Lecture Notes in Computational Science and Engineering

95

Editors:

For further volumes:
http://www.springer.com/series/3527

Mejdi Azaïez • Henda El Fekih
Jan S. Hesthaven
Editors

Spectral and High Order Methods for Partial Differential Equations ICOSAHOM 2012

Selected papers from the ICOSAHOM conference, June 25-29, 2012, Gammarth, Tunisia

Springer

Editors
Mejdi Azaïez
Institut Polytechnique de Bordeaux
Université de Bordeaux
Pessac, France

Henda El Fekih
Ecole Nationale d'Ingénieurs de Tunis
Université de Tunis El Manar
Tunis, Tunisia

Jan S. Hesthaven
Ecole Polytechnique Fédérale de Lausanne
EPFL-SB-MATHICSE-MCSS
Lausanne, Switzerland

ISSN 1439-7358
ISBN 978-3-319-37858-9 ISBN 978-3-319-01601-6 (eBook)
DOI 10.1007/978-3-319-01601-6
Springer Cham Heidelberg New York Dordrecht London

Mathematics Subject Classification (2010): Primary: 65M70; 65N35; 65N30; 74S25; 76M10; 76M22; 78M10; 78M22
Secondary: 33Cxx; 41Axx; 65Cxx; 65Dxx; 65Lxx; 65Mxx; 65Nxx; 74Jxx; 76Mxx; 78Mxx; 80Mxx76

Printed on acid-free paper

Springer is part of Springer Science+Business Media (www.springer.com)

Foreword

This volume presents selected papers from the ninth International Conference on Spectral and High Order Methods (ICOSAHOM'12) conference that was held at Gammarth, Tunisia, during the week June 25–29, 2012. These selected papers were refereed by members of the scientific committee of ICOSAHOM as well as by other leading scientists.

The first ICOSAHOM conference was held in Como, Italy, in 1989 and marked the beginning of an international conference series in Montpelier, France (1992); Houston, Texas, USA (1995); Tel Aviv, Israel (1998); Uppsala, Sweden (2001); Providence, Rhode Island, USA (2004); Beijing, China (2007); and Trondheim, Norway (2009).

The ICOSAHOM conferences have established itself as the main meeting place for researchers with interest in the theoretical, applied and computational aspects of highorder methods for the numerical solution of partial differential equations.

The number of registered participants for this most recent event, the first of its kind to be held at the African continent, exceeded 130, while the total number of talks was 105, comprising nine invited talks, 43 talks in nine different topicspecific mini-symposia, and 53 talks in sessions devoted to contributed papers.

The content of the proceedings is organized as follows. First, contributions from the invited speakers are included, listed in alphabetical order according to name of the invited speaker. Next, contributions from the speakers in mini-symposia and from contributed sessions are included and listed in alphabetical order according to the first author of each paper.

The success of the meeting was ensured through the generous financial support by Institut de Mécanique et d'Ingénierie - Bordeaux, Laboratoire de Modélisation Mathématique et Numérique pour les Sciences de l'Ingénieur (LAMSIN), Ecole Nationale d'Ingénieurs de Tunis, Université de Tunis El Manar, Agence Nationale de Promotion de la Recherche en Tunisie, Laboratoire de Mathématiques Appliquées de Compiègne, Institut Français de Tunisie and European Office of Aerospace Research and Development, Air Force Office of Scientific Research (AFOSR) and US Air Force Research Laboratory (AFRL).

Special thanks go to our colleagues and partners, Faker Ben Belgacem and Nabil Gmati, both central members of the organizing committee, who help enable and share with us this wonderful experience.

Finally, the conference could not have happened without the invaluable support and assistance of the members and graduate students of LAMSIN.

Pessac, France — Mejdi Azaïez
Tunis, Tunisia — Henda El Fekih
Lausanne, Switzerland — Jan S. Hesthaven

Contents

A Quasi-optimal Sparse Grids Procedure for Groundwater Flows 1
Joakim Beck, Fabio Nobile, Lorenzo Tamellini, and Raúl Tempone

The Geometric Basis of Numerical Methods 17
Marc Gerritsma, René Hiemstra, Jasper Kreeft, Artur Palha, Pedro Rebelo, and Deepesh Toshniwal

Spectral Element Methods on Simplicial Meshes 37
Richard Pasquetti and Francesca Rapetti

Exponential Convergence of hp-DGFEM for Elliptic Problems in Polyhedral Domains 57
Dominik Schötzau, Christoph Schwab, Thomas Wihler, and Marcel Wirz

A Contribution to the Outflow Boundary Conditions for Navier-Stokes Time-Splitting Methods 75
E. Ahusborde, M. Azaïez, S. Glockner, and A. Poux

High Order Space-Time Discretization for Elastic Wave Propagation Problems 87
Paola F. Antonietti, Ilario Mazzieri, Alfio Quarteroni, and Francesca Rapetti

Laguerre Simulation of Boundary Layer Flows: Conditions at Large Distance from the Wall 99
F. Auteri and L. Quartapelle

Implementation of an Explicit Algebraic Reynolds Stress Model in an Implicit Very High-Order Discontinuous Galerkin Solver ... 111
F. Bassi, L. Botti, A. Colombo, A. Ghidoni, and S. Rebay

Investigation of Near-Wall Grid Spacing Effect in High-Order Discontinuous Galerkin RANS Computations of Turbomachinery Flows 125
F. Bassi, L. Botti, A. Colombo, A. Ghidoni, and S. Rebay

A Fourth-Order Compact Finite Volume Scheme for the Convection-Diffusion Equation 135
Christine Baur and Michael Schäfer

On the Effect of Flux Functions in Discontinuous Galerkin Simulations of Underresolved Turbulence 145
Andrea D. Beck, Gregor J. Gassner, and Claus-Dieter Munz

Generation of High-Order Polynomial Patches from Scattered Data 157
Karsten Bock and Jörg Stiller

Towards a High Order Fourier-SEM Solver of Fluid Models in Tokamaks 169
A. Bonnement, S. Minjeaud, and R. Pasquetti

Whitney Forms, from Manifolds to Fields 179
Alain Bossavit and Francesca Rapetti

Exponential Convergence of the hp Version of Isogeometric Analysis in 1D 191
Annalisa Buffa, Giancarlo Sangalli, and Christoph Schwab

High-Order Locally Implicit Time Integration Strategies in a Discontinuous Galerkin Method for Maxwell's Equations 205
S. Descombes, S. Lanteri, and L. Moya

High-Order ADI Schemes for Convection-Diffusion Equations with Mixed Derivative Terms 217
B. Düring, M. Fournié, and A. Rigal

A Numerical Study of Averaging Error Indicators in p-FEM 227
Philipp Dörsek and J. Markus Melenk

Coupling of an Exact Transparent Boundary Condition with a DG Method for the Solution of the Time-Harmonic Maxwell Equations 237
M. El Bouajaji, N. Gmati, S. Lanteri, and J. Salhi

A New Proof for Existence of $\mathcal{H}$-Matrix Approximants to the Inverse of FEM Matrices: The Dirichlet Problem for the Laplacian 249
Markus Faustmann, Jens M. Melenk, and Dirk Praetorius

Multidomain Extension of a Pseudospectral Algorithm for the Direct Simulation of Wall-Confined Rotating Flows 261
G. Fontaine, S. Poncet, and E. Serre

A Comparison of High-Order Time Integrators for Highly Supercritical Thermal Convection in Rotating Spherical Shells 273
F. Garcia, M. Net, and J. Sánchez

High Order Methods with Exact Conservation Properties 285
René Hiemstra and Marc Gerritsma

Spectral Element Discretization for the Vorticity, the Velocity and the Pressure Formulation of the Axisymmetric Navier-Stokes Problem 297
Chahira Jerbi and Nahla Abdellatif

Higher-Order Compatible Discretization on Hexahedrals 311
Jasper Kreeft and Marc Gerritsma

Mimetic Spectral Element Advection 325
Artur Palha, Pedro Pinto Rebelo, and Marc Gerritsma

Large Eddy Simulation of a Muffler with the High-Order Spectral Difference Method 337
Matteo Parsani, Michael Bilka, and Chris Lacor

Stability Tools for the Spectral-Element Code Nek5000: Application to Jet-in-Crossflow 349
A. Peplinski, P. Schlatter, P.F. Fischer, and D.S. Henningson

A High-Order Discontinuous Galerkin Method for Viscoelastic Wave Propagation 361
Fabien Peyrusse, Nathalie Glinsky, Céline Gélis, and Stéphane Lanteri

Mixed Mimetic Spectral Element Method Applied to Darcy's Problem ... 373
Pedro Pinto Rebelo, Artur Palha, and Marc Gerritsma

Novel Outflow Boundary Conditions for Spectral Direct Numerical Simulation of Rotating Flows 383
Stéphanie Rodriguez, Bertrand Viaud, and Eric Serre

A Geometric Approach Towards Momentum Conservation 393
Deepesh Toshniwal, R.H.M. Huijsmans, and M.I. Gerritsma

A Spectral Method for Optimal Control Problems Governed by the Time Fractional Diffusion Equation with Control Constraints 403
Xingyang Ye and Chuanju Xu

Two-Phase Flow Solved by High Order Discontinuous Galerkin Method 415
J.S.B. van Zwieten, D.R. van der Heul, R.H.A. IJzermans, R.A.W.M. Henkes, and C. Vuik

A Quasi-optimal Sparse Grids Procedure for Groundwater Flows

Joakim Beck, Fabio Nobile, Lorenzo Tamellini, and Raúl Tempone

Abstract In this work we explore the extension of the quasi-optimal sparse grids method proposed in our previous work "*On the optimal polynomial approximation of stochastic PDEs by Galerkin and Collocation methods*" to a Darcy problem where the permeability is modeled as a lognormal random field. We propose an explicit a-priori/a-posteriori procedure for the construction of such quasi-optimal grid and show its effectiveness on a numerical example. In this approach, the two main ingredients are an estimate of the decay of the Hermite coefficients of the solution and an efficient nested quadrature rule with respect to the Gaussian weight.

1 Introduction

Uncertainty quantification plays a crucial role in the area of groundwater flows where, given the time and length scale of most problems, it is quite common to have partial and fragmented knowledge about most of the system properties, e.g. on the permeability field, forcing terms, boundary conditions. Broad classes of applications of interest could be oil or water reservoir management, see e.g. [4].

Given the complexity of the deterministic solvers for such problems, a non-intrusive computational approach to perform the uncertainty quantification analysis is quite appealing. In this work we consider a Darcy problem with uncertain

J. Beck · R. Tempone (✉)
Applied Mathematics and Computational Science, KAUST, Thuwal, Saudi Arabia
e-mail: joakim.back.09@ucl.ac.uk; raul.tempone@kaust.edu.sa

F. Nobile · L. Tamellini
CSQI – MATHICSE, Ecole Politechnique Fédérale Lausanne, Lausanne, Switzerland

MOX, Department of Mathematics "F. Brioschi", Politecnico di Milano, Milan, Italy
e-mail: fabio.nobile@epfl.ch; lorenzo.tamellini@mail.polimi.it

M. Azaïez et al. (eds.), *Spectral and High Order Methods for Partial Differential Equations - ICOSAHOM 2012*, Lecture Notes in Computational Science and Engineering 95, DOI 10.1007/978-3-319-01601-6_1,

permeability modeled as a lognormal random field, and we explore (rather heuristically) the possibility to extend to this problem the quasi-optimal sparse grid method that we had proposed in [5] for problems depending instead on a set of uniform random variables. See also [18, 19, 21] for other probabilistic collocation approaches to the Darcy problem, where however only isotropic approximations are proposed.

The well-posedness of the lognormal problem has been thoroughly investigated in [9, 13]. The optimal convergence rate of its so-called Polynomial Chaos Expansion approximation has been analyzed theoretically in [16]. Although the deterministic Darcy problem is more commonly approximated numerically in its mixed form (see e.g. [1,6]), in this work we will consider a standard Finite Element discretization of the primal elliptic formulation of the Darcy problem, in which the unknown is the water pressure.

The rest of this work is organized as follows. In Sect. 2 we specify the model assumptions on the random permeability field, on the deterministic problem and on the quantity of interest. Section 3 deals with the finite dimensional Fourier expansion of the random field, and Sect. 4 with the derivation of the quasi-optimal sparse grid for the problem at hand. Finally, we present some numerical results in Sect. 5, and draw some conclusions in Sect. 6.

2 Problem Setting

Let $(\Omega, \mathcal{F}, P)$ be a complete probability space, where Ω denotes the set of outcomes, $\mathcal{F}$ its σ-algebra, and $P : \mathcal{F} \to [0, 1]$ a probability measure. Following a standard notation, we denote with $H^1(D)$ the Sobolev space of square-intergrable functions in D with square integrable derivatives. $L^q_P(\Omega)$ will denote the Banach space of random functions with bounded q-th moment with respect to the probability measure P, and $L^q_P(\Omega; H^1(D))$ the Bochner space of $H^1(D)$-valued random fields with q-th bounded moment with respect to P, that is $f \in L^q_P(\Omega; H^1(D)) \Leftrightarrow \int_\Omega \|f(\cdot,\omega)\|^q_{H^1(D)} + dP(\omega) < \infty$.

As mentioned in the introduction, the permeability field is supposed to be uncertain. Moreover, since in hydrogeological applications the pointwise permeability values can vary within several orders of magnitude it is rather common to model the logarithm of the permeability as a random field, rather than the permeability itself. Observe also that this approach automatically guarantees positivity of the permeability. More in detail, we will make the following assumption.

Assumption 1. *The permeability $a(\mathbf{x}, \omega) : \overline{D} \times \Omega \to \mathbb{R}$ is a lognormal field, that is*

$$a(\mathbf{x}, \cdot) = e^{\gamma(\mathbf{x},\cdot)}, \quad \gamma(\mathbf{x}, \cdot) \sim \mathcal{N}(\mu, \sigma^2) \quad \forall \mathbf{x} \in D,$$

where $\mathcal{N}(\mu, \sigma^2)$ denotes a Gaussian probability distribution with expected value μ and variance σ^2, and $\gamma(\mathbf{x}, \omega) : \overline{D} \times \Omega \to \mathbb{R}$ is such that for $\mathbf{x}, \mathbf{x}' \in D$ the covariance *function* $C_\gamma(\mathbf{x},\mathbf{x}') = \mathbb{C}\mathrm{ov}\left[\gamma(\mathbf{x},\cdot)\gamma(\mathbf{x}',\cdot)\right]$ *depends only on the distance* $\|\mathbf{x}-\mathbf{x}'\|$ *("second-order isotropic stationary field"). Moreover,* $C_\gamma(\mathbf{x},\mathbf{x}') = C_\gamma(\|\mathbf{x}-\mathbf{x}'\|)$ *is Lipschitz continuous, and is a positive definite function.*

As for the choice of C_γ, several models have been proposed in the literature, depending on the specific application (exponential, spherical, gaussian, matern, hole effect and others, see e.g. [4, 22]). Given the exploratory level of this work, we choose here to work with with a simple Gaussian covariance function

Assumption 2. *The Gaussian field* $\gamma(\mathbf{x},\omega)$ *has a Gaussian covariance function with correlation length* $L_c > 0$*,*

$$C_\gamma(\mathbf{x},\mathbf{x}') = \sigma^2 \exp\left(-\frac{\|\mathbf{x}-\mathbf{x}'\|^2}{L_c^2}\right).$$

The Darcy problem will be set in a horizontal square domain $D = (0,L)^2$, $L = 1$, with no forcing terms. We impose a pressure gradient acting on the water by setting $p = 1$ on the left boundary $\mathcal{B}_1 = \{\mathbf{x} \in D : x_1 = 0\}$ and $p = 0$ on the right boundary $\mathcal{B}_2 = \{\mathbf{x} \in D : x_1 = L\}$. Finally, we consider a no-flux Neumann condition on the upper and lower boundaries $\mathcal{B}_3 = \{\mathbf{x} \in D : x_2 = 0\}$ and $\mathcal{B}_4 = \{\mathbf{x} \in D : x_2 = L\}$. The Darcy problem thus reads:

Strong Formulation 1. *Find a random pressure* $p : \overline{D} \times \Omega \to \mathbb{R}$ *such that* P*-almost everywhere the following equation holds*

$$\begin{cases} -\operatorname{div}(a(\mathbf{x},\omega)\nabla p(\mathbf{x},\omega)) = 0 & \mathbf{x} \in D,\\ p(\mathbf{x},\omega) = 1 & \mathbf{x} \in \mathcal{B}_1,\\ p(\mathbf{x},\omega) = 0 & \mathbf{x} \in \mathcal{B}_2,\\ a(\mathbf{x},\omega)\nabla p(\mathbf{x},\omega)\cdot \mathbf{n} = 0 & \mathbf{x} \in \mathcal{B}_3 \cup \mathcal{B}_4. \end{cases} \tag{1}$$

It is straightforward to see that, thanks to the Lax–Milgram lemma, (1) is well-posed for almost every $\omega \in \Omega$. Proving the well-posedness of (1) in the Bochner spaces $L^q_P(\Omega; H^1(D))$ for $q > 0$ is instead not trivial, since a is not uniformly bounded nor uniformly coercive with respect to ω. It is however possible to prove the following result (see e.g. [2, 9, 10, 13]):

Proposition 1. *For every* $q > 0$*, there exists a unique* $H^1(D)$*-valued random pressure* $p = p(\mathbf{x},\omega)$ *in* $L^q_P(\Omega; H^1(D))$ *solving* (1).

As for quantities of interest, we aim at computing the expected value of the total flux crossing the right boundary $\mathcal{B}_2$. This is indeed a random variable,

$$Z_p(\omega) = \int_{\mathcal{B}_2} a(\mathbf{x},\omega)\partial_{\mathbf{n}} p(\mathbf{x},\omega) d\mathbf{x}, \tag{2}$$

and also represents the "effective permeability" of the random medium in D.

3 Series Expansion of the Log-Permeability Random Field

To get to a computable representation of p we need to derive an approximation of a in terms of a finite set of N random variables $y_i(\omega)$, $i = 1, \ldots, N$ ("finite noise approximation"). Such approximation is usually obtained by suitably truncating a series expansion such as the Karhunen-Loève expansion, see e.g. [20]. As an alternative, we consider here a Fourier-based decomposition of γ (see e.g. [15]), which uses trigonometric polynomials as basis functions in the physical space. This choice allows analytical computation of the expansion and highlights the contribution of each spatial frequency to the total field a.

Proposition 2 (Fourier expansion). *Let $\gamma(\mathbf{x}, \omega) : [0, L]^2 \times \Omega \to \mathbb{R}$ be a weakly stationary gaussian random field as in Assumption 1, with pointwise variance σ^2. Then the covariance function can be expanded in cosine-Fourier series*

$$C_\gamma(\|\mathbf{x} - \mathbf{x}'\|) = \sigma^2 \sum_{\mathbf{k}=(k_1,k_2)\in\mathbb{N}_0^2} c_\mathbf{k} \cos(\omega_{k_1}(x_1 - x_1')) \cos(\omega_{k_2}(x_2 - x_2')), \quad (3)$$

with $\omega_{k_1} = \frac{k_1\pi}{L}$, $\omega_{k_2} = \frac{k_2\pi}{L}$, and normalized Fourier coefficient $c_\mathbf{k}$ so that $\sum_{\mathbf{k}\in\mathbb{N}_0^2} c_\mathbf{k} = 1$. In particular, for the Gaussian covariance function in Assumption 2 and sufficiently small values of L_c, $c_\mathbf{k}$ are well approximated by

$$c_\mathbf{k} \approx \lambda_{k_1}\lambda_{k_2}, \text{ where } \lambda_k = \begin{cases} \dfrac{L_c\sqrt{\pi}}{2L} & \text{if } k = 0 \\ \dfrac{L_c\sqrt{\pi}}{L} \exp\left(-\dfrac{(k\pi L_c)^2}{4L^2}\right) & \text{if } k > 0\,. \end{cases} \quad (4)$$

The random field γ admits then the following expansion

$$\gamma(\mathbf{x}, \omega) = \mathbb{E}[\gamma(\mathbf{x}, \cdot)] + \sigma \sum_{\mathbf{k}\in\mathbb{N}^2} \sum_{i=1}^{4} \left(\sqrt{c_\mathbf{k}}\, y_{\mathbf{k},i}(\omega)\phi_{\mathbf{k},i}(\mathbf{x})\right) \quad (5)$$

where $y_{\mathbf{k},i}(\omega)$ are identically distributed and independent standard Gaussian random variables, and $\phi_{\mathbf{k},i}$ are defined as $\phi_{\mathbf{k},1}(\mathbf{x}) = \cos(\omega_{k_1}x_1)\cos(\omega_{k_2}x_2)$, $\phi_{\mathbf{k},2}(\mathbf{x}) = \sin(\omega_{k_1}x_1)\sin(\omega_{k_2}x_2)$, $\phi_{\mathbf{k},3}(\mathbf{x}) = \cos(\omega_{k_1}x_1)\sin(\omega_{k_2}x_2)$, $\phi_{\mathbf{k},4}(\mathbf{x}) = \sin(\omega_{k_1}x_1)\cos(\omega_{k_2}x_2)$.

Proof. See [27, Chap. 4].

A good approximation of γ, γ_N, can be achieved by retaining in (5) only the N random variables corresponding to the frequencies $\mathbf{k}$ in the set

$$\mathcal{K}_\kappa = \left\{\mathbf{k} \in \mathbb{N}_0^2 : k_1^2 + k_2^2 \le \kappa^2\right\}, \quad (6)$$

Table 1 Random variables needed to represent α % of the total variance of a random field with Gaussian covariance function for different correlation lengths L_c

	$\alpha = 0.7$	$\alpha = 0.9$	$\alpha = 0.99$
$L_c = 0.35$	$N = 13$	$N = 25$	$N = 49$
$L_c = 0.25$	$N = 25$	$N = 49$	$N = 97$
$L_c = 0.1$	$N = 161$	$N = 293$	$N = 593$

for a given $\kappa \in \mathbb{N}$. Following the argument of [9], it can be shown in particular that γ_N converges to γ almost surely in $C^0(D)$ as $\kappa \to \infty$.

Example 1. Table 1 shows the number of random variables that need to be included into (5) to take into account a fraction α of the total variance of γ for different correlation lengths L_c. The need to include a high number of random variables in the approximation of the random field γ, and hence the high-dimensionality of the vector $\mathbf{y}$ of input random variables clearly emerges. Observe that less regular covariance functions C_γ will experience a slower eigenvalue decay than (4), see e.g. [25], further enlarging the set of input random variables needed to represent the solution. In practice, the level of truncation should be related to the error in the variance of the solution of the PDE.

Let us now denote $\Gamma_i = \mathbb{R}$ the support of $y_i(\omega)$, $\Gamma = \Gamma_1 \times \ldots \times \Gamma_N$ the support of $\mathbf{y} = [y_1, \ldots, y_N]$, $\rho_i(y_i) : \Gamma_i \to \mathbb{R}$ the probability density function of y_i and $\rho(\mathbf{y}) : \Gamma \to \mathbb{R}$ the joint probability density function of $\mathbf{y}$, with $\rho(\mathbf{y}) = \prod_{n=1}^N \rho_i(y_i), \rho_i(y_i) = \frac{1}{\sqrt{2\pi}} e^{-\frac{y_i^2}{2}}$. Having introduced the random variables y_i, we can replace the abstract probability space $(\Omega, \mathcal{F}, P)$ with $(\Gamma, \mathscr{B}(\Gamma), \rho(\mathbf{y})d\mathbf{y})$, where $\mathscr{B}(\Gamma)$ denotes the Borel σ-algebra, and hence $L^q_P(\Omega)$ with $L^q_\rho(\Gamma)$ and $L^q_P(\Omega; H^1(D))$ with $L^q_\rho(\Gamma; H^1(D))$. Moreover, the permeability and pressure fields can now be seen as functions of $\mathbf{x}$ and $\mathbf{y}$, $a(\mathbf{x}, \omega) \approx a_N(\mathbf{x}, \mathbf{y}(\omega)) = e^{\gamma_N(\mathbf{x}, \mathbf{y}(\omega))}$, $p(\mathbf{x}, \omega) \approx p_N(\mathbf{x}, \mathbf{y}(\omega))$ and the quantity of interest (2) becomes a random function $Z_p : \Gamma \to \mathbb{R}$. We will not however address here the study on the convergence of p_N to p, see e.g. [9] to this end. Here we just mention that, following again the argument in [9], it is possible to show that the almost sure convergence of γ_N to γ guarantees the almost sure convergence of a_N to a in $C^0(D)$, and that for any $q > 0$ there holds $\|a_{N(\kappa)} - a\|_{L^q(\Omega, C^0(D))} \le C_1(q)\kappa e^{-C_2(L,L_c)\kappa^2}$, $N(\kappa)$ being the cardinality of the set $\mathcal{K}_\kappa$ defined in (6). In the rest of this work, with a slight abuse of notation, we will therefore omit subscript $\cdot_N$ if no confusion arise. Moreover, the quasi-optimal Sparse Grid Collocation technique that we will present in the next Section is able to automatically select the "most important" random variables that should be retained for the approximation of p. This would allow us to work without truncating the expansion of the permeability, i.e. with formally $N = \infty$ random variables.

The previous results on the well-posedness of the problem still hold after having replaced ω with $\mathbf{y}$, and we can write the problem in weak form.

Weak Formulation 1. *Find* $p \in H^1(D) \otimes L^2_\rho(\Gamma)$ *such that* $p = 1$ *on* $\mathcal{B}_1$, $p = 0$ *on* $\mathcal{B}_2$ *and* $\forall\, v \in H^1_{\{\mathcal{B}_1 \cup \mathcal{B}_2\}}(D) \otimes L^2_\rho(\Gamma)$

$$\int_\Gamma \int_D a(\mathbf{x},\mathbf{y}) \nabla p(\mathbf{x},\mathbf{y}) \cdot \nabla v(\mathbf{x},\mathbf{y})\, \rho(\mathbf{y})\, d\mathbf{x}\, d\mathbf{y} = 0. \tag{7}$$

where $H^1_{\{\mathcal{B}_1 \cup \mathcal{B}_2\}}(D)$ is the subset of $H^1(D)$ functions that vanish on the Dirichlet boundary $\mathcal{B}_1 \cup \mathcal{B}_2$.

4 Quasi-optimal Sparse Grid Approximation

As highlighted in Example 1, both the permeability a and the pressure p depend on a large number of random variables y_i. To obtain efficiently an approximation of p over Γ we then resort to the sparse grid method [2,3,7,23,28], that allows to obtain an accurate representation of p while keeping the number of interpolation points considerably lower than what would be needed if a full tensor grid approximation was employed. In formulae, the sparse grid approximation of p is written as

$$p_{\mathrm{w}}(\mathbf{y}) = \mathcal{S}^m_{\mathcal{I}(\mathrm{w})}[p](\mathbf{y}) = \sum_{\mathbf{i} \in \mathcal{I}(\mathrm{w})} \bigotimes_{n=1}^{N} \Delta_n^{m(i_n)}[p](\mathbf{y}), \tag{8}$$

where

- $\mathbf{i} \in \mathbb{N}^N_+$ is a multiindex with non-zero components;
- $\Delta_n^{m(i_n)} = \mathcal{U}_n^{m(i_n)} - \mathcal{U}_n^{m(i_n-1)}$ is called "detail operator", and is the difference between two consecutive one-dimensional interpolants, using $m(i)$ and $m(i-1)$ points respectively;
- $\Delta^{m(\mathbf{i})}[p] = \bigotimes_{n=1}^N \Delta^{m(i_n)}[p]$ is called "hierarchical surplus";
- $\{\mathcal{I}(\mathrm{w})\}_{\mathrm{w} \in \mathbb{N}}$ denotes a sequence of index sets. Each of these sets has to be *admissible* in the following sense for the sparse grid to be consistent (see e.g. [12]):

$$\forall\, \mathbf{i} \in \mathcal{I},\ \mathbf{i} - \mathbf{e}_j \in \mathcal{I} \text{ for } 1 \le j \le N,\ i_j > 1, \tag{9}$$

$\mathbf{e}_j$ being the j-th canonical vector. Roughly speaking, the sparse grid approximation of p can be understood as a linear combination of tensor grid approximations of p over Γ, each one built over "few" points.

The efficiency of the sparse grid depends on the choice of the interpolation points used in $\mathcal{U}_n^{m(i)}$ and of the index sets $\mathcal{I}(\mathrm{w})$. The interpolation points should be chosen in agreement with the probability measure over Γ, a good choice being given e.g. by the Gauss-Hermite points (see e.g. [8]).

Regarding the index sets $\mathcal{I}(\mathrm{w})$, the best strategy is to include in (8) only the hierarchical surpluses with the highest profits [5, 12, 14]. The latter is defined as the ratio between the expected error decrease by adding a given hierarchical surplus to the sparse grid approximation and the corresponding cost, quantified here by the number of interpolation points in the hierarchical surplus,

$$\mathcal{I}(\mathrm{w}) = \left\{ \mathbf{i} \in \mathbb{N}_+^N : \frac{\Delta E(\mathbf{i})}{\Delta W(\mathbf{i})} \geq \epsilon_{\mathrm{w}} \right\} \tag{10}$$

with $\{\epsilon_{\mathrm{w}}\}_{\mathrm{w}\in\mathbb{N}}$ decreasing to 0 as $w \to \infty$ and $\Delta E(\mathbf{i})$, $\Delta W(\mathbf{i})$ representing the error and work contribution of each hierarchical surplus respectively. Note that $\mathcal{I}(\mathrm{w})$ in (10) may not satisfy the admissibility condition (9), that has to be explicitly enforced.

This criterion can be implemented in an adaptive procedure (see e.g. [12]) that explores the space of hierarchical surpluses and adds to $\mathcal{I}(\mathrm{w})$ the most profitable according to (10). As an alternative, in [5] we have detailed an a-priori/a-posteriori procedure to detect $\mathcal{I}(\mathrm{w})$ based on estimates of $\Delta E(\mathbf{i})$ and $\Delta W(\mathbf{i})$. On the one hand, the a-priori approach saves the computational cost of the exploration of the space of hierarchical surpluses, but on the other hand it will be effective only if the estimates of $\Delta E(\mathbf{i})$ and $\Delta W(\mathbf{i})$ are sufficiently sharp. In [5] only the case of uniform random variables has been investigated. Deriving sharp estimates for the problem at hand, that depends on Gaussian random variables, is the goal of the present work.

We begin with the estimate of the work contribution corresponding to an additional index $\mathbf{i}$, which can be easily computed if the considered interpolant operators $\mathcal{U}_n^{m(i_n)}$ are nested and the set $\mathcal{I}(\mathrm{w})$ is admissible:

$$\Delta W(\mathbf{i}) = \prod_{n=1}^{N} (m(i_n) - m(i_n - 1)). \tag{11}$$

The estimate of the error contribution requires instead more effort. As a preliminary step, we need to introduce a spectral basis for $L^2_\rho(\Gamma)$. To this end, let $\{H_p(y_n)\}_{p\in\mathbb{N}}$ be the family of orthonormal Hermite polynomials relative to the weight $e^{-y^2/2}/\sqrt{2\pi}$ in the n-th direction [8]. The set of multidimensional Hermite polynomials $\mathcal{H}_{\mathbf{q}}(\mathbf{y}) = \prod_{n=1}^{N} H_{q_n}(y_n)$, $\forall \mathbf{q} \in \mathbb{N}^N$ is an orthonormal basis for $L^2_\rho(\Gamma)$, that can be used to formally construct the spectral expansion of $p(\mathbf{y})$

$$p(\mathbf{y}) = \sum_{\mathbf{q}\in\mathbb{N}^N} p_{\mathbf{q}} \mathcal{H}_{\mathbf{q}}(\mathbf{y}), \quad p_{\mathbf{q}} = \int_\Gamma p(\mathbf{y}) \mathcal{H}_{\mathbf{q}}(\mathbf{y}) \rho(\mathbf{y}) d\mathbf{y}. \tag{12}$$

We can now state a heuristic estimate for the error contribution of the hierarchical surplus $\Delta^{m(\mathbf{i})}$ in the spirit of what was done in [5], Eq. (4.9):

$$\Delta E(\mathbf{i}) \approx B(\mathbf{i}) \left\| p_{m(\mathbf{i}-\mathbf{1})} \right\|_{H^1(D)}, \tag{13}$$

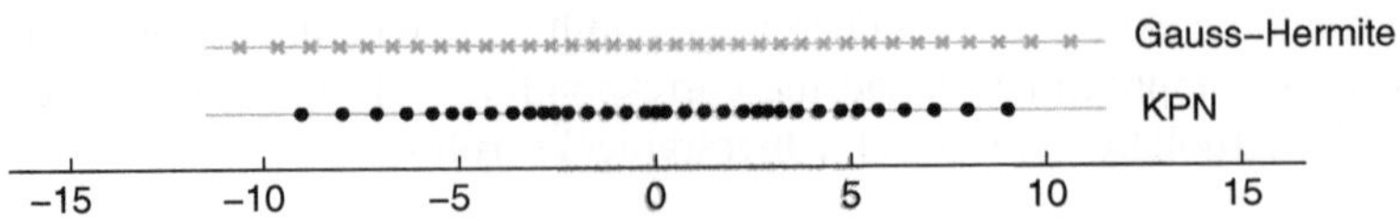

Fig. 1 First 35 *KPN* and Gauss-Hermite knots

where $p_{m(\mathbf{i}-\mathbf{1})}$ is the $m(\mathbf{i}-\mathbf{1})$-th coefficient of the spectral expansion (12), and $B(\mathbf{i})$ is a factor that depends on the interpolation points only, in the spirit of the Lebesgue constant. This is a reasonable heuristic assumption, since in this way the error contribution estimate "encodes" information on both the quality of the solution (through the decay of the spectral coefficients), and the quality of the interpolant operator itself. Numerical results in the next section will also show the effectiveness of (13).

To make estimates (11) and (13) computable we still need to:

1. Choose a family of nested univariate interpolant operators for the Gaussian measure;
2. Provide an estimate for the factor $B(\mathbf{i})$ in (13);
3. Provide an estimate for the coefficients $p_{m(\mathbf{i}-\mathbf{1})}$ in (12) and (13).

4.1 Nested Quadrature Formulae for Gaussian Measure

The family of nested points we choose is the so-called "Kronrod-Patterson-Normal" (*KPN* in short, see Fig. 1). Such family of interpolation/quadrature points is due to Genz and Keister, see [11], that applied the Kronrod-Patterson procedure [17, 24] to the classical Gauss-Hermite quadrature points (i.e. the roots of the Hermite polynomials $H_p(y_n)$). We recall that the Kronrod-Patterson procedure is a way to modify a quadrature rule, by adding new points in a nested fashion retaining the highest degree of exactness possible. The knots and the corresponding quadrature weights are tabulated up to level 5 (35 nodes) and can be found e.g. at http://www.sparse-grids.de/. For such family of points there holds

$$m(i_n) = 1, 3, 9, 19, 35 \quad \text{for } i_n = 1, \ldots, 5 \tag{14}$$

i.e. consecutive interpolants are built over $1, 3, 9, 19, 35$ points respectively.

4.2 Estimate for $B(\mathbf{i})$

In [5] the constant $B(\mathbf{i})$ in Eq. (13) was chosen to be equal to the product of the Lebesgue constants of interpolant operators in each direction, $B(\mathbf{i}) = \prod_{n=1}^{N} \mathbb{L}_n^{m(i_n)}$. Such an estimate is also supported by numerical verification.

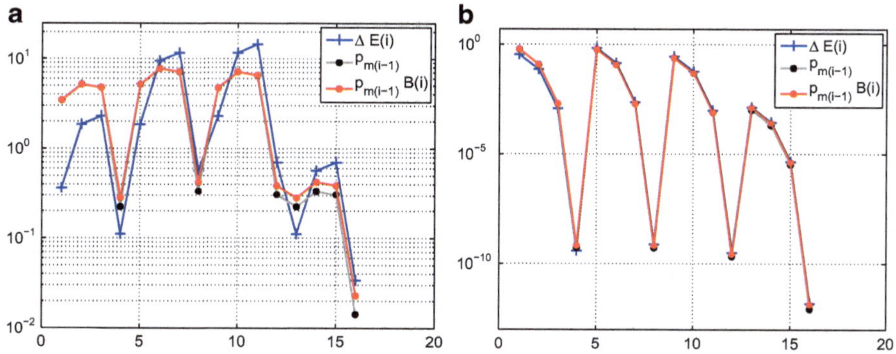

Fig. 2 Numerical comparison between $\Delta E(\mathbf{i})$ and $|p_{m(\mathbf{i}-1)}|$ for p of the form $p(y_1, y_2) = e^{-1-b_1 y-1-b_2 y_2}$. The quantities $\Delta E(\mathbf{i})$ for $\mathbf{i}$ s.t. $\max\{i_1, i_2\} \leq 4$ have been computed with a standard Smolyak sparse grid, with $\mathcal{I}(\mathrm{w}) = \{\mathbf{i} \in \mathbb{N}_+^N : |\mathbf{i} - \mathbf{1}| \leq \mathrm{w}\}$, $\mathrm{w} = 10$, and "doubling" function $m(i)$: $m(0) = 0, m(1) = 1, m(i) = 2^{i-1} + 1$. The Hermite coefficients $|p_{m(\mathbf{i}-1)}|$ have been computed analytically with the formula stated in Lemma 1. (**a**) $p(y_1, y_2) = e^{-1-1.5y_1-1.5y_2}$. (**b**) $p(y_1, y_2) = e^{-1-y_1-0.2y_2}$

However, it is not easy to obtain a sharp bound for the Lebesgue constant in case of interpolation in spaces with Gaussian measure. Thus, we propose here a different estimate for $B(\mathbf{i})$, which on the one hand gives good numerical results when tested on model problems (see Fig. 2) and on the other hand is close to the original choice when applied to a problem with uniform random variables.

To this end, we go back to the definition of error contribution for a hierarchical surplus, and exploit the fact that p admits a Hermite expansion. To improve the readability we will use $\|\cdot\|_\otimes$ to denote the norm $\|\cdot\|_{H^1(D)\otimes L^2_\rho(\Gamma)}$.

$$\begin{aligned} \Delta E(\mathbf{i}) &= \left\| \left(p - \mathcal{S}^m_{\{\mathcal{J}\cup\mathbf{i}\}}[p]\right) - \left(p - \mathcal{S}^m_{\mathcal{J}}[p]\right) \right\|_\otimes = \left\| \Delta^{m(\mathbf{i})}[p] \right\|_\otimes \\ &= \left\| \Delta^{m(\mathbf{i})}\Big[\sum_{\mathbf{q}\in\mathbb{N}^N} p_\mathbf{q} \mathcal{H}_\mathbf{q} \Big] \right\|_\otimes = \left\| \sum_{\mathbf{q}\in\mathbb{N}^N} p_\mathbf{q} \Delta^{m(\mathbf{i})}[\mathcal{H}_\mathbf{q}] \right\|_\otimes . \end{aligned} \tag{15}$$

Observe now that by construction of hierarchical surplus there holds $\Delta^{m(\mathbf{i})}[\mathcal{H}_\mathbf{q}] = 0$ for polynomials such that $\exists n : q_n < m(i_n - 1)$. Next, we apply the triangular inequality and get to

$$\Delta E(\mathbf{i}) \leq \sum_{\mathbf{q}\geq m(\mathbf{i}-1)} \left\| p_\mathbf{q} \right\|_{H^1(D)} \left\| \Delta^{m(\mathbf{i})}[\mathcal{H}_\mathbf{q}] \right\|_{L^2_\rho(\Gamma)} . \tag{16}$$

Therefore, the error estimate (13) is equivalent to assuming that the summation on the right-hand side of (16) is dominated by the first term, with

$$B(\mathbf{i}) = \left\| \Delta^{m(\mathbf{i})}[\mathcal{H}_{m(\mathbf{i}-1)}] \right\|_{L^2_\rho(\Gamma)} = \prod_{n=1}^{N} B_n(i_n) , \tag{17}$$

$$B_n(i_n) = \left\| \Delta^{m(i_n)}[H_{m(i_n-1)}] \right\|_{L^2_{\rho_n}(\Gamma_n)}.$$

The quantity $B_n(i_n)$ can be easily computed numerically, and has a moderate growth with respect to i_n:

$$B_n(i_n) = 1,\ 1,\ 1,\ 1.28,\ 5.46 \quad \text{for } i = 1, \ldots, 5. \tag{18}$$

Finally, we test estimate (13) on the model function $p(y_1, y_2) = 1/\exp(1 + b_1 y_1 + b_2 y_2)$, so that we can compute each $\Delta E(\mathbf{i})$ as

$$\Delta E(\mathbf{i}) = \left\| \Delta^{m(\mathbf{i})}[p] \right\|_{L^2_\rho(\Gamma)} = \left\| \mathcal{S}^m_{\{\mathcal{J}\cup\mathbf{i}\}}[p] - \mathcal{S}^m_{\mathcal{J}}[p] \right\|_{L^2_\rho(\Gamma)}$$

using a sufficiently accurate sparse grid quadrature. The Hermite coefficients of p can be computed either numerically or analytically, see Lemma 1 in the next section. Once such quantities are available, we can verify the accuracy of (13), with $B(\mathbf{i})$ as in (17). The results are shown in Fig. 2: the proposed estimate is thus seen to be quite reasonable.

Remark 1. As mentioned earlier, the procedure used here to derive an estimate for $B(\mathbf{i})$ could be applied to the problems investigated in [5] as well. It can be seen numerically (see [27]) that estimating $B(\mathbf{i})$ in this way would end up in results not significantly different from the original choice, namely $B(\mathbf{i}) = \prod_{n=1}^N \mathbb{L}_n^{m(i_n)}$.

4.3 Convergence of Hermite Expansions

To derive an estimate for $\|p_\mathbf{q}\|_{H^1(D)}$ we first consider a simplified Darcy problem with a lognormal permeability field a constant over D, $a = a(\mathbf{y}) = \exp\left(b_0 + \sum_{i=1}^N b_i y_i\right)$ and with homogeneous Dirichlet boundary conditions,

$$\begin{cases} -\operatorname{div}(a(\mathbf{y})\nabla p(\mathbf{x},\mathbf{y})) = f(\mathbf{x}) & \mathbf{x} \in D, \\ p(\mathbf{x},\mathbf{y}) = 0 & \mathbf{x} \in \partial D. \end{cases} \tag{19}$$

Furthermore, let $h(\mathbf{x})$ be the solution of the Poisson problem $-\Delta h = f$ with homogeneous Dirichlet boundary conditions. We can then write the analytic expression for p solving (19), which is separable with respect to $\mathbf{y}$, $p(\mathbf{x},\mathbf{y}) = h(\mathbf{x})e^{-b_0} \prod_{n=1}^N \exp(-b_n y_n)$, and further derive the exact expression of the coefficients of the Hermite expansion.

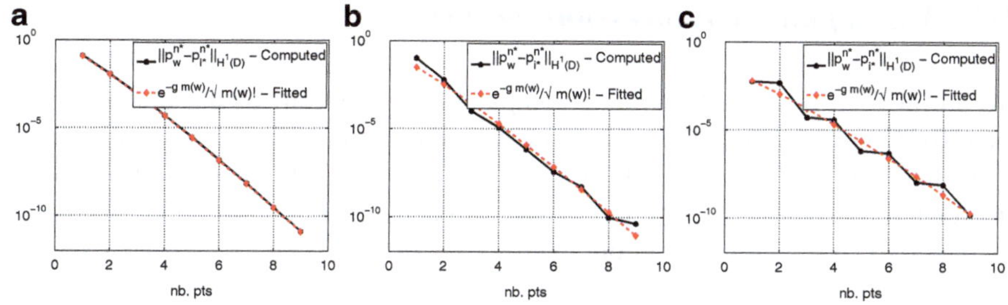

Fig. 3 Assessment of the rates $g_n, n = 0, 2, 5$, used to build the quasi-optimal set (22), estimated according to Eq. (21). For each random variable y_n the corresponding harmonic in the Fourier expansion (23) is specified. The plots show the decay of $\left\| Z^{n^*}_{p,\mathrm{w}} - Z^{n^*}_{p,i^*} \right\|_{H^1(D)}$ as a function of the number of point $m(\mathrm{w})$ and its fitting according to the proposed estimate $e^{-g_n m(\mathrm{w})}/\sqrt{m(\mathrm{w})!}$. (**a**) y_0, constant, $g = 2.07$. (**b**) y_2, $\sin(\pi x/L)$, $g = 1.95$. (**c**) y_5, $\cos(3\pi x/L)$, $g = 1.37$

Lemma 1. *Given problem* (19)*, the* $H^1(D)$ *norm of the Hermite coefficients* (12) *of p can be estimated as*

$$\left\| p_{\mathbf{q}} \right\|_{H^1(D)} = C_{\mathcal{H}} \prod_{n=1}^{N} \frac{e^{-g_n q_n}}{\sqrt{q_n!}}, \tag{20}$$

with $C_{\mathcal{H}} = \|h\|_{H^1(D)} e^{-b_0} \prod_{n=1}^{N} e^{b_n^2/2}$ *and* $g_n = -\log(b_n)$.

Proof. See [27] for details.

Our numerical experience shows that estimate (20) is satisfactory even in the more general case where $a(\mathbf{x}, \mathbf{y}) = e^{\gamma(\mathbf{x},\mathbf{y})}$, and the boundary conditions are those specified in Eq. (1); on the other hand, the more general estimate $\left\| p_{\mathbf{q}} \right\|_{H^1(D)} = C e^{-\sum_n g_n \sqrt{q_n}}$ that applies to analytic (but not entire) functions seems to be too pessimistic in this context.

As pointed out in [5], it is generally better to estimate the rates g_n numerically to get sharper bounds. This is achieved by freezing all the variables y_i but the n^*-th one e.g. at the midpoint of their support, and computing the solution $p^{n^*}_{\mathrm{w}}$ of such reduced problem increasing the sparse grid level w from 1 to i^*. If the quadrature points are accurate enough (i.e. Gaussian quadrature points), then the intermediate solutions $p^{n^*}_{\mathrm{w}}$ will converge to $p^{n^*}_{i^*}$ with the same rate, and the same holds for any quantity of interest $Z_p = Z_p(\mathbf{y})$ depending on p_{w}, that is

$$\left\| p^{n^*}_{\mathrm{w}} - p^{n^*}_{i^*} \right\|_{\otimes} \le C \frac{e^{-g_n m(\mathrm{w})}}{\sqrt{m(\mathrm{w})!}}, \quad \left\| Z^{n^*}_{p,\mathrm{w}} - Z^{n^*}_{p,i^*} \right\|_{L^2_\rho(\Gamma)} \le C \frac{e^{-g_n m(\mathrm{w})}}{\sqrt{m(\mathrm{w})!}}. \tag{21}$$

It is then possible to use a least square fitting on the computed errors to derive an estimated value for g_n. Figure 3 in next Section shows the results of such procedure applied to a test case, and confirms the quality of the method proposed. Alternative estimates for the decay of the Hermite coefficients are available in [16].

4.4 A Computable Expression for $\mathcal{I}(\mathrm{w})$

We are now in position to write a computable expression for the quasi-optimal set (10). Combining together the work contribution (11), the error contribution estimate (13), the estimate (20) for $\|p_{m(\mathbf{i}-1)}\|_{H^1(D)}$ and the numerical values obtained for $m(i_n)$, $B_n(i_n)$ and g_n, see respectively Eqs. (14), (18), and (21), we obtain the following expression

$$\mathcal{I}(\mathrm{w}) = \left\{ \mathbf{i} \in \mathbb{N}_+^N : \frac{\prod_{n=1}^{N} B_n(i_n) \frac{e^{-g_n m(i_n-1)}}{\sqrt{m(i_n-1)!}}}{\prod_{n=1}^{N} (m(i_n) - m(i_n-1))} \geq \epsilon_{\mathrm{w}} \right\}. \tag{22}$$

Again, note that (22) may not satisfy the admissibility condition (9), that has to be enforced by adding the missing multiindices.

5 Numerical Results

In this section we test on an example the effectiveness of the proposed sparse grid. We consider the case of a stratified material in the direction transversal to the flow: that is, the log-permeability field γ depends only on x_1 and is constant along x_2. Thus the covariance function is

$$C_\gamma(s,t) = \sigma^2 \exp\left(-\frac{|s-t|^2}{L_c^2}\right), \quad s,t \in [0,1],$$

and the truncated Fourier expansion of γ (5) simplifies to

$$\gamma(x_1, \mathbf{y}) = \overline{\gamma}(x_1) + \sigma\sqrt{c_0}y_0 + \sigma \sum_{k=1}^{K} \sqrt{c_k}\,[\, y_{2k-1}\cos(\omega_k x_1) + y_{2k}\sin(\omega_k x_1)] \tag{23}$$

with $\overline{\gamma} = \mathbb{E}[\gamma]$. As in Proposition 2, we have $y_k \sim \mathcal{N}(0,1)$, $\omega_k = k\pi/L$, $L = 1$, and λ_k as in Eq. (4). Obviously, in this case it holds $c_k \approx \lambda_k$ rather than $c_{\mathbf{k}} \approx \lambda_{k_1}\lambda_{k_2}$, due to the layer structure of γ. We set the correlation length to $L_c = 0.2$, the pointwise standard deviation to $\sigma = 0.3$ and $\overline{\gamma}(x_1) = 0$.

We consider three different levels of truncation for γ in (23): $K = 6, 10, 16$ corresponding to $N = 11, 21, 33$ random variables. With these truncation we take into account up to 2, 10^{-2} and 10^{-9} % respectively of the total variance of γ. For each truncation we compute the quasi-optimal sparse grid approximation $p_{N,\mathrm{w}}$ using the sets (22) with $\epsilon_{\mathrm{w}} = e^{-\mathrm{w}}$, $\mathrm{w} = 0, 1, 2, \ldots$, and then compute the expected value

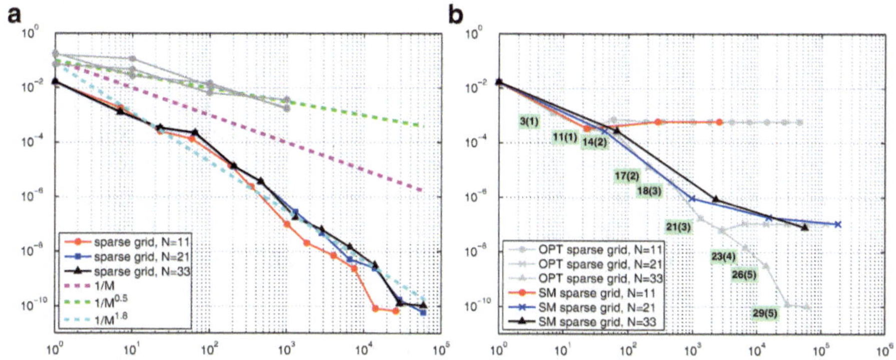

Fig. 4 Convergence for MC and sparse grid methods. (**a**) Convergence of quasi-optimal sparse grid approximations. (**b**) Convergence with respect to the reference solution with $N = 33$ random variables

for the total outgoing flux Z (see Eq. (2)), using the resulting sparse grid quadrature rule. We also perform a classical Monte Carlo simulation, repeated three times. The deterministic problems are solved with $\mathbb{P}1$ finite elements on an unstructured regular mesh with approximately 1,400 vertices.

We first fix the number of random variables N and study the convergence of the sparse grid approximation as the number of points in the sparse grid increases. Since we do not have an exact solution, we compute errors with respect to a reference solution, i.e. we measure the error as $\left|\mathbb{E}\left[Z_{p_{N,\mathrm{w}}}\right] - \mathbb{E}\left[Z_{p_{N,\mathrm{w}^*}}\right]\right|$, with $\mathrm{w}^* = 13$. Results are shown in Fig. 4a. The Monte Carlo simulations converge with the expected rate $1/2$; we also show the convergence rate 1 that would be obtained with a quasi-Monte Carlo method, like Sobol' sequences (see e.g. [26]). As for the sparse grids approximation, it is important to observe that not only they all converge with a rate higher than $1/2$, but such rate seems to be almost independent of the truncation level N. This would mean that the strategy detailed in Sect. 4 is quite effective in reducing the deterioration of the performance of the standard sparse grids as the number of random variables increases. Indeed, the selection of the most profitable hierarchical surpluses manages to "activate" (i.e. to put interpolation points) only in those directions that are most useful in explaining the total variance of the solution, so that the less influent random variables get activated only for small approximation errors. Beside the number of "active" variables, another interesting indicator is the number of "interacting" variables in the sparse grid. As was previously mentioned, a sparse grid is indeed a linear combinations of a number of "small" tensor grids, that put interpolation points only in some of the directions $y_1, \ldots, y_N$ at a time, say $\bar{n}$ directions out of N. We call the largest $\bar{n}$ in a sparse grid the number of interacting variables, that could be much lower than the number of active ones. The convergence curve for $N = 33$ is also shown in Fig. 4b (grey line), where for each level w we show the number of active variables followed by the number of interacting variables in parenthesis. For instance, the sparse grid labeled 18(3) places collocation points in 18 variables, but each tensor grid covers 3 dimensions at most.

We then repeat the analysis by computing the error for all the three approximations with respect to the *same* reference solution, i.e. p_{33,w^*}, again with $w^* = 13$. Results are shown in Fig. 4b, where we also show the convergence of a standard isotropic sparse grid obtained using the "doubling" function $m(i)$: $m(0) = 0, m(1) = 1, m(i) = 2^{i-1} + 1$ and Gauss-Hermite quadrature points. As expected, the convergence of the solutions with $N = 11$ and 21 stagnates when the error due to the truncation of the random field becomes predominant. Moreover, the convergence rate of the optimized grid is independent of the number of random variables up to the stagnation, while the convergence of the isotropic grid shows the well-known "curse of dimensionality" effect. Observe however that the performances of the isotropic and optimized sparse grids are equivalent up to about 1,000 collocation points (18 random variables), after which using the optimized sparse grid outperforms the standard one. Using a sharper profit estimate to build the optimized set (22) would further improve the efficiency of the optimized sparse grid. We finally remark that, even if the performances are similar for low tolerances, the advantage of using the optimized sparse grid with respect to the isotropic one is that it allows to work with virtually $N = \infty$ random variables, and provides an automatic truncation the permeability field at each tolerance level.

6 Conclusions

In this work we have considered a Darcy problem with uncertain permeability, modeled as a lognormal random field with Gaussian covariance function, and we have applied the quasi-optimal sparse grid paradigm derived in [5] to the problem at hand: we have introduced a nested quadrature/interpolation rule and we have estimated the proportionality constant $B(\mathbf{i})$ between error contribution of the sparse grids and the coefficients of the Hermite expansion of the solution, for which we have derived an estimate.

We have applied our quasi-optimal sparse grid thus obtained to a test case describing a layered material, that has been discretized with a Fourier expansion with $N = 11, 21$ and 33 random variables. Numerical results on this preliminary test seem to suggest that the quasi-optimal sparse grid procedure achieves a convergence rate higher than the ones of the most common sampling methods. Moreover, it is quite effective in reducing considerably the degradation of the performance suffered by the standard sparse grids approach when the number of input random variables increases.

Acknowledgements The second and third authors have been supported by the Italian grant FIRB-IDEAS (Project n. RBID08223Z) "Advanced numerical techniques for uncertainty quantification in engineering and life science problems". Support from the VR project "Effektiva numeriska metoder för stokastiska differentialekvationer med tillämpningar" and King Abdullah University of Science and Technology (KAUST) AEA project "Predictability and Uncertainty Quantification for Models of Porous Media" is also acknowledged. The fourth author is a member of the KAUST SRI Center for Uncertainty Quantification in Computational Science and Engineering.

References

1. T. Arbogast, M. F. Wheeler, and I. Yotov. Mixed finite elements for elliptic problems with tensor coefficients as cell-centered finite differences. *SIAM J. Numer. Anal.*, 34(2):828–852, 1997.
2. I. Babuška, F. Nobile, and R. Tempone. A stochastic collocation method for elliptic partial differential equations with random input data. *SIAM Review*, 52(2):317–355, June 2010.
3. V. Barthelmann, E. Novak, and K. Ritter. High dimensional polynomial interpolation on sparse grids. *Adv. Comput. Math.*, 12(4):273–288, 2000.
4. J. Bear and A.H.D. Cheng. *Modeling Groundwater Flow and Contaminant Transport*. Theory and Applications of Transport in Porous Media. Springer, 2010.
5. J. Beck, F. Nobile, L. Tamellini, and R. Tempone. On the optimal polynomial approximation of stochastic PDEs by Galerkin and collocation methods. *Mathematical Models and Methods in Applied Sciences*, 22(09), 2012.
6. F. Brezzi and M. Fortin. *Mixed and hybrid finite element methods*, volume 15 of *Springer Series in Computational Mathematics*. Springer-Verlag, New York, 1991.
7. H.J Bungartz and M. Griebel. Sparse grids. *Acta Numer.*, 13:147–269, 2004.
8. C.G. Canuto, Y. Hussaini, A. Quarteroni, and T.A. Zang. *Spectral Methods: Fundamentals in Single Domains*. Springer, 2006.
9. J. Charrier. Strong and weak error estimates for elliptic partial differential equations with random coefficients. *SIAM J. Numer. Anal.*, 50(1), 2012.
10. J. Galvis and M. Sarkis. Approximating infinity-dimensional stochastic Darcy's equations without uniform ellipticity. *SIAM J. Numer. Anal.*, 47(5):3624–3651, 2009.
11. A. Genz and B. D. Keister. Fully symmetric interpolatory rules for multiple integrals over infinite regions with Gaussian weight. *J. Comput. Appl. Math.*, 71(2), 1996.
12. T. Gerstner and M. Griebel. Dimension-adaptive tensor-product quadrature. *Computing*, 71(1):65–87, 2003.
13. C. J. Gittelson. Stochastic Galerkin discretization of the log-normal isotropic diffusion problem. *Math. Models Methods Appl. Sci.*, 20(2):237–263, 2010.
14. M. Griebel and S. Knapek. Optimized general sparse grid approximation spaces for operator equations. *Math. Comp.*, 78(268):2223–2257, 2009.
15. M. Grigoriu. *Stochastic Calculus: Applications in Science and Engineering*. Birkhäuser Boston, 2002.
16. V. H. Hoang and C. Schwab. N-term galerkin wiener chaos approximations of elliptic pdes with lognormal gaussian random inputs. SAM-Report 2011–59, ETHZ, 2011.
17. A. S. Kronrod. *Nodes and weights of quadrature formulas. Sixteen-place tables*. Authorized translation from the Russian. Consultants Bureau, New York, 1965.
18. H. Li and D. Zhang. Probabilistic collocation method for flow in porous media: Comparisons with other stochastic methods. *Water Resources Research*, 43(9), 2007.
19. G. Lin and A.M. Tartakovsky. An efficient, high-order probabilistic collocation method on sparse grids for three-dimensional flow and solute transport in randomly heterogeneous porous media. *Advances in Water Resources*, 32(5):712–722, 2009.
20. M. Loève. *Probability theory. II*. Springer-Verlag, New York, fourth edition, 1978. Graduate Texts in Mathematics, Vol. 46.
21. F. Müller, P. Jenny, and D.W. Meyer. Probabilistic collocation and lagrangian sampling for advective tracer transport in randomly heterogeneous porous media. *Advances in Water Resources*, 34(12):1527–1538, 2011.

22. S.P. Neuman, M. Riva, and A. Guadagnini. On the geostatistical characterization of hierarchical media. *Water Resources Research*, 44(2), 2008.
23. F. Nobile, R. Tempone, and C.G. Webster. An anisotropic sparse grid stochastic collocation method for partial differential equations with random input data. *SIAM J. Numer. Anal.*, 46(5):2411–2442, 2008.
24. T. N. L. Patterson. The optimum addition of points to quadrature formulae. *Math. Comp. 22 (1968), 847–856; addendum, ibid.*, 22(104):C1–C11, 1968.
25. C. Schwab and R. A. Todor. Karhunen-Loève approximation of random fields by generalized fast multipole methods. *Journal of Computational Physics*, 217(1):100 – 122, 2006.
26. I. H. Sloan and H. Woźniakowski. When are quasi-Monte Carlo algorithms efficient for high-dimensional integrals? *J. Complexity*, 14(1):1–33, 1998.
27. L. Tamellini. *Polynomial approximation of PDEs with stochastic coefficients*. PhD thesis, Politecnico di Milano, 2012.
28. D. Xiu and J.S. Hesthaven. High-order collocation methods for differential equations with random inputs. *SIAM J. Sci. Comput.*, 27(3):1118–1139, 2005.

The Geometric Basis of Numerical Methods

Marc Gerritsma, René Hiemstra, Jasper Kreeft, Artur Palha, Pedro Rebelo, and Deepesh Toshniwal

Abstract The relation between physics, its description in terms of partial differential equations and geometry is explored in this paper. Geometry determines the correct weak formulation in finite element methods and also dictates which basis functions should be employed to obtain discrete well-posedness.

1 Introduction

Tonti in 1972 [43], made the classification for a very large number of physical theories based on geometry. Reading Tonti is a fascinating experience, because his work sheds a completely new light on how we perceive physical equations.

Mattiussi [33], wrote in 2000 a very clear paper based on Tonti's work. Mattiussi relates the geometric concepts to numerical methods. Bossavit explains the whole geometric structure underlying the Maxwell equations in [6]. See also [4, 27] for a description in terms of differential forms and cochains.

M. Gerritsma (✉) · A. Palha · P. Rebelo
Aerospace Engineering, TU Delft, Delft, The Netherlands
e-mail: m.i.gerritsma@tudelf.nl; a.palha@tudelf.nl; p.rebelo@tudelf.nl

D. Toshniwal
EPFL, Lausanne, Switzerland
e-mail: deepesh.toshniwal@epfl.ch

R. Hiemstra
ICES, University of Austin, Austin, TX, USA
e-mail: Hiemstra@ices.utexas.edu

J. Kreeft
Shell Global Solution, Amsterdam, The Netherlands
e-mail: Jasper.Kreeft@shell.com

M. Azaïez et al. (eds.), *Spectral and High Order Methods for Partial Differential Equations - ICOSAHOM 2012*, Lecture Notes in Computational Science and Engineering 95, DOI 10.1007/978-3-319-01601-6_2,

Originally, the reconstruction of differential forms from discrete cochains was established by means of Whitney forms [6, 7, 24, 39, 40]. Later other basis functions were developed. In the finite element setting important contributions are made by Arnold, Falk and Winther [1, 2] and Hiptmair [23]. Application of these ideas to finite difference/finite volume methods can be found in [8, 9, 28, 42]. For the application of these ideas to isogeometric methods, see [11, 15, 16, 21, 22]. A very geometric approach was developed at Caltech [12, 25, 26]. Application to fluid dynamics and a motivation to use these structure-preserving techniques is described by Perot [37, 38] and the description of incompressible flow in terms of the Lie group of volume-preserving diffeomorphisms with associated Lie algebra of divergence-free vector fields can be found in [17, 36].

The above cited references serve as an excellent introduction into the relation between physical models, partial differential equations, numerical methods and the underlying geometrical structure.

Understanding the geometry associated with physical variables also has profound implications for numerical methods. If one faithfully respects the underlying geometrical structure, some of the discrete relations become exact. This exactness is preserved under a large class of transformations and should be independent of the basis functions.

2 Physics and Geometry

All physical variables are associated with geometric objects. Mass is associated to a volume. A flux is associated to a surface and circulation is associated to a contour. Although these examples are trivial, there are situations where this association is less clear. Consider, for instance, the perfect fluid given by

$$\operatorname{div}\mathbf{u} = 0\,, \tag{1}$$

$$\frac{\partial \mathbf{u}}{\partial t} + (\mathbf{u}, \nabla)\,\mathbf{u} + \nabla p = 0\,, \tag{2}$$

where p is the pressure and $\mathbf{u}$ is the velocity. But what we commonly call 'velocity' is slightly more difficult when we take geometry into account. In (1) the velocity is associated with a surface, while in (2) the convective velocity (the $\mathbf{u}$ between brackets) is associated with a line. The remaining velocities (momentum) in (2) is probably the most difficult one in terms of geometry and it is associated with *both lines and volumes*, see [44] for a treatment of momentum. In the remainder of this paper we will focus on variables which are easier to associate with geometry.

Mass and mass density The physical variable 'mass' is associated with three dimensional volumes. It is important to note that the relation between mass and volume *is independent of the actual shape and size of the volume.*

In engineering we usually do not work with mass directly, but prefer *mass density*. If we have a mass occupying a certain volume, we can calculate its average density, $\bar{\rho}$, using $\bar{\rho} = M/V$.

The average mass density is considered to contain too little detail to be useful, it tells us nothing about the *mass distribution*. So what one can do is to divide the brick in two and calculate the average mass densities for the two halves, $\bar{\rho}_1$ and $\bar{\rho}_2$. If these two average densities are equal, one might be tempted to conclude that the mass distribution was uniform before cutting the brick in two. This remains a hypothesis which cannot be corroborated, but only falsified by successively cutting the brick in smaller parts. Mathematicians are less inhibited by physical reality and define *the* mass density as

$$\rho = \lim_{V \to 0} \frac{M}{V} .$$

Physically this limiting process is not feasible.

Given the physical limitations of mass density, it is easier to work with the global quantity mass and the volume it is associated with. Another advantage is that mass and volume are additive, whereas mass densities are not. These rather trivial observations are very relevant for numerical methods, where, no matter how small we make our mesh, we always work with finite cells or elements.

Flux and flux density The instantaneous flow through a surface is generally called *the flux*. This suggests that the physical variable flux is associated with two dimensional objects, i.e. surfaces. If the area or the orientation of the surface changes, the flux will generally also change. In order to remove the dependence on the area of the surface, the flux density normal to the surface is defined as the limit of the average flux density for the area to zero.

$$\rho_{flux} = \lim_{A \to 0} \frac{F}{A} .$$

Similar considerations as for mass density apply to flux density. Mathematically this might be a well-defined concept, but physically it is questionable. We might be able to measure the flux through a very small surface, but never the flux *in a point*.

Again, for numerical methods it is usually preferable to work with the global quantities such as mass and flux, then the limiting variables mass density and flux density, because our meshes will be far too coarse (no matter what your computational resources are) to come close to these limiting values. The only thing we can hope for is that we *approximate* the mass density and the flux density, whereas mass and flux can be represented *exactly* on a mesh.

Displacement and velocity The final example concerns the measurement of velocity. One way to do this, is to release tracer particles in the flow. Make two snapshots of the flow while the tracer particles are illuminated by a laser beam. This allows one to determine the position, r_1, of a tracer particle at t_1 and the position of the tracer particle, r_2, at time $t_2 > t_1$. This is in practice more difficult then described here, but for the moment we assume that we can determine t_1, t_2, r_1 and r_2 with infinite precision. Then we know that

$$r_2 - r_1 = \int_{t_1}^{t_2} \frac{dr}{dt}\, dt = \int_{t_1}^{t_2} \mathbf{v}\, dt \,.$$

Here $r_2 - r_1$ denotes the *displacement* of the particle in the time interval $\Delta t = t_2 - t_1$ and $\mathbf{v}$ is the velocity of the particle in the time interval Δt. This relation between displacement and velocity is exact. We have no idea what the particle did between t_1 and t_2. It could go in a straight line from r_1 to r_2, but it could also have oscillated several times between the points r_1 and r_2, or the particle could have taken a grand detour to get from r_1 to r_2. So we cannot say much about the velocity in the time interval Δt, but we do know that the time integral of the velocity equals the displacement.

The measurement described above is called particle image velocimetry (PIV) and the velocity measurement from PIV is usually done by *approximating* the velocity by

$$\mathbf{v} \approx \frac{r_2 - r_1}{t_2 - t_1} \,.$$

This approximation is exact if the direction and speed of the particle is constant during the time interval Δt, in which case we have

$$r_2 - r_1 = \int_{t_1}^{t_2} \mathbf{v}\, dt = \mathbf{v} \int_{t_1}^{t_2} dt = \mathbf{v}(t_2 - t_1) \,.$$

However, this is based on the assumption that the direction and speed of the particle is constant. But this assumption is unknowable. We could decrease the time interval to test our assumption, but we can never corroborate this assumption. Basis functions and interpolation in numerical methods play the same role as the assumption made by the experimentalist about the behaviour of the system between the observations.

In this example, velocity is a physical quantity associated with curves, the particle path or the streamline for steady flows. Any attempt to define the velocity in the PIV experiment at a certain point and time instant may be mathematically correct, but physically not realistic.

These three simple examples serve to illustrate the connection between physical variables and geometric objects such as volumes, surfaces, lines and points. In fact, all physical variables are associated with geometric objects, but due to a twist of fate we have been working with point-wise defined quantities in engineering and have tried to apply this nodal approach in our numerical schemes; instead of using surface forces we use the stress tensor and mass density instead of mass. The reader may easily find more examples.

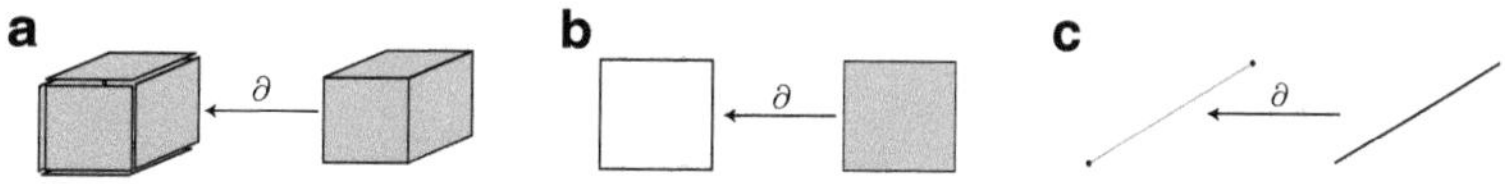

Fig. 1 Application of boundary operator. (**a**) The boundary operator ∂ which maps a volume onto its six faces. (**b**) The boundary operator ∂ which maps a surface onto its four edges. (**c**) The boundary operator ∂ which maps a curve onto its two endpoints

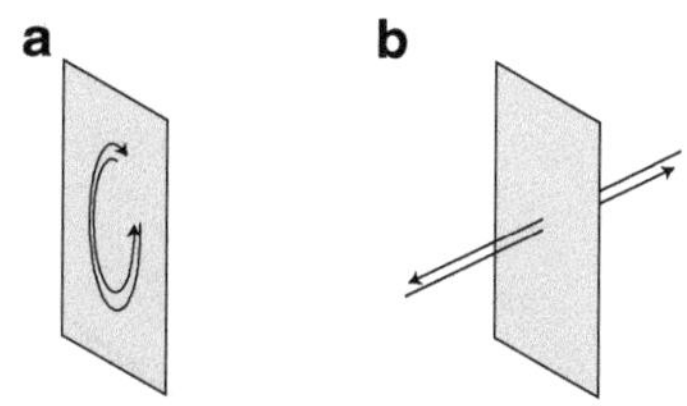

Fig. 2 The two types or orientation for a surface. Inner-orientation (*left*), outer-orientation (*right*). (**a**) The two ways of choosing an orientation *in* the surface. (**b**) Selecting a positive orientation *through* a surface

3 Geometry, Orientation and Basic Operations

The most important operator that we need to discuss is the boundary operator.

The picture given in Fig. 1a we see the boundary operator, ∂, applied to a volume. It maps the volume onto its six faces. Similarly, the boundary operator applied to a surface yields its four edges as shown in Fig. 1b. We also have the boundary operator applied to a curve which returns its two endpoints as shown in Fig. 1c. Note that although in these three figures we use an orthogonal volume and square for a surface and a straight line segment for the curve, nothing changes if we deform the volume, surface or curve. This indicates that the boundary operator is a topological operator which does not depend on a specific shape or size.

The boundary operator maps a k-dimensional object onto a $(k-1)$-dimensional object as the above examples show. Now that we use higher dimensional objects than points, we also need to take orientation into account. In Fig. 2a we see a surface with the two possible ways of orientation. We can either choose a clockwise rotation to be the default orientation or the counter-clockwise rotation. There is no preferred orientation, but once you choose a default orientation, you need to stick to it during further analysis.

Besides orienting a surface by choosing a sense of rotation *in* the surface, there is a second type of orientation where you choose a positive direction *through* the surface, as shown in Fig. 2b. The first type of orientation, shown in Fig. 2a, is called *inner orientation*, whereas the second type of orientation, shown in Fig. 2b, is referred to as *outer orientation*.

It turns out that the representation of physical variables are not only associated with geometric objects, but to geometric objects with *a specific type of orientation*, inner- or outer-oriented. Mass and flux are associated to outer-oriented objects,

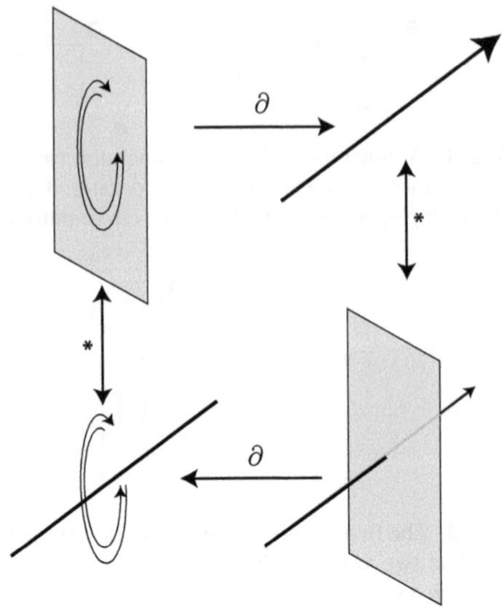

Fig. 3 The boundary operator, ∂, and the switch between inner- and outer orientation, *

while velocity, in the PIV example, is associated to inner-oriented curves. So a full geometric description needs to take this aspect also into account. Note that inner-orientation can be defined without reference to the ambient space in which the geometric object is embedded, whereas outer-orientation depends explicitly on the dimension of the ambient space, see [6, 32, 34].

Now that we have inner and outer orientation, we can set up the loop displayed in Fig. 3. We see that the boundary operator maps inner-oriented k-dimensional objects onto inner-oriented $(k-1)$-dimensional objects and outer-oriented k-dimensional objects onto outer-oriented $(k-1)$-dimensional objects. The boundary operator does not change the type of orientation. In Fig. 3 we also included an operation, *, which leaves the 'sense of orientation invariant' but changes the associated geometric object. On the left we see a sense of rotation *in* a plane (top left) and a sense of rotation *around* a line (bottom left), while on the right, we have the direction *along* a line (top right) and the same direction, but now considered as *through* a plane (bottom right). In general, the *-operator associates a k-dimensional object to an $(n-k)$-dimensional object and thereby changes the type of orientation (from inner to outer and vice versa). Here n is the dimension of the space in which the objects are embedded. Whereas the boundary operator is a purely topological operator, the *-operator is metric-dependent. See, Harrison, [20], for a more formal description of the geometric *-operator.

With this *-operator and the boundary operator we can define a map from $(k-1)$-dimensional objects to k-dimensional objects (of the same type of orientation), by

$$\partial^* := *\partial* \;:\; (k-1)\text{-dim} \rightarrow k\text{-dim}\;.$$

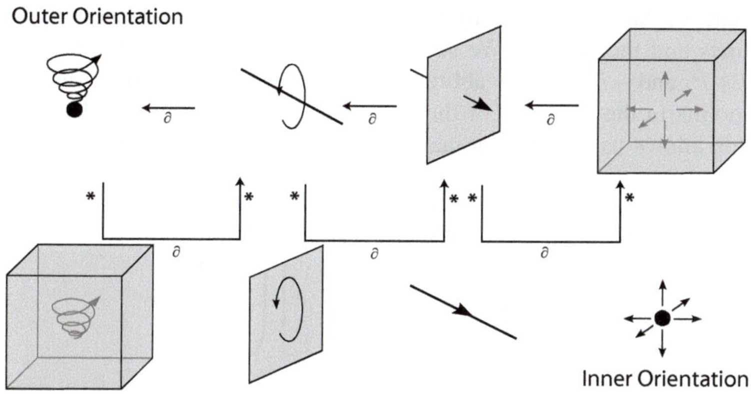

Fig. 4 Inner- and outer-oriented geometric objects and the operators ∂ and $\partial^* = *\partial*$

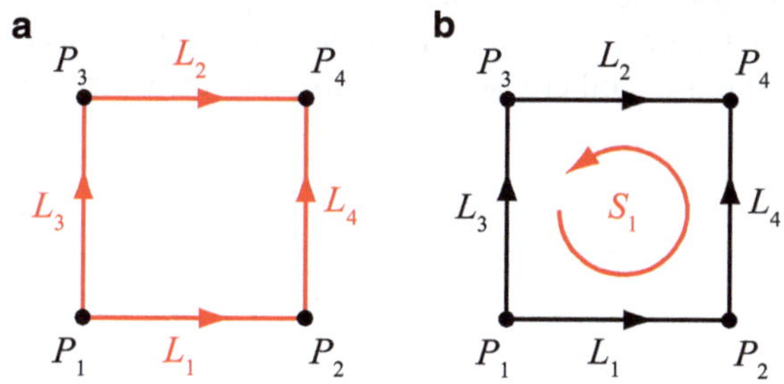

Fig. 5 Oriented grid. Relation between line segments and points (*left*) and surface and line segments (*right*). (**a**) The numbering of inner-oriented points and surfaces to set up the incidence matrix. (**b**) Orientation of the surface and the boundary for the incidence matrix

If we start with the inner-oriented line in the top right of Fig. 3, apply the * to get the traverse direction *through* the plane, then apply the boundary operator ∂ to get the circulation *around* the line and finally the *-operator again to get a sense of orientation *in* the plane. Since this *-operator is metric-dependent, ∂^* will depend on the metric as well.

The loop in Fig. 3 can be extended to include all k-dimensional objects in 3D and this is depicted in Fig. 4. Inner-oriented volumes are usually oriented with right hand rules and inner-oriented points are oriented as source or sinks. The geometric framework displayed in Fig. 4 forms the geometric basis for the double DeRham complex. The boundary operator ∂ and its metric-dependent antagonist ∂^* will play a crucial role in numerical methods. Many operations we perform in numerical analysis find their origin in these two operations, as we will see.

The boundary operator also has a very natural matrix representation in terms of *incidence matrices*.

In Fig. 5 we have a two-dimensional 'grid' consisting of four points, four line segments and one surface. Although all the lines are straight and all angles rectangular, we can deform the figure without changing the connectivity between points, lines and the surface. We see that the boundary of the line labeled L_1 is given by $+P_2$ and $-P_1$, usually abbreviated as $\partial L_1 = P_2 - P_1$. We write a $+$ when the line points in the direction of the point and a $-$ when the line departs from a point. This allows us to set up a matrix representation for the boundary operator given by

$$\begin{pmatrix} \partial L_1 \\ \partial L_2 \\ \partial L_3 \\ \partial L_4 \end{pmatrix} = \begin{bmatrix} -1 & 1 & 0 & 0 \\ 0 & 0 & -1 & 1 \\ -1 & 0 & 1 & 0 \\ 0 & -1 & 0 & 1 \end{bmatrix} \begin{pmatrix} P_1 \\ P_2 \\ P_3 \\ P_4 \end{pmatrix} .$$

Similarly, we can set up an incidence matrix which connects the oriented surface, S_1 with the four bounding line segments, see Fig. 5b. The incidence matrix, which connects the oriented surface to the bounding line segments, is constructed such that if the orientation of the line segment L_i is in the same direction as the orientation of the surface, we write $+L_i$, otherwise $-L_i$. This gives the matrix representation

$$\partial S = [1 \quad -1 \quad -1 \quad 1] \begin{pmatrix} L_1 \\ L_2 \\ L_3 \\ L_4 \end{pmatrix} .$$

So we have a matrix representation for the boundary operator. If we deform our 'grid' without changing the connectivity between points, lines and the surface, the incidence matrices remain the same. Note that $\partial \circ \partial S \equiv \emptyset$. So the boundary of the boundary is empty. This result is not specific for this example, but is generally true. For all geometric objects, the boundary of the boundary is the empty set. The reverse is generally not true, i.e. if the boundary of an object is empty, the object does not need to be a boundary itself.

4 Assigning Values to Objects

We started this paper with the fact that physical variables are associated to geometric objects. We then discussed geometric objects briefly and now we will add values to geometric objects. To make life easier, we are going to introduce some simple notation: Suppose we have a collection of points, lines which connect them, surfaces bounded by lines and volumes bounded by surfaces. We will call the points 0-cells and denote them by $\sigma_{(0),i}$ where the subscript (0) indicates that it is a point and i is just a label to distinguish different points, just as we did in the example for the incidence matrices. Similarly, $\sigma_{(1),i}$ will refer to the oriented line segments (not necessarily straight) in our mesh, $\sigma_{(2),i}$ the oriented and labeled surfaces and finally

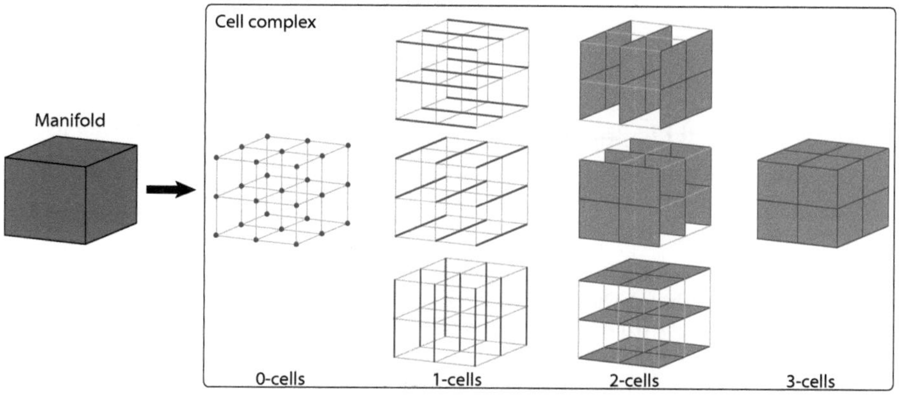

Fig. 6 Subdivision of the domain (manifold) in points (0-cells), line segments (1-cells), faces (2-cells) and volumes (3-cells)

$\sigma_{(3),i}$ will refer to oriented volumes, all of the same type of orientation (either inner-oriented or outer-oriented). Together these building blocks will constitute a so-called *cell complex*, but in computational science we usually refer to such a collection as a *grid* or a *mesh*, see Fig. 6. The main difference is that a grid or mesh is usually not oriented whereas a cell complex is.

A collection of oriented k-dimensional cells will be called a k-chain, $\mathbf{c}_{(k)}$, and is usually written as a formal sum

$$\mathbf{c}_{(k)} = \sum_{i=1}^{\#k} m^i \sigma_{(k),i} \ ,$$

where $\#k$ denotes the number of k-cells in the complex and m^i is 0, when the cell $\sigma_{(k),i}$ is not part of the chain, is equal to 1 when the cell $\sigma_{(k),i}$ is in the chain and $m^i = -1$ when $\sigma_{(k),i}$ is in the chain but the orientation is opposite to its default orientation.

In the examples given above (mass, flux and velocity) we assigned values to geometric objects. Now we are going to assign values to the k-cells. Let $\sigma^{(k),j}$ be the operator which assigns the value 1 to the k-cell $\sigma_{(k),j}$ and 0 to all the other k-cells. This will be denoted by

$$\langle \sigma^{(k),j}, \sigma_{(k),i} \rangle = \delta_i^j = \begin{cases} 1 & \text{if } i = j \\ 0 & \text{if } i \neq j \end{cases} .$$

If we want to assign a different value to a k-cell, say the value c_j, then we apply $c_j \sigma^{(k),j}$ to the k-cells. We can collect all these assignments into a formal sum and write

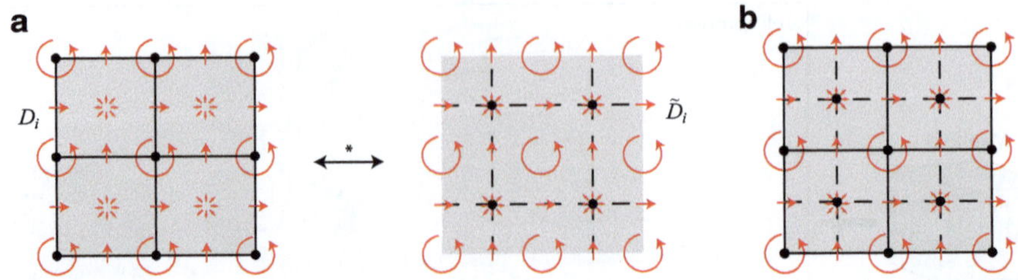

Fig. 7 The geometric construction of a staggered grid consisting of an overlapping inner-oriented and outer-oriented mesh. (**a**) Outer-oriented 2D mesh on the *left* and the corresponding inner-oriented mesh on the *right*. (**b**) A staggered grid is obtained by putting the two dual cell-complexes displayed in Fig. 7a on top of each other

$$\mathbf{c}^{(k)} = \sum_{i=1}^{\#k} c_i \sigma^{(k),i} \ , \quad c_i \in \mathbb{R} \ .$$

So if we have mass, $k = n = 3$ and c_i denotes the mass assigned to the 3-cell $\sigma_{(3),i}$. If we have a k-chain $c_{(k)}$, we can assign a value to the whole chain by

$$\langle \mathbf{c}^{(k)}, \mathbf{c}_{(k)} \rangle = \left\langle \sum_{i=1}^{\#k} c_i \sigma^{(k),i}, \sum_{i=j}^{\#k} \sigma_{(k),j} \right\rangle = \sum_{i=1}^{\#k} \sum_{i=j}^{\#k} c_i \langle \sigma^{(k),i}, \sigma_{(k),j} \rangle = \sum_{i=1}^{\#k} c_i \ .$$

The operator $\mathbf{c}^{(k)}$ is called the *k-cochain*. So we now have a discrete description of geometry and a way of assigning values to geometric objects. Note that this assignment of a number to a collection of geometric objects is the discrete analogue of integration over a k-dimensional manifold Ω_k, as given by

$$\left\langle a^{(k)}, \Omega_k \right\rangle := \int_{\Omega_k} a^{(k)} \in \mathbb{R} \ ,$$

where $a^{(k)}$ is differential k-form and Ω_k the k-dimensional submanifold. So integration can be considered as the continuous analogue of assigning numbers to geometric objects. Both the duality pairing of cochains and chains as well as integration of differential forms are *metric-free operations*.

Typical examples of these abstract constructions are the association of mass to a volume (integration of mass density over a volume in the continuous case), assigning a flux to a surface (integration of flux density in the continuous case) or assigning circulation to a curve (integration of velocity along a curve at the continuous level).

We considered inner-orientation and outer-orientation and in order to represent both types of orientation we need to use two meshes. One representing the inner-oriented variables the other the outer-oriented variables, see Fig. 7a.

The two meshes on which we represent inner- and outer orientation constitute a *staggered grid* as shown in Fig. 7b. This configuration is quite common in finite volume methods for incompressible flow.

5 Discrete Derivative

Duality pairing between cochains and chains allows us to introduce the discrete derivative as the formal adjoint of the boundary operator.

$$\forall \mathbf{c}_{(k+1)},\ \forall \mathbf{c}^{(k)}\ \exists ! \mathbf{c}^{(k+1)}\ s.t.\ \langle \mathbf{c}^{(k+1)}, \mathbf{c}_{(k+1)} \rangle = \langle \mathbf{c}^{(k)}, \partial \mathbf{c}_{(k+1)} \rangle \ .$$

The map which associates the unique $\mathbf{c}^{(k+1)}$ to each $\mathbf{c}^{(k)}$ in the above relation, is called the *coboundary operator*, δ

$$\delta\, C^k \rightarrow C^{k+1}\ s.t.\ \ \langle \delta \mathbf{c}^{(k)}, \mathbf{c}_{(k+1)} \rangle = \langle \mathbf{c}^{(k)}, \partial \mathbf{c}_{(k+1)} \rangle \quad \forall \mathbf{c}_{(k+1)}\ ,$$

where C^k denotes the space of all k-cochains on our grid. Note that we have here the discrete analogue of well-known integral relations: Let C be an arbitrary curve going from the point A to the point B

$$k = 0\ :\ \int_C \operatorname{grad} \phi \cdot \mathrm{d}\mathbf{s} = \int_{\partial C} \phi = \phi(B) - \phi(A)\ .$$

Let S be a surface bounded by ∂S then

$$k = 1\ :\ \int_S \operatorname{curl} \cdot \mathbf{A}\, \mathrm{d}\mathbf{S} = \int_{\partial S} \mathbf{A} \cdot \mathrm{d}\mathbf{s}\ .$$

Let V be a volume, bounded by ∂V then

$$k = 2\ :\ \int_V \operatorname{div} \mathbf{F}\, \mathrm{d}V = \int_{\partial V} \mathbf{F} \cdot \mathrm{d}\mathbf{S}\ .$$

So the coboundary is the discrete analogue of the gradient, curl and divergence operator. Note that duality pairing is metric-free and the boundary operator is metric-free, therefore we have a metric-free representation of grad, curl and div, see also [42]. This implies that if we use basis functions in a finite element method or interpolations in finite difference or finite volume methods, the above integral should hold ALWAYS. On nice orthogonal grids this needs to hold, but also on highly curved grids (even self-overlapping grids). It should hold for low order approximations (linear basis functions), but also for high order methods (spectral elements). In fact, these relations can be satisfied without the introduction of basis

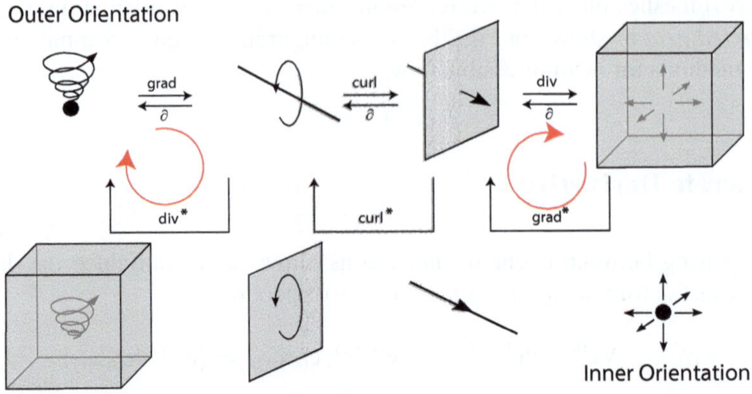

Fig. 8 The double DeRham complex with the metric-free operations in the top row and the metric-dependent operations via the detour along the dual cell-complex. The scalar Laplace operator which maps scalars defined in points to scalars defined in points (*red arrow on the left*) and scalars associated with volumes to volumes (*red arrow on the right*)

functions and if you do introduce basis functions, they should cancel from the equation, see [18, 19, 32].

Since we have a matrix representation for the boundary operator in terms of the incidence matrices as we described above, we also have a matrix representation for the coboundary operator. The discrete derivative is therefore completely determined by the mesh.

We can also define – formally – the adjoint of the *-operator which switched the type of orientation. For every $\mathbf{c}^{(n-k)}$ there exists a unique $\mathbf{c}^{(k)}$ such that

$$\langle \mathbf{c}^{(k)}, \mathbf{c}_{(k)} \rangle = \langle \mathbf{c}^{(n-k)}, *\mathbf{c}_{(k)} \rangle \,, \quad \forall \mathbf{c}_{(k)}$$

We will denote the map which associates this unique $\mathbf{c}^{(k)}$, by $\star\mathbf{c}^{(n-k)}$, i.e. $\star : C^{n-k} \rightarrow C^k$. This duality relation allows us to write the formal adjoint of ∂^*

$$\langle \delta^* c^{(k+1)}, c_{(k)} \rangle = \langle c^{(k+1)}, \partial^* c_{(k)} \rangle \,.$$

It is easy to verify that, just like the coboundary operator, δ^* is a discrete derivative operator which represents the grad, curl and div. However, it cannot be the same grad, curl and div as represented by δ, because δ is purely topological, independent of angles, length, curvature, while δ^* is a metric operator because it is composed of the metric-dependent $\star$-operator. In order to make this distinction explicit, we will refer to grad*, curl* and div* whenever we refer to the vector operator represented by δ^*. The associated geometric picture is displayed in Fig. 8. This is one of the reasons to use a more geometric approach, because now we can see the distinction between the topological vector operations, grad, curl and div, and the metric-dependent vector operations, grad*, curl* and div*; a distinction which is completely obscured by vector calculus. We noted that discretization of topological operators needs to be independent of basis functions, while the discrete version of the metric-dependent vector operators will always depend on the basis functions.

As a consequence of the fact that the boundary of the boundary is empty, we have that $\delta \circ \delta \equiv 0$ and $\delta^* \circ \delta^* \equiv 0$, which in vector calculus is represented by curl∘grad$\equiv 0$, div∘curl$\equiv 0$ and curl*∘grad$^* \equiv 0$, div*∘curl$^* \equiv 0$. So geometric identities immediately provide well-known vector identities [42].

Figure 8 also gives us the various Laplace operators in 3D. For instance, the well-known scalar Laplace operator defined in points is illustrated by the left red arrow in Fig. 8. Following the red arrow in Fig. 8 shows that we first apply the topological grad followed by the metric-dependent div*, so $\Delta = \text{div}^* \,\text{grad}$. Alternatively, we can write the Laplace operator for scalars associated with volumes as shown by the rightmost red arrow in Fig. 8. The Laplace operator associated with volumes is then given by $\Delta = \text{div}\,\text{grad}^*$. At first sight there is no difference with the Laplace operator defined on points, however closer inspection reveals that now the grad is the metric dependent operator, while for the scalar Laplace defined in points is was the div. This seemingly minor change has consequences for finite element methods, see Sect. 7.

6 Going from Continuous to Discrete and Back Again

In the previous sections we mainly focused on the discrete setting in terms of chains (geometry) and cochains (variables). At the continuous level we have k-dimensional manifolds (geometry) and differential k-forms (variables). Generally a physical problem is given in terms of continuous variables. These need to be converted to discrete values (cochains) to apply the above ideas. However, not everything can be accomplished fully at the discrete level. Once in a while we need to reconstruct a continuous representation from the discrete values. This is for instance the case with metric-dependent operations such as the $\star$-operator and when presenting the final solution. So here we will briefly outline the reduction operation, $\mathscr{R}$ which maps continuous variables to cochains and the reconstruction operator, $\mathscr{I}$, which reconstructs continuous differential forms from discrete cochains. For more details, see [32].

Reduction [4] Given a differential k-form, $a^{(k)}$, and a discrete k-cells, $\sigma_{(k),i}$ then the associated k-cochains is defined as

$$\mathscr{R}(a^{(k)}) = c^k = \sum_i \left[\int_{\sigma_{(k),i}} a^{(k)} \right] \sigma^{(k),i} .$$

So if the continuous variable is the mass density ρ, then integration over all volumes in the mesh gives us the mass contained in all volumes and this discrete value associated to a volume is a cochain. The reduction operator is denoted by $\mathscr{R}$.

Reduction commutes with differentiation, i.e. the following diagram commutes

$$\mathscr{R}\mathrm{d} = \delta\mathscr{R} \quad \Leftrightarrow \quad \begin{array}{ccc} \Lambda^k & \xrightarrow{\mathrm{d}} & \Lambda^{k+1} \\ \downarrow \mathscr{R} & & \downarrow \mathscr{R} \\ C^k & \xrightarrow{\delta} & C^{k+1} \end{array}$$

In plain vector calculus, we can take grad, curl or div and then reduce or first reduce to discrete values and then apply the coboundary which acts as grad, curl and div as was shown previously. Here Λ^k is the space of differential k-forms and C^k is the space of k-cochains.

Reconstruction [4] Reconstruction, denoted by $\mathscr{I}$, is a representation of the discrete variables at the continuous level. While reduction $\mathscr{R}$ is more or less fixed, reconstruction allows more freedom and essentially the various ways in which one can reconstruct discrete data has lead to the plethora of numerical methods. The reconstruction operator should satisfy the following criteria:

1. $\mathscr{I}$ should be the right inverse of the reduction operator $\mathscr{R}$, i.e. $\mathscr{R} \circ \mathscr{I} \equiv \mathbb{I}$;
2. $\mathscr{I}$ should be an *approximate left inverse* of the reduction operator $\mathscr{R}$, i.e. $\mathscr{I} \circ \mathscr{R} = \mathbb{I} + \mathscr{O}(h^p)$, where h denotes a characteristic mesh size and p the order of the method;
3. $\mathscr{I}$ should commute with continuous and discrete differentiation, i.e.

$$\mathscr{I}\delta = \mathrm{d}\mathscr{I} \quad \Leftrightarrow \quad \begin{array}{ccc} C^k & \xrightarrow{\delta} & C^{k+1} \\ \downarrow \mathscr{I} & & \downarrow \mathscr{I} \\ \Lambda^k & \xrightarrow{\mathrm{d}} & \Lambda^{k+1} \end{array}$$

The first condition is obvious and when applied to 0-cochains it is just nodal interpolation. For k-cochains, other than $k = 0$, things are less obvious. In [18, 32] spectral element basis functions are given which interpolate 1-cochains in 1D, see Fig. 9, and Hiemstra [21, 22] discusses B-spline reconstructions for 0- and 1-cochains in 1D. Higher-dimensional reconstructions are obtained using tensor products, which restricts the reconstruction to quadrilateral elements. On triangular grids the well-known Whitney forms have the desired interpolation property, see for instance [6, 39] and the families $\mathscr{P}_r^-\Lambda^k(\mathscr{T}_h)$ and $\mathscr{P}_r\Lambda^k(\mathscr{T}_h)$ developed by Arnold, Falk and Winter [1, 2], which apply to regular simplicial triangulations of polyhedra in any dimension. Recently, Arnold and Awanou have extended finite element exterior calculus to cubical meshes [3]. The use of quadrilateral elements could be beneficial in dealing with thin solids or boundary layer resolving meshes in viscous, incompressible flow.

The basis functions in Fig. 9 have the property that

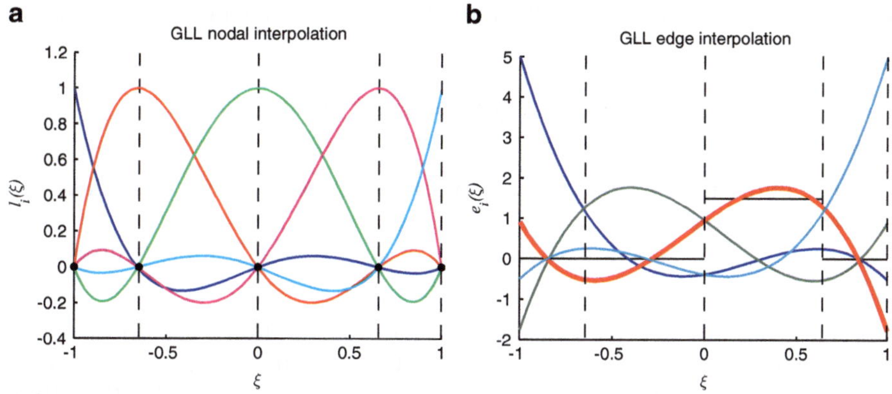

Fig. 9 (**a**) Nodal interpolation for the reconstruction of 0-forms from 0-cochains and (**b**) edge functions to reconstruct 1-forms from 1-cochains. These spectral basis functions are defined for any order

$$h_i(\xi_j) = \delta_{ij} \qquad \int_{\xi_{j-1}}^{\xi_j} e_j(\xi) = \delta_{ij} \ ,$$

which is another way of writing $\mathscr{R} \circ \mathscr{I} = \mathbb{I}$ as described in [18] and [32, Lemma 7, p. 52]

The second condition is an approximability condition necessary for convergence. Spectral and high-order methods focus on reconstructions for high values of p. The third requirement states that reconstruction commutes with differentiation. So, we can take the discrete derivative (coboundary) and then reconstruct or first reconstruct and then take the continuous derivative, $\mathscr{I}\delta = \mathrm{d}\mathscr{I}$. When these basis functions are used, the expansion coefficients are precisely the cochains discussed in the previous section.

Mimetic discretization We define the discretization as $\pi = \mathscr{I} \circ \mathscr{R}$. It follows that

$$\mathrm{d}\pi = \mathrm{d}\mathscr{I}\mathscr{R} = \underline{\mathscr{I}\delta\mathscr{R}} = \mathscr{I}\mathscr{R}\mathrm{d} = \pi\mathrm{d} \quad \Leftrightarrow \quad \begin{array}{ccc} \Lambda^k & \xrightarrow{\ \mathrm{d}\ } & \Lambda^{k+1} \\ \downarrow{\scriptstyle \pi} & & \downarrow{\scriptstyle \pi} \\ \Lambda^k_h & \xrightarrow{\ \mathrm{d}\ } & \Lambda^{k+1}_h \end{array}$$

Here we have underlined the expression in the middle, because this is the form we actually use in numerical computations. Note that we have a matrix expression for the coboundary operator in terms of the incidence matrices, so this whole expression only depends on the choice of $\mathscr{I}$. This commutation relation has some immediate consequences. For instance, we satisfy vector identities at the discrete level (not only on orthogonal grids but also on curvilinear grids). For example

$$\mathrm{div}_h\, \mathrm{curl}_h\, \mathbf{u}_h = \mathrm{div}_h\, \mathrm{curl}_h\, (\pi\mathbf{u}) = \mathrm{div}_h\, \pi\, (\mathrm{curl}\, \mathbf{u}) = \pi\, (\mathrm{div}\, \mathrm{curl}\, \mathbf{u}) = 0 \ .$$

If we have a conservation law of the form $\operatorname{div}\mathbf{u} = f$, then at the discrete level we have

$$\operatorname{div}_h \mathbf{u}_h - f_h = \operatorname{div}_h (\pi \mathbf{u}) - \pi f = \pi (\operatorname{div}\mathbf{u} - f) = 0 \, .$$

- If $f \in \operatorname{Im}(\operatorname{div})$, then $f_h \in \operatorname{Im}(\operatorname{div}_h)$, therefore existence of a discrete solution.
- $\operatorname{Ker}(\operatorname{div}_h) \subset \operatorname{Ker}(\operatorname{div})$.
- $\operatorname{div}_h u_h = f_h$ holds pointwise, in particular, $\operatorname{div}_h u_h = 0$ is strongly divergence free.

Especially, the fact that the discrete null space of the divergence operator (but also the other vector operators) is a proper subspace of the continuous divergence operator is an important property in mixed formulations [10, 29, 30]. Mixed formulations are in another aspect closely related to geometry as we will show in §7 for the Laplace equations for volume forms.

If operations commute many nice properties follow. However, this is all based on the coboundary operator, δ, i.e. the formal adjoint of boundary operator, ∂. But these relations do NOT hold for δ^*. So everything works fine for grad, curl and div, but these properties are lost when we consider grad*, curl* or div*. This is another reason to make a distinction between the metric-free vector operations and the metric-dependent vector operations. It would be nice if we could remove the metric dependent operations from our models, but as the examples of the scalar Laplacian above show, they need to be incorporated.

7 Discretization of Metric in Finite Element Methods

In finite volume methods one needs to explicitly construct the $\star$-operator, while in finite element methods the $\star$-operator is hidden in the inner product. Let a^k and b^k be differential forms (the continuous analogues of the cochains), then we have that [13, p. 16/17] and [14, Eq. (14.6), p. 362]

$$(a^k, b^k)\omega^n := a^k \wedge \star b^k \, .$$

Here ω^n is a standard volume form. (a^k, b^k) is the inner-product of k-forms and $\wedge$ is the wedge product of differential forms. We see by this relation that the inner-product and the wedge define the $\star$-operator. So the inner-product employed in finite element methods implicitly defines the metric and orientation. The exterior derivate d is the continuous analogue of the coboundary operator δ, whereas the codifferential d* is the continuous analogue of the discrete metric-dependent operator δ^*. The inner-product provides a relation between the metric-dependent vector operations and the topological vector operations by *integration by parts*. On a domain without boundary we have

$$
\begin{aligned}
(\mathrm{d}a^k, b^{k+1})\omega^n = \mathrm{d}a^k \wedge \star b^{k+1} &= (-1)^{k+1} a^k \wedge \mathrm{d} \star b^{k+1} \\
&= a^k \wedge \star \mathrm{d}^* b^{k+1} = (a^k, \mathrm{d}^* b^{k+1})\omega^n
\end{aligned}
$$

The whole purpose of integration by parts is to convert metric-dependent vector operators represented by d^*, into metric-free vector operators, represented by d, which have all the nice commuting properties. In order to do so, one has to have a sound geometric understanding which vector operations depend on the metric and which ones do not.

The scalar Laplace equation associated with points, see Fig. 8, leads to the scalar Laplace equation $\mathrm{div}^* \, \mathrm{grad}\varphi = f$. Taking the inner-product with a test function ψ, integrating over the domain Ω and applying integration by parts gives

$$
-(\mathrm{grad}\varphi, \mathrm{grad}\psi) = (f, \psi) \;, \quad \forall \psi \in \Lambda^0(\Omega) \;,
$$

modulo a boundary integral. This is the standard weak formulation of the Laplacian. The main purpose of integration by parts is to convert the metric-dependent div^* into the topological grad using the metric of the inner-product.

Next consider the scalar Laplace equation associated with volumes as shown in Fig. 8. This equation is given by $\mathrm{div}\,\mathrm{grad}^*\varphi = f$. We see that in Galerkin methods integration by parts converts the metric-free div to the metric-dependent grad^*, so instead of removing metric dependence we introduce even more metric-dependence. One way to resolve this, is to rewrite the Laplace equation as an equivalent first order system by introducing $q = \mathrm{grad}^*\varphi$ which gives

$$
q - \mathrm{grad}^*\varphi = 0 \;, \quad \mathrm{div} q = f \;.
$$

The second equation does not depend on the metric and only the first equation requires integration by parts to convert grad^* into div.

$$
(q, p) + (\varphi, \mathrm{div}\, p) = 0 \;, \quad (\mathrm{div} q, \psi) = (f, \psi) \;, \quad \forall p \in \Lambda^2(\Omega) \;, \; \psi \in \Lambda^3(\Omega) \;,
$$

modulo a boundary integral. And here we have the weak formulation in which all metric-dependent operations have been removed. This is a well-known mixed formulation [10]. When appropriate reconstructions are employed, this formulation satisfies automatically the inf-sup relation for q and φ if at the discrete level the Poincaré inequality is satisfied, see [5, § 7.1][30, Lemma 3, p. 12]. The important message of these two examples is that one is *not* free to choose between a direct weak formulation or a mixed formulation. This is determined by the geometry. In the first case we had scalars associated with points, in the second scalars associated with volumes. Whether we consider a variable to be associated to points or volumes is not up to us, but this is determined by the physics. Applications of these ideas can be found in these proceedings [7, 29–31, 35, 41].

Acknowledgements The authors want to thank the reviewers for their critical remarks and useful suggestions.

References

1. D. Arnold, R. Falk & R. Winther, *Finite element exterior calculus, homological techniques, and applications*, Acta Numerica,15, pp. 1–155, 2006.
2. D. Arnold, R. Falk & R. Winther, *Finite element exterior calculus: from Hodge theory to numerical stability*, Bull. Amer. Math. Soc., 47, pp. 281–354, 2010.
3. D. Arnold & G. Awanou, *Finite element differential forms on cubical meshes* , Arxiv preprint math/1204.2595, to appear in: Mathematics of Computation, 2013. http://arxiv.org/abs/1204.2595
4. P. Bochev & J. Hyman, *Principles of mimetic discretizations of differential operators*, IMA Volumes In Mathematics and its Applications, Springer, 142, pp. 89–114, 2006.
5. P. Bochev, *A discourse on variational and geometric aspects of stability of discretizations*, in VKI Lecture Series: 33rd computational fluid dynamics course - novel methods for solving convection dominated systems March 24–28, 2003.
6. A.Bossavit, *Discretization of electromagnetic problems* in Handbook of Numerical Analysis, Vol. 13, Elsevier, pp. 105–197, 2005.
7. A. Bossavit & F. Rapetti, *Whitney elements, from manifolds to fields*, proceedings ICOSAHOM 2012, 2013.
8. F. Brezzi & A. Buffa, *Innovative mimetic discretizations for electromagnetic problems*, Journal of Computational and Applied Mathematics, 234, pp. 1980–1987, 2010.
9. F. Brezzi, A. Buffa & K. Lipnikov, *Mimetic finite differences for elliptic problems*, Mathematical Modelling and Numerical Analysis, 43, pp. 277–296, 2009.
10. F. Brezzi & M. Fortin, Mixed and Hybrid Finite Element Methods, Springer Verlag, 1991.
11. A. Buffa, G. Sangalli & R. Vázquez, *Isogeometric analysis in electromagnetics: B-splines approximation*, Computer Methods in Applied Mechanics and Engineering 199 (17–20), pp. 1143–1152, 2010.
12. M. Desbrun, A. Hirani, M. Leok, J. Marsden, *Discrete exterior calculus*, Arxiv preprint math/0508341, 2005.
13. H. Flanders, Differential forms with applications to the physical sciences, Dover Publications, New York, 1989.
14. Th. Frankel, The Geometry of Physics. An Introduction. 2nd edition, Cambridge University Press, 2011.
15. J.A. Evans & T.J.R. Hughes, *Isogeometric divergence-conforming B-splines for the unsteady Navier-Stokes equations*, Journal of Computational Physics 241 , pp. 141–167, 2013.
16. J.A. Evans & T.J.R. Hughes, *Isogeometric divergence-conforming B-splines for the Darcy-Stokes-Brinkman equations*, Mathematical Models and Methods in Applied Sciences 23 (4), pp. 671–741, 2013.
17. E.S. Gawlik, P. Mullen, D. Pavlov, J.E. Marsden & M. Desbrun, *Geometric, variational discretization of continuum theories*, Physica D: Nonlinear Phenomena 240 (21) , pp. 1724–1760, 2011.
18. M.I. Gerritsma, *Edge functions for spectral element methods*, Spectral and High Order Methods for Partial differential equations, Eds J.S. Hesthaven & E.M. Rønquist, Lecture Notes in Computational Science and Engineering, 76, pp. 199–207, 2011.
19. M.I. Gerritsma, *An Introduction to a Compatible Spectral Discretization Method*, Mechanics of Advanced Materials and Structures, 19 pp. 48–67, 2012.
20. J. Harrison, *Geometric Hodge star operator with applications to the theorems of Gauss and Green* , Math. Proc. Camb Phil. Soc., 140, pp. 135–155, 2006.
21. R.R. Hiemstra, R.H.M. Huijsmans & M.I. Gerritsma, *High order gradient, curl and divergence conforming spaces, with an application to compatible IsoGeometric Analysis*, submitted to JCP, 2012.
22. R.R. Hiemstra, & M.I. Gerritsma, *High order methods with exact conservation properties*, proceedings ICOSAHOM 2012
23. R. Hiptmair, *Discrete hodge operators*, Numerische Mathematik, 90, pp. 265–289, 2001.

24. R. Hiptmair, *Higher order Whitney forms*, Geometric Methods for Computational Electromagnetics, 42, pp. 271–299, 2001.
25. A.N. Hirani, *Discrete Exterior Calculus*, PhD thesis, California Institute of Technology, 2003. http://thesis.library.caltech.edu/1885/3/thesis_hirani.pdf
26. A.N. Hirani, K. Kalyanaraman & E.B.van der Zee, *Delaunay Hodge star*, CAD Computer Aided Design 45 (2), pp. 540–544, 2013.
27. J.M. Hyman & J.C. Scovel, *Deriving mimetic difference approximations to differential operators using algebraic topology*, Math. Comp. 52, No.186, pp. 471–494, 1989.
28. J.M. Hyman, M. Shashkov & S. Steinberg, *The numerical solution of diffusion problems in strongly heterogeous non-isotropic materials*, Journal of Computational Physics, 132, 1, pp. 30–148, 1997.
29. J.J. Kreeft & M.I. Gerritsma, *Mixed Mimetic Spectral Element Method for Stokes Flow: A Pointwise Divergence-Free Solution*, Journal of Computational Physics, 240, pp. 284–309, 2013. .
30. J.J. Kreeft & M.I. Gerritsma, *A priori error estimates for compatible spectral discretization of the Stokes problem for all admissible boundary conditions*, arXiv preprint arXiv:1206.2812, 2012. http://arxiv.org/abs/1206.2812
31. J.J. Kreeft & M.I. Gerritsma, *Higher-order compatible discretization on hexahedrals*, proceedings ICOSAHOM 2012. http://arxiv.org/abs/1304.7018
32. J.J. Kreeft, A. Palha & M.I. Gerritsma, *Mimetic framework on curvilinear quadrilaterals of arbitrary order*, Arxiv preprint arXiv:1111.4304, pp. 1–69, 2011. http://arxiv.org/abs/1111.4304
33. C. Mattiussi, *The finite volume, finite element, and finite difference methods as numerical methods for physical field problems*, Advances in Imaging and electron physics, 133, pp. 1–147, 2000.
34. A. Palha, P. Rebelo, R.R. Hiemstra, J.J. Kreeft & M.I. Gerritsma, *Physics-compatible discretization techniques on single and dual grids, with application to the Poisson equation of volume forms*, submitted to JCP, http://arxiv.org/abs/1304.6908, 2012.
35. A. Palha, P. Rebelo & M.I. Gerritsma, *Mimetic Spectral Element advection*, proceedings ICOSAHOM 2012. http://arxiv.org/abs/1304.6926
36. D. Pavlov, P. Mullen, Y. Tong, E. Kanso, J.E. Marsden & M. Desbrun, *Structure-preserving discretization of incompressible fluids*, Physica D: Nonlinear Phenomena 240 (6), pp. 443–458, 2011.
37. J. Blair Perot, *Conservation properties of unstructured staggered mesh schemes*, Journal of Computational Physics, 159, pp. 58–89 , 2000.
38. J. Blair Perot, *Discrete Conservation Properties of Unstructured Mesh Schemes*, Annual Review of Fluid Mechanics, Annual Reviews, 43, pp. 299–318, 2011.
39. F. Rapetti & A. Bossavit, *Whitney forms of higher degree*, SIAM J. Numer. Anal., 47, pp. 2369–2386, 2009.
40. F. Rapetti & A. Bossavit, *Geometrical localisation of the degrees of freedom for Whitney elements of higher order*, IET Science, Measurement and Technology 1 (1) , pp. 63–66, 2007.
41. P. Rebelo, A. Palha & M.I. Gerritsma, *Mixed Mimetic Spectral Element method applied to Darcy's problem*, proceedings ICOSAHOM 2012 http://arxiv.org/abs/1304.7147
42. N. Robidoux & S. Steinberg, *A discrete vector calculus in tensor grids*, Computational Methods in Applied Mathematics 11 (1) , pp. 23–66, 2011.
43. Tonti, E, *On the formal structure of physical theories*, preprint of the Italian National Research Council, 1975. http://www.dic.univ.trieste.it/perspage/tonti/DEPOSITO/CNR.pdf
44. D. Toshniwal, R.H.M. Huijsmans & M.I. Gerritsma, *A Geometric approach towards momentum conservation*, proceedings ICOSAHOM 2012. http://arxiv.org/abs/1304.6991

Spectral Element Methods on Simplicial Meshes

Richard Pasquetti and Francesca Rapetti

Abstract We present a review in the construction of accurate and efficient multivariate polynomial approximations on elementary domains that are not Cartesian products of intervals, such as triangles and tetrahedra. After the generalities for high-order nodal interpolation of a function over an interval, we introduce collapsed coordinates and warped tensor product expansions. We then discuss about the choice of interpolation and quadrature points together with the assembling of the final system matrices in the case of elliptic operators. We also present two efficient ways of solving the associated linear system, namely a Schur complement strategy and a p-multigrid solver. Two applications to Computational Fluid Dynamics problems conclude this contribution.

1 Introduction

Spectral element methods (SEMs) combine the flexibility of finite element methods (FEMs) with some basic facts from polynomial approximation theory which allow to construct a good interpolant. Applied in a domain subdivided into elements, they can provide accurate approximations to solutions of many problems with fewer degrees of freedom (dof) than low-order approaches. High accuracy results from the use of orthogonal polynomial basis to construct the interpolating functions over the elements. The Galerkin projection operators link the differential to the algebraic problem and keep the global system matrices sparse by imposing minimal continuity requirement on the approximated solution.

R. Pasquetti · F. Rapetti (✉)
Laboratoire de Mathématiques "J.A. Dieudonné" UMR CNRS 7351, Université de Nice Sophia-Antipolis, Parc Valrose, 06108 Nice Cedex 02, France
e-mail: rpas@unice.fr; frapetti@unice.fr

M. Azaïez et al. (eds.), *Spectral and High Order Methods for Partial Differential Equations - ICOSAHOM 2012*, Lecture Notes in Computational Science and Engineering 95, DOI 10.1007/978-3-319-01601-6_3,

In this work we focus on SEMs based on Jacobi polynomials. These polynomials are defined as the eigenfunctions of a particular singular Sturm-Liouville problem, which constitute a basis for the expansion of square integrable functions. It is known that eigenfunction expansions of a function u based on singular Sturm-Liouville problems converge to u with a rate which depends only on the regularity of u. Approximated solutions of partial differential equations (PDEs) based on these expansions enjoy the same property. Namely, if u is sufficiently smooth, the discretization error between u and its SE approximation u_N decays exponentially fast to zero, asymptotically with respect to the polynomial degree N. However, exponential convergence is always at risk in simulations of complex phenomena since the non-uniformity of the meshes, the singularities of the geometry, the discontinuities of the boundary conditions and the jumps in the physical parameters of the problem degrade the convergence. Hopefully, the accuracy of SE approximations can be improved in two ways, by increasing either the number of mesh elements (h-refinement, where h denotes the maximal diameter of the mesh elements) or the interpolation polynomial degree in the elements (N-refinement), and this makes the SE approach robust. Other advantages of such methods are low dissipation and dispersion errors (see an example in [34]), their possible generalization to include non-conforming elements and their efficient/scalable implementation on modern computer architectures. Classical references on SEMs are [1, 7–9, 14, 22, 25, 33, 48], and many others therein. Early work with SEMs focused on meshes composed of quadrilateral or hexahedral elements. More recent advances concern the formulation of a nodal SEM for meshes composed of Triangles or Tetrahedra (TSEM) that will be emphasized in these pages.

2 Three Aspects of the TSEM

The key ingredients for the success of the TSEM are: (i) the definition of an orthogonal polynomial basis on non tensorial domains T through the warped tensor product (Sect. 2.1); (ii) the nodal interpolation of functions in T at suitable points which result from either a minimization procedure or a generating formula (Sect. 2.2); (iii) the computation of integrals over T by high-order quadrature formulas (Sect. 2.3).

2.1 *Warped Tensor Product*

Let $\mathscr{P}_N(I)$ be the space of polynomials of maximal degree N over an interval $I \subset I\!R$. The polynomial fitting of a function f at $n = N + 1$ points $\{x_k\}$ of the interval I consists in finding a polynomial $I_N f \in \mathscr{P}_N(I)$ such that $I_N f(x_k) = f_k := f(x_k)$, for $k = 1, \dots, n$. In terms of the (canonical) functions $\{x^{j-1}\}_{j=1,\dots,n}$ of $\mathscr{P}_N(I)$, we may write $I_N f(x) = \sum_{j=1}^n a_j x^{j-1}$ with

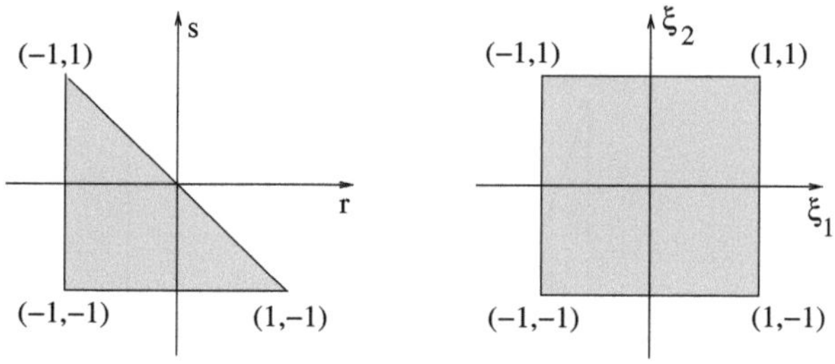

Fig. 1 (ξ_1, ξ_2) in Q are the collapsed Cartesian coordinates of the point (r, s) in T, with $\xi_1 = 2\frac{(1+r)}{(1-s)} - 1$, $\xi_2 = s$ and inversely $r = \frac{(1+\xi_1)(1-\xi_2)}{2} - 1$, $s = \xi_2$. Moreover, $\nabla_{r,s} = \begin{pmatrix} \frac{2}{(1-\xi_2)} & 0 \\ \frac{2(1+\xi_1)}{(1-\xi_2)} & 1 \end{pmatrix} \nabla_{\xi_1, \xi_2}$

the coefficients a_j solution of the Vandermonde linear system $V\mathbf{a} = \mathbf{f}$ where $V = (V_{kj}) = (x_k^{j-1})$, $\mathbf{a} = (a_j)$ and $\mathbf{f} = (f_k)$. Adopting a different basis $\{\psi_j\}$ for $\mathscr{P}_N(I)$ and representing f in terms of the (cardinal) functions $\{\phi_i\}$ of $\mathscr{P}_N(I)$ defined by $\phi_i(x_j) = \delta_{ij}$, we may write $I_N(f)(x) = \sum_{k=1}^n f_k\, \phi_k(x)$ with $\phi_k(x) = \sum_{j=1}^n c_{kj}\psi_j(x)$. For each k, the vector $\mathbf{c}_k = (c_{k1}, c_{k2}, \ldots, c_{kn})^t$ is solution of the generalized Vandermonde linear system $V\, \mathbf{c}_k = \mathbf{e}_k$ where $V = (V_{kj}) = (\psi_j(x_k))$ and $(\mathbf{e}_k)_j = \delta_{kj}$. The conditioning of the Vandermonde matrix V is sensitive to the choice of the basis $\{\psi_j\}$ in $\mathscr{P}_N(I)$: the best condition numbers are obtained when the basis $\{\psi_j\}$ is L^2-orthogonal. Yes, but which orthogonal basis could be defined on simplices? Earlier in the literature, the answer has been given in [28, 42] but highly determinant in the spectral method community has been the work [15]. Dubiner introduces the warped tensor product expansion to marry the tensor product idea with the flexibility of simplices. Warped tensor product expansions exploit collapsed coordinate systems in simplices, see Fig. 1 for $d = 2$ and Fig. 2 for $d = 3$, to transform a simplicial region T into a region with constant bounds (as the d-dimensional cube $Q = [-1, 1]^d$). In 2D for example, integrals in the variables (r, s) over $T = \{(r, s) \in I\!R^2 : -1 \le r, s \le 1,\ r + s \le 0\}$ are transformed into integrals in the variables (ξ_1, ξ_2) over $Q = [-1, 1]^2$:

$$\int_T u(r,s)dr\,ds = \int_{-1}^{1} \int_{-1}^{-s} u(r,s)\,dr\,ds$$
$$= \int_{-1}^{1} \int_{-1}^{1} u(\xi_1, \xi_2) \left| \frac{\partial(r,s)}{\partial(\xi_1, \xi_2)} \right| d\xi_1\, d\xi_2 = \int_Q u(\xi_1, \xi_2) \left(\frac{1-\xi_2}{2}\right) d\xi_1\, d\xi_2.$$

In 3D, we have $T = \{(r, s, t) \in I\!R^3 : -1 \le r, s, t \le 1,\ r+s+t \le 0\}$, $Q = [-1, 1]^3$ and

$$\int_T u(r,s,t)dr\,ds\,dt = \int_Q u(\xi_1, \xi_2, \xi_3) \left(\frac{1-\xi_2}{2}\right) \left(\frac{1-\xi_3}{2}\right)^2 d\xi_1\, d\xi_2\, d\xi_3.$$

We now introduce the Koornwinder-Dubiner (KD) orthogonal polynomials on the reference triangle (and similarly on the tetrahedron). Let $\{P_i^{\alpha,\beta}(x)\}_{i\ge 0}$ be the family of Jacobi polynomials evaluated at x, orthogonal in $L_w^2([-1, 1])$, with the weight $w = (1-x)^\alpha(1+x)^\beta$, being α, $\beta > -1$ reals. Just define the warped tensor product basis functions on $[-1, 1]^2$ as follows

$$\psi_\ell(r,s) = \varphi_{\sigma(i,j)}(\xi_1, \xi_2) = \phi_i(\xi_1)\phi_{i,j}(\xi_2) = P_i^{(0,0)}(\xi_1) \left[(\frac{1-\xi_2}{2})^i P_j^{(2i+1,0)}(\xi_2) \right],$$

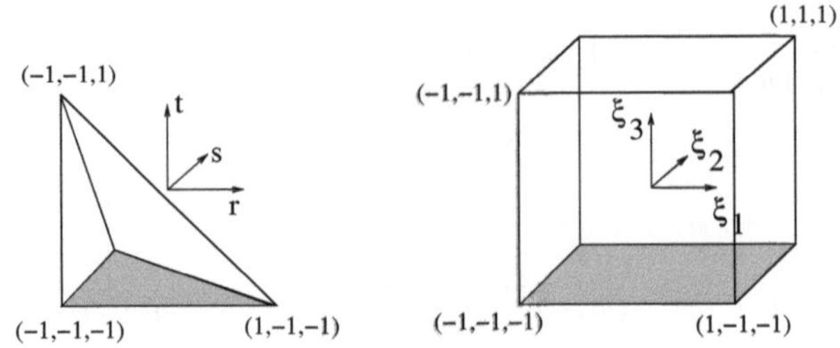

Fig. 2 (ξ_1, ξ_2, ξ_3) in Q are the collapsed Cartesian coordinates of the point (r, s, t) in T, with $\xi_1 = 2\frac{(1+r)}{(-s-t)} - 1$, $\xi_2 = 2\frac{(1+s)}{(1-t)} - 1$, $\xi_3 = t$ and $r = \frac{(1+\xi_1)(1-\xi_3)}{2} - 1$, $s = \frac{(1+\xi_2)(1-\xi_3)}{2} - 1$, $t = \xi_3$. Moreover, $\nabla_{r,s,t} = \begin{pmatrix} \frac{4}{(1-\xi_2)(1-\xi_3)} & 0 & 0 \\ \frac{2(1+\xi_1)}{(1-\xi_2)(1-\xi_3)} & \frac{2}{(1-\xi_3)} & 0 \\ \frac{2(1+\xi_1)}{(1-\xi_2)(1-\xi_3)} & \frac{(1+\xi_2)}{(1-\xi_3)} & 1 \end{pmatrix} \nabla_{\xi_1,\xi_2,\xi_3}$.

with the variables ξ_1 and ξ_2 replaced by their expression in r, s given in Fig. 1. Here, $\ell = \sigma(i, j)$ denotes a bijective index mapping, and ϕ_i, $\phi_{i,j}$ are the principal functions in the collapsed coordinates (ξ_1 and ξ_2, resp.). Note that $\varphi_{\sigma(i,j)}$ (resp. ψ_ℓ) is a polynomial of degree i in ξ_1 (resp. r) and $i + j$ in ξ_2 (resp. s). Moreover, if $s \to 1$ then $r \to -1$, so that $(2r + s + 1)/(1 - s) \to -1$, and we have $\varphi_{\sigma(0,j)}(\xi_1, \xi_2) = c_{ij} P_j^{1,0}(\xi_2)$ with c_{ij} a normalizing factor. It can be proven that Legendre polynomials result in KD ones when $Q \to T$, that KD polynomials are orthonormal in $L^2(T)$ and that $\{\psi_\ell, \ell = \sigma(i, j), i, j = 0, \ldots, N, i + j \leq N\}$ constitutes a basis of $\mathscr{P}_N(T)$, the space of polynomials defined in T of maximal degree N. In a tetrahedron, we have

$$\begin{aligned}\psi_\ell(r, s, t) &= \varphi_{\sigma(i,j,k)}(\xi_1, \xi_2, \xi_3) = \phi_i(\xi_1)\phi_{i,j}(\xi_2)\phi_{i,j,k}(\xi_3) \\ &= P_i^{0,0}(\xi_1) \left[\left(\tfrac{1-\xi_2}{2}\right)^i P_j^{2i+1,0}(\xi_2) \right] \left[\left(\tfrac{1-\xi_3}{2}\right)^{i+j} P_k^{2i+2j+2,0}(\xi_3) \right],\end{aligned}$$

with $\ell = \sigma(i, j, k)$ and the variables ξ_1, ξ_2, ξ_3 replaced by their expression in r, s, t given in Fig. 2 (more details are in [25]). Just remark that if $t \to 1$ then $r, s \to -1$, whereas if $(-s - t) \to 0$ then $r \to -1$ and $s \to 1$. Note that the expression of $\varphi_{\sigma(i,j,k)}(\xi_1, \xi_2, \xi_3)$ has to be updated for $\sigma(0, j, k)$, $\sigma(i, 0, k)$ and $\sigma(0, 0, k)$. The cardinality of $\mathscr{P}_N(T)$ is $n = \frac{(N+1)(N+2)}{2}$ in 2D and $n = \frac{(N+1)(N+2)(N+3)}{6}$ in 3D.

2.2 *Set of Interpolation Points*

The question of generating reasonable sets of points in a simplex has been extensively discussed in the literature (see Figs. 3 and 4 for some simple ideas). We ask the set of points $\{x_j\}$ to be unisolvent for $\mathscr{P}_N(T)$, i.e., given any set of n real values $\{f_j\}$ at points $\{x_j\}$ there exists a unique polynomial $I_N f \in \mathscr{P}_N(T)$

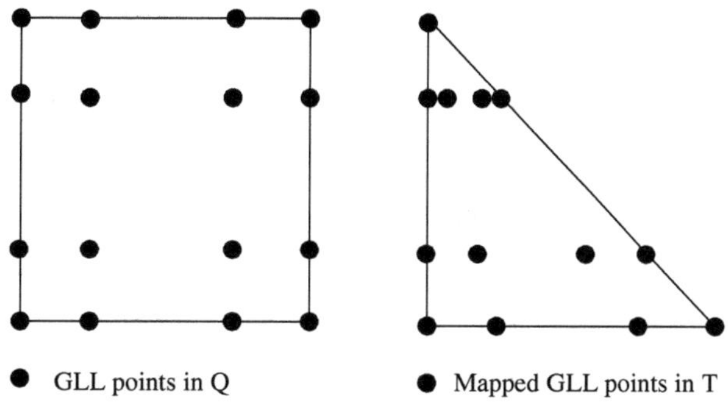

Fig. 3 We may create a set of interpolation points in T by mapping the GLL defined in Q. We preserve the accuracy but we generate more points than necessary and useless accumulation of points at $(-1, 1)$

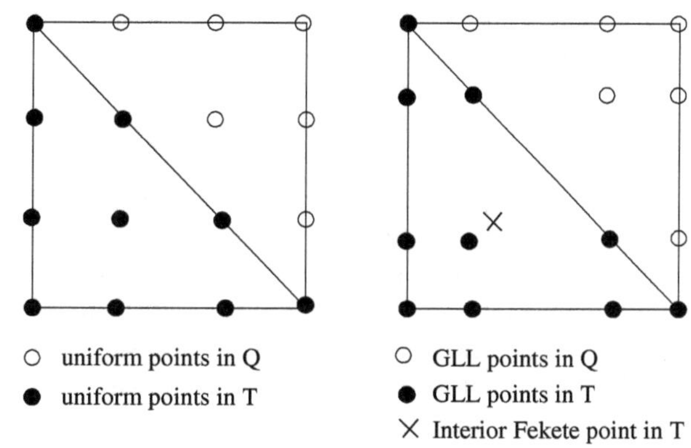

Fig. 4 We may create a set of interpolation points in T by truncating the GLL distribution in Q, as it occurs for equally-spaced points in FEMs. We have the correct number of points in T but low accuracy (apart from the case where internal • points are replaced by others ×)

such that $I_N f(x_j) = f_j, j = 1, \ldots, n$ which is equivalent to ask that the associated generalized Vandermonde matrix V is invertible. Moreover, we may impose to the set of points $\{x_j\}$ to fulfill certain symmetries and boundary constraints. Most of all, we ask to the interpolation operator $I_N : \mathscr{C}^0(\bar{T}) \to \mathscr{P}_N(T)$, which exists due to the unisolvence requirement, to have good approximation properties. The unisolvence condition yields the existence of a nodal basis in $\mathscr{P}_N(T)$ associated with the points $\{x_j\}$ given by the characteristic Lagrange polynomials $\{\varphi_j\}$ at these points (i.e., $\varphi_j \in \mathscr{P}_N(T)$ such that $\varphi_j(x_k) = \delta_{jk}$). The symmetry condition enhances the efficiency of the interpolation process in the sense that the number of degrees of freedom (dofs), represented by the unknown coordinates of the points $\{x_j\}$, can be drastically reduced. As an example, in 2D, while searching for rotationally invariant configurations, we may pass from $2n$ to $\lfloor n/3 \rfloor$ dofs which can be further reduced by imposing $N + 1$ constraints on each side to ensure inter-element continuity. When we ask for good approximation properties for the operator I_N, we are implicitly thinking to the Runge's phenomenon [47] and to the growth with N of the so-called Lebesgue's constant Λ_N. Runge [47] remarked that polynomial fitting on equally spaced (ES) nodes $\{x_j\}$ in a domain may lead to unbounded and oscillatory interpolations even for smooth functions. A classical example is given by the *averisera* of Agnesi, that in 1D reads $f(x) = 1/(x^2 + 1)$, $x \in IR$. Indeed, if one constructs the polynomial $I_N^{ES} f$ which fits the exact values of f at $N + 1$ ES nodes in $[-1, 1]$, one gets $lim_{N\to\infty}|f - I_N^{ES} f| \neq 0$. Wild oscillations appear in a neighborhood of ± 1 suggesting that the regions near the interval extremities are not enough resolved. To reduce oscillations, modern methods combine a partitioning (called mesh) of the domain into elements (images of Q or T) with a clever choice of the interpolation points in each element. As an example, when the interpolation polynomial I_N^{GL} is built at Gauss-Lobatto (GL) points of Chebyshev or Legendre type, it verifies $||f(x) - I_N^{GL} f(x)||_{L^2_w} \leq N^{-s}||f||_{H^s_w}$ where s denotes the regularity of f. We remark in fact that GL points are spaced $O(1/N^2)$ close to the interval extremities and $O(1/N)$ at the center of the interval. We know that GL points can be straightforwardly extended to higher dimension on tensor product domains (Cartesian products of 1D intervals) but on other domains there is no equivalent of GL points defined as zeroes or extrema of suitable orthogonal polynomials, thus solving the Runge's phenomenon problem in a simplex.

The quality of the interpolation in a domain Ω is measured by the Lebesgue's constant $\Lambda_N = \max_{x\in\Omega} \sum_{j=1,n} |\varphi_j(x)|$, a positive real number which depends on the interpolation points $\{x_j\}$. The Lebesgue's function $\Lambda(x) = \sum_{j=1,n} |\varphi_j(x)|$ takes value 1 at the nodes x_j and reaches maximal values where nodal coverage is poor. Indeed, due to the Runge's phenomenon, the Lagrangians φ_j can wildly oscillate when going from a zero value at a point x_k, $k \neq j$, to 1 at the point x_j, and these oscillations accumulate in the constant Λ_N. The Lebesgue's constant Λ_N plays an important role in the Lebesgue's lemma, linking the interpolation error in the maximum norm to the best approximation error in the same norm. This lemma states that $\|f - I_N f\| \leq (1 + \|I_N\|)\|f - f^*\|$ for all $f \in \mathscr{C}^0(\bar{T})$ where f^* is the best fit of f in the norm $\|f\| = \max_{x\in\Omega} |f(x)|$ (that is, $\|f - f^*\| = \inf_{p\in\mathscr{P}_N(T)} \|f - p\|$) and $\|I_N\| = \max_{\|f\|=1} \|I_N(f)\| = \Lambda_N$. If Λ_N increases faster with N than $\|f - f^*\|$ goes to zero, uniform convergence to f can never be attained. In 1D, for Chebyshev (C) points the Lebesgue's constant increases slowly with $N \to +\infty$, namely $\Lambda_N < \frac{2}{\pi} \log(N + 1) + 1$, whereas for ES points we have an exponential grow, indeed $\Lambda_N \sim 2^{N+1}/(e\, N \log N)$. The Lebesgue's constant is defined in terms of the Lagrangians which depend on the points $\{x_j\}$ regardless of the basis functions ψ_j of $\mathscr{P}_N(T)$ used to express them in T (the latter instead influence the conditioning of V). If one wants to limit Λ_N, one has to optimize the position of nodes x_j in T. The points which give the best Lebesgue's constant (thus called Lebesgue's points) are not known even in 1D and what researchers try to do is to select the properties of the GL points in 1D and extend them to higher dimension while keeping an eye on the Lebesgue's constant (see for example [43] and the references therein).

In [50] it has been established that GL points over an interval or tensor product domains can be determined as the steady-state, minimum energy solution to an electrostatic problem and in [52] that this minimum is unique. This idea led to the definition of the so-called electrostatic points in a triangle [23] and in a tetrahedron [24]. Another approach to the construction of a set of points in a simplex with the required conditions is based on the minimization of the Lebesgue's constant

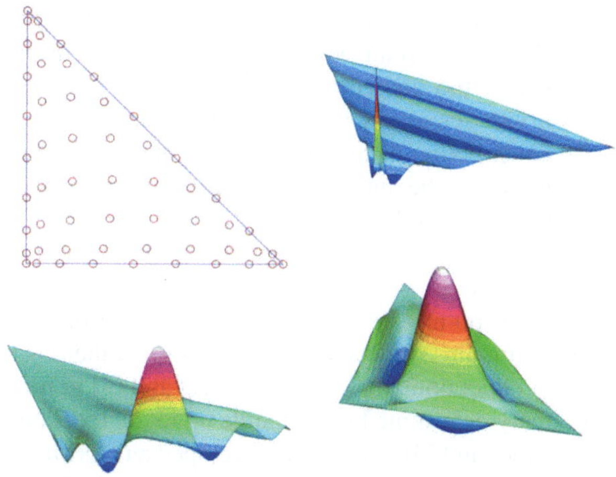

Fig. 5 Fekete nodes in a triangle T for $N = 9$ and Lagrangians associated to nodes located in a vertex, on a side, at the interior of T. Note that Fekete nodes on the *triangle sides* are distributed as GLL ones, allowing in 2D conforming meshes of triangles and quadrilaterals

or the optimization of related quantities. The Lebesgue's function is not convex in 2D or 3D, thus in [10, 11] it has been proposed to minimize the L^2-average $(\int_T \sum_j |\varphi_j(x)|^2 dx)^{1/2}$. A slightly different strategy, which leads to Fekete's points (after the Hungarian mathematician Michael Fekete, 1886–1957), relies on the generalized Vandermonde matrix $V = V(x_1, \ldots, x_n) = (\psi_j(x_i))$ where $\{\psi_j\}$ is any (better if L^2-orthogonal) basis in $\mathscr{P}_N(T)$. It exists a relation between the Lagrangians φ_j and the basis functions ψ_i of $\mathscr{P}_N(T)$ through the Vandermonde matrix, namely $V_{ij}\varphi_i = \psi_j$. By Cramer's rule, one has

$$\varphi_j(x) = \det[\psi(x_1), \ldots, \psi(x_{i-1}), \psi(x), \psi(x_{i+1}), \ldots, \psi(x_n)] / \det[\psi(x_1), \ldots, \psi(x_n)] \tag{1}$$

where the vector $\psi(.)$ stands for $(\psi_i(.))_{i=1,n}$. To minimize Runge's phenomenon, thus reducing the Lebesgue's constant, one should choose points $\{x_j\}$ that maximizes $\det[V]$.

2.2.1 Fekete Nodes

For a fixed set of basis functions $\{\psi_i\}$ of $\mathscr{P}_N(T)$, Fekete [17] nodes $\{x_j\}$ are those which maximize over $\bar{T}$ the determinant of the Vandermonde matrix V, with $V_{ij} = \psi_j(x_i)$. If $\det[V]$ has been really maximized, the numerator in (1) can never be larger than the denominator and thus, by construction, $|\varphi_j(x)| \leq 1$ for all $x \in \bar{T}$ (see Fig. 4). This yields a bound on the Lebesgue's constant, that is $\Lambda_N \leq n$ (numerical simulations indicate that $\Lambda_N \leq N$) and by bounding the Lebesgue norm of the interpolation polynomial, Fekete nodes ensure spectral convergence (Fig. 5).

Theoretically, Fekete points $\{x_j\}$ do not depend on the choice of the basis of $\mathscr{P}_N(T)$. The Vandermonde matrices constructed from two different bases of

$\mathscr{P}_N(T)$ at a given set of points are such that their determinants differ by a multiplicative constant which has no influence in the maximization process. But numerically, the choice of the basis matters a lot, since the inversion of V is necessary to get, at any point in T, the values of the Lagrangians that are involved in the construction of the interpolation operator. From (1) we may understand that the success of the interpolation procedure depends on the selection of a basis such that V is well-conditioned (a possible choice is the $L^2(T)$-orthogonal KD basis introduced in Sect. 1) and of a set of points $\{x_j\}$ that maximize det$[V]$ (i.e., Fekete's).

Fekete's points are a possible generalization of GLL points over not tensor product domains. Indeed, it has been shown that GLL points are Fekete points on tensor product domains (see [16] in 1D and [5] in higher dimension). Fekete points were firstly determined in a triangle up to degree 7 (and analytically up to degree 4) in [4]. Later, in [53] it has been developed an algorithm to approximate sets of points that locally maximize det$[V]$ up to degree 19. Through a coupling of a modified version of this algorithm with a simple global perturbation method, computations have been run up to degree 30 in [46]. For large values of N, the resulting (symmetric) Fekete-like configurations are the best, measured by the Lebesgue norm, found to date. The difficulty related to the computation of Fekete-like nodes in 3D suggested the researchers to look for other more direct strategies where the nodes' construction has a generating formula.

2.2.2 Generating Formulas

Recently, some authors introduced node distributions that have a generating formula for their construction. This is of great practical interest. Moreover, it reveals basic structural properties of the distributions that are responsible for the interpolation quality demonstrated by the optimization-based node sets discussed in Sect. 2.2.1. Among these node distributions, we recall the Lobatto grid introduced in a triangle [3] where the nodes are generated by deploying Lobatto interpolation points along the three edges of the triangle and computing interior nodes by averaged intersections to achieve a three-fold rotational symmetry. This construction has been extended to the tetrahedron in [31]. In [30], interpolation points in the triangle T are obtained from the GLL ones in Q through a new rectangle-to-triangle mapping, which pulls one edge (at the middle point) of T to two edges of Q. The idea of defining interpolation nodes on concentric triangles was firstly analyzed in [4]. An alternative to this construction, which results in a very easy recursive generation algorithm, has been later generalized to arbitrarily shaped domains in [19]. A very efficient approach is the one presented in [54], where the task of creating a nodal distribution is replaced with the closely related task of building a coordinate-warping transformation for the triangle/tetrahedron. This is a familiar problem faced when using curvilinear finite elements [21]. New contributions of this work include the application of the warp & blend transform to create nodal elements, and the construction of coordinate transforms that do not actually change the overall geometry of the triangle. A summary of these nodal distributions is given in [40] where we compare them numerically in terms of Lebesgue constants, generalized Vandermonde matrix conditioning and accuracy when adopted as approximation points in a TSEM approach applied to a model problem.

2.3 High-Order Quadrature Formulas

The cornerstone of nodal spectral-element methods is the co-location of the interpolation and integration points, yielding a diagonal mass matrix that is efficient for explicit time-integration. In classical SEMs over meshes composed of tensor product elements, GLL points are both good interpolation and integration points; on simplices, analogous points have not yet been found. Moreover, quadrature formulas centered at Fekete nodes as well as at all other sets of interpolation nodes described in Sect. 2 for a given polynomial degree N are precise only in $\mathscr{P}_N(T)$, yielding to a loss of exponential convergence [37]. Therefore, either we try to find another set of points with both interpolation and quadrature good properties or we separate the interpolation nodes from the quadrature ones. The first possibility has been considered in [20] (see also references therein) where cubature points have been used for interpolation up to degree 7 through a careful filtering of the basis. The other possibility, presented in [38], involves two sets of points: the set of n points $\{x_i\}$ for a polynomial interpolation of degree $\leq N$ over T and the set of m points $\{y_i\}$ allowing for an exact quadrature in $\mathscr{P}_M(T)$. Given the values at the approximation points of any polynomial in $\mathscr{P}_N(T)$, one can set up interpolation and differentiation matrices to compute its values and derivatives at the quadrature points.

Let $u_N \in \mathscr{P}_N(T)$. Knowing the $u_N(x_i)$, $i = 1, \ldots, n$, one can easily compute the $u_N(y_i)$, $i = 1, \ldots, m$. To this end we use the KD polynomials. With u_i for $u_N(x_i)$, we have:

$$u_i = \sum_{j=1}^{n} \hat{u}_j \psi_j(x_i) = \sum_{j=1}^{n} V_{ij}\hat{u}_j,$$

where the $\hat{u}_j$ are the components of u_N in the KD basis. In matrix form, with $\hat{\mathbf{u}}$ the vector of the $\hat{u}_j$, we have $\mathbf{u} = V\hat{\mathbf{u}}$. Similarly, with $\mathbf{u}'$ for the vector gathering the $u(y_i)$ and $V'_{ij} = \psi_j(y_i)$, we obtain $\mathbf{u}' = V'\hat{\mathbf{u}} = V'V^{-1}\mathbf{u}$. To differentiate, e.g., with respect to r, we use again the KD polynomials:

$$\psi_j(x) = \sum_{k=1}^{n} \psi_j(x_k)\phi_k(x), \qquad \partial_r\psi_j(y_i) = \sum_{k=1}^{n} \psi_j(x_k)\partial_r\phi_k(y_i),$$

thus for the differentiation matrix: $D'^r = V'^r V^{-1}$, with $(V'^r)_{ij} = \partial_r\psi_j(y_i)$.

Once knowing such differentiation matrices, D'^r and D'^s, it is an easy task, by applying the chain rule, to compute derivatives at the quadrature points. Note that if the mapping g between the reference triangle T and the current mesh triangle is a non-linear mapping, such an approach is not equivalent to the computation of the derivatives at the Fekete points followed by an interpolation at the Gauss points.

The TSEM is thus efficient, (i) in terms of computational time, especially to set up the stiffness matrix, (ii) because it does not require much memory for deformed triangles and finally (iii) because it is very flexible with respect to the choice of interpolation and quadrature nodes.

The TSEM requires of course the use of highly accurate quadrature rules based on Gauss points, as proposed in [12, 51]. Unfortunately, in practice such quadrature rules are not yet known (or accessible) for large values of N. One possibility is to use Gauss points based quadrature rules for the quadrilateral and map them to the triangle through a mapping, say $b : [-1, 1]^d \to T$. Such a mapping is, e.g., recalled in Sect. 1 and detailed in [25] where it is moreover suggested to use the quadrature points and weights associated to the tensor product of Jacobi polynomials $P_i^{1,0}$ and $P_j^{0,0}$ (Legendre polynomials) of suitable degree i, j. Such polynomials show indeed the property to be orthogonal with a weight proportional to the Jacobian determinant of b (2D case). Within polynomial Galerkin methods, it is a general practice to use quadrature rules that allow to integrate linear differential terms exactly. We have seen that, for integrands resulting from the TSEM applied to discretize linear differential terms, we may enforce exact quadratures in $\mathscr{P}_M(T)$ with $M = 2N - 1$ or $M = 2N - 2$ rather than $\mathscr{P}_{2N}(T)$, without loosing in accuracy. However an opposite strategy should be adopted in presence of nonlinear terms [26, 27]. Indeed, an insufficient quadrature rule for nonlinear terms leads to an aliasing pollution that degrades the accuracy of the solution and in some cases lead to numerical instabilities. To avoid these numerical problems, the use of more quadrature points than what would be necessary to integrate linear differential terms exactly is required. Quadrature formulas in T with minimal number of nodes for a prescribed degree of precision q are known for several values of q but not for all. Note that for high values of q, some weights are negative and some quadrature points can be out of the triangle. The latter aspect is manageable with the TSEM since the interpolation points are different from the quadrature ones. One can define a quadrature rule in T by mapping a quadrature rule defined in Q. However, when using such a quadrature rule, the quadrature points are no-longer symmetric in T and their number is maximal: $(N + 1)^2$ quadrature points are required for an exact quadrature of polynomials of maximum degree $2N + 1$ in each variable. Moreover, an a priori useless accumulation of points occurs at the upper vertex. To avoid the singularity in the simplex top vertex, quadrature formulas of Gauss type, rather than of Gauss-Lobatto type, should be adopted. On a generic triangle $T_k = g_k(T)$ of a simplicial mesh, the approximated L^2-inner product in $\mathscr{P}_N(T)$ is defined as

$$(u, v)_{T_k,N} = \sum_{i=1}^{m} (u \circ g_k)(y_i)(v \circ g_k)(y_i)|J_{g_k}(y_i)|w_i, \tag{2}$$

where w_i are the original quadrature weights. In terms of grid-point values at the interpolation set, we may write:

$$(u,v)_{T_k,N} = \mathbf{v}^t \mathscr{M}\mathbf{u}, \qquad \mathscr{M} = (V^{-1})^t (V')^t W V' V^{-1}, \qquad W = \mathrm{diag}(w_i).$$

3 Solving the TSEM Systems

Another non-negligible area of research covers the efficient resolution of the algebraic linear systems resulting from the TSEM application to discretize PDEs. The matrices resulting from applying the TSEM to second-order elliptic PDEs are indeed less sparse than the corresponding SEM ones and more ill-conditioned, since its condition number grows as $O(N^4h^{-2})$ rather than $O(N^3h^{-2})$ in 2D, where N is the total degree of the polynomial approximation in each element and h is the maximal diameter of the mesh elements.

Finite-element matrices constructed on piecewise linear (in 1D), bilinear (in 2D) or trilinear (in 3D) shape functions centered at GLL nodes of the reference cube $[-1,1]^d$ provide optimal preconditioners for SEM matrices built on $[-1,1]^d$. A simple Rayleigh quotient argument shows that the spectral condition number is optimal, that is, uniformly bounded with respect to both the polynomial degrees N and the element sizes h in the case of quadrilateral (or parallelepiped) elements. This is known as the FEM-SEM equivalence. Conversely, it is shown that the finite-element preconditioners are not optimal on simplices when constructed on Fekete points or other interpolation points (see details in [8, 55]). Thus, other types of preconditioners or multi-level solvers have to be considered (for a general introduction to domain decomposition methods and preconditioners we refer to [49]).

The following methods have been considered. (i) Neumann-Neumann Schur complement methods [35], with each spectral element being considered as a subdomain: Addressing the Schur complement with Balancing Neumann-Neumann (BNN) type preconditioners has yielded promising results. Without any numerical manipulation, the condition number of the Schur complement matrix only shows a $O(Nh^{-2})$ behavior. (ii) Overlapping Schwarz (OS) methods [36], with subdomains containing more than one spectral element: Impressive results can be obtained but with the drawback that, differently to the SEM, a "generous overlap" (overlap of one entire mesh element) must be adopted due to the not tensor product distribution of the Fekete points in the element.

On the basis of these considerations, we present two strategies. On the one hand, the Schur complement approach with a simple local implementation, as in [29]. On the other hand, the p-multigrid approach which makes use of a fixed simplicial mesh and of different approximation levels, each of them associated with a different polynomial degree to solve elliptic problems [39]. Let us consider the 2D model problem

$$-\nabla\cdot(\nu\nabla u) + \sigma u = f \quad \text{in } \Omega, \qquad u|_\Gamma = 0 \tag{3}$$

where ν, $\sigma > 0$ are given in Ω and where f is a given function in $L^2(\Omega)$. For simplicity, homogeneous Dirichlet conditions have been assumed. The weak formulation of problem (3) reads: Given $f \in L^2(\Omega)$, find $u \in E = H_0^1(\Omega)$ such that

$$a(u, v) := \int_\Omega (\nu \nabla u \cdot \nabla v + u v)\, d\Omega = \int_\Omega f v\, d\Omega\,, \qquad \forall v \in E. \tag{4}$$

The variational formulation (4) is discretized by the conforming TSEM over a mesh of K simplices T_k, where the approximation space, say $E_{K,N}$, contains continuous, piecewise polynomials of total degree $\leq N$,

$$E_{K,N} = \{v \in E : v|_{T_k} \circ g_k \in \mathscr{P}_N(T),\ 1 \leq k \leq K\}. \tag{5}$$

Knowing how to compute derivatives and integrals, we can use the usual FEM methodology to set the discrete problem

$$\sum_{k=1}^{K} a_{k,N}(u, v) = \sum_{k=1}^{K} (f, v)_{k,N} \qquad \forall v \in E_{K,N}\,, \tag{6}$$

where $a_{k,N}(\cdot,\cdot)$ is obtained from $a(\cdot,\cdot)$ by replacing each integral with the quadrature rule (2). By using for the test functions all the involved (Lagrangian) basis functions, Eq. (6) can be written in matrix form as a linear system $A\mathbf{u} = \mathbf{b}$.

3.1 Schur Complement with Local Implementation

Let us consider the matrix form of Eq. (6) restricted to each element T_k, $k = 1, \ldots, K$. Using for the test functions $v \circ g_k$ the Lagrangians based on the Fekete points we obtain, $A_k\mathbf{u}_k = \mathbf{b}_k + \mathbf{r}_k$ where $\mathbf{u}_k$ is the vector of the unknowns at the interpolation nodes in the element T_k whereas $\mathbf{r}_k$ stands for the contribution of an (unknown) Neumann condition at the edges shared by two elements. Note however that such terms compensate when assembled, i.e. $\sum_k{}' \mathbf{r}_k = 0$, where $\sum'$ is used to denote the assembling procedure.

By reordering (if necessary) the boundary nodes and then the interior ones, and since $\mathbf{r}_k$ has no contribution to the inner nodes, the matrix system $A_k\mathbf{u}_k = \mathbf{b}_k + \mathbf{r}_k$ can be rewritten as

$$\begin{pmatrix} A_{k,\gamma\gamma} & A_{k,\gamma I} \\ A_{k,I\gamma} & A_{k,II} \end{pmatrix} \begin{pmatrix} \mathbf{u}_{k,\gamma} \\ \mathbf{u}_{k,I} \end{pmatrix} = \begin{pmatrix} \mathbf{b}_{k,\gamma} + \mathbf{r}_{k,\gamma} \\ \mathbf{b}_{k,I} \end{pmatrix}.$$

where the subscript (k, γ) (resp. (I, γ)) refers to the boundary (resp. inner) nodes of element T_k. Assuming now that $A_{k,II}$ is not singular, we can eliminate the variables $\mathbf{u}_{k,I}$ and set up the following equation for $\mathbf{u}_{k,\gamma}$:

$$S_k \mathbf{u}_{k,\gamma} = \mathbf{g}_k \quad \text{with} \quad S_k = (A_{k,\gamma\gamma} - A_{k,\gamma I} A_{k,II}^{-1} A_{k,I\gamma})$$
$$\mathbf{g}_k = \mathbf{b}_{k,\gamma} + \mathbf{r}_{k,\gamma} - A_{k,\gamma I} A_{k,II}^{-1} \mathbf{b}_{k,I} \,.$$

The Schur complement matrix S_k is of smaller dimension than matrix A_k, i.e. $3N$ rather than $n = (N+1)(N+2)/2$. Moreover, since A_k is symmetric S_k is also symmetric.

By the assembling procedure and taking into account the compensatory equation $\sum_k{}' \mathbf{r}_k = 0$, one obtains:

$$S\mathbf{u}_\gamma = \mathbf{g}_\gamma \quad \text{where} \quad S = \sum_k{}' S_k \tag{7}$$

$$\mathbf{g}_\gamma = \sum_k{}' \mathbf{b}_{k,\gamma} - A_{k,\gamma I} A_{k,II}^{-1} \mathbf{b}_{k,I} \tag{8}$$

where γ refers to the union of all element boundaries, included those elements that touch the domain boundary Γ (in the present formulation of the Schur complement method, $\Gamma \subset \gamma$). The dimension of the Schur complement matrix S is $O(N)$, thus smaller than the dimension $O(N^2)$ of the matrix A. Indeed, S results from the assembling of elemental matrices of dimension $O(N)$. Moreover, S is better conditioned than A since its condition number is $O(Nh^{-2})$, as proved in [35] (we recall that the Schur complement method has been here considered in the case where each mesh element is a subdomain).

In practice we assemble the source term $\mathbf{g}_\gamma$ but do not assemble the Schur complement matrix S, for memory space reasons. Because the Schur complement system is solved by using a PCG method, we indeed only need to realize matrix vector product, which is easy from:

$$S\mathbf{u}_\gamma = \sum_k{}' S_k \mathbf{u}_{k,\gamma} \,.$$

This is an alternative to the more common approach based on the use of low storage algorithms for sparse matrices. For the preconditioner, we simply use the diagonal term of S, which is also assembled to this end. Details concerning the imposition of boundary conditions with a non assembled system are given in [29].

In the implementation all operators specific to each element, i.e. $A_{k,II}^{-1}$, $A_{k,I\gamma}$, $A_{k,\gamma I} A_{k,II}^{-1}$ and S_k, are computed and stored in a preliminary calculation. The storage requirement is then $O(KN^2)$, with K the number of mesh elements. Such storage capacity remains reasonable and provides the guaranty of an efficient resolution.

3.2 p-Multigrid for TSEM

In the frame of classical SEMs, the p-multigrid solver was initially proposed in [32, 44, 45] and recently used in conjunction with Overlapping Schwarz preconditioners for CFD purposes in [18]. To develop a p-multigrid strategy we have to define the prolongation/restriction operators between different approximation levels, an efficient way to set up the coarse level algebraic systems and the smoothing procedure.

For the sake of simplicity we address the two level case, but the approach can be easily extended to an arbitrary number of levels. Each level corresponds to a given approximation polynomial degree N over T and is associated to the set of Fekete points $\{x_i\}$ in T for the assigned N at that level. The superscripts c and f are hereafter used to denote the coarse and fine polynomial approximations. Thus, N_c is the polynomial approximation degree, $\{x_i^c\}_{i=1,\dots,n_c}$ the set of Fekete points and $\{\varphi_i^c\}_{i=1,\dots,n_c}$ the corresponding Lagrange polynomials. Accordingly, we use N_f, $\{x_i^f\}_{i=1,\dots,n_f}$ and $\{\varphi_i^f\}_{i=1,\dots,n_f}$, with $N_f > N_c$.

In the frame of spectral methods, defining the prolongation operator is natural. Using the polynomial interpolant yields :

$$u_f(x_i^f) = \sum_{j=1}^{n_c} u_c(x_j^c)\varphi_j^c(x_i^f)\,, \quad 1 \le i \le n_f$$

where u_c (resp. u_f) denotes u_{N_c} (resp. u_{N_f}). In matrix notation we thus obtain the prolongation operator P, such that, for the element T_k:

$$\mathbf{u}_f = P\,\mathbf{u}_c\,, \quad [P]_{ij} = \varphi_j^c(x_i^f)\,.$$

Note that the side point values of u_f only depend on the side point values of u_c, so that the approximation remains conforming. There are indeed $N+1$ Fekete points on each side of T, so that the Lagrange polynomials based on points on the other sides or inside T vanish at this side. As a result, the operator P shows a special structure.

Defining the restriction operator, say R, is less trivial. As just done for the prolongation operator, one may proceed by *interpolation*, but a clever handling of the highest frequencies cannot really be expected from the interpolation strategy. One may then prefer to proceed by *projection* or more generally by *filtering*, in, e.g., the KD orthogonal hierarchical basis, so that $u_c(x_i^c) = \sum_{j=1}^{n_f} Q_j \hat{u}_j \psi_j(x_i^c)$. For a projection, one simply uses $Q_j = 1$ if $j \le n_c$ and $Q_j = 0$ if $n_c < j \le n_f$. For a filtering, the values of the Q_j should be associated with the total degree of the polynomial ψ_j, say $N(j)$. Using, e.g., the raised cosine filter $Q_j = (\cos(N(j)/N_f)\pi + 1)/2$.

In the frame of variational methods, weak formulations with $L^2(T)$ inner products are involved. Taking this into account yields to set up a restriction operator by *transposition* of the prolongation operator. Indeed,

$$(u_f, \varphi_i^c) = (u_f, \sum_{j=1}^{n_f} \varphi_i^c(x_j^f)\varphi_j^f) = \sum_{j=1}^{n_f} \varphi_i^c(x_j^f)(u_f, \varphi_j^f)$$

so that $R = P^t$. Note that if R is obtained by transposition of P, then its structure is such that the inner point values of the right-hand side $\mathbf{b}_c$ only depend on the inner point values of the residual $\mathbf{r}_f$.

The prolongation and restriction operators being chosen, it remains to set up the coarse level matrix, say A_c. Matrix A_c may be set up *directly*, i.e., like the fine level matrix A, or by *aggregation* of A : $(A_c)_k = RA_k P$, with A_k for the elementary matrix of element T_k. Concerning the boundary conditions to be implemented in A_c, or more generally at all sublevels if more than two approximation levels are considered, they must be taken homogeneous and of the same type, Dirichlet, Neumann or Robin, involved in the initial problem.

The aggregation approach is generally coupled to the definition of the restriction operator by transposition. One can indeed observe that, because of the previously mentioned properties of the prolongation and restriction operators, one obtains for the full system something similar to what was obtained for each element. Keeping unchanged the notations, for the sake of simplicity, one has then $A_c = RAP$ and $R = P^t$, where R and P are easily identifiable matrices. One can check that if A is symmetric and positive definite, the coarse level error $\mathbf{e}_c$ such that $A_c\mathbf{e}_c = R\mathbf{r}_f$ solves the constrained optimization problem : Minimize

$$\begin{aligned} \phi(\mathbf{u}^*) &= \frac{1}{2}(A\mathbf{u}^*, \mathbf{u}^*) - (\mathbf{b}, \mathbf{u}^*) \quad \textit{constrained by} \qquad (9) \\ \mathbf{u}^* &= \mathbf{u}_f + P\mathbf{e}_c \,. \end{aligned}$$

On the basis of a standard Gauss-Seidel smoothing, tests have shown that the most satisfactory results were obtained by using the transposition strategy, $R = P^t$, to set up the restriction operator and by adopting the aggregation of the system matrix A to set up the coarse grid matrix A_c, as detailed in [39].

4 Two Numerical Results in CFD and Concluding Remarks

The Fekete-Gauss TSEM methodology has been implemented in a numerical solver for the incompressible Navier-Stokes equations. The solver is based on a projection method, to enforce the divergence free velocity constraint, and a second order time approximation, with an implicit/explicit treatment of the diffusion/advection terms.

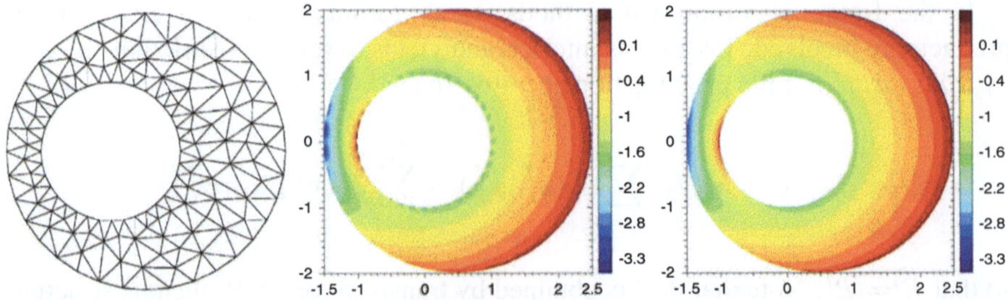

Fig. 6 Flow between eccentric cylinders at Reynold number $Re = 37.2$ [29]. Here, $N = 9$ and $M = 2N$, over a mesh (*left*) of 222 triangles, the vorticity field is presented when using either linear elements (*center*) or isoparametric elements (*right*)

The Schur complement approach is used to solve the resulting elliptic problems. Two applications are here briefly presented (see [29] for complete details).

Results obtained for the flow between eccentric cylinders, the inner one being rotating, are presented in Fig. 6. This test case allows to point out the requirement of using isoparametric elements when a curved boundary is involved, that means rely on a polynomial parametrization of the sides of triangles touching the curved boundary of the same order as the TSEM basis functions. For some interesting remarks on standard mapping techniques based on conformal transformations to go from $[-1, 1]^d$ into a domain with curved boundaries, see [2]. In Fig. 6, we show the TSEM mesh and compare the vorticity fields obtained with straight triangles to the one obtained with deformed triangles at the boundary. Thanks to a correct treatment of the circular boundaries, the vorticity peaks that appear at the element vertices in the former case have disappeared in the latter.

Results obtained for the driven cavity flow are presented in Fig. 7. The boundary condition is $\mathbf{u} = (-1, 0)$ at the upper boundary and $\mathbf{u} = (0, 0)$ elsewhere. Such a driven cavity flow is challenging to be captured with a high-order method as it involves singularities in the upper corners of Ω. Indeed, u_x is not continuous at those corners and consequently the vorticity $\omega = \partial_x u_y - \partial_y u_x$ blows up. At these points we simply enforce the no-slip condition, i.e. no sophisticated singularity treatment is implemented.

Finally, it should be mentioned that here we have used a $P_N - P_N$ approximation, i.e., equal polynomial degrees for the velocity components and for the pressure, which contrasts with the usual $P_N - P_{N-2}$ SEM formulation, see [33]. Despite the fact that our Navier-Stokes solver is based on a projection method, one may conjecture that the TSEM formulation developed here, i.e., based on different sets of points for interpolation and quadrature, yields no spurious modes, see e.g. [1], and so has a filtering effect. This conjecture has been validated numerically for the Stokes problem, by verifying that the dimension of $\ker(B^t)$ is 1, where B is the matrix associated to the constraint of incompressibility, for $N \in \{3, 6, 9, 12\}$ and either $2N$ or $2N - 1$ as maximal polynomial degree for which the integration is exact. We hope to be more precise on this point in near future.

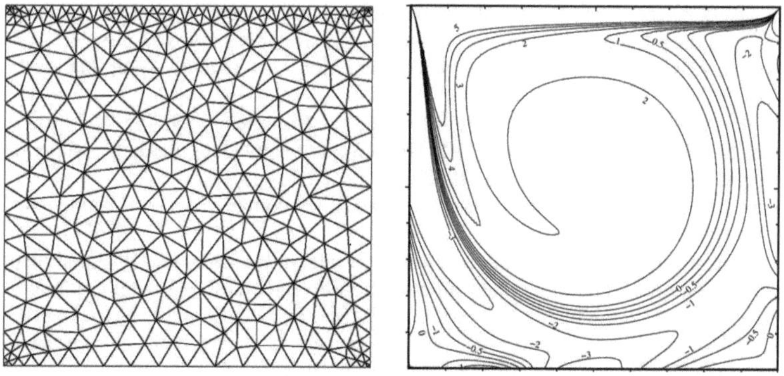

Fig. 7 Driven cavity flow at $Re = 1{,}000$ [29]. Mesh (at *left*) and vorticity (at *right*). The computation has been done with $N = 9$, $M = 2N$ and 682 elements, so that $dof = 28{,}090$. Such a result compares well with the reference one of [6]. The computed vorticity at the *upper left* corner is $\omega \approx 1{,}000$

Concerning the theoretical analysis, little is known to date about the approximation properties of the interpolation operators at Fekete's or other points in T. The proposed TSEM has proved to be a high-order approach to PDEs that enjoys the SEM properties of high-accuracy, low dispersion and dissipation, efficiency and scalability on modern computer architectures and with possible extension to include non-conforming elements.

References

1. Bernardi, C., Maday, Y.: Spectral methods. In: Ciarlet, P.J., Lions, J.L. (eds.) Handbook of numerical analysis, vol.5, pp. 209–486. North Holland, Amsterdam (1997)
2. Bjøntegaard, T., Rønquist, E.M., Tråsdahl, Ø.: Spectral approximation of partial differential equations in highly distorted domains. J. Sci. Comput. **52**, 603–618 (2012)
3. Blyth, M.G., Pozrikidis, C.: A Lobatto interpolation grid over the triangle. IMA J. Appl. Math. **71**, 153–169 (2005)
4. Bos, L.: Bounding the Lebesgue function for Lagrange interpolation in the simplex. J. Approx. Theory **38**, 43–59 (1983)
5. Bos, L., Taylor, M.A., Wingate, B.A.: Tensor product Gauss-Lobatto points are Fekete points for the cube. Math. Comp. **70** 1543–1547 (2001)
6. Botella, O., Peyret, R.: Benchmark spectral results on the lid driven cavity flow. Comput. Fluids **27**(4), 421–433 (1998)
7. Canuto, C., Hussaini, M.Y., Quarteroni, A., Zang, T.A.: Spectral methods in fluid dynamics. Springer, New York (1988)
8. Canuto, C., Hussaini, M.Y., Quarteroni, A., Zang, T.A.: Spectral methods. Fundamentals in single domains. Springer, New York (2007)
9. Canuto, C., Hussaini, M.Y., Quarteroni, A., Zang, T.A.: Spectral methods. Evolution to complex domains and applications to fluid dynamics. Springer, New York (2007)
10. Chen, Q., Babuska, I.: Approximate optimal points for polynomial interpolation of real functions in an interval and in a triangle. Comput. Methods Appl. Mech. Engrg. **128**, 405–417 (1995)

11. Chen, Q., Babuska, I.: The optimal symmetrical points for polynomial interpolation of real functions in the tetrahedron. Comput. Methods Appl. Mech. Engrg. **137**, 89–94 (1996)
12. Cools, R.: Advances in multidimensional integration. J. Comput. Appl. Math., **149**, 1–12 (2002)
13. Cools, R.: An encyclopaedia of cubature formulas. J. Complexity **19**, 445–453 (2003)
14. Deville, M., Fischer, P.F., Mund, E.H.: High-order methods for incompressible fluid flow. Cambridge Univ. Press, Cambridge (2002)
15. Dubiner, M.: Spectral methods on triangles and other domains. J. Sci. Comput. **6**, 345–390 (1991)
16. Fejér, L.: Bestimmung derjenigen Abszissen eines Intervalles für welche die Qradratsumme der Grundfunktionen der Lagrangeschen Interpolation im Intervalle $[-1, 1]$ ein möglichst kleines Maximum besitzt. Ann. Scuola Norm. Sup. Pisa, Sci. Fis. Mt. Ser. II **1**, 263–276 (1932)
17. Fekete, M.: Über die Verteilung der Wurzeln bei gewissen algebraischen Gleichungen mit ganzzahligen Koeffizienten. Mathematische Zeitschrift **17**, 228–249 (1923)
18. Fischer, P.F., Lottes, J.W.: Hybrid Schwarz-multigrid methods for the spectral element method: Extensions to Navier-Stokes. J. Sci. Comput. **6**, 345–390 (2005)
19. Gassner, G., Lörcher, F., Munz, C.-D., Hestaven, J.: Polymorphic nodal elements for discontinuous spectral/hp element methods. J. Comput. Phys. **228**(5), 1573–1590 (2009)
20. Giraldo, F.X., Taylor, M.A.: A diagonal-mass-matrix triangular-spectral-element method based on cubature points. J. Engrng. Math. **56**(3), 307–322 (2006)
21. Gordon, W.N., Hall, C.A.: Construction of curvilinear coordinate systems and application to mesh generation. Int. J. Num. Meth. Eng. **7**, 461–477 (1973)
22. Gottlieb, D., Orszag, S.A.: Numerical analysis of spectral methods: theory and applications. SIAM, Philadelphia (1977)
23. Hestaven, J.: From electrostatic to almost optimal nodal sets for polynomial interpolation in a simplex. SIAM J. Numer. Anal. **35**, 655–676 (1998)
24. Hestaven, J., Teng, C.H.: Stable spectral methods on tetrahedral elements. SIAM J. Sci. Comput. **21**, 2352–2380 (2000)
25. Karniadakis, G.E., Sherwin, S.J.: Spectral/*hp* element methods for CFD. Oxford Univ. Press, New York (1999)
26. Kirby, R.M., Karniadakis, G.E.: De-aliasing on non-uniform grids: algorithms and applications. J. Comput. Phys. **191**, 249–264 (2003)
27. Kirby, R.M., Sherwin, S.J.: Aliasing errors due to quadratic nonlinearities on triangular spectral/hp element discretisations. J. Eng. Math. **56**, 273–288 (2006)
28. Koornwinder, T.: Two-variable analogues of the classical orthogonal polynomials. In: Askey, R.A. (ed.) Theory and application of special functions, 435–495. Academic Press, New York (1975)
29. Lazar, L., Pasquetti, R., Rapetti, F.: Fekete-Gauss Spectral Elements for Incompressible Navier-Stokes Flows: The Two-Dimensional Case. Commun. Comput. Phys. **13**(5), 1309–1329 (2013)
30. Li, Y., Wang, L.-L., Li, H., Ma, H.: A new spectral method on triangles. In: J.S.Hestaven, E.M. Rønquist (eds.) Spectral and High Order Methods for PDEs, LNCSE **76**, 237–246, Springer-Verlag (2011)
31. Luo, H., Pozrikidis, C.: A Lobatto interpolation grid in the tretrahedron. IMA J. Appl. Math. **71**, 298–313 (2006)
32. Maday, Y., Munoz, R.: Spectral element multigrid. II. Theoretical justification. J. Sci. Comput. **3**, 323–353 (1988)
33. Maday, Y., Patera, A.T.: Spectral element methods for the Navier-Stokes equations. In: Noor, A.K., Oden, J.T. (eds.) State-Of-The-Art Surveys in Computational Mechanics, chap.3. ASME (1989)
34. Mazzieri, I., Rapetti, F.: Dispersion analysis of triangle-based spectral element methods for elastic wave propagation. Numerical Algorithms **60**(4), 631–650 (2012)
35. Pasquetti, P., Pavarino, L.F., Rapetti, F., Zampieri, E.: Neumann-Neumann-Schur complement methods for Fekete spectral elements. J. Engrg. Math. **56**(3), 323–335 (2006)

36. Pasquetti, P., Pavarino, L.F., Rapetti, F., Zampieri, E.: Overlapping Schwarz methods for Fekete and Gauss-Lobatto spectral elements SIAM J. on Sci. Comp. **29**(3), 1073–1092 (2007)
37. Pasquetti, R., Rapetti, F.: Spectral element methods on triangles and quadrilaterals: comparisons and applications. J. Comput. Phys. **198**(1), 349–362 (2004)
38. Pasquetti, R., Rapetti, F.: Spectral element methods on unstructured meshes: comparisons and recent advances. J. Sci. Comput. **27**(1–3), 377–387 (2006)
39. Pasquetti, R., Rapetti, F.: p-multigrid method for Fekete-Gauss spectral element approximations of elliptic problems. Commun. in Comput. Phys. **5**, 667–682 (2009)
40. Pasquetti, R., Rapetti, F.: Spectral element methods on unstructured meshes: Which interpolation points ? Num. Alg. **55**(2), 349–366 (2010)
41. Patera, A.T.: A spectral element method for fluid dynamics: laminar flows in a channel expansion. J. Comput. Phys. **54**, 468–488 (1984)
42. Proriol, J.: Sur une famille de polynômes à deux variables orthogonaux dans un triangle. C. R. Acad. Sci. Paris **257**, 2459–2461 (1957)
43. Rapetti, F., Sommariva, A., Vianello, M.: Computing symmetric Lebesgue-type points on the triangle. J. Comput. Appl. Math. **236**(18), 4925–4932 (2012)
44. Rønquist, E.M.: Optimal spectral element methods for the unsteady three-dimensional incompressible Navier-Stokes equations. Ph.D. Thesis at Massachusetts Institute of Technology (1988)
45. Rønquist, E.M., Patera, A.T.: Spectral element multigrid. 1- Formulation and numerical results. J. Sci. Comput. **2**, 389–405 (1987)
46. Roth, M.J.: Nodal configurations and Voronoi tessellations for triangular spectral elements. PhD Thesis, University of Victoria (2005)
47. Runge, C.: Überempirische functionen und die interpolation zwischen äquidistanten ordinaten. Zeitschrift für Mathematik und Physik **46**, 224–243 (1901)
48. Schwab, Ch.: p- and hp-finite element methods. Oxford Univ. Press, Oxford (1998)
49. Smith, B.F., Bjørstad, P.E., Gropp, W.D.: Domain Decomposition. Parallel multilevel methods for elliptic partial differential equations. Cambridge University Press (1996)
50. Stieltjes, T.J.: Sur les polynômes de Jacobi. C. R. Acad. Sci. Paris **100**, 620–622 (1885)
51. Stroud, A.H.: Approximate calculations of multiple integrals. Prentice Hall (1971).
52. Szegö, G.: Orthogonal Polynomials. Vol. 23 in AMS Coll. Publ., New York (1939)
53. Taylor, M.A., Wingate, B.A., Vincent, R.E.: An algorithm for computing Fekete points in the triangle. SIAM J. Numer. Anal. **38**, 1707–1720 (2000)
54. Warburton, T.: An explicit construction for interpolation nodes on the simplex. J. Eng. Math. **56**(3), 247–262 (2006)
55. Warburton, T., Pavarino, L.F., Hesthaven, J.S.: A pseudo-spectral scheme for the incompressible NavierStokes equations using unstructured spectral elements. J. Comput. Phys. **164**(1), 1–21 (2000)

Exponential Convergence of *hp*-DGFEM for Elliptic Problems in Polyhedral Domains

Dominik Schötzau, Christoph Schwab, Thomas Wihler, and Marcel Wirz

Abstract We review the recent results of D. Schötzau et al. (*hp*-dGFEM for elliptic problems in polyhedra. I: Stability and quasioptimality on geometric meshes. Technical report 2009-28, Seminar for applied mathematics, ETH Zürich, 2009. To appear in SIAM J Numer Anal, 2013; *hp*-dGFEM for elliptic problems in polyhedra. II: Exponential convergence. Technical report 2009-29, Seminar for applied mathematics, ETH Zürich, 2009. To appear in SIAM J Numer Anal, 2013), and establish the exponential convergence of *hp*-version discontinuous Galerkin finite element methods for the numerical approximation of linear second-order elliptic boundary-value problems with homogeneous Dirichlet boundary conditions and constant coefficients in three-dimsional and axiparallel polyhedra. The exponential rates are confirmed in a series of numerical tests.

1 Introduction

A key feature of the *hp*-version finite element method (FEM) is the possibility to achieve exponential convergence rates in terms of the number of degrees of freedom. Indeed, in the mid eighties, Babuška and Guo proved that using *hp*-FEM for the numerical approximation of elliptic boundary-value problems with piecewise

---

D. Schötzau (✉)
Department of Mathematics, University of British Columbia, Vancouver, BC, V6T 1Z2, Canada
e-mail: schoetzau@math.ubc.ca

C. Schwab
Seminar for Applied Mathematics, ETH, 8092 Zürich, Switzerland
e-mail: schwab@math.ethz.ch

T. Wihler · M. Wirz
Mathematisches Institut, Universität Bern, 3012 Bern, Switzerland
e-mail: wihler@math.unibe.ch; wirz@math.unibe.ch

M. Azaïez et al. (eds.), *Spectral and High Order Methods for Partial Differential Equations - ICOSAHOM 2012*, Lecture Notes in Computational Science and Engineering 95, DOI 10.1007/978-3-319-01601-6_4,

analytic data in a polygonal domain Ω leads to energy norm error bounds of the form $C\exp(-bN^{1/3})$, where N is the dimension of the hp-version finite element space, and C and b are constants independent of N; see [2, 10, 11] and the references therein. Exponential convergence is achieved by employing geometric mesh refinement towards the singular support $\mathscr{S}$ of the solutions (i.e., the set of vertices of Ω), and nonuniform elemental polynomial degrees which increase linearly with the elements' distance from $\mathscr{S}$. The proof of elliptic regularity in countably weighted Sobolev spaces of the solutions, which constitutes an essential ingredient of the convergence proof, has been a major technical achievement. Let us mention that generalizations to conforming methods for higher-order elliptic problems and hp-version mixed methods for Stokes flow in polygons can be found in [8, 14, 18].

In the 1990s, steps to extend the analytic regularity and the hp-convergence analysis to polyhedral domains in three dimensions were undertaken in [3, 9, 12, 13] and the references therein. The difficulty in this case is the appearance of anisotropic edge and corner-edge singularities. It was claimed and confirmed numerically that the energy norm errors decay exponentially as $C\exp(-bN^{1/5})$, i.e., with an exponent containing the fifth root of N.

The discontinuous Galerkin finite element method (DGFEM) emerged in the seventies as a stable discretization of first-order transport-dominated problems, and as a nonconforming discretization of second-order elliptic problems; see the historical survey [1] and the references therein. Later, in the 1990s, DGFEM was employed to realize hp-version methods for first-order transport and for advection-reaction-diffusion problems in two- and three-dimensional domains (see [15, 16]). Exponential convergence rates were established for piecewise analytic solutions excluding, in particular, corner singularities as occurring in polygonal domains. In the context of purely elliptic problems, the well-posedness of hp-version local discontinuous Galerkin methods was shown in [17]. Finally, exponential convergence of hp-DGFEM in polygonal domains was proved in [25] for diffusion problems, and in [24] for Stokes flow, thereby extending the results of Babuška and Guo to the discontinuous Galerkin framework. In the recent articles [21, 22], the hp-DGFEM for the approximation of three-dimensional elliptic problems in polyhedra was considered. In addition, the paper [26] addresses mixed hp-DGFEM discretizations of the linear elasticity and Stokes equations in polyhedral domains; this work is based on the inf-sup stability of mixed hp-DGFEM (based on uniform isotropic, but variable polynomial degrees) for our class of hp-discretizations, which has been established in [19, 20].

In this paper, we will review the recent results of [21, 22], and, in particular, the proof of exponential convergence with a fifth root in N for an hp-version DGFEM for elliptic problems with constant coefficients in axiparallel polyhedra. Our proof is based on the recent analytic regularity results of [5], which measure corner, edge and corner-edge singularities in analytic classes of anisotropically weighted Sobolev spaces. We begin by introducing hp-version DG approximations on general meshes consisting of axiparallel and possibly anisotropic cuboids, along with elemental degree vectors which may also be anisotropic. Moreover, we review the well-posedness of the resulting finite element methods, show the Galerkin orthogonality property, and derive abstract error estimates for the DG energy errors. To resolve singularities, we shall then construct a family of anisotropically and geometrically refined meshes, characterized by a subdivision ratio $\sigma \in (0,1)$ and a number ℓ of refinements. The corresponding degree vectors are linearly increasing with slope $\mathfrak{s} > 0$ away from corners and edges. This family of hp-discretizations contains, in particular, three-dimensional and anisotropic generalizations of all mesh-degree combinations which were found to be optimal in the univariate case in [7]. By proceeding as in [22], we will then specify suitable polynomial interpolation operators, and show that our estimates lead to the exponential convergence bound $C\exp(-bN^{1/5})$ in the DG energy norm. We will also present a series of new numerical tests which verify the exponential convergence in three dimensions. In particular, we confirm the fifth root in N for corner-edge singularities.

The outline of the article is as follows: In Sect. 2, we introduce a model problem, and recapitulate its analytic regularity in the weighted Sobolev spaces of [5]. In Sect. 3, we introduce and analyze an hp-version interior penalty DGFEM with anisotropic elemental polynomial degrees on meshes of anisotropic and axiparallel elements. Section 4 is devoted to proving our exponential convergence estimate on geometric mesh families. Finally, Sect. 5 contains a series of numerical results.

2 Model Problem and Analytic Regularity

We consider the boundary-value problem

$$-\nabla\cdot(A\nabla u) = f \qquad \text{in } \Omega \subset \mathbb{R}^3, \tag{1}$$

$$u = 0 \qquad \text{on } \Gamma = \partial\Omega, \tag{2}$$

where Ω is an axiparallel Lipschitz polyhedron, A a constant symmetric positive definite coefficient matrix, and f an analytic right-hand side (more precise assumptions will be made in Proposition 1 below).

We specify the precise regularity of the solution u of (1)–(2) in countably normed weighted Sobolev spaces. To that end, we follow [5], but mention the papers [9, 12, 13] where alternative definitions of countably normed weighted Sobolev spaces in terms of local spherical coordinates have originally been defined and studied.

Let us denote by $\mathscr{C}$ the set of corners c, and by $\mathscr{E}$ the set of edges e of Ω. The singular support of the solution u is given by

$$\mathscr{S} = \Big(\bigcup_{c\in\mathscr{C}} c\Big) \cup \Big(\bigcup_{e\in\mathscr{E}} e\Big) \subset \Gamma. \tag{3}$$

For $c \in \mathscr{C}$, $e \in \mathscr{E}$ and $x \in \Omega$, we define the distance functions:

$$r_c(x) = \operatorname{dist}(x,c), \qquad r_e(x) = \operatorname{dist}(x,e), \qquad \rho_{ce}(x) = r_e(x)/r_c(x). \tag{4}$$

For each corner $c \in \mathscr{C}$, we define by $\mathscr{E}_c = \{e \in \mathscr{E} : c \cap \overline{e} \neq \emptyset\}$ the set of all edges of Ω which meet at c. For any $e \in \mathscr{E}$, the set of corners of e is given by $\mathscr{C}_e = \{c \in \mathscr{C} : c \cap \overline{e} \neq \emptyset\}$. Then, for $c \in \mathscr{C}$, $e \in \mathscr{E}$ respectively $e \in \mathscr{E}_c$, and a parameter $\varepsilon > 0$, we define the neighborhoods

$$\begin{aligned} \omega_c &= \{x \in \Omega : r_c(x) < \varepsilon \wedge \rho_{ce}(x) > \varepsilon \quad \forall e \in \mathscr{E}_c\}, \\ \omega_e &= \{x \in \Omega : r_e(x) < \varepsilon \wedge r_c(x) > \varepsilon \quad \forall c \in \mathscr{C}_e\}, \\ \omega_{ce} &= \{x \in \Omega : r_c(x) < \varepsilon \wedge \rho_{ce}(x) < \varepsilon\}. \end{aligned} \tag{5}$$

By choosing ε sufficiently small, we may then partition the domain Ω into four disjoint parts,

$$\overline{\Omega} = \overline{\Omega_0 \,\dot\cup\, \Omega_{\mathscr{C}} \,\dot\cup\, \Omega_{\mathscr{E}} \,\dot\cup\, \Omega_{\mathscr{C}\mathscr{E}}}, \tag{6}$$

where

$$\Omega_{\mathscr{C}} = \bigcup_{c \in \mathscr{C}} \omega_c, \qquad \Omega_{\mathscr{E}} = \bigcup_{e \in \mathscr{E}} \omega_e, \qquad \Omega_{\mathscr{C}\mathscr{E}} = \bigcup_{c \in \mathscr{C}} \bigcup_{e \in \mathscr{E}_c} \omega_{ce}. \tag{7}$$

We shall refer to the subdomains $\Omega_{\mathscr{C}}$, $\Omega_{\mathscr{E}}$ and $\Omega_{\mathscr{C}\mathscr{E}}$ as corner, edge and corner-edge neighborhoods of Ω, respectively, and define the remaining interior part of the domain Ω by $\Omega_0 := \Omega \setminus \overline{\Omega_{\mathscr{C}} \cup \Omega_{\mathscr{E}} \cup \Omega_{\mathscr{C}\mathscr{E}}}$.

To each $c \in \mathscr{C}$ and $e \in \mathscr{E}$, we associate a corner and an edge exponent $\beta_c, \beta_e \in \mathbb{R}$, respectively. We collect these quantities in the multi-exponent

$$\underline{\beta} = \{\beta_c : c \in \mathscr{C}\} \cup \{\beta_e : e \in \mathscr{E}\} \in \mathbb{R}^{|\mathscr{C}|+|\mathscr{E}|}. \tag{8}$$

Inequalities of the form $\underline{\beta} < 1$ and expressions like $\underline{\beta} \pm s$ are to be understood componentwise.

Near corners $c \in \mathscr{C}$ and edges $e \in \mathscr{E}$, we shall use local coordinate systems in ω_e and ω_{ce}, which are chosen such that e corresponds to the direction $(0, 0, 1)$. Then, we denote quantities that are transversal to e by $(\cdot)^\perp$, and quantities parallel to e by $(\cdot)^\parallel$. In particular, if $\alpha \in \mathbb{N}_0^3$ is a multi-index corresponding to the three local coordinate directions in ω_e or ω_{ce}, then we have $\alpha = (\alpha^\perp, \alpha^\parallel)$, where $\alpha^\perp = (\alpha_1, \alpha_2)$ and $\alpha^\parallel = \alpha_3$. Following [5, Definition 6.2 and Eq. (6.9)], we introduce the anisotropically weighted semi-norm

$$\begin{aligned} |u|^2_{M^m_{\underline{\beta}}(\Omega)} &= |u|^2_{H^m(\Omega_0)} + \sum_{e \in \mathscr{E}} \sum_{\substack{\alpha \in \mathbb{N}_0^3 \\ |\alpha| = m}} \left\| r_e^{\beta_e + |\alpha^\perp|} \mathsf{D}^\alpha u \right\|^2_{L^2(\omega_e)} \\ &\quad + \sum_{c \in \mathscr{C}} \sum_{\substack{\alpha \in \mathbb{N}_0^3 \\ |\alpha| = m}} \left(\left\| r_c^{\beta_c + |\alpha|} \mathsf{D}^\alpha u \right\|^2_{L^2(\omega_c)} + \sum_{e \in \mathscr{E}_c} \left\| r_c^{\beta_c + |\alpha|} \rho_{ce}^{\beta_e + |\alpha^\perp|} \mathsf{D}^\alpha u \right\|^2_{L^2(\omega_{ce})} \right), \end{aligned} \tag{9}$$

for $m \in \mathbb{N}_0$, and define the norm $\|u\|_{M^m_{\underline{\beta}}(\Omega)}$ by $\|u\|^2_{M^m_{\underline{\beta}}(\Omega)} = \sum_{k=0}^m |u|^2_{M^k_{\underline{\beta}}(\Omega)}$. Here, $|u|^2_{H^m(\Omega_0)}$ is the usual Sobolev semi-norm of order m on Ω_0, and the operator D^α denotes the partial derivative in the local coordinate directions corresponding to the multi-index α. The space $M^m_{\underline{\beta}}(\Omega)$ is the weighted Sobolev space obtained as the closure of $C_0^\infty(\Omega)$ with respect to the norm $\|\cdot\|_{M^m_{\underline{\beta}}(\Omega)}$. Finally, for a weight $\underline{\gamma} \in \mathbb{R}^{|\mathscr{C}|+|\mathscr{E}|}$, we define the analytic class

$$A_{\underline{\gamma}}(\Omega) = \left\{ u \in \bigcap_{m\geq 0} M^m_{\underline{\gamma}}(\Omega) \,:\, \exists\, C_u > 0 \text{ s.t. } |u|_{M^m_{\underline{\gamma}}(\Omega)} \leq C_u^{m+1} m! \;\forall\, m \in \mathbb{N}_0 \right\}; \tag{10}$$

cf. [5, Definition 6.3]. The following shift theorem from [5, Corollary 7.1] now establishes the analytic regularity of solutions to problem (1)–(2).

Proposition 1. *There exist bounds $\beta_{\mathscr{E}}, \beta_{\mathscr{C}} > 0$ (depending on Ω and the coefficients in (1)) such that, for all weight vectors $\underline{\beta}$ satisfying*

$$0 < \beta_e < \beta_{\mathscr{E}}, \quad 0 < \beta_c < \frac{1}{2} + \beta_{\mathscr{C}}, \qquad e \in \mathscr{E},\ c \in \mathscr{C}, \tag{11}$$

the following property holds: if the right-hand side f in (1) belongs to $A_{1-\underline{\beta}}(\Omega)$, then the solution u of (1)–(2) belongs to $A_{-1-\underline{\beta}}(\Omega)$.

3 Discretization

3.1 Finite Element Spaces

We consider (a family of) meshes $\mathscr{M}$ consisting of axiparallel cuboids $\{K\}$. Hence, each element K is the image of the reference cube $\hat{Q} = (-1,1)^3$ under a composition $\Phi_K : \hat{Q} \to K$ of a translation and a dilation. We allow for anisotropic elements and irregular meshes. Additional assumptions will be introduced in (22) below. With each cuboid $K \in \mathscr{M}$, we associate a polynomial degree vector $\underline{p}_K = (p_{K,1}, p_{K,2}, p_{K,3}) \in \mathbb{N}^3$, whose components correspond to the coordinate directions in $\hat{Q} = \Phi_K^{-1}(K)$. For technical reasons, we shall assume throughout that $p_{K,i} \geq 3$. The polynomial degree is called isotropic if $p_{K,1} = p_{K,2} = p_{K,3} = p_K$. We combine the elemental polynomial degrees $\underline{p}_K$ into the polynomial degree vector $\underline{p} = \{\, \underline{p}_K : K \in \mathscr{M} \,\}$, and introduce the hp-version finite element space

$$S^{\underline{p}}(\mathscr{M}) = \left\{ u \in L^2(\Omega) \,:\, u|_K \in \mathbb{Q}^{\underline{p}_K}(K),\ K \in \mathscr{M} \right\}. \tag{12}$$

The local polynomial approximation space $\mathbb{Q}^{\underline{p}_K}(K)$ is defined as follows: first, on the reference element $\hat{Q}$ and for a polynomial degree vector $\underline{p} = (p_1, p_2, p_3) \in \mathbb{N}_0^3$, we introduce the tensor product polynomial space:

$$\mathbb{Q}^{\underline{p}}(\hat{Q}) = \mathbb{P}^{p_1}(\hat{I}) \otimes \mathbb{P}^{p_2}(\hat{I}) \otimes \mathbb{P}^{p_3}(\hat{I}) = \text{span}\left\{ \hat{x}_1^{\alpha_1}\hat{x}_2^{\alpha_2}\hat{x}_3^{\alpha_3} \; : \; \alpha_i \le p_i, \; 1 \le i \le 3 \right\}. \tag{13}$$

Here, for $p \in \mathbb{N}_0$, we denote by $\mathbb{P}^p(\hat{I})$ the space of all polynomials of degree at most p on the reference interval $\hat{I} = (-1, 1)$. Then, if K is an axiparallel element of $\mathscr{M}$ with associated elemental mapping $\Phi_K : \hat{Q} \to K$ and polynomial degree vector $\underline{p}_K = (p_{K,1}, p_{K,2}, p_{K,3})$, we set

$$\mathbb{Q}^{\underline{p}_K}(K) = \left\{ u \in L^2(K) : (u|_K \circ \Phi_K) \in \mathbb{Q}^{\underline{p}_K}(\hat{Q}) \right\}. \tag{14}$$

If the polynomial degrees are uniform and isotropic, i.e., $p_{K,1} = p_{K,2} = p_{K,3} = p_K = p \ge 1$ for all $K \in \mathscr{M}$, we simply write $S^p(\mathscr{M})$ instead of $S^{\underline{p}}(\mathscr{M})$.

3.2 Element Boundary Operators

We denote the set of all interior faces in $\mathscr{M}$ by $\mathscr{F}_I(\mathscr{M})$, and the set of all boundary faces by $\mathscr{F}_B(\mathscr{M})$. In addition, let $\mathscr{F}(\mathscr{M}) = \mathscr{F}_I(\mathscr{M}) \cup \mathscr{F}_B(\mathscr{M})$ signify the set of all (smallest) faces of $\mathscr{M}$. Furthermore, for an element $K \in \mathscr{M}$, we denote the set of its faces by $\mathscr{F}_K = \{ f \in \mathscr{F} : f \subset \partial K \}$. If $f \in \mathscr{F}_K$, then we denote by $h_{K,f}^{\perp}$ the diameter of K perpendicular to the face f. Similarly, if $\underline{p}_K$ is the polynomial degree vector on K, we denote by $p_{K,f}^{\perp}$ the polynomial degree perpendicular to f.

Next, we recall the standard DG trace operators. For this purpose, consider an interior face $f = \partial K^{\sharp} \cap \partial K^{\flat} \in \mathscr{F}_I(\mathscr{M})$ shared by two neighboring elements $K^{\sharp}, K^{\flat} \in \mathscr{M}$. Furthermore, let v and w be a scalar-valued function and a vector-valued function, respectively, both sufficiently smooth inside the elements $K^{\sharp}, K^{\flat}$. Then we define the following trace operators along f:

$$[\![v]\!] = v|_{K^{\sharp}} n_{K^{\sharp}} + v|_{K^{\flat}} n_{K^{\flat}}, \qquad \langle\!\langle w \rangle\!\rangle = 1/2 \left(w|_{K^{\sharp}} + w|_{K^{\flat}} \right). \tag{15}$$

Here, for an element $K \in \mathscr{M}$, we denote by n_K the outward unit normal vector on ∂K. For a boundary face $f = \partial K \cap \partial \Omega \in \mathscr{F}_B(\mathscr{M})$ for $K \in \mathscr{M}$, and sufficiently smooth functions v, w on K, we let $[\![v]\!] = v|_K n_\Omega$, $\langle\!\langle w \rangle\!\rangle = w|_K$, where n_Ω is the outward unit normal vector on $\partial\Omega$.

3.3 Discontinuous Galerkin Discretizations

For a given mesh $\mathscr{M}$ and associated polynomial degree distribution $\underline{p}$, we define the hp-version symmetric interior penalty DG solution $u_{DG} \in S^{\underline{p}}(\mathscr{M})$ by

$$u_{\mathrm{DG}} \in S^{\underline{p}}(\mathscr{M}): \qquad a_{\mathrm{DG}}(u_{\mathrm{DG}}, v) = \int_{\Omega} f v \,\mathrm{d}x \qquad \forall\, v \in S^{\underline{p}}(\mathscr{M}), \tag{16}$$

where the bilinear form $a_{\mathrm{DG}}(u, v)$ is given by

$$\begin{aligned} a_{\mathrm{DG}}(u, v) &= \int_{\Omega} (A\nabla_h u) \cdot \nabla_h v \,\mathrm{d}x - \int_{\mathscr{F}(\mathscr{M})} \langle\!\langle A\nabla_h v\rangle\!\rangle \cdot [\![u]\!] \,\mathrm{d}s \\ &\quad - \int_{\mathscr{F}(\mathscr{M})} \langle\!\langle A\nabla_h u\rangle\!\rangle \cdot [\![v]\!] \,\mathrm{d}s + \gamma \int_{\mathscr{F}(\mathscr{M})} \mathrm{j}\, [\![u]\!] \cdot [\![v]\!] \,\mathrm{d}s. \end{aligned} \tag{17}$$

Here, ∇_h is the elementwise gradient, and $\gamma > 0$ is the interior penalty parameter that will be chosen sufficiently large. Furthermore, $\mathrm{j} \in L^\infty(\mathscr{F}(\mathscr{M}))$ is the facewise constant function given by

$$\mathrm{j}|_f = \begin{cases} \dfrac{\max\left(p^{\perp}_{K_\sharp,f}, p^{\perp}_{K_\flat,f}\right)^2}{\min\left(h^{\perp}_{K_\sharp,f}, h^{\perp}_{K_\flat,f}\right)} & \text{if } f = \partial K_\sharp \cap \partial K_\flat \in \mathscr{F}_I(\mathscr{M}), \\ \dfrac{(p^{\perp}_{K,f})^2}{h^{\perp}_{K,f}} & \text{if } f = \partial K \cap \partial\Omega \in \mathscr{F}_B(\mathscr{M}). \end{cases} \tag{18}$$

We remark that we have omitted an explicit dependence of the penalty jump terms on the diffusion tensor A.

3.4 Well-Posedness

We show the well-posedness of the hp-DGFEM in the standard DG energy norm defined by

$$|\!|\!| v |\!|\!|^2_{\mathrm{DG}} = \int_{\Omega} |\nabla_h v|^2 \,\mathrm{d}x + \gamma \int_{\mathscr{F}} \mathrm{j}\, |[\![v]\!]|^2 \,\mathrm{d}s, \tag{19}$$

for any $v \in S^{\underline{p}}(\mathscr{M}) + H^1(\Omega)$. To that end, we recall the anisotropic polynomial trace inequality from [21, Lemma 4.3 (a)]: let $K = (0, h_1) \times (0, h_2) \times (0, h_3)$ be an axiparallel element, then there exists a constant $C_I > 0$ only depending on the reference element such that

$$\|q\|_{L^2(f)} \leq C_I (p^{\perp}_{K,f})^2 (h^{\perp}_{K,f})^{-1} \|q\|^2_{L^2(K)} \tag{20}$$

for all $f \in \mathscr{F}_K$, $K \in \mathscr{M}$, and $q \in \mathbb{Q}^{\underline{p}_K}(K)$.

Proceeding as in [21, Theorem 4.4], the following result can be shown.

Proposition 2. *There is a threshold parameter $\gamma_{\min} > 0$ such that for $\gamma \geq \gamma_{\min}$ the DG bilinear form $a_{DG}(\cdot,\cdot)$ is continuous and coercive over $S^{\underline{p}}(\mathcal{M})$. That is, we have*

$$|a_{DG}(v,w)| \leq C_1 |||v|||_{DG} |||w|||_{DG} \qquad \forall\, v,w \in S^{\underline{p}}(\mathcal{M}),$$

$$a_{DG}(v,v) \geq C_2 |||v|||_{DG}^2 \qquad \forall\, v \in S^{\underline{p}}(\mathcal{M}).$$

The constants $\gamma_{\min}$, C_1 and C_2 only depend on γ appearing in (17) *and* (19)*, the coefficient matrix A, and the constant C_I in the trace inequality* (20).

Next, we discuss the Galerkin orthogonality of the DG scheme (16) under the assumption that the solution u of (1)–(2) belongs to $M^2_{-1-\underline{\beta}}(\Omega)$ for a weight vector $\underline{\beta}$ as in (11). We notice that, in this case, it is not obvious that the expression $a_{DG}(u,v)$ is well defined for $v \in S^{\underline{p}}(\mathcal{M})$, since u may exhibit corner and/or edge singularities. Here, integrals of the form

$$\int_f A\nabla u \cdot v \,\mathrm{d}s, \qquad f \in \mathcal{F}_K \cap \mathcal{F}_B(\mathcal{M}), \tag{21}$$

appearing in the bilinear form a_{DG} require some special care, in particular, for faces f which abut at the singular support $\mathcal{S}$. However, in [21, Sect. 4.5], it is shown that the regularity $u \in M^2_{-1-\underline{\beta}}(\Omega)$ implies $A\nabla u \in L^1(f)$, $f \in \mathcal{F}_K \cap \mathcal{F}_B(\mathcal{M})$. Thus, the above integrals are in fact properly defined as bilinear forms on $L^1(f) \times L^\infty(f)$. Consequently, a Green's formula can be established which leads to the following result; see [21, Theorem 4.9].

Proposition 3. *Let the solution u of* (1)–(2) *satisfy $u \in M^2_{-1-\underline{\beta}}(\Omega)$ for $\underline{\beta}$ as in* (11)*, and let u_{DG} be the DG approximation of* (16) *obtained with $\gamma \geq \gamma_{\min}$ (cf. Proposition 2). Then we have the Galerkin orthogonality property $a_{DG}(u - u_{DG}, v) = 0$ for all $v \in S^{\underline{p}}(\mathcal{M})$.*

3.5 Error Estimates

To derive error estimates, we shall now assume the following bounded variation property in the mesh size: there is a constant $\lambda \in (0,1)$ such that

$$\lambda \leq h^{\perp}_{K^\flat,f} / h^{\perp}_{K^\sharp,f} \leq \lambda^{-1}, \tag{22}$$

for all interior faces $f = \partial K^\flat \cap K^\sharp \in \mathcal{F}_I(\mathcal{M})$, uniformly in the mesh family.

To account for the singular solution behavior near corner and edges, we disjointly partition $\mathcal{M}$ into

$$\mathcal{M} = \mathfrak{O} \,\dot{\cup}\, \mathfrak{T}, \tag{23}$$

where elements in $\mathfrak{O}$ are bounded away from $\mathscr{S}$, and elements in the terminal layer $\mathfrak{T}$ have a nontrivial intersection with the singular support $\mathscr{S}$.

Let now $u \in M^2_{-1-\underline{\beta}}(\Omega)$ be the solution of problem (1)–(2), and $u_{\mathrm{DG}} \in S^{\underline{p}}(\mathscr{M})$ be the DG approximation from (16). As usual we split the error into the two parts

$$u - u_{\mathrm{DG}} = \eta + \xi, \quad \text{with } \eta = u - \Pi u \text{ and } \xi = \Pi u - u_{\mathrm{DG}} \in S^{\underline{p}}(\mathscr{M}), \tag{24}$$

for an appropriate hp-version (quasi)interpolation operator $\Pi u \in S^{\underline{p}}(\mathscr{M})$ of u.

To bound $|||\xi|||_{\mathrm{DG}}$ in terms of quantities involving η, we apply the coercivity of the DG form in Proposition 2, the Galerkin orthogonality property in Proposition 3, and the anisotropic trace inequality: given a cuboid $K = (0, h_1) \times (0, h_2) \times (0, h_3)$, $v \in W^{1,t}(K)$ for $t \geq 1$, there exists a constant $C_t > 0$ only depending on t and the reference element such that

$$\|v\|^t_{L^t(f)} \leq C_t (h^{\perp}_{K,f})^{-1} \left(\|v\|^t_{L^t(K)} + (h^{\perp}_{K,f})^t \left\| \mathsf{D}_{K,f,\perp} v \right\|^t_{L^t(K)} \right), \tag{25}$$

for any $f \in \mathscr{F}_K$; cf. [21, Lemma 4.2]. Here, the operator $\mathsf{D}_{K,f,\perp}$ signifies the partial derivative on element K in direction perpendicular to f.

Consequently, we find the following generic error bound; see [21, Theorem 4.10] for details.

Theorem 1. *Assume* (22) *and let* $u \in M^2_{-1-\underline{\beta}}(\Omega)$ *with* $\underline{\beta}$ *as in* (11). *Then we have the error estimate*

$$|||u - u_{\mathrm{DG}}|||^2_{\mathrm{DG}} \leq C \left|\underline{p}\right|^4 \left(E_{\mathfrak{O}}[\eta] + E_{\mathfrak{T}}[\eta]\right), \tag{26}$$

where

$$\begin{aligned} E_{\mathfrak{O}}[\eta] = &\sum_{K \in \mathfrak{O}} \left(\max_{f \in \mathscr{F}_K} \left(h^{\perp}_{K,f}\right)^{-2} \|\eta\|^2_{L^2(K)} + \|\mathsf{D}\eta\|^2_{L^2(K)} \right) \\ &+ \sum_{K \in \mathfrak{O}} \sum_{f \in \mathscr{F}_K} \left(h^{\perp}_{K,f}\right)^2 \left\|\mathsf{D}_{K,f,\perp} \mathsf{D}\eta\right\|^2_{L^2(K)}, \end{aligned} \tag{27}$$

and

$$\begin{aligned} E_{\mathfrak{T}}[\eta] = &\sum_{K \in \mathfrak{T}} \left(\max_{f \in \mathscr{F}_K} \left(h^{\perp}_{K,f}\right)^{-2} \|\eta\|^2_{L^2(K)} + \|\mathsf{D}\eta\|^2_{L^2(K)} \right) \\ &+ \sum_{K \in \mathfrak{T}} \sum_{f \in \mathscr{F}_K} |f|^{-1} h^{\perp}_{K,f} \|\mathsf{D}\eta\|^2_{L^1(f)}. \end{aligned} \tag{28}$$

The constant $C > 0$ *is independent of the elemental aspect ratios, mesh sizes, and polynomial degree vectors. The quantity* $|f|$ *is the surface measure of a face* f, *and* $|\underline{p}| = \max_{K \in \mathscr{M}} \max\{p_{K,1}, p_{K,2}, p_{K,3}\}$ *is the maximal polynomial degree.*

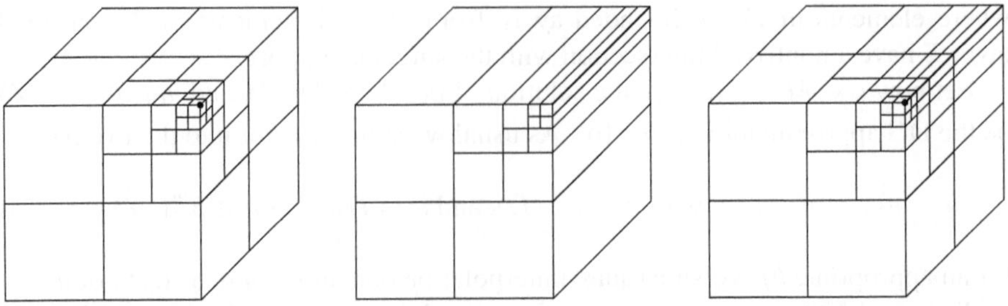

Fig. 1 Canical geometric refinements in $\tilde{Q}$ with subdivision ratio $\sigma = \frac{1}{2}$. (Ex2): isotropic towards the corner c (*left*), (Ex3): anisotropic towards the edge e (*center*), (Ex4): anisotropic towards the edge-corner pair ce (*right*). The sets c, e, ce are shown in boldface

4 Exponential Convergence on hp-Version Subspaces

4.1 Geometric Meshes

To construct geometrically refined meshes, we start from a coarse regular and shape-regular, quasi-uniform partition $\mathscr{M}^0 = \{Q_j\}_{j=1}^J$ of Ω into J convex axiparallel hexahedra. Each of these elements $Q_j \in \mathscr{M}^0$ is the image under an affine mapping G_j of the reference patch $\tilde{Q} = (-1,1)^3$, i.e., $Q_j = G_j(\tilde{Q})$. The mappings G_j are again compositions of (isotropic) dilations and translations.

We then introduce three canonical geometric refinements towards corners, edges and corner-edges of $\tilde{Q}$, which are referred to as extensions (Ex2), (Ex3), and (Ex4) in [21], and which are illustrated in Fig. 1. The extension (Ex1) introduced in [21] corresponds to the case where no refinement is considered on $\tilde{Q}$.

Geometric meshes in Ω are now obtained by applying the patch mappings G_j to transform these canonical geometric mesh patches on the reference patch $\tilde{Q}$ to the macro-elements $Q_j \in \mathscr{M}^0$. More precisely, we denote by $\tilde{\mathscr{M}}_j = \{\tilde{K}\}_{\tilde{K}\in\tilde{\mathscr{M}}_j}$ the elements in the canonical geometric mesh patch associated with $Q_j \in \mathscr{M}^0$. The patches Q_j away from the singular support $\mathscr{S}$ (i.e., with $\overline{Q}_j \cap \mathscr{S} = \emptyset$) are left unrefined by taking $\tilde{\mathscr{M}}_j = \{\tilde{Q}\}$. Then, we denote by $\mathscr{M}_j = \{\, K = G_j(\tilde{K}) \,:\, \tilde{K} \in \tilde{\mathscr{M}}_j\}$ the patch mesh on Q_j, and a geometric mesh in Ω is given by

$$\mathscr{M} = \bigcup_{j=1}^{J} \mathscr{M}_j. \tag{29}$$

It is important to note that the geometric refinements in the canonical patches have to be suitably selected, oriented and combined in order to achieve a proper geometric refinement towards corners and edges of Ω. By construction, each element $K \in \mathscr{M}$ is the image of the reference cube $\hat{Q} = (-1,1)^3$ under an element mapping

$\Phi_K = G_{j(K)} \circ H_K : \hat{Q} \to K \in \mathcal{M}$, where $H_K : \hat{Q} \to \tilde{K}$, $\tilde{K} \subset \tilde{M}_j$, is a possibly anisotropic dilation combined with a translation, and $G_j(K) : \tilde{Q} \to Q_j$ is the patch map.

In what follows, we will consider a sequence of σ-geometrically refined meshes denoted by $\mathfrak{M}_\sigma = \{\mathcal{M}_\sigma^{(\ell)}\}_{\ell\geq 1}$. Here, $\sigma \in (0,1)$ is a fixed parameter defining the ratio of the geometric subdivisions in the canonical refinements in Fig. 1. The index ℓ is the refinement level. There holds: if $K \in \mathcal{M}_\sigma^{(\ell)}$, then there exists $K' \in \mathcal{M}_\sigma^{(\ell-1)}$ such that $K \subset K'$. We shall refer to the sequence $\mathfrak{M}_\sigma$ as a σ-geometric mesh family; see [21, Sect. 3].

In addition to the mesh refinements, the extensions (Ex1)–(Ex4) in [21, Sect. 3] also provide appropriate polynomial degree distributions $\{\underline{p}^{(\ell)}\}_{\ell\geq 1}$. They increase $\mathfrak{s}$-linearly away from the singular set $\mathcal{S}$. In particular, in the edge and corner-edge patches the polynomial degrees are anisotropic. On elements in the interior of the domain, they are uniform, isotropic and proportional to the number ℓ of geometric refinements.

Let us point out that the geometric mesh family constructed in [21, Sect. 3] satisfies the bounded variation property (22) with a constant $\lambda \in (0,1)$ depending on σ and $\mathcal{M}^0$. In addition, the associated family of polynomial degree vectors $\{\underline{p}^{(\ell)}\}_{\ell\geq 1}$ satisfies a similar property: there is a constant $\mu \in (0,1)$ depending on the slope parameter $\mathfrak{s}$ such that

$$\mu \leq p^{\perp}_{K^\flat,f} / p^{\perp}_{K^\sharp,f} \leq \mu^{-1}, \tag{30}$$

for all interior faces $f = \partial K^\flat \cap K^\sharp \in \mathcal{F}_I(\mathcal{M})$ and $\ell \geq 1$.

4.2 Exponential Convergence Rates

The main result of this review is the following exponential convergence result from [22, Theorem 6.1].

Theorem 2. *Assume that the right-hand side f of the boundary-value problem* (1)–(2) *belongs to $A_{1-\underline{\beta}}(\Omega)$ for a weight vector $\underline{\beta}$ as in* (11) *(hence, the solution u is in $A_{-1-\underline{\beta}}(\Omega)$ due to Proposition 1).*

Let $\mathfrak{M}_\sigma = \{\mathcal{M}_\sigma^{(\ell)}\}_{\ell\geq 1}$ be a σ-geometric mesh family with a geometric refinement factor $\sigma \in (0,1)$ and $\{\underline{p}^{(\ell)}\}_{\ell\geq 1}$ the associated (possibly anisotropic) $\mathfrak{s}$-linear degree distribution vectors with a slope parameter $\mathfrak{s} > 0$, generated by the hp-extensions (Ex1)–(Ex4) in [21, Sect. 3]. Consider the resulting hp-version finite element spaces

$$V_{\sigma,\mathfrak{s}}^{(\ell)} := S^{\underline{p}^{(\ell)}}(\mathcal{M}_\sigma^{(\ell)}), \qquad \ell \geq 1. \tag{31}$$

Then, for each $\ell \geq 1$, the DG approximation $u_{DG} \in V_{\sigma,\mathfrak{s}}^{(\ell)}$ is well defined for $\gamma \geq \gamma_{\min}$ (see Proposition 2), and we have the error estimate

$$|||u - u_{\mathrm{DG}}|||_{\mathrm{DG}} \leq C \exp(-bN^{1/5}), \tag{32}$$

where $N = \dim(V_{\sigma,\mathfrak{s}}^{(\ell)})$*. The constants* $C > 0$ *and* $b > 0$ *are independent of* N*, and solely depend on the constants in the trace inequalities* (20) *and* (25)*, respectively, the parameters* σ*,* $\mathfrak{s}$*, the initial mesh* $\mathcal{M}^0$*, the analyticity constant* C_u *in* (10) *of the solution* u*, the weight vector* $\underline{\beta}$*, the diffusion tensor* A*, and the penalty parameter* γ*.*

Remark 1. As proved in [22, Theorem 6.1 and Corollary 5.19], the exponential convergence result (32) also holds for spaces with uniform and isotropic polynomial degrees, i.e., for the family

$$V_\sigma^\ell = S^{p^{(\ell)}}(\mathcal{M}_\sigma^{(\ell)}), \qquad \ell \geq 1, \tag{33}$$

provided that $p^{(\ell)} \simeq \max\{3, \ell\}$. However, in this case, the constant b in the exponent has to be replaced by a smaller constant $\overline{b} > 0$.

Remark 2. The discontinuous hp-version interpolant constructed to prove Theorem 2 yields an exponential DG norm approximation bound of the form (32) for any function $u \in A_{-1-\underline{\beta}}(\Omega)$ with weights given by (11). In particular, for more general diffusion-reaction problems with non-constant coefficients as in [21] or for second-order elliptic systems as in [26], exponential convergence of hp-DGFEM can be achieved for solutions in $A_{-1-\underline{\beta}}(\Omega)$.

4.3 Ingredients of the Proof

Let us give some insights into the proof of Theorem 2. We apply the error estimates of Theorem 1. To that end and according to (23), we subdivide the geometric mesh $\mathcal{M}_\sigma^{(\ell)}$ into

$$\mathcal{M}_\sigma^{(\ell)} = \mathfrak{O}_\sigma^{(\ell)} \,\dot{\cup}\, \mathfrak{T}_\sigma^{(\ell)}. \tag{34}$$

After specification of the hp-version interpolation operator Πu in (24), Theorem 1 requires bounding the two terms $E_{\mathfrak{O}_\sigma^{(\ell)}}[\eta]$ and $E_{\mathfrak{T}_\sigma^{(\ell)}}[\eta]$ in (27) and (28), respectively. Since the approximation spaces are discontinuous, we can choose different interpolation operators in the two submeshes $\mathfrak{O}_\sigma^{(\ell)}$ and $\mathfrak{T}_\sigma^{(\ell)}$.

Bounding $E_{\mathfrak{O}_\sigma^{(\ell)}}[\eta]$**:** In the elements away from $\mathcal{S}$, we choose Πu to be an elementwise tensorized operator of univariate hp-interpolation operators: for an element $K \in \mathfrak{O}_\sigma^{(\ell)}$ and a polynomial degree $\underline{p}_K = (p_1, p_2, p_3)$, we set

$$(\Pi u)|_K = \pi_{p_1,2}^1 \otimes \pi_{p_2,2}^2 \otimes \pi_{p_3,2}^3 u|_K, \qquad K \in \mathfrak{O}_\sigma^{(\ell)}, \tag{35}$$

where $\pi^i_{p_i,2}$ is a properly scaled version of the C^1-conforming univariate projector into polynomials of degree p_i, constructed and analyzed in [6, Sect. 8] and acting in coordinate direction x_i.

To take into account different weighting of the singularities in the different neighborhoods, we shall further subdivide $\mathfrak{O}^{(\ell)}_\sigma$ into discrete corner, edge, corner-edge and interior neighborhoods of the form

$$\mathfrak{O}^\ell_\sigma = \mathfrak{O}^\ell_{\mathscr{C}} \,\dot{\cup}\, \mathfrak{O}^\ell_{\mathscr{E}} \,\dot{\cup}\, \mathfrak{O}^\ell_{\mathscr{C}\mathscr{E}} \,\dot{\cup}\, \mathfrak{O}^\ell_{\mathrm{int}}. \tag{36}$$

In each of these neighborhoods, the geometrically refined elements can be grouped into certain subsets of elements with identical scaling properties in terms of their relative distance to the sets $\mathscr{C}$ and $\mathscr{E}$. Hence, combining the analytic regularity properties in each of the discrete neighborhoods with classical hp-approximation techniques for $u - \Pi u$ (similarly to the two-dimensional case) yields

$$E_{\mathfrak{O}^{(\ell)}_\sigma}[\eta] \le C \exp(-2b\ell), \tag{37}$$

with constants $C > 0$ and $b > 0$ independent of ℓ. We refer to [22, Sects. 5.2 and 5.3] for details.

Bounding $E_{\mathfrak{T}^{(\ell)}_\sigma}[\eta]$: For elements $K \in \mathfrak{T}^{(\ell)}_\sigma$ at the boundary of Ω, the zero interpolation operator $\Pi u \equiv 0$ is sufficient; indeed, this may be motivated by the fact that the exact solution u satisfies homogeneous Dirichlet boundary conditions. In addition, the weights appearing in the $\|.\|_{M^2_{-1-\underline{\beta}}}$-norm from (9) carry negative exponents for $|\alpha| = 0, 1$, which results in exponentially small scaled element contributions in $\mathfrak{T}^{(\ell)}_\sigma$. Thence, the following bound can be obtained:

$$E_{\mathfrak{T}^{(\ell)}_\sigma}[\eta] \le C \exp(-2b\ell), \tag{38}$$

with constants $C > 0$ and $b > 0$ independent of ℓ, see [22, Sect. 5.4].

Counting the degrees of freedom: From Theorem 1 and the bounds (37), (38), we conclude that

$$|\!|\!| u - u_{\mathrm{DG}} |\!|\!|_{\mathrm{DG}} \le C \exp(-b\ell).$$

Then we note that $N = \dim(V^{(\ell)}_{\sigma,\mathfrak{s}}) \simeq \ell^5 + O(\ell^4)$ for $\ell \to \infty$, which implies the bound (32).

5 Numerical Experiments

For the numerical experiments the software library deal.ii [4] is employed. Our computations are based on the geometrically (with ratio $\sigma = 1/2$) refined hp-spaces V^ℓ_σ from (33) featuring uniform and isotropic polynomial degree $p \simeq \ell$,

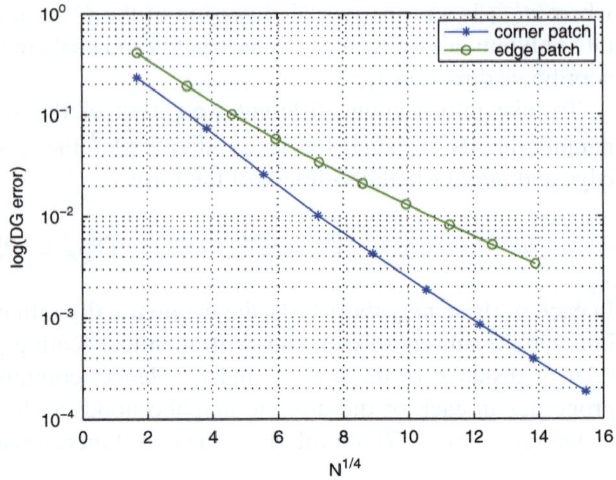

Fig. 2 Performance of the hp-DGFEM in corner and edge patches

where ℓ is the number of mesh layers. We test the hp-DGFEM (16) for the three reference situations 'corner patch', 'edge patch', 'corner-edge patch', as displayed in Fig. 1 (and scaled to the unit cube $(0, 1)^3$). They correspond to the hp-extensions (Ex2), (Ex3), (Ex4) in [21], respectively. In all experiments, the penalty parameter in (17) is chosen to be $\gamma = 10$. Then, we monitor the decay of the error measured in the DG-energy norm (19) as the number of refinements ℓ is increased. In all experiments we prescribe the exact solution u, and choose the right-hand side f in (1) (as well as the (nonhomogeneous) Dirichlet boundary conditions) accordingly.

- Corner patch (Ex2): The exact solution is chosen to be $u_c(r_c) = r_c^{\gamma_c}$, with $\gamma_c = 1/3$, where r_c denotes the distance to the origin. This solution has an isotropic singularity at 0, and is resolved using the isotropic geometric corner mesh shown in Fig. 1 (left). The number of degrees of freedom N in the corresponding hp-spaces is proportional to $\ell^4 \simeq p^4$. In Fig. 2, we observe that the DG energy error decays with a nearly constant slope in a semi-logarithmic plot, thereby confirming exponential convergence (with respect to $N^{1/4}$).
- Edge patch (Ex3): Here, we choose $u_e(r_e) = r_e^{\gamma_e}$, for $\gamma_e = 1/2$, with r_e signifying the distance to the edge $e = \{x_1 = 0\} \times \{x_2 = 0\} \times \{0 < x_3 < 1\}$. The solution exhibits an anisotropic and non-local edge singularity along e, and is refined by means of the anisotropic geometric edge-mesh depicted in Fig. 1 (center). Again, the number of degrees of freedom is proportional to $\ell^4 \simeq p^4$, and the exponential decay of the DG energy error is clearly visible in Fig. 2.
- Corner-edge patch (Ex4): Finally, we consider the anisotropic corner-edge singularity solution $u_{ce}(r_c, r_e) = r_c^{\gamma_c} r_e^{\gamma_e}$, with $\gamma_c = 1/3$ and $\gamma_e = 1/2$. It is refined by employing the anisotropic corner-edge mesh presented in Fig. 1 (right). This is the most complex of the three model cases discussed here; in fact, in contrast to the previous examples, it features $N \simeq \ell^5 \simeq p^5$ degrees of

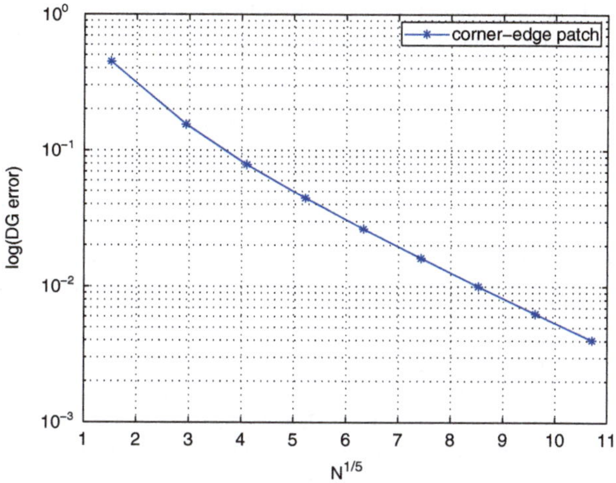

Fig. 3 Performance of the hp-DGFEM in corner-edge patch

freedom. Correspondingly, the DG energy error is plotted against the fifth root of N. As before, exponential convergence is achieved already after a few initial refinements (Fig. 3).

Our experiments show that the hp-DGFEM (16) on the proposed geometric hp-meshes is able to resolve isotropic as well as anisotropic singularities, and, in particular, that exponential rates of convergence are attained in all the reference configurations shown in Fig. 1.

6 Concluding Remarks

Ongoing research is concerned with extensions of the exponential convergence theory for hp-DGFEM in three dimensions to elliptic problems with mixed and Neumann boundary conditions, see [23], to problems with more complicated geometries and non-constant coefficients, as well as to more general elliptic systems.

Acknowledgements This work was supported in part by the Natural Sciences and Engineering Research Council of Canada (NSERC), the European Research Council AdG grant STAHDPDE 247277, and the Swiss National Science Foundation (SNF, Grant 200020_144442/1).

References

1. D. N. Arnold, F. Brezzi, B. Cockburn, and L. D. Marini. Unified analysis of discontinuous Galerkin methods for elliptic problems. *SIAM J. Numer. Anal.*, 39:1749–1779, 2001.

2. I. Babuška and B. Q. Guo. Regularity of the solution of elliptic problems with piecewise analytic data. I. Boundary value problems for linear elliptic equation of second order. *SIAM J. Math. Anal.*, 19(1):172–203, 1988.
3. I. Babuška and B. Q. Guo. Approximation properties of the h-p version of the finite element method. *Comput. Methods Appl. Mech. Engrg.*, 133(3–4):319–346, 1996.
4. W. Bangerth, R. Hartmann, and G. Kanschat. deal.II – a general purpose object oriented finite element library. *ACM Trans. Math. Softw.*, 33(4):24/1–24/27, 2007.
5. M. Costabel, M. Dauge, and S. Nicaise. Analytic regularity for linear elliptic systems in polygons and polyhedra. *Math. Models Methods Appl. Sci.*, 22(8), 2012.
6. M. Costabel, M. Dauge, and C. Schwab. Exponential convergence of hp-FEM for Maxwell's equations with weighted regularization in polygonal domains. *Math. Models Methods Appl. Sci.*, 15(4):575–622, 2005.
7. W. Gui and I. Babuška. The h, p and h-p versions of the finite element method in 1 dimension. II. The error analysis of the h- and h-p versions. *Numer. Math.*, 49(6):613–657, 1986.
8. B. Q. Guo. The h-p version of the finite element method for elliptic equations of order $2m$. *Numer. Math.*, 53(1–2):199–224, 1988.
9. B. Q. Guo. The h-p version of the finite element method for solving boundary value problems in polyhedral domains. In *Boundary Value Problems and Integral Equations in Nonsmooth Domains*, volume 167 of *Lecture Notes in Pure and Applied Mathematics*, pages 101–120. Dekker, New York, 1995.
10. B. Q. Guo and I. Babuška. The hp-version of the finite element method. Part I: The basic approximation results. *Comp. Mech.*, 1:21–41, 1986.
11. B. Q. Guo and I. Babuška. The hp-version of the finite element method. Part II: General results and applications. *Comp. Mech.*, 1:203–220, 1986.
12. B. Q. Guo and I. Babuška. Regularity of the solutions for elliptic problems on nonsmooth domains in $\mathbb{R}^3$. I. Countably normed spaces on polyhedral domains. *Proc. Roy. Soc. Edinburgh Sect. A*, 127(1):77–126, 1997.
13. B. Q. Guo and I. Babuška. Regularity of the solutions for elliptic problems on nonsmooth domains in $\mathbb{R}^3$. II. Regularity in neighbourhoods of edges. *Proc. Roy. Soc. Edinburgh Sect. A*, 127(3):517–545, 1997.
14. B. Q. Guo and C. Schwab. Analytic regularity of Stokes flow on polygonal domains in countably weighted Sobolev spaces. *J. Comp. Appl. Math.*, 119:487–519, 2006.
15. P. Houston, C. Schwab, and E. Süli. Stabilized hp-finite element methods for first-order hyperbolic problems. *SIAM J. Numer. Anal.*, 37:1618–1643, 2000.
16. P. Houston, C. Schwab, and E. Süli. Discontinuous hp-finite element methods for advection–diffusion–reaction problems. *SIAM J. Numer. Anal.*, 39:2133–2163, 2002.
17. I. Perugia and D. Schötzau. An hp-analysis of the local discontinuous Galerkin method for diffusion problems. *J. Sci. Comput.*, 17:561–571, 2002.
18. D. Schötzau and C. Schwab. Exponential convergence in a Galerkin least squares hp-FEM for Stokes flow. *IMA J. Numer. Anal.*, 21:53–80, 2001.
19. D. Schötzau, C. Schwab, and A. Toselli. Stabilized hp-DGFEM for incompressible flow. *Math. Models Methods Appl. Sci.*, 13(10):1413–1436, 2003.
20. D. Schötzau, C. Schwab, and A. Toselli. Mixed hp-DGFEM for incompressible flows. II. Geometric edge meshes. *IMA J. Numer. Anal.*, 24(2):273–308, 2004.
21. D. Schötzau, C. Schwab, and T. P. Wihler. hp-dGFEM for elliptic problems in polyhedra. I: Stability and quasioptimality on geometric meshes. Technical Report 2009-28, Seminar for Applied Mathematics, ETH Zürich, 2009. To appear in SIAM J. Numer. Anal., 2013.
22. D. Schötzau, C. Schwab, and T. P. Wihler. hp-dGFEM for elliptic problems in polyhedra. II: Exponential convergence. Technical Report 2009-29, Seminar for Applied Mathematics, ETH Zürich, 2009. To appear in SIAM J. Numer. Anal., 2013.
23. D. Schötzau, C. Schwab, and T. P. Wihler. Exponential convergence of hp-dGFEM for elliptic problems with mixed and Neumann boundary conditions in polyhedral domains. In preparation, 2013.

24. D. Schötzau and T. P. Wihler. Exponential convergence of mixed hp-DGFEM for Stokes flow in polygons. *Numer. Math.*, 96:339–361, 2003.
25. T. P. Wihler, P. Frauenfelder, and C. Schwab. Exponential convergence of the hp-DGFEM for diffusion problems. *Comput. Math. Appl.*, 46:183–205, 2003.
26. T. P. Wihler and M. Wirz. Mixed hp-Discontinuous Galerkin FEM for linear elasticity in three dimensions. *Math. Models Methods Appl. Sci.*, 22(8), 2012.

A Contribution to the Outflow Boundary Conditions for Navier-Stokes Time-Splitting Methods

E. Ahusborde, M. Azaïez, S. Glockner, and A. Poux

Abstract We present in this paper a numerical scheme for incompressible Navier-Stokes equations with open boundary conditions, in the framework of the pressure and velocity correction schemes. In Poux et al. (J Comput Phys 230:4011–4027, 2011), the authors presented an almost second-order accurate version of the open boundary condition with a pressure-correction scheme in finite volume framework. This paper proposes an extension of this method in spectral element method framework for both pressure- and velocity-correction schemes. A new way to enforce this type of boundary condition is proposed and provides a pressure and velocity convergence rate in space and time higher than with the present state of the art. We illustrate this result by computing some numerical tests.

1 Introduction

A difficulty in obtaining the numerical solution of the incompressible Navier-Stokes equations, lies in the Stokes stage and specifically in the determination of the pressure field which will ensure a solenoidal velocity field. Several approaches are possible. We can for instance consider exact methods as the Uzawa [1] and augmented lagrangian [5] ones. In complex geometries or three dimensional methods, theses techniques are inappropriate since their computational time costs are very high. An alternative consists in decoupling the pressure from the velocity by means of a time splitting scheme. A large number of theoretical and numerical

E. Ahusborde (✉)
University of Pau, LMAP UMR 5142 CNRS, PAU Cedex, France
e-mail: etienne.ahusborde@univ-pau.fr

M. Azaïez · S. Glockner · A. Poux
University of Bordeaux, IPB-I2M UMR 5295, Bordeaux, France
e-mail: azaiez@enscbp.fr; glockner@enscbp.fr; alexandre.poux@enscbp.fr

M. Azaïez et al. (eds.), *Spectral and High Order Methods for Partial Differential Equations - ICOSAHOM 2012*, Lecture Notes in Computational Science and Engineering 95, DOI 10.1007/978-3-319-01601-6_5,

studies have been published that discuss the accuracy and the stability properties of such approaches. The most popular methods are pressure-correction schemes. They were first introduced by Chorin-Temam [2, 18], and improved by Goda (the standard incremental scheme) in [6], and later by Timmermans in [19] (the rotational incremental scheme). They require the solution of two sub-steps: the pressure is treated explicitly in the first one, and is corrected in the second one by projecting the predicted velocity onto an ad-hoc space. A less studied alternative technique known as the velocity-correction scheme, developed by Orszag et al. in [15], Karniadakis et al. in [11], Leriche et al. in [12] and more recently by Guermond et al. in [10], consists in switching the two sub-steps.

In [17] and [7], the authors proved the reliability of such approaches from the stability and the convergence rate points of view. A series of numerical issues related to the analysis and implementation of fractional step methods for incompressible flows are addressed in the review paper of Guermond et al. [9]. In this reference the authors describe the state of the art for both theoretical and numerical results related to the time splitting approach.

Another difficulty consists in the treatment of outflow boundary conditions. Indeed the majority of the studies made on these methods consider only Dirichlet boundary conditions. We are interested here in outflow boundary conditions. A large variety of boundary conditions of this type exists, such as the non reflecting boundary condition developed by Orlanski [14] or Engquist [4]. Here we present some results on the open and traction boundary condition, Liu [13] and Guermond [8].

With open or traction boundary conditions, while no studies have been reported with a velocity-correction scheme, a few have been done with pressure-correction schemes. Guermond et al. proved in [9] that only spatial and time convergence rates between $O(\Delta x+\Delta t)$ and $O(\Delta x^{3/2}+\Delta t^{3/2})$ on the velocity and $O(\Delta x^{1/2}+\Delta t^{1/2})$ on the pressure are to be expected with the standard incremental scheme, and between $O(\Delta x + \Delta t)$ and $O(\Delta x^{3/2} + \Delta t^{3/2})$ on the velocity and pressure for the rotational incremental scheme. In [16], the authors presented a new version of the boundary condition for the pressure-correction scheme in the finite volume framework. They obtained a second-order accuracy for the velocity and rates between $O(\Delta x^{3/2} + \Delta t^{3/2})$ and $O(\Delta x^2 + \Delta t^2)$ with the standard incremental scheme while with the rotational version, a second order convergence is reached for both velocity and pressure. The goal of this paper is to extend this method in spectral element method framework for both pressure-correction and velocity-correction schemes.

2 Pressure-Correction Scheme for Open Boundary Condition

2.1 *Governing Equations*

Let Ω be a regular bounded domain in $\mathbb{R}^d$ with $\mathbf{n}$ a unit vector on the outward normal along the boundary $\Gamma = \partial\Omega$ oriented outward. We suppose that Γ is partitioned into two portions Γ_D and Γ_N.

Our study consists, for a given finite time interval $]0, t^*]$ in computing velocity $\mathbf{u} = \mathbf{u}(\mathbf{x}, t)$ and pressure $p = p(\mathbf{x}, t)$ fields satisfying:

$$\rho \frac{\partial \mathbf{u}}{\partial t} - \mu \Delta \mathbf{u} + \nabla p = \mathbf{f} \quad \text{in } \Omega \times]0, t^*], \tag{1}$$

$$\nabla \cdot \mathbf{u} = 0 \quad \text{in } \Omega \times]0, t^*], \tag{2}$$

$$\mathbf{u} = \mathbf{g} \quad \text{on } \Gamma_D \times]0, t^*], \tag{3}$$

$$(\mu \nabla \mathbf{u} - p\mathbf{I})\, \mathbf{n} = \mathbf{t} \quad \text{on } \Gamma_N \times]0, t^*], \tag{4}$$

where ρ and μ are the density and the dynamic viscosity of the flow respectively and $\mathbf{I}$ the unit tensor. The body force $\mathbf{f} = \mathbf{f}(\mathbf{x}, t)$, the constraint $\mathbf{t} = \mathbf{t}(\mathbf{x}, t)$ and the boundary condition $\mathbf{g} = \mathbf{g}(\mathbf{x}, t)$ are known. For the sake of simplicity, we chose $\mathbf{g} = \mathbf{0}$. Finally, the initial state is characterised by a given $\mathbf{u}(., 0)$.

We shall compute two sequences $(\mathbf{u}^n)_{0 \le n \le N}$ and $(p^n)_{0 \le n \le N}$ in a recurrent way that approximates in some sense the quantities $(\mathbf{u}(., t^n))_{0 \le n \le N}$ and $(p(., t^n))_{0 \le n \le N}$, solutions of unsteady Stokes problem (1)–(4). Using a second order backward difference formula (BDF) time scheme, its semi-discrete version reads:

$$\rho \frac{\alpha \mathbf{u}^{n+1} + \beta \mathbf{u}^n + \gamma \mathbf{u}^{n-1}}{\Delta t} - \mu \Delta \mathbf{u}^{n+1} + \nabla p^{n+1} = \mathbf{f}^{n+1} \text{ in } \Omega, \tag{5}$$

$$\nabla \cdot \mathbf{u}^{n+1} = 0 \quad \text{in } \Omega, \tag{6}$$

$$\mathbf{u}^{n+1} = \mathbf{g} \quad \text{on } \Gamma_D, \tag{7}$$

$$\left(\mu \nabla \mathbf{u}^{n+1} - p^{n+1}\mathbf{I}\right) \mathbf{n} = \mathbf{t}^{n+1} \text{ on } \Gamma_N. \tag{8}$$

Values of parameters α, β, γ depend on the temporal scheme used. Namely:

- $\alpha = 1, \beta = -1, \gamma = 0$ for the first order Euler time scheme,
- $\alpha = \frac{3}{2}, \beta = -2, \gamma = \frac{1}{2}$ for the second order Backward Difference Formulae time scheme.

Equations (5)–(8) are split into two sub-problems. The first one is a prediction diffusion problem that computes a predicted velocity field: *Find* $\mathbf{u}^{n+1/2}$ *such that*

$$\rho \frac{\alpha \mathbf{u}^{n+1/2} + \beta \mathbf{u}^n + \gamma \mathbf{u}^{n-1}}{\Delta t} - \mu \Delta \mathbf{u}^{n+1/2} + \nabla p^n = \mathbf{f}^{n+1} \text{ in } \Omega, \tag{9}$$

$$\mathbf{u}^{n+1/2} = 0 \quad \text{on } \Gamma_D, \tag{10}$$

$$\left(\mu \nabla \mathbf{u}^{n+1/2} - \tilde{p}^{n+1}\mathrm{Id}\right) \mathbf{n} = \mathbf{t}^{n+1} \text{ on } \Gamma_N. \tag{11}$$

Expression of $\tilde{p}^{n+1}$ depends on the time scheme:

- For the first order time scheme

$$\tilde{p}^{n+1} = p^n, \tag{12}$$

- For the second order time scheme

$$\tilde{p}^{n+1} = 2p^n - p^{n-1}. \tag{13}$$

The second step is a correction pressure-continuity: *Find* $(\mathbf{u}^{n+1}, \varphi^{n+1})$ *such that*

$$\frac{\rho\alpha}{\Delta t}(\mathbf{u}^{n+1} - \mathbf{u}^{n+1/2}) + \nabla\varphi^{n+1} = \mathbf{0} \quad \text{in } \Omega, \tag{14}$$

$$\nabla \cdot \mathbf{u}^{n+1} = 0 \quad \text{in } \Omega, \tag{15}$$

$$\mathbf{u}^{n+1} \cdot \mathbf{n} = 0 \quad \text{on } \Gamma_D, \tag{16}$$

$$B.C.\,(\varphi^{n+1}) \qquad \text{on } \Gamma_N. \tag{17}$$

The pressure is upgraded via:

$$p^{n+1} = p^n + \varphi^{n+1} - \chi\mu\nabla \cdot \mathbf{u}^{n+1/2} \quad \text{in } \Omega. \tag{18}$$

The parameter χ is used to switch between the standard incremental scheme and the rotational one:

- $\chi = 0$ for the standard incremental scheme,
- $\chi = 0.7$ for the rotational incremental scheme.[1]

In practice, this second step is replaced by a Poisson problem on φ^{n+1}:

$$\frac{\Delta t}{\alpha\rho}\Delta\varphi^{n+1} = \nabla \cdot \mathbf{u}^{n+1/2} \quad \text{in } \Omega, \tag{19}$$

$$\frac{\partial\varphi^{n+1}}{\partial n} = 0 \qquad \text{on } \Gamma_D, \tag{20}$$

$$B.C.\,(\varphi^{n+1}) \qquad \text{on } \Gamma_N, \tag{21}$$

completed by:

$$p^{n+1} = p^n + \varphi^{n+1} - \chi\mu\nabla \cdot \mathbf{u}^{n+1/2} \quad \text{in } \Omega, \tag{22}$$

[1] Ideally, $\chi = 1$ but as Guermond proved [8], for stability issues, χ is necessarily strictly lower than $2\mu/d$.

$$\mathbf{u}^{n+1} = \mathbf{u}^{n+1/2} - \frac{\Delta t}{\alpha\rho}\nabla\varphi^{n+1} \qquad \text{in } \Omega. \tag{23}$$

The natural choice for B.C. (φ^{n+1}) consists in choosing $\varphi^{n+1} = 0$ on Γ_N. Such a choice involves a numerical locking for $\chi = 0$ since the boundary condition on the pressure increment causes the pressure on the limit to be equal to its initial value. A real improvement is obtained for $\chi = 0.7$ but the expected rates of convergence are not reached.

In the next section we will keep the nature of the boundary condition of φ^{n+1} and will suggest a value of it allowing the reduction of the boundary layer effect mentioned previously.

2.2 *Improvement of the Pressure Boundary Conditions*

For the sake of simplicity we choose a square domain Ω with Γ_N at its right boundary. The starting point of our approach is the derivation on x_1 of the first component of (14):

$$-\frac{\Delta t}{\alpha\rho}\frac{\partial^2\varphi^{n+1}}{\partial x_1^2} = \frac{\partial {u_{x_1}}^{n+1}}{\partial x_1} - \frac{\partial {u_{x_1}}^{n+\frac{1}{2}}}{\partial x_1}. \tag{24}$$

Then, we project on direction x_1 the Eqs. (4) and (8):

$$\mu\frac{\partial u_{x_1}^{n+1}}{\partial x_1} - p^{n+1} = t_{x_1}^{n+1}, \tag{25}$$

$$\mu\frac{\partial u_{x_1}^{n+\frac{1}{2}}}{\partial x_1} - \tilde{p}^{n+1} = t_{x_1}^{n+1}. \tag{26}$$

The combination of those three last Eqs. (24)–(26) gives:

$$-\frac{\Delta t}{\alpha\rho}\frac{\partial^2\varphi^{n+1}}{\partial x_1^2} = \frac{1}{\mu}(p^{n+1} - \tilde{p}^{n+1}). \tag{27}$$

Replacing $\tilde{p}^{n+1}$ by its expressions (12) gives for the first order scheme:

$$\left(\frac{\Delta t}{\alpha\rho}\frac{\partial^2}{\partial x_1^2} + \frac{1}{\mu}\right)\varphi^{n+1} = +\chi\nabla\cdot\mathbf{u}^{n+\frac{1}{2}}. \tag{28}$$

Or for a second order scheme using (13) :

$$\left(\frac{\Delta t}{\alpha\rho}\frac{\partial^2}{\partial x_1^2}+\frac{1}{\mu}\right)\varphi^{n+1}=\frac{\varphi^n}{\mu}+\chi\nabla\cdot\left(\mathbf{u}^{n+\frac{1}{2}}-\mathbf{u}^{n-\frac{1}{2}}\right). \tag{29}$$

Moreover taking into account the Poisson problem (19):

$$\frac{\partial^2\varphi^{n+1}}{\partial x_1^2}+\frac{\partial^2\varphi^{n+1}}{\partial x_2^2}=\nabla\cdot\mathbf{u}^{n+1/2}, \tag{30}$$

and subtracting (28) or (29) in (30) one obtains:

- First-order open boundary condition (OBC1):

$$\left(\frac{\Delta t}{\alpha\rho}\frac{\partial^2}{\partial x_2^2}-\frac{1}{\mu}\right)\varphi^{n+1}=(1-\chi)\nabla\cdot\mathbf{u}^{n+\frac{1}{2}}, \tag{31}$$

- Second-order open boundary condition (OBC2):

$$\left(\frac{\Delta t}{\alpha\rho}\frac{\partial^2}{\partial x_2^2}-\frac{1}{\mu}\right)\varphi^{n+1}=(1-\chi)\nabla\cdot\mathbf{u}^{n+\frac{1}{2}}-\frac{\varphi^n}{\mu}+\chi\nabla\cdot\mathbf{u}^{n-\frac{1}{2}}. \tag{32}$$

To summarize, we propose a pressure-correction step that writes: *Find* φ^{n+1} *such that*

$$\frac{\Delta t}{\alpha\rho}\Delta\varphi^{n+1}=\nabla\cdot\mathbf{u}^{n+1/2}\quad\text{in }\Omega, \tag{33}$$

$$\frac{\partial\varphi^{n+1}}{\partial n}=0\qquad\text{on }\Gamma_D, \tag{34}$$

$$\varphi^{n+1}=\varphi^*\qquad\text{on }\Gamma_N, \tag{35}$$

where φ^* is solution of:

$$\left(\frac{\Delta t}{\alpha\rho}\frac{\partial^2}{\partial x_2^2}-\frac{1}{\mu}\right)\varphi^*=(1-\chi)\nabla\cdot\mathbf{u}^{n+\frac{1}{2}}-2\gamma\left(\frac{\varphi^n}{\mu}-\chi\nabla\cdot\mathbf{u}^{n-\frac{1}{2}}\right)\ \text{on }\Gamma_N, \tag{36}$$

$$\frac{\partial\varphi^*}{\partial x_2}(\pm 1)=0. \tag{37}$$

2.3 Numerical Experiments

2.3.1 Spectral Element Method Implementation

The domain Ω is the union of quadrangular elements $\overline{\Omega} = \cup_{k=1}^{K} \overline{\Omega}_k$.
For simplification, we consider only rectilinear elements with edges collinear to the axis x and y, that is:

$$\Omega_k =]c_k, c'_k[\times]d_k, d'_k[.$$

The partition is conforming in the sense that the intersection of two adjacent elements is either a verte or a whole edge.
The discrete and stable subspaces to approximate the velocity and the pressure, $\mathbf{X}_p \subset (H_0^1(\Omega))^2$ and $M_p \subset L_0^2(\Omega)$ are chosen to be:

$$\mathbf{X}_p = \Big\{\mathbf{w}_p \in (H_0^1(\Omega))^2, \qquad \mathbf{w}_p^k = \mathbf{w}_{p|\Omega_k} \in (\mathbb{P}_p(\Omega))^2\Big\}, \tag{38}$$

$$M_p = \Big\{q_p \in L^2(\Omega), \qquad q_p^k = q_{p|\Omega_k} \in \mathbb{P}_{p-2}(\Omega_k), \quad \int_\Omega q_p \, d\mathbf{x} = 0\Big\}. \tag{39}$$

The spectral Legendre approach consists in using the Legendre-Galerkin methods introduced in [3] applied to the variational formulation of elliptic problems introduced in our algorithms.

2.3.2 Numerical Results for the Stokes Problem

Exact solutions for $\mathbf{u}^{ex} = \left(u_{x_1}^{ex}, u_{x_2}^{ex}\right)$ and p^{ex} correspond to these data:

$$u_{x_1}^{ex}(x_1, x_2, t) = \sin(x_1)\sin(x_2)\cos(2\pi\omega t), \tag{40}$$

$$u_{x_2}^{ex}(x_1, x_2, t) = \cos(x_1)\cos(x_2)\cos(2\pi\omega t), \tag{41}$$

$$p^{ex}(x_1, x_2, t) = -2\cos(1)\sin(2(x_1 - 1) - x_2)\cos(2\pi\omega t). \tag{42}$$

To study the time splitting error, we consider the unsteady case $\omega = 0.7$ and the errors at $t^* = 1$ with a second order time discretization for a range of time steps $5 \times 10^{-4} \leq \Delta t \leq 10^{-1}$.

Figure 1 depicts results when we use the natural choice for the boundary conditions for φ^{n+1} that is $\varphi^{n+1} = 0$ on Γ_N. The left part of the figure displays the error in L^2-norm for both pressure and velocity when using the standard incremental scheme ($\chi = 0$). We can see that the results are very bad and no order of convergence can be calculated. The right part exhibits the same quantities when using the rotational scheme with $\chi = 0.7$. We can see that only rates close to 1 are obtained while order 2 is expected.

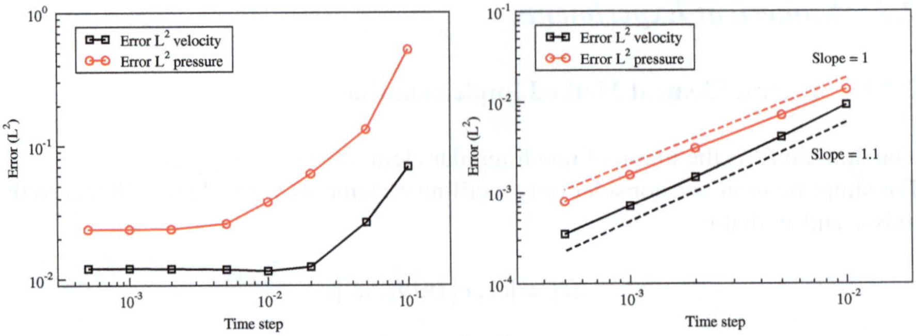

Fig. 1 Time convergence rates with the standard incremental scheme (*left*) and the rotational scheme (*right*) at $t^* = 1$ with $K = 1$ and $p = 18$ with standard open boundary conditions and spectral element method

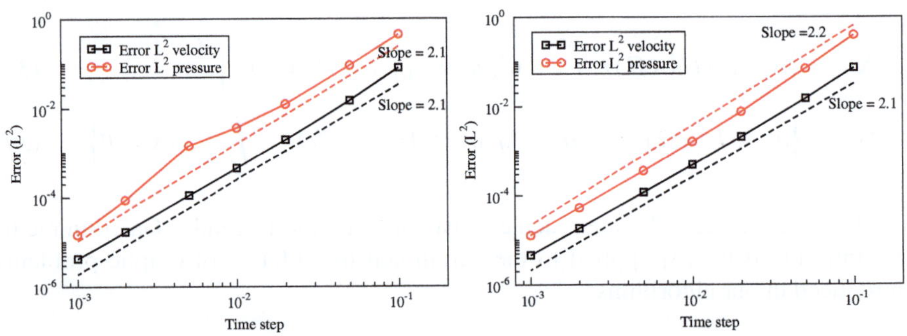

Fig. 2 Time convergence rates with the standard incremental scheme (*left*) and the rotational scheme (*right*) at $t^* = 1$ with $K = 1$ and $p = 18$ with the proposed open boundary conditions and spectral element method

Figure 2 displays the same results using our boundary condition (32). Again, the left part of the figure depicts the errors with standard incremental scheme whereas the right part depicts the errors with the rotational scheme. Contrary to [16], where the authors obtained an almost second-order for the standard incremental scheme and a full second-order for the rotational scheme, we obtain here, in both cases, convergence rates equal to 2.

3 Velocity-Correction Scheme for Open Boundary Condition

3.1 Governing Equations

We propose now to extend our boundary condition for the velocity-correction scheme. The scheme developed by Guermond and Shen in [10] consists on two sub-steps.

The first one is the prediction problem that computes a pressure increment and a solenoidal velocity: *find* φ^{n+1} *and* $\mathbf{u}^{n+1}$ *such that:*

$$\rho\frac{\alpha\mathbf{u}^{n+1} + (\beta-\alpha)\tilde{\mathbf{u}}^{n} + (\gamma-\beta)\tilde{\mathbf{u}}^{n-1} - \gamma\tilde{\mathbf{u}}^{n-2}}{\Delta t} + \nabla\varphi^{n+1} = \mathbf{f}^{n+1} - \mathbf{f}^{n} \quad \text{in } \Omega, \tag{43}$$

$$\nabla\cdot\mathbf{u}^{n+1} = 0 \quad \text{in } \Omega, \tag{44}$$

$$\mathbf{u}^{n+1}\cdot\mathbf{n} = 0 \quad \text{on } \Gamma_D, \tag{45}$$

$$\mu\partial_n(\mathbf{u}^{n+1}\cdot\mathbf{n}) - p^{n+1} = \mathbf{t}^{n+1}\cdot\mathbf{n} \quad \text{on } \Gamma_N, \tag{46}$$

where φ is the pressure increment defined as:

$$\varphi^{n+1} = p^{n+1} - p^{n} + \chi\mu\nabla\cdot\tilde{\mathbf{u}}^{n}. \tag{47}$$

In practice, this step is processed by solving the following problem: *find* φ^{n+1} *such that:*

$$\nabla\cdot\left(\frac{\Delta t}{\rho}\nabla\varphi^{n+1}\right) = \nabla\cdot\left(\frac{\Delta t}{\rho}(\mathbf{f}^{n+1} - \mathbf{f}^{n}) - (\beta-\alpha)\tilde{\mathbf{u}}^{n} - (\gamma-\beta)\tilde{\mathbf{u}}^{n-1} + \gamma\tilde{\mathbf{u}}^{n-2}\right) \quad \text{in } \Omega, \tag{48}$$

$$\partial_n\varphi^{n+1} = (\mathbf{f}^{n+1} - \mathbf{f}^{n})\cdot\mathbf{n} \quad \text{on } \Gamma_D, \tag{49}$$

$$B.C.\,(\varphi^{n+1}) \quad \text{on } \Gamma_N, \tag{50}$$

and upgrading the pressure and the solenoidal velocity via (47) and (43).

The second step is a correction-diffusion problem: *find* $\tilde{\mathbf{u}}^{n+1}$ *such that:*

$$\rho\frac{\alpha\tilde{\mathbf{u}}^{n+1} + \beta\tilde{\mathbf{u}}^{n} + \gamma\tilde{\mathbf{u}}^{n-1}}{\Delta t} - \mu\Delta\tilde{\mathbf{u}}^{n+1} = \mathbf{f}^{n+1} - \nabla p^{n+1} \quad \text{in } \Omega, \tag{51}$$

$$\tilde{\mathbf{u}}^{n+1} = \mathbf{0} \quad \text{on } \Gamma_D, \tag{52}$$

$$\mu\partial_n(\tilde{\mathbf{u}}\cdot\mathbf{n})^{n+1} = \mathbf{t}^{n+1}\cdot\mathbf{n} + p^{n+1} \quad \text{on } \Gamma_N, \tag{53}$$

$$\mu\partial_n(\tilde{\mathbf{u}}\cdot\boldsymbol{\tau})^{n+1} = \mathbf{t}^{n+1}\cdot\boldsymbol{\tau} \quad \text{on } \Gamma_N. \tag{54}$$

Again the main difficulty lies on the boundary condition (50). The natural choice consisting in choosing $\varphi^* = 0$ leads to the same issues as for the pressure-correction scheme since rates of convergences are lower that the expected ones. We have carried out the same reasoning as for the pressure-correction scheme and we propose this formulation for the pressure computation step : *Find* φ^{n+1} *such that*

$$\frac{\Delta t}{\rho}\Delta\varphi^{n+1} = \nabla\cdot\left(\frac{\Delta t}{\rho}\left(\mathbf{f}^{n+1}-\mathbf{f}^{n}\right)-(\beta-\alpha)\tilde{\mathbf{u}}^{n}-(\gamma-\beta)\tilde{\mathbf{u}}^{n-1}+\gamma\tilde{\mathbf{u}}^{n-2}\right) \quad \text{in } \Omega, \tag{55}$$

$$\partial_n\varphi^{n+1} = \left(\mathbf{f}^{n+1}-\mathbf{f}^{n}\right)\cdot\mathbf{n} \quad \text{on } \Gamma_D, \tag{56}$$

$$\varphi^{n+1} = \varphi^{*} \quad \text{on } \Gamma_N, \tag{57}$$

where φ^* is solution of:

$$\left(\frac{\Delta t}{\rho}\partial_{x_2^2}-\frac{\alpha}{\mu}\right)\varphi^{n+1} = \partial_{x_2}\frac{\Delta t}{\rho}\left(f_{x2}^{n+1}-f_{x2}^{n}\right)-\nabla\cdot\left((\beta-\alpha)\tilde{\mathbf{u}}^{n}+(\gamma-\beta)\tilde{\mathbf{u}}^{n-1}-\gamma\tilde{\mathbf{u}}^{n-2}\right)-H^{n}, \tag{58}$$

with:

$$H^n = \chi\nabla\cdot\left(\alpha\tilde{\mathbf{u}}^{n}+\beta\tilde{\mathbf{u}}^{n-1}+\gamma\tilde{\mathbf{u}}^{n-2}\right)-\frac{1}{\mu}\left(\beta\varphi^{n}+\gamma\varphi^{n-1}\right)$$
$$-\frac{1}{\mu}\left(\alpha\,(t_{x_1}^{n+1}-t_{x_1}^{n})+\beta(t_{x_1}^{n}-t_{x_1}^{n-1})+\gamma\,(t_{x_1}^{n-1}-t_{x_1}^{n-2})\right). \tag{59}$$

3.2 *Numerical Experiments*

The same numerical experiments as for the pressure-correction scheme are carried out. Again we present firstly in Fig. 3 the results using the natural choice for φ^* that is $\varphi^* = 0$ on Γ_N. The left part of the figure displays the error in L^2-norm for the pressure and velocity when we use the standard incremental scheme. We can see that the results are very bad and no order of convergence can be calculated. The right part exhibits the same quantities when using the rotational scheme with $\chi = 0.7$. We can see that only rates close to 1 for the pressure and $\frac{3}{2}$ for the velocity are obtained.

In Fig. 4, results corresponding to the proposed boundary condition are exhibited. Again, the left part of the figure depicts the errors with standard incremental scheme whereas the right part depicts the errors with the rotational scheme. We can see that for the standard incremental scheme rates of convergence close to 2 are obtained as expected. For the rotational scheme, we can remark that unlike the pressure-correction scheme for which the standard and rotational schemes give the same results with a slight improvement with the rotational scheme, the results show

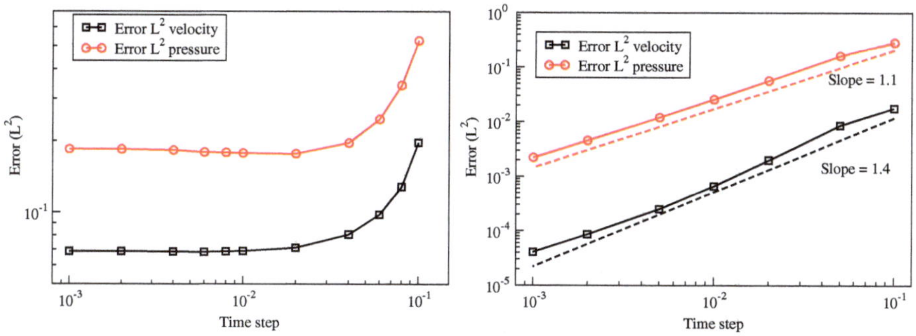

Fig. 3 Time convergence rates with the standard incremental scheme (*left*) and the rotational scheme (*right*) at $t^* = 1$ with $K = 1$ and $p = 18$ with standard open boundary conditions and spectral element method

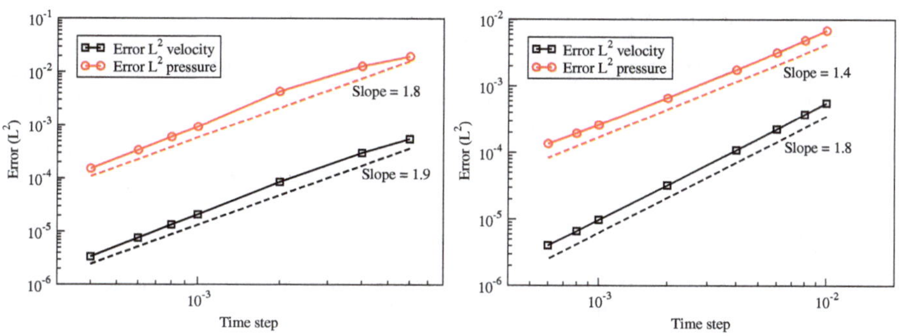

Fig. 4 Time convergence rates with the standard incremental scheme (*left*) and the rotational scheme (*right*) at $t^* = 1$ with $K = 1$ and $p = 18$ with the proposed open boundary conditions and spectral element method

here a distinct advantage for the standard version. Indeed, for the pressure, the convergence rate is now 1.4. This conclusion is confirmed by several numerical tests. A similar observation can be found in the paper of Guermond et al. [10] where the Dirichlet boundary condition is considered for the Stokes problem (on the right part of figure [3]).

Acknowledgements The second author would like to thank Professor Claudio Canuto for many discussions which allowed improving the presentation of this method.

References

1. Arrow K. J., Hurwicz L. , Uzawa H., Studies in linear and non-linear programming, Stanford University Press, Stanford, (1958).
2. Chorin A., Numerical solution of the Navier-Stokes equations, Mathematics of Computation, **22**, 745–762, (1968).
3. Deville M. O., Fischer P. F., Mund E.H., High-Order Methods for Incompressible Fluid Flow, Cambridge University Press, Cambridge, (2002).
4. Engquist B., Absorbing Boundary Conditions for Numerical Simulation of Waves, Proceedings of the National Academy of Sciences, **74**, 1765–1766, (1977).
5. Fortin M., Glowinski R., Méthodes de Lagrangien Augmenté - Applications à la résolution numérique de problémes aux limites, Dunod, Paris, (1982).
6. Goda K., A multistep technique with implicit difference schemes for calculating two- or three-dimensional cavity flows, Journal of Computational Physics, **30**, 76–95, (1979).
7. Guermond J. L., Calculation of Incompressible Viscous Flows by an Unconditionally Stable Projection FEM, Journal of Computational Physics, **132**, 12–33, (1997).
8. Guermond J. L., Minev P., Shen J., Error Analysis of Pressure-Correction Schemes for the Time-Dependent Stokes Equations with Open Boundary Conditions. SIAM Journal on Numerical Analysis, **43**, 239–258, (2005).
9. Guermond J. L., Minev P., Shen J., An overview of projection methods for incompressible flows, Computer Methods in Applied Mechanics and Engineering, **195**, 6011–6045, (2006).
10. Guermond J. L., Shen J., Velocity-correction projection methods for incompressible flows, SIAM Journal on Numerical Analysis, **41**, 112–134, (2004).
11. Karniadakis G. E., Israeli M., Orszag S. A., High-order splitting methods for the incompressible Navier-Stokes equation, Journal of Computational Physics, **97**, 414–443, (1991).
12. Leriche E., Labrosse G., High-order direct Stokes solvers with or without temporal splitting: numerical investigations of their comparative properties, SIAM Journal on Scientific Computing, **22**, 1386–1410, (2000).
13. Liu J., Open and traction boundary conditions for the incompressible NavierStokes equations. Journal of Computational Physics, **228**, 7250–7267, (2009).
14. Orlanski I., A simple boundary condition for unbounded hyperbolic flows, Journal of Computational Physics, **21**, 251–269,1976.
15. Orszag S. A., Israeli M., Deville M. O., Boundary conditions for incompressible flows. Journal of Scientific Computing, **1**, 75–111, (1986).
16. Poux A., Glockner S., Azaïez M., Improvements on open and traction boundary conditions for NavierStokes time-splitting methods, Journal of Computational Physics, **230**, 4011–4027, (2011).
17. Shen J., On error estimates of the projection methods for the Navier-Stokes equations: Second-order schemes, Mathematics of Computation, **65**, 1039–1066, (1996).
18. Témam R., Navier Stokes Equations: Theory and Numerical Analysis, North-Holland Publishing Company, Amsterdam, (1984).
19. Timmermans L. J. P., Minev P. D., Van De Voss F. N., An approximate projection scheme for incompressible flow using spectral elements, International Journal for Numerical Methods in Fluids, **22**, 673–688, (1996).

High Order Space-Time Discretization for Elastic Wave Propagation Problems

Paola F. Antonietti, Ilario Mazzieri, Alfio Quarteroni, and Francesca Rapetti

Abstract In this work we consider the numerical solution of elastic wave propagation problems in heterogeneous media. Our approximation is based on a Discontinuous Galerkin spectral element method coupled with a fourth stage Runge-Kutta time integration scheme. We partition the computational domain into non-overlapping subregions, according to the involved materials, and in each subdomain a spectral finite element discretization is employed. The partitions do not need to be geometrically conforming; furthermore, different polynomial approximation degrees are allowed within each subdomain. The numerical results show that the proposed method is accurate, flexible and well suited for wave propagation analysis.

1 Introduction

The possibility of inferring the physical parameter distribution of the Earth's substratum, from information provided by elastic wave propagations, has increased the interest for computational seismology. The rapid development of efficient

P.F. Antonietti · I. Mazzieri (✉)
Department of Mathematics, Politecnico di Milano, 20133 Milan, Italy
e-mail: paola.antonietti@polimi.it; ilario.mazzieri@mail.polimi.it

A. Quarteroni
CMCS-MATHICSE, École Polytechnique Fédérale de Lausanne, Station 8, 1015 Lausanne, Switzerland
e-mail: alfio.quarteroni@epfl.ch

F. Rapetti
Université de Nice Sophia Antipolis, Laboratoire de Mathématiques J.A. Dieudonné, Parc Valrose, 06108 Nice, Cedex 02, France
e-mail: frapetti@unice.fr

M. Azaïez et al. (eds.), *Spectral and High Order Methods for Partial Differential Equations - ICOSAHOM 2012*, Lecture Notes in Computational Science and Engineering 95, DOI 10.1007/978-3-319-01601-6_6,

numerical tools makes it possible to simulate with high accuracy the complete seismic wavefront field in highly heterogeneous media, even in complex geometries. Recent developments on computational seismology have been based on high-order spectral element (SE) methods (see, for example, [2, 6, 8–10, 15]). Spectral element methods, which stem from a weak variational formulation, allow a flexible treatment of boundaries, or subdomain interfaces, and deal with free-surface boundary conditions naturally. They feature both combine the geometrical flexibility typical of low-order methods and the exponential convergence rate associated with spectral techniques; as a matter of fact, they yield minimal numerical dispersion and dissipation errors. Moreover, they retain a high level parallel structure, and are therefore well suited for parallel computations.

In this paper we consider a non-conforming high-order technique, namely the Discontinuous Galerkin (DG) spectral element method to simulate seismic wave propagation in heterogeneous media. In contrast to standard conforming discretizations, like the SE method, DG techniques can accommodate discontinuities, not only in the parameters, but also in the wave-field, and they are energy conservative.

The paper is organized as follows. In Sect. 2 we describe the model of linear elastodynamics. In Sect. 3 we introduce the semi-discrete formulation obtained by using the DG spectral element approximation. The corresponding algebraic formulation and the time integration scheme are described in Sect. 4. Numerical results are presented and discussed in Sect. 5. Finally in Sect. 6 we draw some conclusions. All along the paper, matrix or tensor quantities are denoted by Greek or capital letters, while vectors are typed in bold. Moreover, we adopt the standard notation $(\cdot,\cdot)_\Omega$ to denote the L^2-inner product for regular enough scalar, vectorial and tensorial functions defined in Ω.

2 Problem Formulation

We consider an elastic heterogeneous medium occupying an open and bounded region $\Omega \subset \mathbb{R}^d$, for $d = 2, 3$, with boundary $\Gamma := \partial\Omega$. The boundary can be composed by three disjoint parts: Γ_D where displacements are prescribed, Γ_N where external loads are applied and Γ_{NR} where suitable non-reflecting conditions are imposed. To simplify, we assume that the measure of Γ_D is positive. For a given displacement vector $\mathbf{v}$, let $\boldsymbol{\sigma}(\mathbf{v})$ be the Cauchy stress tensor $\boldsymbol{\sigma}(\mathbf{v}) := \lambda(\nabla \cdot \mathbf{v})\mathrm{I} + 2\mu\boldsymbol{\epsilon}(\mathbf{v})$, where $\boldsymbol{\epsilon}(\mathbf{v}) := 1/2(\nabla\mathbf{v} + \nabla\mathbf{v}^\top)$ is the strain tensor, I is the identity tensor and λ, μ are the Lamé parameters. For a given density of body forces $\mathbf{f}$, and a given vector field $\mathbf{t}$, we consider the linear elastodynamics system:

$$\rho\mathbf{u}_{tt} - \nabla \cdot \boldsymbol{\sigma}(\mathbf{u}) = \mathbf{f}, \qquad \text{in } \Omega \times (0, T], \tag{1}$$

coupled with boundary conditions

$$\mathbf{u} = \mathbf{0} \quad \text{on } \Gamma_D, \qquad \boldsymbol{\sigma}(\mathbf{u})\mathbf{n} = \mathbf{t}^* \quad \text{on } \Gamma_N \cup \Gamma_{NR}, \tag{2}$$

where $\mathbf{n}$ is the unit outward normal vector to Γ and $\mathbf{t}^* := \mathbf{t}$ on Γ_N and

$$\mathbf{t}^* := \rho(c_P - c_S)(\mathbf{u}_t \cdot \mathbf{n})\mathbf{n} + \rho c_S \mathbf{u}_t \quad \text{on } \Gamma_{NR}. \tag{3}$$

Here, $c_P := \sqrt{(\lambda + 2\mu)/\rho}$ and $c_S := \sqrt{\mu/\rho}$ are the propagation velocities of P and S waves, respectively. The representation of the radiation condition associated with external boundary is a difficult problem, and numerous numerical schemes have been proposed in literature (see for instance [7]). In (3) we adopt a first order approximation close to the one proposed by Stacey [16] that is based upon a one-way treatment that perfectly absorbs waves impinging at right angles to the boundary, but that is less effective for waves that graze the boundary [3]. Finally, to complete the system (1)–(3) we prescribe initial conditions $\mathbf{u} = \mathbf{u}_0$ and $\mathbf{u}_t = \mathbf{u}_1$ for the displacement and the velocity, respectively. We remark that for heterogeneous media, ρ, λ and μ are bounded functions of the spatial variable, not necessarily continuous, i.e., ρ, λ and $\mu \in L^\infty(\Omega)$.

3 Discontinuous Galerkin Spectral Element Method

In this section we describe the non-conforming technique adopted to approximate (1)–(3). To simplify the discussion, we present the method in the case Ω is decomposed into two non-overlapping subdomains; the extension to any partition of Ω into a fixed number of subdomains is straightforward [1, 14]. Moreover we suppose $\Gamma = \Gamma_D \cup \Gamma_N$, the general case is obtained similarly.

Let Ω_1 and Ω_2 be two open and connected subsets of Ω, with Lipschitz boundaries $\partial\Omega_i$ such that $\overline{\Omega} = \overline{\Omega_1} \cup \overline{\Omega_2}$ with $\Omega_1 \cap \Omega_2 = \emptyset$. Let $\mathbf{n}_i$ denote the unit normal to $\partial\Omega_i$, exterior to Ω_i and let Γ_{12} denote the interface between Ω_1 and Ω_2, that is $\Gamma_{12} := \partial\overline{\Omega_1} \cap \partial\overline{\Omega_2}$. For any smooth enough functions $\mathbf{v}$ and $\boldsymbol{\sigma}$, we set $\mathbf{v}_i := \mathbf{v}_{|\Omega_i}$ and $\boldsymbol{\sigma}_i := \boldsymbol{\sigma}_{|\Omega_i}$, $i = 1, 2$. The jump and average trace of $\mathbf{v}$ and $\boldsymbol{\sigma}$ through Γ_{12} are defined as $\{\mathbf{v}\} := \frac{1}{2}(\mathbf{v}_1 + \mathbf{v}_2)$, $[\![\mathbf{v}]\!] := \mathbf{v}_1 \otimes \mathbf{n}_1 + \mathbf{v}_2 \otimes \mathbf{n}_2$, and $\{\boldsymbol{\sigma}\} := \frac{1}{2}(\boldsymbol{\sigma}_1 + \boldsymbol{\sigma}_2)$, $[\![\boldsymbol{\sigma}]\!] := \boldsymbol{\sigma}_1\mathbf{n}_1 + \boldsymbol{\sigma}_2\mathbf{n}_2$, where $\mathbf{a} \otimes \mathbf{b} \in \mathbb{R}^{d\times d}$ is the tensor with entries $(\mathbf{a} \otimes \mathbf{b})_{ij} := a_i b_j$, $1 \le i, j \le d$, for all $\mathbf{a}, \mathbf{b} \in \mathbb{R}^d$. We remark that in this setting, problem (1) can be reformulated equivalently as

$$\rho(\mathbf{u}_i)_{tt} - \nabla \cdot \boldsymbol{\sigma}(\mathbf{u}_i) = \mathbf{f}, \qquad \text{in } \Omega_i \times (0, T], \tag{4}$$

for $i = 1, 2$ coupled with transmission conditions $[\![\mathbf{u}]\!] = \mathbf{0}$ and $[\![\boldsymbol{\sigma}]\!] = \mathbf{0}$ on Γ_{12}.

Next, in each Ω_i we introduce a conforming partition $\mathscr{T}_{h_i}$, made by quadrilateral elements Ω_i^j with typical linear size h_i and $\overline{\Omega}_i = \bigcup_{j=1} \overline{\Omega}_i^j$. Each element Ω_i^j is the image of the reference square $\hat{\Omega} = (-1, 1)^d$ by means of a suitable map $\mathbf{F}_i^j$ with Jacobian matrix $\mathbf{J}_i^j$, i.e., $\Omega_i^j = \mathbf{F}_i^j(\hat{\Omega})$. We define an interior edge (face for $d = 3$) as the non-empty interior of the intersection of two neighbouring elements belonging to different subdomains. More precisely, let $\Omega_1^i \in \Omega_1$ and $\Omega_2^k \in \Omega_2$ be two neighbouring elements, we set $\overline{\gamma} := \partial\overline{\Omega}_1^i \cap \partial\overline{\Omega}_2^k$. We remark that in this setting the skeleton Γ_{12} can be expressed as $\overline{\Gamma}_{12} = \bigcup_{j=1} \overline{\gamma}_j$.

Now, in each Ω_i^j, we introduce the space $\mathbf{Q}_{N_i}(\Omega_i^j) := \{\mathbf{v} = \hat{\mathbf{v}} \circ (F_i^j)^{-1} : \hat{\mathbf{v}} \in \mathbf{Q}_{N_i}(\hat{\Omega})\}$, where $\mathbf{Q}_{N_i}(\hat{\Omega})$ is the space of vectorial functions defined over $\hat{\Omega}$ and such that each component is an algebraic polynomial of degree less than or equal to $N_i \geq 2$ in each space variable. Finally we define the finite dimensional spaces $\mathbf{X}_\delta(\Omega_i) := \{\mathbf{v}_\delta \in \mathbf{C}^0(\overline{\Omega}_i) : \mathbf{v}_{\delta|\Omega_i^j} \in \mathbf{Q}_{N_i}(\Omega_i^j),\ \forall \Omega_i^j \in \mathscr{T}_{h_i}\}$, and $\mathbf{V}_\delta := \{\mathbf{v}_\delta \in \mathbf{L}^2(\Omega) : \mathbf{v}_{\delta|\Omega_i} \in \mathbf{X}_\delta(\Omega_i),\ i = 1,2 : \mathbf{v}_{\delta|\Gamma_D} = \mathbf{0}\}$, where $\delta := \{\mathbf{h}, \mathbf{N}\}$ with $\mathbf{h} := (h_1, h_2)$ and $\mathbf{N} := (N_1, N_2)$ are couplets of discretization parameters. Each component h_i and N_i represents the mesh size and the degree of the polynomial interpolation in the region Ω_i, respectively.

A nodal basis for $\mathbf{V}_\delta$ is obtained introducing on each element Ω_i^j a set of interpolation points $\{\mathbf{p}_i\}$ (Gauss-Legendre-Lobatto (GLL) points) and corresponding degrees of freedom which allow to identify uniquely a generic function in $\mathbf{V}_\delta$. In the SE approach the interpolation points are used as quadrature points. Thus, we have

$$(\mathbf{f}, \mathbf{g})_{\Omega_i^j} \approx (\mathbf{f}, \mathbf{g})_{NI,\Omega_i^j} := \sum_{k=1}^{(N_i+1)^d} (\mathbf{f} \circ \mathbf{F}_i^j)(\mathbf{p}_k) \cdot (\mathbf{g} \circ \mathbf{F}_i^j)(\mathbf{p}_k) |det(\mathbf{F}_i^j)| \omega_k, \tag{5}$$

where ω_k are the weights of the GLL quadrature formula [5]. The spectral shape functions $\boldsymbol{\Phi}_i \in \mathbf{V}_\delta$ are defined as $\boldsymbol{\Phi}_i(\mathbf{p}_j) = \delta_{ij}$, $i, j = 1, \ldots, (N_i + 1)^2$, where δ_{ij} is the Kronecker symbol. It is straightforward to see that by the definition of $\mathbf{V}_\delta$, the basis functions will not be globally continuous on the whole domain and that the restriction of any spectral function to Ω_i^j either coincides with a Lagrange polynomial or vanishes. Moreover, the support of any shape function is limited to the neighbouring elements if the spectral node lies on the interface between two or more elements, while it is limited to only one element for internal nodes.

To introduce the non-conforming semi-discrete DG variational formulation, we multiply by a generic test function $\mathbf{v} \in \mathbf{V}_\delta$ Eq. (4), integrate it by parts and sum over the elements $\Omega_i^j \in \mathscr{T}_{h_i}$. For each $t \in (0, T]$, we now seek for $(\mathbf{u}_{1,\delta}, \mathbf{u}_{2,\delta}) \in \mathbf{V}_\delta$ such that

$$\sum_{i=1}^{2} (\rho \mathrm{d}_{tt} \mathbf{u}_{i,\delta}, \mathbf{v}_i)_{\Omega_i} + \mathscr{A}(\mathbf{u}_{i,\delta}, \mathbf{v}_i)_{\Omega_i} = \sum_{i=1}^{2} \mathscr{L}(\mathbf{v}_i)_{\Omega_i} \qquad \forall\, (\mathbf{v}_1, \mathbf{v}_2) \in \mathbf{V}_\delta, \tag{6}$$

where $\mathscr{L}(\mathbf{v}_i)_{\Omega_i} := (\mathbf{f}, \mathbf{v}_i)_{\Omega_i} + (\mathbf{t}, \mathbf{v}_i)_{\Gamma_N}$ and

$$\sum_{i=1}^{2} \mathscr{A}(\mathbf{u}, \mathbf{v})_{\Omega_i} := \sum_{i=1}^{2} (\boldsymbol{\sigma}(\mathbf{u}), \boldsymbol{\epsilon}(\mathbf{v}))_{\Omega_i} + \sum_{j:\gamma_j \in \Gamma_{12}} \theta\, ([\![\mathbf{u}]\!], \{\boldsymbol{\sigma}(\mathbf{v})\})_{\gamma_j} - (\{\boldsymbol{\sigma}(\mathbf{u})\}, [\![\mathbf{v}]\!])_{\gamma_j} + \eta_j\, ([\![\mathbf{u}]\!], [\![\mathbf{v}]\!])_{\gamma_j}\,. \tag{7}$$

Here $\theta \in \{-1, 0, 1\}$ and η_j are positive constants depending on the discretization parameters $\mathbf{h}$ and $\mathbf{N}$ and on the Lamé coefficients. More precisely, $\eta_j := \alpha\{\lambda + 2\mu\}_A \mathtt{N_j}^2/\mathtt{h_j}$, where $\{q\}_A$ represents the harmonic average of the quantity q, defined by $\{q\}_A := (2q_1q_2)/(q_1 + q_2)$, $\mathtt{N_j} := \max(N_1, N_2)$, $\mathtt{h_j} := \min(h_1, h_2)$ and α is a positive constant at disposal.

It is possible to prove that problem (6) admits a unique solution $\mathbf{u}_{DG}(t) \in \mathbf{V}_\delta$, satisfying optimal a-priori error bound with respect to a suitable DG-norm. See [1, 13] for further details.

4 Algebraic Formulation and Time Integration Scheme

In this section we discuss the algebraic formulation of the DG spectral element method presented in the previous section and we introduce the time integration scheme. We start by introducing a basis $\{\boldsymbol{\Phi}_i^1, \boldsymbol{\Phi}_i^2\}_{i=1}^D$, for the finite dimensional space $\mathbf{V}_\delta$, where D represents the degrees of freedom of the problem. Omitting the subscript δ, we write the discrete function $\mathbf{u} \in \mathbf{V}_\delta$ as

$$\mathbf{u}(\mathbf{x}, t) := \sum_{j=1}^{D} \boldsymbol{\Phi}_j^1(\mathbf{x}) U_j^1(t) + \boldsymbol{\Phi}_j^2(\mathbf{x}) U_j^2(t).$$

Then, using the above expression, we recast Eq. (6) for any test function $\boldsymbol{\Phi}_j^\ell(\mathbf{x}) \in \mathbf{V}_\delta$, for $\ell = 1, 2$, obtain the following set of discrete ordinary differential equations for the nodal displacement $\mathbf{U} := [\mathbf{U}^1, \mathbf{U}^2]^\top$:

$$\mathrm{M}\ddot{\mathbf{U}} + \mathrm{A}\mathbf{U} = \mathbf{F}, \tag{8}$$

where $\ddot{\mathbf{U}}$ represents the vector of nodal acceleration and $\mathbf{F}$ the vector of externally applied loads:

$$\mathbf{F}_i^\ell := (\mathbf{f}, \boldsymbol{\Phi}_i^\ell)_\Omega + (\mathbf{t}, \boldsymbol{\Phi}_i^\ell)_{\Gamma_N}, \quad \text{for } i = 1, \ldots D.$$

Equation (8) can be written equivalently as

$$\begin{bmatrix} \mathrm{M}^1 & 0 \\ 0 & \mathrm{M}^2 \end{bmatrix} \begin{bmatrix} \ddot{\mathbf{U}}^1 \\ \ddot{\mathbf{U}}^2 \end{bmatrix} + \begin{bmatrix} \mathrm{A}^{11} & \mathrm{A}^{12} \\ \mathrm{A}^{21} & \mathrm{A}^{22} \end{bmatrix} \begin{bmatrix} \mathbf{U}^1 \\ \mathbf{U}^2 \end{bmatrix} = \begin{bmatrix} \mathbf{F}^1 \\ \mathbf{F}^2 \end{bmatrix}.$$

As a consequence of (5) and of assumptions on the basis functions, the mass matrix M has a diagonal structure with elements

$$\mathrm{M}_{ij}^\ell := (\rho \boldsymbol{\Phi}_j^\ell, \boldsymbol{\Phi}_i^\ell)_\Omega, \quad \text{for } i, j = 1, \ldots, D,$$

The matrix A associated to the bilinear form $\mathscr{A}(\cdot, \cdot)$ defined in (7) is such that for $i, j = 1, \ldots, D$ it holds

$$\mathrm{A}_{ij}^{11} := \mathscr{A}(\boldsymbol{\Phi}_j^1, \boldsymbol{\Phi}_i^1)_\Omega, \quad \mathrm{A}_{ij}^{12} := \mathscr{A}(\boldsymbol{\Phi}_j^2, \boldsymbol{\Phi}_i^1)_\Omega,$$
$$\mathrm{A}_{ij}^{21} := \mathscr{A}(\boldsymbol{\Phi}_j^1, \boldsymbol{\Phi}_i^2)_\Omega, \quad \mathrm{A}_{ij}^{22} := \mathscr{A}(\boldsymbol{\Phi}_j^2, \boldsymbol{\Phi}_i^2)_\Omega.$$

In order to apply the four-stage Runge-Kutta method to system (8) we define $\mathbf{V} := \dot{\mathbf{U}}$ the vector of nodal velocities and we prescribe initial conditions $\mathbf{U}(0) := \mathbf{u}_0$ and $\mathbf{V}(0) := \mathbf{u}_1$. Then, we subdivide the interval $(0, T]$ into N subintervals of amplitude $\Delta t = T/N$ and set $t_n = n\Delta t$, for $n = 1, \dots, N$. Using the notation just introduced, we rewrite (8) as the following first order system of equations

$$\begin{bmatrix} \mathrm{I} & 0 \\ 0 & \mathrm{M} \end{bmatrix} \begin{bmatrix} \dot{\mathbf{U}}(t) \\ \dot{\mathbf{V}}(t) \end{bmatrix} = \begin{bmatrix} 0 & \mathrm{I} \\ -\mathrm{A} & 0 \end{bmatrix} \begin{bmatrix} \mathbf{U}(t) \\ \mathbf{V}(t) \end{bmatrix} + \begin{bmatrix} \mathbf{0} \\ \mathbf{F}(t) \end{bmatrix},$$

that is equivalent to

$$\dot{\mathbf{W}}(t) = \mathbf{g}(t, \mathbf{W}(t)),$$

where $\mathbf{W}(t) := [\mathbf{U}(t)\ \ \mathbf{V}(t)]^\top$, and

$$\mathbf{g}(t, \mathbf{W}(t)) := \begin{bmatrix} 0 & \mathrm{I} \\ -\mathrm{M}^{-1}A & 0 \end{bmatrix} \mathbf{W}(t) + \begin{bmatrix} \mathbf{0} \\ \mathrm{M}^{-1}\mathbf{F}(t) \end{bmatrix}.$$

Then, the four stage Runge-Kutta method [11] takes the form

$$\mathbf{W}(t_{n+1}) = \mathbf{W}(t_n) + \frac{\Delta t}{6}\left(\mathbf{K}_1 + 2\mathbf{K}_2 + 2\mathbf{K}_3 + \mathbf{K}_4\right), \tag{9}$$

where

$$\mathbf{K}_1 := \mathbf{g}(t_n, \mathbf{W}(t_n)), \qquad \mathbf{K}_2 := \mathbf{g}(t_{n+\frac{1}{2}}, \mathbf{W}(t_n) + \frac{\Delta t}{2}\mathbf{K}_1),$$

$$\mathbf{K}_3 := \mathbf{g}(t_{n+\frac{1}{2}}, \mathbf{W}(t_n) + \frac{\Delta t}{2}\mathbf{K}_2), \quad \mathbf{K}_4 := \mathbf{g}(t_{n+1}, \mathbf{W}(t_n) + \Delta t\mathbf{K}_3).$$

We remark that the above scheme is fourth order accurate, explicit and conditionally stable. In order to guarantee boundedness of the discrete solution for all the observation time the Courant-Friedrich-Levy condition (CFL) prescribes a restriction on the time step Δt that reads $\Delta t \leq C_{CFL}\Delta x/c_P$, where Δx is the shortest distance between two GLL nodes and C_{CFL} is a constant depending on the dimension, the order of the scheme, the mesh geometry and the polynomial order. Since $\Delta x \approx N^{-2}$ (see [5]) it follows that $C_{CFL} \approx N^{-2}$ as for the well-known (and widely used in seismic applications) leap-frog scheme [4]. Moreover, a direct computation of the C_{CFL} constant shows that the stability bounds for the Runge-Kutta scheme (9) are less restrictive of the leap-frog one, see [13] for details. The grid dispersion and dissipation properties of the four stage Runge Kutta method coupled with DG discretization has been analyzed in [13], in particular it is shown that, heuristically, 5 points per wavelength are sufficient for providing negligible errors (less that 10^{-6}). This makes the proposed scheme well suited for the approximation of wave propagation problems.

5 Elastic Scattering by Circular Inclusion

This example illustrates the scattering of a plane wave by an elastic circle included in a homogeneous halfspace, see Fig. 1. Such an example appears frequently in non-destructive testing and near-surface seismic studies (cavity, gas inclusions). From the numerical point of view, the main difficulty lies in meshing the curved internal interface, specially in the case of two media with high velocity contrast where both an accurate approximation of the body and of the interface waves are mandatory.

In order to illustrate the flexibility of the proposed DG approximation we consider the case of a compliant inclusion buried in a stiffer elastic space since we want to test the accuracy of the DG spectral element method for an extremely high velocity contrast.

The physical model that we are considering consists in a circular inclusion (CI) of diameter 500 m embedded in the square elastic half space (ES) of dimension $(0, 2{,}000) \times (0, -2{,}000)$ m. In Table 1 we report the different material properties for the case considered. The set of parameters is borrowed form [12] and was used firstly in [15] to study a two quarter space problem, that is two elastic half-spaces in contact along a vertical material discontinuity, with special emphasis in the simulation of the interface waves travelling along the vertical interface, a geometry well suited for classical spectral element methods based on quadrilateral meshes. Here, we consider the scattering by an elastic circle of a plane wave travelling upwards.

In this example, an interface wave is expected to travel along the circular boundary of the buried circle.

We set the mesh sizes $h_{CI} = 35$ m and $h_{ES} = 70$ m, for the circular inclusion (CI) and the elastic space (ES), respectively. The partition within the circle and the halfspace is designed in order to have at least 5 points per wavelength with polynomial degree $N_{CI} = N_{ES} = 4$ and do not match at the interface.

We apply free-surface boundary conditions, i.e. $\mathbf{t} = \mathbf{0}$, on the top of the domain, while non-reflecting boundary conditions are imposed on the remaining parts of the boundary. A body force $\mathbf{f}(\mathbf{x}, t)$ is prescribed along the bottom boundary by an initial displacement $\mathbf{u}_0$ and velocity $\mathbf{u}_1$. The force time history is represented by Ricker plane wave $R(t)$ with incident angle $\vartheta = 0$ (see Fig. 1) and modulation of 5 Hz central frequency: $R(t) = [1 - 2\beta(t - t_0)^2]\exp[-\beta(t - t_0)^2]$, where $t_0 = 1$ s and $\beta = 246.7401\,\text{s}^{-1}$.

Along the curved interface, starting from the point $R_1 = (1{,}000, -1{,}250)$ m, 50 receivers are placed in a counter clockwise order. The wave field is propagated using the Runge-Kutta scheme, described in the previous section, for $T = 7$ s using

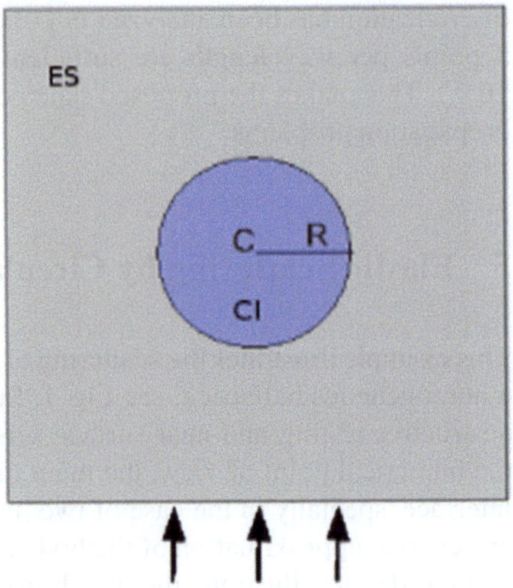

Fig. 1 Circular inclusion with center $C = (0, -1{,}000)$ m and radius $R = 500$ m in the elastic space $(0, 2{,}000) \times (0, -2{,}000)$ m

Table 1 Dynamic and mechanical parameters for the circular inclusion (CI) and the elastic space (ES)

Layers	ρ [kg/m^3]	c_P [m/s]	c_S [m/s]
CI	1	600	1,400
ES	2	2,310	4,000

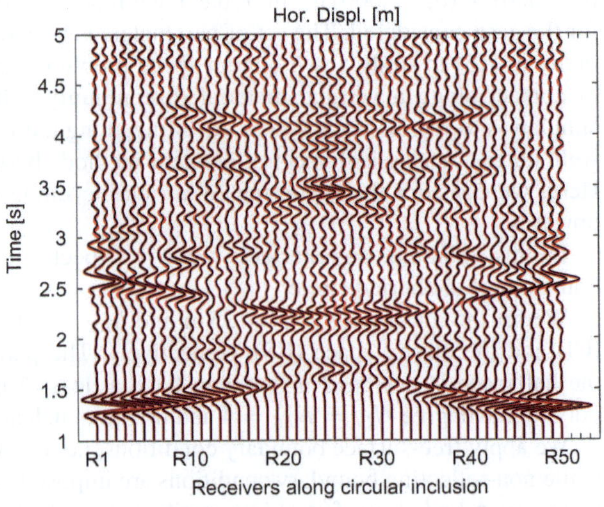

Fig. 2 Horizontal displacement along the circle's boundary computed with the DG approach (*black line*) and the conforming spectral element approach (*red line*)

a time step $\Delta t = 10^{-4}$ s. All the results are compared with those obtained with a conforming spectral element approximation choosing fourth order polynomials and $h_{CI} = h_{ES} = 35$ m. In Fig. 2 we report the horizontal displacement recorded by the receivers along the interface. Notice that as expected the displacement turns out to be symmetric with respect to the vertical axis.

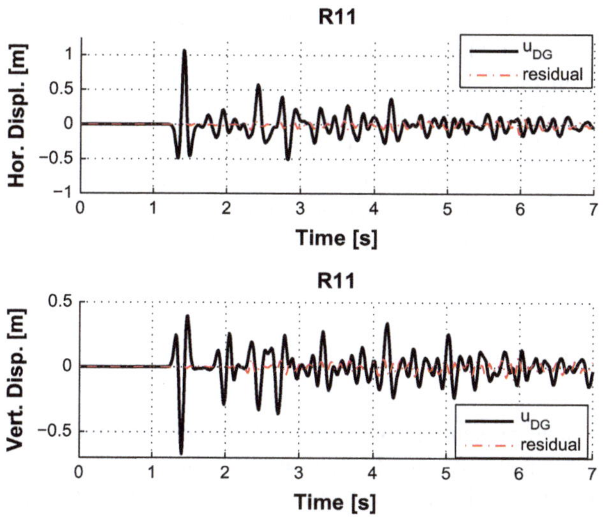

Fig. 3 Displacement recorded by R_{11} and corresponding residual

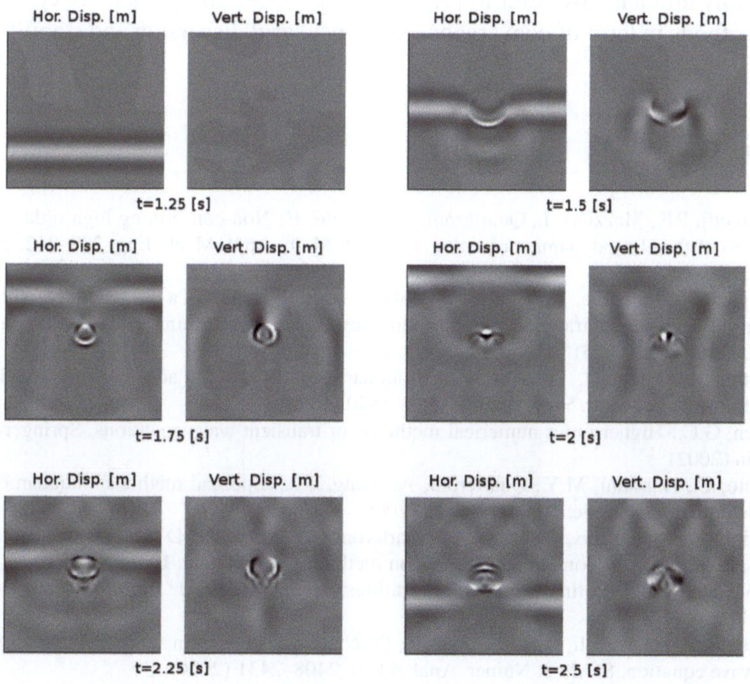

Fig. 4 Snapshots of the horizontal and vertical displacements obtained with the DG method

In Fig. 3 we report in the same graphics the displacement recorded by the receiver $R_{11} = (239.67, -1{,}071.1)$ m and the difference (magnified by 10 for visualization purposes) between the solution obtained with the DG scheme and the reference one obtained with the conforming SE method. It can be observed that DG method reproduces accurately the wave front field for all observation times. It is evident the effect on the wavefield induced by the softer inclusion: the waves that travel across the circle are trapped within it and then phenomena of reflection and refraction arise. This is more evident from the snapshots of the computed solution shown in Fig. 4.

6 Conclusions

In this note we have presented a DG spectral element method combined with the fourth order Runge-Kutta scheme for the discretization of elastic wave propagation in heterogeneous materials. The test case analyzed showed the flexibility of the proposed method when complex mesh geometries are considered (curved internal interfaces). The results, compared with those obtained with the conforming spectral element method, show that the non-conforming strategy is both accurate and computationally efficient. We refer to [1, 13, 14] for a more comprehensive comparison of the methods in term of convergence, accuracy, grid dispersion and stability.

References

1. Antonietti, P.F., Mazzieri, I., Quarteroni, A., Rapetti, F.: Non-conforming high order approximations of the elastodynamics equation. Comput. Meth. Appl. Mech. Eng. **209–212**, 212–238 (2012)
2. Chaljub, E., Capdeville, Y., Vilotte, J.P.: Solving elastodynamics in a fluid-solid heterogeneous sphere: a parallel spectral element approximation on non-conforming grids. J. Comput. Phys. **187**(2), 457–491 (2003)
3. Clayton, R., Engquist, B.: Absorbing boundary conditions for acoustic and elastic wave equations, Bull. Seism. Soc. Am. **67**, 1529–1540 (1977)
4. Cohen, G.C.: Higher-order numerical methods for transient wave equations. Springer-Verlag, Berlin (2002)
5. Canuto, C., Hussaini, M.Y., Quarteroni, A., Zang, T.A.: Spectral methods - Fundamentals in single domains, Springer-Verlag, Berlin, (2006)
6. Faccioli, E., Maggio, F., Paolucci, R., Quarteroni, A.: 2-D and 3-D elastic wave propagation by a pseudo-spectral domain decomposition method. J. of Seismol. **1**, 237–251 (1997)
7. Givoli, D.: Non-reflecting boundary conditions: review article. J. Comput. Phys. **94**, 1–29 (1991)
8. Grote, M.J., Schneebeli, A., Schotzau, D.: Discontinuous Galerkin finite element method for the wave equation. SIAM J. Numer. Anal. **44**(6), 2408–2431 (2006)
9. de la Puente, J., Kaser, M., Dumbser, M., Igel, H.: An arbitrary high-order discontinuous Galerkin method for elastic waves on unstructured meshes – IV. Anisotropy. Geophys. J. Int. **169**(3), 1210–1228 (2007)

10. Komatitsch, D., Tromp, J.: Introduction to the spectral-element method for 3-D seismic wave propagation. Geophys. J. Int. **139**, 806–822 (1999)
11. Lambert, J.D.: Numerical methods for ordinary differential systems: the initial value problem. John Wiley & Sons Inc., New York, USA (1991)
12. Mercerat, E.D., Vilotte, J.P., Sanchez-Sesma, F.J.: Triangular spectral-element simulation of two-dimensional elastic wave propagation using unstructured triangular grids. Geophys. J. Int. **166**(2), 679–698 (2006)
13. Mazzieri, I.: Non-conforming high order methods for the elastodynamics equation. PhD. Thesis, Politecnico di Milano (2012)
14. Mazzieri, I., Smerzini, C., Antonietti, P.F., Rapetti, F., Stupazzini, M., Paolucci, R., Quarteroni, A.: Non-conforming spectral approximations for the elastic wave equation in heterogeneous media, Proceedings of COMPDYN 2011, 3rd International Conference in Computational Methods in Structural Dynamics and Earthquake Engineering (2011)
15. Priolo, E., Carcione, J.M., Seriani, G.: Numerical simulation of interface waves by high-order spectral modeling techniques. J. Acoust. Soc. Am. **95**(2), 681–693 (1994)
16. Stacey, R.: Improved transparent boundary formulations for the elastic-wave equation. Bull. Seismol. Soc. Am. **78**(6), 2089–2097 (1988)

Laguerre Simulation of Boundary Layer Flows: Conditions at Large Distance from the Wall

F. Auteri and L. Quartapelle

Abstract In this contribution, a fully spectral projection method for simulating the flow over a flat plate is presented. The incompressible Navier–Stokes equations are integrated in time using a second order fractional step method, while suitable Legendre and Laguerre polynomial basis functions are combined with a Fourier expansion in the spanwise direction to represent the spatial dependence in the truly unbounded domain. An original feature of the proposed method is the treatment of the asymptotic free-stream condition far from the wall on the normal velocity component by a Petrov-Galerkin method. Convergence tests assess the spectral accuracy of the proposed method in space and its second order accuracy in time. Results from the first large scale, with $\approx$30 millions of unknowns, simulation of boundary layer flow exploiting Laguerre polynomials are also reported.

1 Introduction

The active control of turbulent flows over a flat plate is a very active current research area, see, e.g., [1]. Amongst the various numerical techniques employed so far to investigate this kind of flows and boundary layer transition, spectral methods are the most convenient by virtue of their accuracy and efficiency. Owing to the semi-infinite character of the flow domain in the direction normal to the wall, truncation of the corresponding unbounded coordinate is typically adopted in connection with the use of Chebyshev polynomials as basis functions [2].

An alternative and more sound approach to solve elliptic equations in unbounded domains with spectral accuracy relies on Laguerre functions [3]. The aim of the

F. Auteri (✉) · L. Quartapelle
Dipartimento di Ingegneria Aerospaziale, Politecnico di Milano, Via La Masa 34, 20156 Milano, Italy
e-mail: auteri@aero.polimi.it

M. Azaïez et al. (eds.), *Spectral and High Order Methods for Partial Differential Equations - ICOSAHOM 2012*, Lecture Notes in Computational Science and Engineering 95, DOI 10.1007/978-3-319-01601-6_7,

present contribution is to describe an innovative fully spectral DNS code for the simulation of boundary layer flows which uses different bases in the three spatial directions: Laguerre functions are employed in the wall normal direction (y) together with Legendre polynomials for the streamwise coordinate (x) and the Fourier representation for the periodic spanwise direction (z).

In Blasius boundary layer simulations, of interest here, satisfying the asymptotic condition at large distance from the wall is particularly difficult. In fact, the asymptotic behaviour of the velocity component v normal to the wall is such that $v \to a$ as $y \to \infty$, where the quantity a is unknown and part of the solution.

We have devised an original procedure that, in the context of a Laguerre spectral approximation which is naturally homogeneous at infinity, accounts for both the asymptotic Neumann specification and the unknown embedded level of the solution variable.

A second order projection method is employed to advance the solution in time [10] and a Galerkin formulation has been adopted for the spectral discretization in space. The reported results show the spectral accuracy of the method with respect to the spatial discretization and second order accuracy with respect to the time discretization.

2 Problem Definition

This paper deals with the simulation of the incompressible flow of a Newtonian fluid over a flat plate. The governing partial differential equations are the Navier–Stokes equations

$$\begin{cases} \dfrac{\partial \mathbf{u}}{\partial t} + (\mathbf{u}\cdot\nabla)\mathbf{u} - \nu\nabla^2\mathbf{u} + \nabla p = \mathbf{f}(\mathbf{r},t), \\ \nabla\cdot\mathbf{u} = 0, \end{cases} \tag{1}$$

over the computational domain consisting in an infinite region bounded by five planes, which can be defined as $\Omega = [\overline{x}, \overline{x} + L] \times [0, \infty) \times [-S, S]$, see Fig. 1. Our simulations will assume a flow field with uniform velocity and zero pressure gradient at large distance from the plate. Thus, the initial and boundary conditions for the flow field will based on the similar 2D flow over a semi-infinite flat plate as given by the solution $f(\eta)$ to the Blasius equation, with η representing the similarity variable. It is therefore convenient to introduce the Blasius velocity field $\mathbf{u}_\mathrm{B} = u_\mathrm{B}\,\hat{\mathbf{x}} + v_\mathrm{B}\,\hat{\mathbf{y}}$ with components u_B parallel and v_B perpendicular to the wall defined in terms of $f(\eta)$ and $g(x) = \sqrt{\nu x/U}$ by the relations

$$\begin{aligned} u_\mathrm{B}(x,y) &= Uf'(\eta) \\ v_\mathrm{B}(x,y) &= Ug'(x)[\eta f'(\eta) - f(\eta)] \end{aligned} \tag{2}$$

Fig. 1 Geometry and computational domain sketch

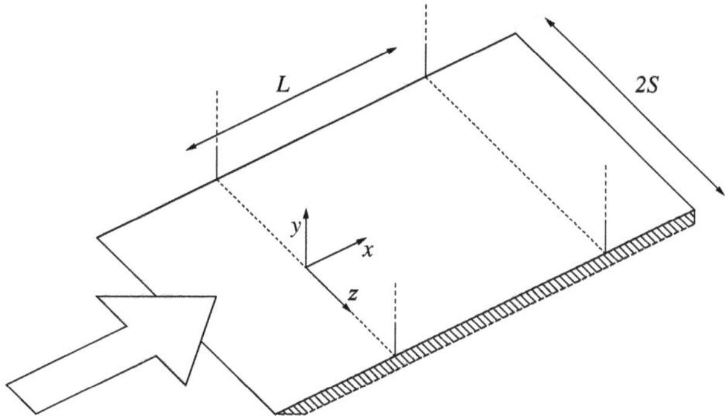

where $\eta = \eta(x, y) = y/g(x)$ and U is the uniform velocity at large distance from the wall.

The initial condition is assumed to be

$$\mathbf{u}(x, y, z, 0) = \mathbf{u}_{\mathrm{B}}(x, y). \tag{3}$$

Five different kinds of boundary conditions are enforced in the present problem.

1. On the plate surface $\partial\Omega_{\mathrm{w}}$, for $y = 0$, the velocity is prescribed including the possibility of wall oscillations parallel to its plane and of normal suction, namely, $\mathbf{u}(x, 0, z, t) = \mathbf{a}_{\mathrm{w}}(x, z, t)$, $t > 0$. A typical suction condition will be $v(x, 0, z, t) = a_{y,\mathrm{w}}(x, z, t)$, together with the no-slip conditions $u(x, 0, z, t) = 0$ and $w(x, 0, z, t) = 0$. No condition on pressure.
2. On the inlet section $\partial\Omega_{\mathrm{i}}$, for $x = \overline{x}$, the vector velocity is imposed $\mathbf{u}(\overline{x}, y, z, t) = \mathbf{u}_{\mathrm{B}}(\overline{x}, y)$, $t > 0$, (thus $w(\overline{x}, y, z, t) = 0$), and again no condition on pressure.
3. For $y \to \infty$, asymptotic conditions are imposed on the velocity components and on pressure. In detail:

$$\lim_{y\to\infty} u(x, y, z, t) = \lim_{y\to\infty} u_{\mathrm{B}}(x, y) = U$$

$$\lim_{y\to\infty} \partial_y v(x, y, z, t) = 0$$

$$\lim_{y\to\infty} w(x, y, z, t) = 0$$

$$\lim_{y\to\infty} p(x, y, z, t) = c_\infty$$

c_∞ being an arbitrary constant. The derivative condition for the normal velocity v is a consequence of enforcing asymptotically the incompressibility constraint, and can be imposed only in conjunction with a condition for pressure (the last one) imposing a zero gradient, with the constant value c_∞ remaining completely arbitrary.

4. The downstream surface, $\partial\Omega_o$, at $x = \overline{x} + L$ is an outflow boundary for the considered problem. Owing to the ellipticity and nonlinearity of the problem, the velocity and pressure fields on this surface depend on the fluid conditions outside the computational domain. Normally, on outflow surfaces non-reflective boundary conditions ought to be imposed capable of mimicking pure transport phenomena. In this case, for the sake of simplicity, a condition of flow alignment is imposed, prescribing the tangent components of velocity,

$$\begin{aligned}
\partial_x u(\overline{x} + L, y, z, t) &= b_o(y, z, t) \ (= 0)\\
v(\overline{x} + L, y, z, t) &= a_{y,o}(y, z, t) \ (= 0)\\
w(\overline{x} + L, y, z, t) &= a_{z,o}(y, z, t) \ (= 0)\\
p(\overline{x} + L, y, z, t) &= c_o(y, z).
\end{aligned}$$

Typically, when the external pressure is prescribed uniformly, the condition respecting compatibility is $p(\overline{x} + L, y, z, t) = c_\infty$, with the same constant c_∞ of the asymptotic pressure condition (with for instance $c_\infty = 0$, without any loss of generality).
5. On the lateral sides $\partial\Omega_l$, for $z = \pm S$, periodicity conditions on all the variables are imposed, namely, $\mathbf{u}(x, y, -S, t) = \mathbf{u}(x, y, S, t)$ and $p(x, y, -S, t) = p(x, y, S, t)$.

3 Time Discretization: Incremental Projection Method

The problem is discretized in time by a fractional-step projection method introduced by Chorin [4, 5] and Temam [6]. The incremental version of the method was first proposed in [7] and was subsequently modified by eliminating the end-of-step velocity from the final solution algorithm in [8] and [9]. We consider the second order BDF incremental method, which has been fully analysed in [10] and implemented in a spectral context for a bounded domain in [11].

The first step is performed using a first order, semi-implicit, Euler scheme without the pressure in the momentum equation and the first pressure field is found as solution of a Poisson equation, for details see [11].

For all the subsequent (incremental) steps a linear extrapolation of the pressure is employed, so that the viscous equation, reads, after the end-of-step velocity has been eliminated, for $k \geq 1$,

$$\frac{3\mathbf{u}^{k+1} - 4\mathbf{u}^k + \mathbf{u}^{k-1}}{2\Delta t} - \nu\nabla^2\mathbf{u}^{k+1} = \mathbf{f}^{k+1} - (\mathbf{u}_\star^{k+1}\cdot\nabla)\mathbf{u}_\star^{k+1} - \nabla p_\star^k \qquad (4)$$

where $\mathbf{u}_\star^{k+1} = 2\mathbf{u}^k - \mathbf{u}^{k-1}$ and $p_\star^k = \frac{1}{3}(7p^k - 5p^{k-1} + p^{k-2})$, for $k \geq 3$. The projection half-step is written as a Poisson equation for the pressure increment and reads, for $k \geq 1$,

$$-\nabla^2(p^{k+1} - p^k) = -\frac{3}{2\Delta t}\nabla\cdot\mathbf{u}^{k+1}, \quad k \geq 1. \tag{5}$$

The boundary conditions for $p^{k+1} - p^k$ are homogeneous Neumann conditions on the wall and the inflow portion of the boundary, together with the asymptotically uniform condition $p^{k+1} - p^k \to 0$ as $y \to \infty$ and the homogeneous Dirichlet condition at $x = \overline{x} + L$, which is artificially requested by the domain truncation.

In order to enforce the asymptotic condition on velocity as $y \to \infty$, the unknown variables of the velocity components are redefined as follows

$$\mathbf{u} = (U + u)\,\hat{\mathbf{x}} + v\,\hat{\mathbf{y}} + w\,\hat{\mathbf{z}} \tag{6}$$

so that the new parallel velocity $u \to 0$ as $y \to \infty$. The semi-discrete momentum equation (4) is recast for the new unknowns u, v and w by expanding the linear and nonlinear terms appropriately.

4 Spatial Discretization: Legendre–Laguerre–Fourier Method

The flow variables are described by a fully spectral approximation which is based, first, on the Fourier representation for the dependence on the periodic coordinate z. The dimensionless spatial coordinates

$$\xi = \frac{2(x - \overline{x})}{L} - 1, \qquad \eta = \frac{y}{H}, \qquad \zeta = 2\pi\frac{z}{S}, \tag{7}$$

are first introduced, where H is a length scale in direction normal to the plane $y = 0$. Then the velocity and pressure fields $\mathbf{u}(\xi, \eta, \zeta)$ and $p(\xi, \eta, \zeta)$ at a given time t are expanded in the truncated Fourier series

$$\begin{pmatrix}\mathbf{u}\\ p\end{pmatrix} = \begin{pmatrix}\mathbf{u}^0\\ p^0\end{pmatrix} + 2\sum_{m=1}^{N-1}\left[\begin{pmatrix}\mathbf{u}^m\\ p^m\end{pmatrix}\cos(m\zeta) - \begin{pmatrix}\mathbf{u}^{-m}\\ p^{-m}\end{pmatrix}\sin(m\zeta)\right] + \begin{pmatrix}\mathbf{u}^N\\ p^N\end{pmatrix}\cos(N\zeta) \tag{8}$$

since the solution is assumed to be periodic in ζ, with period 2π.

Focusing on the 3D Helmholtz equation in Cartesian coordinates

$$(-\nabla^2 + \gamma)u = f(x, y, z) \tag{9}$$

for a scalar unknown u, it is reduced by the Fourier expansion to a set of 2D elliptic equations for the Fourier coefficients $u^m(\xi,\eta)$, with $-N+1,\dots,N-1,N$, of the form, after multiplication by $L^2/4$,

$$\left(-\partial_\xi^2 - r_y^2\,\partial_\eta^2 + \kappa_\gamma^{|m|}\right)u^m = \frac{L^2}{4} f^m(\xi,\eta), \tag{10}$$

By multiplying the equation by $L^2/4$, one obtains where $r_y = L/2H$,

$$\kappa_\gamma^{|m|} = \left(\frac{\pi L}{S}\right)^2 m^2 + \frac{\gamma L^2}{4} = r_z^2\, m^2 + \frac{\gamma L^2}{4} \tag{11}$$

and $f^m(\xi,\eta)$ is the Fourier transform of function $f\left(\overline{x} + \frac{L}{2}(1+\xi), H\eta, z\right)$.

The spectral approximation of the velocity and pressure Fourier expansion coefficients $\mathbf{u}^m(\xi,\eta)$ and $p^m(\xi,\eta)$ is based on Legendre polynomials in the longitudinal direction and on Laguerre polynomials in the normal direction. Due to the different boundary conditions, different bases are required for each unknown variable. In particular, along the ξ coordinate, Dirichlet conditions are imposed at both extremes $\xi = \mp 1$ for the two velocity components v^m and w^m, while mixed conditions of Dirichlet–Neumann and Neumann–Dirichlet type are imposed on the velocity component u^m and on the pressure p^m, respectively. In the direction η normal to the wall, the boundary condition on it is Dirichlet for $\mathbf{u}^m$ while Neumann for p^m. Finally, the asymptotic condition as $y \to \infty$ is homogeneous Dirichlet for all variables, except for a Neumann condition on the velocity component v^m normal to the wall.

Thus, the Legendre basis used for variables v^m and w^m will be

$$L_0^*(\xi) = 1, \quad L_1^*(\xi) = \frac{\xi}{\sqrt{2}}, \quad L_i^*(\xi) = \frac{L_{i-2}(\xi) - L_i(\xi)}{\sqrt{2(2i-1)}}, \; i = 2,3,\dots \tag{12}$$

$L_i(\xi)$ being the Legendre polynomial of degree i. The two bases for coping with the alternate mixed Dirichlet–Neumann conditions on the variables u^m and p^m will be denoted by $L_i^{\mathrm{DN}}(\xi)$ and $L_i^{\mathrm{ND}}(\xi)$, respectively. The functions of these bases are all coincident with $L_i^*(\xi)$ except for the second one, which is defined by

$$L_1^{\mathrm{DN}}(\xi) = \frac{1+\xi}{\sqrt{2}} \qquad \text{and} \qquad L_1^{\mathrm{ND}}(\xi) = \frac{1-\xi}{\sqrt{2}} \tag{13}$$

in the two cases.

As the normal coordinate η is concerned, the basis function will be defined by

$$\mathscr{B}_j(\eta) \equiv \begin{cases} e^{-\eta/2} & \text{for } j = 0 \\ \eta e^{-\eta/2} \frac{\mathscr{L}_j^{(1)}(\eta)}{j+1} & \text{for } j \geq 1 \end{cases} \tag{14}$$

where $\mathscr{L}_j^{(1)}(\eta)$ are the generalized Laguerre polynomials of degree j. The normal component of velocity v is supplemented by Dirichlet conditions at both extremes $x = \overline{x}$ and $x = \overline{x} + L$, and also on the plate $y = 0$, and by the asymptotic condition of being constant as $y \to \infty$, $v \to a$. The unknown asymptotic constant a is taken into account in the weak formulation by introducing a special expansion function endowed with the appropriate asymptotic behaviour and satisfying also the homogeneous boundary condition on the wall, namely

$$\mathscr{B}^\star(\eta) = 1 - e^{-\eta/2}. \tag{15}$$

This function will replace the last basis function $\mathscr{B}_J(\eta)$ of the Laguerre basis (14). The modified basis is denoted by $\big(\mathscr{B}_j^\star(\eta); j = 0, 1, 2, \ldots, J\big)$, with $\mathscr{B}_j^\star(\eta) = \mathscr{B}_j(\eta)$, for $0 \le j \le J - 1$, and $\mathscr{B}_J^\star(\eta) = \mathscr{B}^\star(\eta)$.

We list the spectral expansion of the four unknown variables u^m, v^m, w^m and p^m:

$$\begin{aligned}
u^m(\xi,\eta) &= \sum_{i=0}^{I}\sum_{j=0}^{J} L_i^{\mathrm{DN}}(\xi)\, u_{i,j}^m\, \mathscr{B}_j(\eta),\\
v^m(\xi,\eta) &= \sum_{i=0}^{I}\sum_{j=0}^{J} L_i^{*}(\xi)\, v_{i,j}^m\, \mathscr{B}_j^\star(\eta),\\
w^m(\xi,\eta) &= \sum_{i=0}^{I}\sum_{j=0}^{J} L_i^{*}(\xi)\, w_{i,j}^m\, \mathscr{B}_j(\eta),\\
p^m(\xi,\eta) &= \sum_{\hat{\imath}=0}^{\hat{I}}\sum_{\hat{\jmath}=0}^{\hat{J}} L_{\hat{\imath}}^{\mathrm{ND}}(\xi)\, \hat{p}_{\hat{\imath},\hat{\jmath}}^m\, \mathscr{B}_{\hat{\jmath}}(\eta).
\end{aligned} \tag{16}$$

Here, the hats in the pressure expansion remind one that the basis employed to represent this variable have an order different (smaller by one or two modes) from those of the bases for the velocity, in order to satisfy the LBB stability condition.

These expansions are introduced in the momentum and pressure equations written in weak form. Exploiting the Galerkin method and imposing the Dirichlet conditions by a lifting, the following four linear systems for the three velocity components and the pressure are obtained, respectively

$$\begin{aligned}
\mathscr{U}^m \mathsf{M}_\infty + r_y^2 \mathscr{M}\, \mathscr{U}^m &+ \chi_\gamma^{|m|} \mathscr{M}\, \mathscr{U}^m \mathsf{M}_\infty = \mathscr{F}^m,\\
\boldsymbol{V}^m \mathsf{M}_\infty^{\star T} + r_y^2\, \mathsf{M}\, \boldsymbol{V}^m \mathsf{D}_\infty^{\star T} &+ \kappa_\gamma^{|m|}\, \mathsf{M}\, \boldsymbol{V}^m \mathsf{M}_\infty^{\star T} = \boldsymbol{G}^m,\\
\boldsymbol{W}^m \mathsf{M}_\infty + r_y^2\, \mathsf{M}\, \boldsymbol{W}^m &+ \chi_\gamma^{|m|}\, \mathsf{M}\, \boldsymbol{W}^m \mathsf{M}_\infty = \boldsymbol{H}^m,\\
\hat{\mathscr{P}}^m \hat{\mathsf{M}}_\infty + r_y^2\, \hat{\mathscr{M}}\, \hat{\mathscr{P}}^m \hat{\mathsf{D}}_\infty &+ \kappa_0^{|m|}\, \hat{\mathscr{M}}\, \hat{\mathscr{P}}^m \hat{\mathsf{M}}_\infty = \hat{\mathscr{K}}^m,
\end{aligned} \tag{17}$$

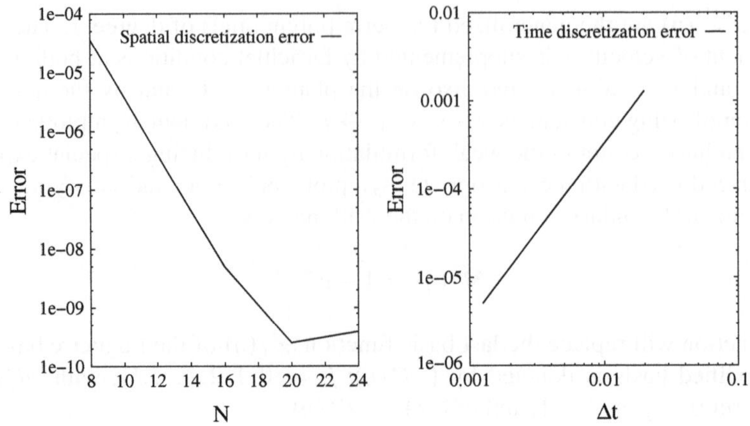

Fig. 2 Space (*left*) and time (*right*) convergence for the test with closed form solution

where $\chi_\gamma^{|m|} = r_z^2 m^2 + \frac{1}{4}(\gamma L^2 - r_y^2)$. The boldface symbols $\boldsymbol{\mathscr{U}}^m$, $\boldsymbol{V}^m$, $\boldsymbol{W}^m$ and $\hat{\boldsymbol{\mathscr{P}}}^m$ denote rectangular arrays of the expansion coefficients of the unknowns. The M letters, in various faces, denote the mass matrices, whose elements are computed by the following inner products: $\mathscr{M}_{i,j} = (L_i^{\mathrm{DN}}, L_j^{\mathrm{DN}})$, $\mathsf{M}_{i,j} = (L_i^*, L_j^*)$, $\hat{\mathscr{M}}_{i,j} = (L_{\hat\imath}^{\mathrm{ND}}, L_{\hat\jmath}^{\mathrm{ND}})$ $\mathsf{M}_{\infty;i,j} = (\mathscr{B}_i, \mathscr{B}_j)$, $\mathsf{M}^\star_{\infty;i,j} = (\mathscr{B}_i, \mathscr{B}_j^\star)$ and $\hat{M}_{\infty;\hat\imath,\hat\jmath} = (\mathscr{B}_{\hat\imath}, \mathscr{B}_{\hat\jmath})$. The D letters denote stiffness matrices, which are all sparse, most tridiagonal, defined as: $\mathsf{D}^\star_{\infty;i,j} = (\mathscr{B}'_i, \mathscr{B}'^\star_j)$ and $\hat{D}_{\infty;\hat\imath,\hat\jmath} = (\mathscr{B}'_{\hat\imath}, \mathscr{B}'_{\hat\jmath})$.

For each time step, the algorithm requires to compute the right hand side of the momentum and pressure equations. This can be done efficiently since the contribution of the pressure gradient and of the velocity divergence can be computed by multiplying the pressure and velocity coefficients by sparse matrices. Moreover, the nonlinear terms are evaluated pseudospectrally. Finally, all the linear systems above are solved very efficiently by the double diagonalization method.

5 Numerical Results

To assess the spectral accuracy of the spatial discretization and the second order accuracy in time, the Navier–Stokes equation with a closed form solution, obtained by introducing a suitable forcing term, have been solved. The steady velocity field is defined by:

$$
\begin{aligned}
u_{\mathrm{s}}(x, y, z) &= (\cos x)\sin(2\pi z/S)\, e^{-y} \\
v_{\mathrm{s}}(x, y, z) &= (\sin x)\sin(2\pi z/S)\, e^{-y} + (\cos x)\cos(2\pi z/S) \\
w_{\mathrm{s}}(x, y, z) &= -(S/\pi)(\sin x)\cos(2\pi z/S)\, e^{-y}
\end{aligned}
$$

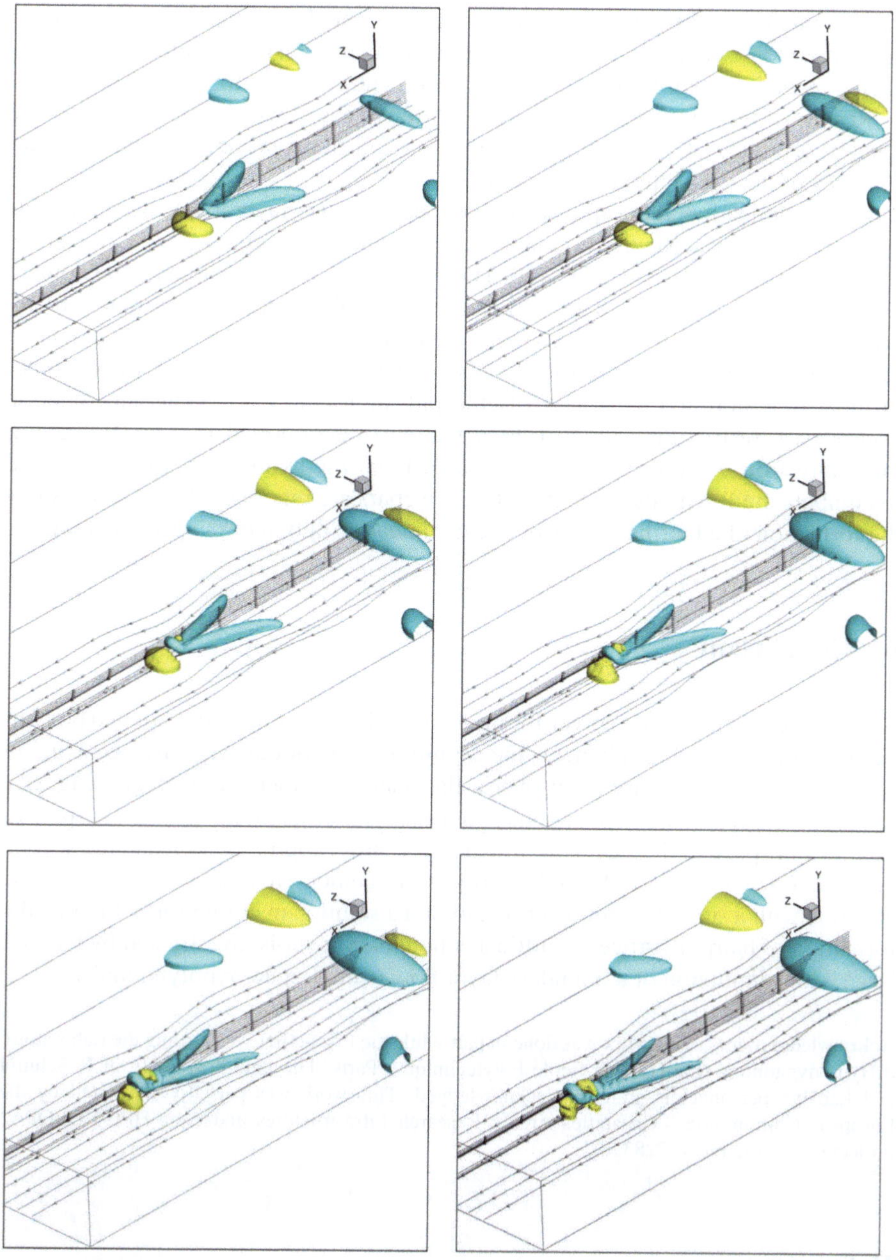

Fig. 3 Evolution of a hairpin vortex in a Blasius boundary layer, streamlines and isobars. $\mathrm{Re}_{\overline{x}} = 92{,}400$, discretization (dealiased): $512 \times 128 \times 128$ modes, $\Delta t = 3.9 \times 10^{-6}$. In lexicographic order: $t_1 = 0.4630$, $t_2 = 0.4834$, $t_3 = 0.5033$, $t_4 = 0.5234$, $t_5 = 0.5435$, $t_6 = 0.5637$. Plotted box size: $1.30 \times 0.13 \times 0.41$

in conjunction with the following pressure field $p_s(x, y, z) = (\sin x) \cos(2\pi z/S)\, e^{-y}$. For the time convergence test, the unsteady velocity field has been assumed in the form $\mathbf{u}(\mathbf{r}, t) = \mathbf{u}_s(\mathbf{r}) \sin t$, and similarly for the pressure.

Results of the convergence tests are reported in Fig. 2 whose left plot shows the spectral accuracy of the spatial discretization and the right one the second order accuracy of the time integration scheme.

As an application, the response of a Blasius boundary layer to a wall forcing has been computed. The Reynolds number based on the inlet abscissa $\overline{x} = 1$ and on the freestream velocity $U = 1$ is 92,400. The dimensions of the computational domain are $L = 2$ and $S = 0.4$. The wall forcing is implemented as a normal velocity boundary condition which mimics the presence of a bumper. The normal velocity is given by $v = 0.0325\, f(x) \cos(\beta z) \sin(\omega t)$, where $\beta = 15.4$ is the wavenumber in the z direction and $\omega = 1.16$ is the pulsation. $f(x)$ is a Gaussian function necessary to restrict the action of the actuation to the interval $x \in (1.19, 1.21)$ in the x direction. All quantities are nondimensionalized by the same quantities used to compute the Reynolds number. In Fig. 3 isobars are reported which clearly show the formation of a hairpin vortex by the nonlinear evolution of the perturbation.

6 Conclusion

A new spectral-projection solver for incompressible flows in an unbounded box has been presented which implements the asymptotic conditions typically occurring in simulations of a boundary layer over a flat plate. The method leverages Laguerre functions to expand the flow variables in direction normal to the plate. The spectral accuracy in space of the Legendre–Laguerre–Fourier approximation and the second order accuracy in time of the BDF2 projection method are assessed by solving the equation with a suitable forcing term to obtain a solution in closed form. Finally, the formation of hairpin vortices in a Blasius boundary layer is investigated by the first large scale simulation of a boundary layer flow using Laguerre polynomials.

Acknowledgements This work was done in part while the first author was visiting the Laboratoire d'Hydrodynamique (LadHyx) at École Polytechnique, Paris. The kind hospitality of P. Schmid and LadHyx personnel is gratefully acknowledged. This work was partially supported by the European Commission – Capacities Area – Research Infrastructures under the HPC-EUROPA2 project (project number: 228398).

References

1. Skote, M.: Turbulent boundary layer flow subject to streamwise oscillation of spanwise wall-velocity. Phys. Fluids **23**, 081703 (2011).
2. Chevalier, M., Schlatter, P., Lundbladh, A., Henningson, D.: A pseudo-spectral solver for incompressible boundary-layer flows. Tech. Rep. TRITA-Mech. KTH, (2007).

3. Shen, Jie: Stable and efficient spectral methods in unbounded domains using Laguerre functions. SIAM J. Numer. Anal., **38**, 1113–1133 (2000).
4. Chorin, A.J.: Numerical solution of the Navier–Stokes equations. Math. Comp. **22**, 745–762 (1968).
5. Chorin, A.J.: On the convergence of discrete approximations to the Navier–Stokes equations. Math. Comp. **23**, 341–353 (1969).
6. Temam, R.: Sur l'approximation de la solution des équations de Navier–Stokes par la méthode de pas fractionnaires. Arch. Rat. Mech. Anal. **33**, 377–385 (1969).
7. Goda, K.: A multistep technique for the with implicit difference schemes for calculating two- or three-dimensional cavity flows. J. Comput. Phys. **30**, 76–95 (1979).
8. Guermond, J.-L.: Sur l'approximation des équations de Navier–Stokes par une méthode de projection. C. R. Acad. Sci. Paris, Série I **319**, 887–892 (1994).
9. Guermond, J.-L., Quartapelle, L.: Calculation of incompressible viscous flows by an unconditionally stable projection FEM. J. Comput. Phys. **132**, 12–33 (1997).
10. Guermond, J.-L.: Un résultat de convergence à l'ordre deux en temps pour l'approximation des équations de Navier–Stokes par une technique de projection. Modél. Math. Anal. Numér. (M^2AN) **33**, 169–189 (1999).
11. Auteri, F., Parolini, N.: A mixed-basis spectral projection method. J. Comput. Phys. **175**, 1–23 (2002).
12. Shen, Jie and Wang, Li-Lian: Some recent advances on spectral methods for unbounded domains. Comm. Comput. Phys. **5**, 195–241 (2009).
13. Guo, Ben-Yu, Shen, Jie and Xu, Cheng-Long: Generalized Laguerre approximation and its application to exterior problems. J. Comput. Math. **23**, 113–130 (2005)

Implementation of an Explicit Algebraic Reynolds Stress Model in an Implicit Very High-Order Discontinuous Galerkin Solver

F. Bassi, L. Botti, A. Colombo, A. Ghidoni, and S. Rebay

Abstract In this work we present the main features of an implicit implementation of the explicit algebraic Reynolds stress model (EARSM) of Wallin and Johansson (J Fluid Mech 403:89–132, 2000) in the high-order Discontinuous Galerkin (DG) solver named MIGALE (Bassi et al. (2011) Discontinuous Galerkin for turbulent flows. In: Wang ZJ (ed) Adaptive high-order methods in computational fluid dynamics. Volume 2 of Advances in computational fluid dynamics. World Scientific). Explicit Algebraic Reynolds stress models replace the linear Boussinesq hypothesis by an algebraic approximation of the anisotropy transport equations, resulting in a non-linear constitutive relation for the Reynolds stress tensor in terms of mean flow strain-rate and rate-of-rotation tensors. The EARSM model has been implemented in the existing k-ω model of the DG code MIGALE without any recalibration of the constants and a basic assessment and validation of its near-near wall behaviour has been done on a turbulent flat plate test case (Slater et al. (2000) The NPARC verification and validation archive. ASME Paper 2000-FED-11233, ASME). Consistently with the mean-flow equations, the turbulence model equations have been discretized to a high-order spatial accuracy on hybrid type elements by using hierarchical and orthonormal polynomial basis functions, local to each element and defined in the physical space. Such discretization preserves its accuracy also for highly-stretched elements with curved boundaries as those used within turbulent boundaries layers. For steady-state computations, the time integration of

F. Bassi · L. Botti · A. Colombo (✉)
Department of Industrial Engineering, University of Bergamo, Viale Marconi 5, 24044 Dalmine (BG), Italy
e-mail: francesco.bassi@unibg.it; lorenzo.botti@unibg.it; alessandro.colombo@unibg.it

A. Ghidoni · S. Rebay
Department of Industrial and Mechanical Engineering, University of Brescia, via Branze 38, 25123 Brescia, Italy
e-mail: antonio.ghidoni@ing.unibs.it; stefano.rebay@ing.unibs.it

M. Azaïez et al. (eds.), *Spectral and High Order Methods for Partial Differential Equations - ICOSAHOM 2012*, Lecture Notes in Computational Science and Engineering 95, DOI 10.1007/978-3-319-01601-6_8,

the fully coupled system of governing equations is performed implicitly by means of the linearized backward Euler method where the Jacobian is derived analytically and a pseudo-transient continuation strategy is employed (Bassi et al. (2010) Very high-order accurate discontinuous Galerkin computation of transonic turbulent flows on aeronautical configurations. In: Norbert Kroll, Heribert Bieler, Herman Deconinck, Vincent Couaillier, Harmen van der Ven, and Kaare Sørensen (eds) ADIGMA – A European initiative on the development of adaptive higher-order variational methods for aerospace applications. Volume 113 of Notes on numerical fluid mechanics and multidisciplinary design. Springer, Berlin/Heidelberg, pp 25–38). The capabilities of the present version of the code will be demonstrated by computing an external aerodynamic problem proposed within the EU-funded project IDIHOM (Project IDIHOM (2012) Industrialisation of high-order methods a top-down approach).

1 Introduction

Computational fluid dynamics is nowadays a key tool for a wide range of industrial design processes. The Reynolds-Averaged Navier-Stokes (RANS) simulation technology plays an important role in spreading the use of numerical simulations within industry since it can provide accurate solutions in a reasonable computational time and with moderate hardware requirements when compared to more involved simulation techniques for turbulent flows (e.g. DES, LES, DNS). However, most of the RANS simulations currently performed in industry relays on standard turbulence models that assume a linear relation between the Reynolds stress and the mean flow strain-rate tensor, which is the Boussinesq hypothesis. It is well known that such hypothesis suffers from several limitations as for example the capability to capture the secondary flows that develop at wall junctions [21, 22]. To improve the prediction capabilities of turbulence models several authors replaced the Boussinesq linear constitutive law with a non-linear relation that could embed the modeling level proper of the Reynolds stress models, see e.g. [17, 18, 24, 25]. In this context Wallin and Johansson [30] proposed a non-linear explicit algebraic constitutive law involving the mean flow strain-rate and rate-of-rotation tensors. The good properties of the model, in terms of compatibility with the Navier–Stokes equations, have been demonstrated in the a priori analysis reported in [23], comparing the model results with those obtained from DNS. An interesting feature of the Wallin and Johansson model is that it can easily fit in the already existing turbulence models implementations.

Improved modelling of turbulence effects coupled with an high-order accurate numerical discretization can provide a reasonably cheap compromise between standard RANS plus two equations turbulence models simulations and more resolved but computational intensive simulations of turbulent flows.

This paper outlines the main features of an implicit fully coupled implementation of the Wallin and Johansson EARSM in a high-order DG solver for the RANS and k-ω equations.

2 Governing Equations

The governing equations can be written as

$$\frac{\partial \rho}{\partial t} + \frac{\partial}{\partial x_j}(\rho u_j) = 0, \tag{1}$$

$$\frac{\partial}{\partial t}(\rho u_i) + \frac{\partial}{\partial x_j}(\rho u_j u_i) = -\frac{\partial p}{\partial x_i} + \frac{\partial \hat{\tau}_{ji}}{\partial x_j}, \tag{2}$$

$$\frac{\partial}{\partial t}(\rho e_0) + \frac{\partial}{\partial x_j}(\rho u_j h_0) = \frac{\partial}{\partial x_j}\left[u_i \hat{\tau}_{ij} - q_j\right] - \mathscr{P}_k + \beta^* \rho \overline{k} e^{\tilde{\omega}}, \tag{3}$$

$$\frac{\partial}{\partial t}(\rho k) + \frac{\partial}{\partial x_j}(\rho u_j k) = \frac{\partial}{\partial x_j}\left[(\mu + \sigma^* \overline{\mu}_t)\frac{\partial k}{\partial x_j}\right] + \mathscr{P}_k - \beta^* \rho \overline{k} e^{\tilde{\omega}}, \tag{4}$$

$$\frac{\partial}{\partial t}(\rho \tilde{\omega}) + \frac{\partial}{\partial x_j}(\rho u_j \tilde{\omega}) = \frac{\partial}{\partial x_j}\left[(\mu + \sigma \overline{\mu}_t)\frac{\partial \tilde{\omega}}{\partial x_j}\right] + \mathscr{P}_\omega - \beta \rho e^{\tilde{\omega}}$$
$$+ (\mu + \sigma \overline{\mu}_t)\frac{\partial \tilde{\omega}}{\partial x_k}\frac{\partial \tilde{\omega}}{\partial x_k} + \sigma_d \frac{\rho}{e^{\tilde{\omega}}} \max\left(\frac{\partial k}{\partial x_k}\frac{\partial \tilde{\omega}}{\partial x_k}; 0\right), \tag{5}$$

where the pressure, the total stress tensor, the heat flux vector, the production terms $\mathscr{P}_k$ and $\mathscr{P}_\omega$ and the limited value of turbulent kinetic energy $\overline{k}$ are defined as

$$p = (\gamma - 1)\rho\,(e_0 - u_k u_k/2)\,, \qquad \hat{\tau}_{ij} = 2\mu S_{ij} + \tau_{ij}, \tag{6}$$

$$q_j = -\left(\frac{\mu}{\mathrm{Pr}} + \frac{\overline{\mu}_t}{\mathrm{Pr}_t}\right)\frac{\partial h}{\partial x_j}, \tag{7}$$

$$\mathscr{P}_k = \tau_{ij}\frac{\partial u_i}{\partial x_j}, \qquad \mathscr{P}_\omega = \alpha \frac{\tau_{ij}}{\overline{k}}\frac{\partial u_i}{\partial x_j}, \tag{8}$$

$$\overline{k} = \max\,(0, k)\,. \tag{9}$$

Here γ is the ratio of gas specific heats, Pr and Pr_t are the molecular and turbulent Prandtl numbers and S_{ij} and Ωij are the mean strain-rate and the rate-of-rotation tensors. The closure coefficients α, α^*, β, β^*, σ, σ^*, σ_d are those of the high- or low-Reynolds number k-ω model of Wilcox [32]. Notice that Eq. (5) of the k-ω turbulence model is not in standard form since the variable $\tilde{\omega} = \log \omega$ is used, see [8]. In the framework of the EARSM of Wallin and Johansson [30], the constitutive relation for the turbulent stress tensor can be written as

$$\frac{\tau_{ij}}{\rho \overline{k}} = -\frac{\overline{u_i u_j}}{\overline{k}} = -\alpha^* a_{ij} - \frac{2}{3}\delta_{ij} = \alpha^*\left(2C_\mu \tau S_{ij} - a_{ij}^{(\mathrm{ex})}\right) - \frac{2}{3}\delta_{ij}, \tag{10}$$

where, for implementation convenience, the anisotropy tensor a_{ij} has been split in a linear part and a non-linear extra anisotropy contribution. The time scale τ and the variable coefficient C_μ are given by

$$\tau = \frac{1}{\beta^* e^{\tilde{\omega}}}, \qquad C_\mu = -\frac{1}{2}\left(\beta_1 + II_\Omega \beta_6\right). \tag{11}$$

We remark that, the time scale τ does not include the near-wall lower bound, based on the Kolmogorov time scale, usually employed in k-ϵ implementations of EARSM since this limitation is actually provided by the finite value of ω set at wall. The eddy viscosity $\overline{\mu}_t$ and the extra-anisotropy tensor $a_{ij}^{(\mathrm{ex})}$ are given by

$$\overline{\mu}_t = \alpha^* C_\mu \tau \rho \overline{k}, \tag{12}$$

$$\begin{aligned} a_{ij}^{(\mathrm{ex})} &= \beta_3 \tau^2 \left(\Omega_{ik}\Omega_{kj} - \frac{1}{3} II_\Omega \delta_{ij}\right) + \beta_4 \tau^2 \left(S_{ik}\Omega_{kj} - \Omega_{ik}S_{kj}\right) \\ &+ \beta_6 \tau^3 \left(S_{ik}\Omega_{kl}\Omega_{lj} + \Omega_{ik}\Omega_{kl}S_{lj} - II_\Omega S_{ij} - \frac{2}{3} IV \delta_{ij}\right) \\ &+ \beta_9 \tau^4 \left(\Omega_{ik}S_{kl}\Omega_{lm}\Omega_{mj} + \Omega_{ik}\Omega_{kl}S_{lm}\Omega_{mj}\right), \end{aligned} \tag{13}$$

where the coefficients $\beta_{i\in\{1,3,4,6,9\}}$ are functions of the invariants II_S, II_Ω and IV

$$II_S = \mathrm{tr}\{\boldsymbol{S}^2\}, \qquad II_\Omega = \mathrm{tr}\{\boldsymbol{\Omega}^2\}, \qquad IV = \mathrm{tr}\{\boldsymbol{S}\boldsymbol{\Omega}^2\}. \tag{14}$$

In this paper, starting from the linear part of the Reynolds stress tensor formulation of Eq. (10), we will evaluate the influence of the non-linear terms of EARSM by computing some turbulent test cases. In the following sections, the notation EARSMx will indicate the EARSM model including anisotropy terms up to the x-th degree. In the 2d case, only linear and quadratic terms of anisotropy are non-zero. At present, the implementation of the EARSM in the 3d code MIGALE has been completed up to the cubic terms of extra-anisotropy.

3 The DG Discrete Setting

Let $\mathscr{T}_h = \{T\}$ denotes a mesh of the domain $\Omega \in \mathbb{R}^d, d \in \{2, 3\}$ consisting of non-overlapping arbitrarily shaped elements T such that

$$\overline{\Omega}_h = \bigcup_{T\in\mathscr{T}_h} \overline{T}. \tag{15}$$

Following the idea to define discrete polynomial spaces in physical coordinates, see e.g. [7, 9–11, 15, 16], we consider DG approximations based on the space

$$\mathbb{P}_d^k(\mathscr{T}_h) \stackrel{\text{def}}{=} \left\{v_h \in L^2(\Omega) \,|\, v_{h|T} \in \mathbb{P}_d^k(T),\ \forall T \in \mathscr{T}_h\right\}, \tag{16}$$

where k is a non-negative integer and $\mathbb{P}_d^k(T)$ denotes the restriction to T of the polynomial functions of d variables and total degree $\leq k$. To build a satisfactory basis for the space (16) we rely on the procedure presented in [28], see also [5, 14], allowing to obtain orthonormal and hierarchical basis functions by means of the modified Gram-Schmidt (MGS) algorithm. The starting set of basis functions for the MGS algorithm are the monomials defined over each elementary space $\mathbb{P}_d^k(T)$, $T \in \mathscr{T}_h$, in an element frame relocated in the barycenter and aligned with the principal axis of inertia of T. For the sake of presenting the DG discretization, we introduce the set $\mathscr{F}_h$ of the mesh faces, partitioned into $\mathscr{F}_h \stackrel{\text{def}}{=} \mathscr{F}_h^{\text{i}} \cup \mathscr{F}_h^{\text{b}}$, where $\mathscr{F}_h^{\text{b}}$ collects the faces located on the boundary of Ω_h and for any $F \in \mathscr{F}_h^{\text{i}}$ there exist two elements $T^+, T^- \in \mathscr{T}_h$ such that $F \in \partial T^+ \cap \partial T^-$. Moreover, for all $F \in \mathscr{F}_h^{\text{b}}$, $\mathbf{n}_F$ denotes the unit outward normal to Ω_h, whereas, for all $F \in \mathscr{F}_h^{\text{i}}$, $\mathbf{n}_F^-$ and $\mathbf{n}_F^+$ are the unit normals pointing exterior to T^- and T^+ respectively.

Since a function $v_h \in \mathbb{P}_d^k(\mathscr{T}_h)$ is double valued over an internal face $F \in \mathscr{F}_h^{\text{i}}$ we introduce the jump $[\![\cdot]\!]$ and average $\{\cdot\}$ trace operators, that is

$$[\![v_h]\!] \stackrel{\text{def}}{=} v_{h|T^+}\mathbf{n}_F^+ + v_{h|T^-}\mathbf{n}_F^-, \qquad \{v_h\} \stackrel{\text{def}}{=} \frac{v_{h|T^+} + v_{h|T^-}}{2}, \tag{17}$$

when applied to a vector function, the average and jump operators act componentwise. Finally, the DG discretization of second-order viscous terms employs the lifting operators $\mathbf{r}_F$ and $\mathbf{R}$. For all $F \in \mathscr{F}_h$, the local lifting operator $\mathbf{r}_F : \left[L^2(F)\right]^d \to [\mathbb{P}_d^k(\mathscr{T}_h)]^d$, is defined so that, for all $\mathbf{v} \in \left[L^2(F)\right]^d$

$$\int_\Omega \mathbf{r}_F(\mathbf{v}) \cdot \boldsymbol{\tau}_h \,\mathrm{d}\mathbf{x} = -\int_F \{\boldsymbol{\tau}_h\} \cdot \mathbf{v} \,\mathrm{d}F \quad \forall \boldsymbol{\tau}_h \in [\mathbb{P}_d^k(\mathscr{T}_h)]^d, \tag{18}$$

and the definition of the global lifting operator $\mathbf{R}$ follows

$$\mathbf{R}(\mathbf{v}) \stackrel{\text{def}}{=} \sum_{F \in \mathscr{F}_h} \mathbf{r}_F(\mathbf{v}). \tag{19}$$

3.1 DG Space Discretization

RANS and turbulence model equations can be written in compact form as

$$\frac{\partial \mathbf{u}}{\partial t} + \nabla \cdot \mathbf{F}_c(\mathbf{u}) + \nabla \cdot \mathbf{F}_v(\mathbf{u}, \nabla \mathbf{u}) + \mathbf{s}(\mathbf{u}, \nabla \mathbf{u}) = \mathbf{0}, \tag{20}$$

where $\mathbf{u}$ and $\mathbf{s}$ are the vectors of the m variables and source terms, and $\mathbf{F}_c, \mathbf{F}_v \in \mathbb{R}^m \otimes \mathbb{R}^d$ are defined as the arrays of the inviscid and viscous flux vectors. A weak formulation of the RANS equations is obtained multiplying each scalar law in Eq. (20) by an arbitrary smooth test function $v_j \in \mathbf{v}$, $1 \leq j \leq m$, and integrating by parts, that is

$$\int_{\Omega} v_j \frac{\partial u_j}{\partial t}\,\mathrm{d}\mathbf{x} - \int_{\Omega} \nabla v_j \cdot \mathbf{F}_j(\mathbf{u}, \nabla \mathbf{u})\,\mathrm{d}\mathbf{x} + \int_{\partial\Omega} v_j \mathbf{F}_j(\mathbf{u}, \nabla \mathbf{u}) \cdot \mathbf{n}\,\mathrm{d}\sigma + \int_{\Omega} v_j \mathbf{s}_j(\mathbf{u}, \nabla \mathbf{u})\,\mathrm{d}\mathbf{x} = 0, \tag{21}$$

where $\mathbf{F}_j$ is the sum of the inviscid and viscous flux vectors of the j-th equation.

To discretize Eq. (21) we replace the solution $\mathbf{u}$ and the test function $\mathbf{v}$ with a finite element approximation $\mathbf{u}_h$ and a discrete test function $\mathbf{v}_h$ respectively, where $\mathbf{u}_h$ and $\mathbf{v}_h$ belong to the space $V_h \stackrel{\text{def}}{=} [\mathbb{P}_d^k(\mathscr{T}_h)]^m$. The discontinuous approximation of the numerical solution requires to introduce a specific treatment of the inviscid and viscous interface fluxes. In order to ensure conservation and correctly account for wave propagation the former is based on the Godunov flux computed with an exact Riemann solver. For the latter we employ the BR2 scheme, proposed in [10] and theoretically analyzed in [2, 12].

Accounting for these aspects, the DG formulation of the compressible RANS and k-ω equations consists in seeking $\mathbf{u}_h \in V_h$ such that

$$\begin{aligned}
&\sum_{T\in\mathscr{T}_h} \int_T v_{h,j} \frac{\partial u_{h,j}}{\partial t}\,\mathrm{d}\mathbf{x} - \sum_{T\in\mathscr{T}_h} \int_T \nabla_h v_{h,j} \cdot \mathbf{F}_j\left(\mathbf{u}_h, \nabla_h \mathbf{u}_h + \mathbf{R}\left(\llbracket \mathbf{u}_h \rrbracket\right)\right)\mathrm{d}\mathbf{x} \\
&\quad + \sum_{F\in\mathscr{F}_h} \int_F \llbracket v_{h,j} \rrbracket \cdot \hat{\mathbf{F}}_j\left(\mathbf{u}_h^{\pm}, \left(\nabla_h \mathbf{u}_h + \eta_F \mathbf{r}_F\left(\llbracket \mathbf{u}_h \rrbracket\right)\right)^{\pm}\right)\mathrm{d}\sigma \\
&\quad + \sum_{T\in\mathscr{T}_h} \int_T v_{h,j} \mathbf{s}_j\left(\mathbf{u}_h, \nabla_h \mathbf{u}_h + \mathbf{R}\left(\llbracket \mathbf{u}_h \rrbracket\right)\right)\mathrm{d}\mathbf{x} = 0 \qquad \forall \mathbf{v}_h \in V_h,
\end{aligned} \tag{22}$$

where boundary conditions are weakly imposed [10].

3.2 Time Integration

The DG space discretization of Eq. (22) results in the following system of (nonlinear) ODEs in time

$$\mathbf{M}\frac{\mathrm{d}\mathbf{U}}{\mathrm{d}t} + \mathbf{R}(\mathbf{U}) = \mathbf{0}, \tag{23}$$

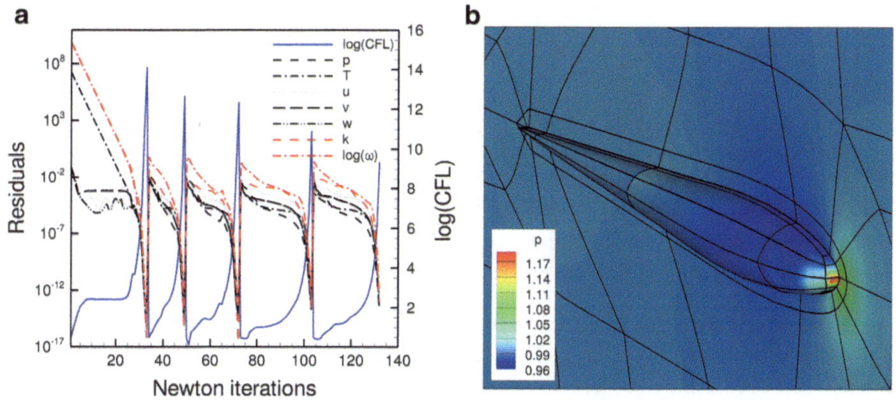

Fig. 1 BTC0: 832 50-node hexahedral elements, $M_\infty = 0.5$, $Re = 10^7$, ¸ $= 5°$. (**a**) Residual history, $\mathbb{P}^{0\rightarrow 4}$. (**b**) Pressure contours, $\mathbb{P}^4$ (EARSM3)

where **U** is the global vector of unknown degrees of freedom, **M** is a global block diagonal matrix and **R** (**U**) is the vector of residuals. Thanks to the use of orthonormal basis functions, the matrix **M** reduces to the identity matrix. In the case of steady-state computations the semi-discrete problem in Eq. (22) is discretized in time by means of the classical backward Euler scheme coupled with the pseudo-transient continuation strategy proposed in [4]. The Jacobian is derived analytically and takes full account of the dependence of the residuals on the unknown vector and on its derivatives, including the implicit treatment of the boundary conditions. To solve the resulting linear system at each time step the matrix-explicit or the matrix-free GMRES algorithm can be used. For this purpose linear algebra and parallelization are handled through PETSc library [3]. In both cases system preconditioning is required to make the convergence of the GMRES solver acceptable in problems of practical interest. The block Jacobi method with one block per process, each of which is solved with ILU(0), or the Additive Schwarz Method (ASM) are usually employed. For simple steady test cases implicit time integration combined with the aforementioned CFL evolution rule provides quadratic Newton convergence to machine accuracy as displayed in Fig. 1.

4 Results

The following section deals with the high-order numerical simulation of turbulent flows of common external aerodynamic configurations. To evaluate the influence of the different EARSM terms on the solution we compare the standard k-ω model [8] with the EARSM in its linear and non-linear formulations. All the computations have been run in parallel, initializing the $\mathbb{P}^0$ solution from uniform flow at freestream conditions and the higher-order solutions from the lower-order ones.

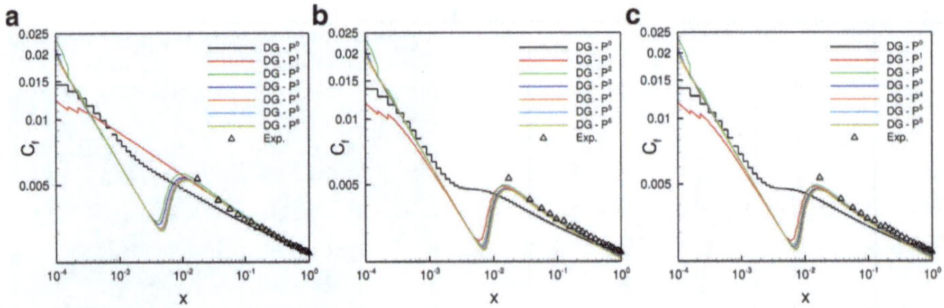

Fig. 2 Wieghardt flat plate: Skin friction coefficient along the plate, C_f. (**a**) k-ω. (**b**) EARSM1. (**c**) EARSM2

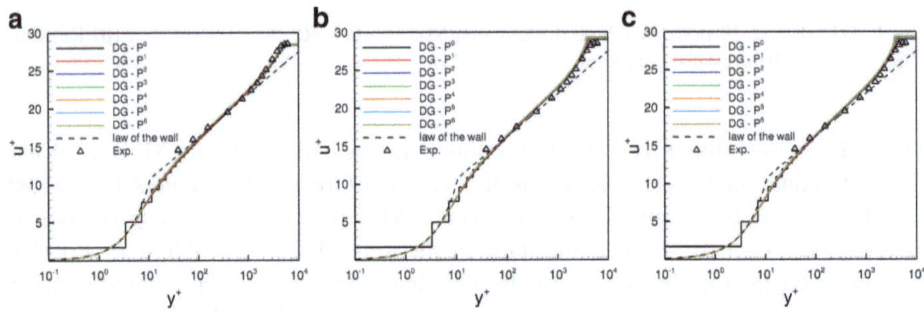

Fig. 3 Wieghardt flat plate: Non-dimensional tangential velocity profile, u^+. (**a**) k-ω. (**b**) EARSM1. (**c**) EARSM2

4.1 Turbulent Flat Plate

The flat plate flow here considered is that reported by Wieghardt [31, 33]. This test case was primarily intended to validate the implementation of the model in the very high-order code MIGALE rather than to assess the modelling capabilities of EARSM. The flow has been computed up to $\mathbb{P}^6$ polynomial approximation with a farfield Mach number $M_\infty = 0.2$ and Reynolds number $Re_\infty = 11.1 \times 10^6$ based on the plate length. The mesh of 8,800 quadrilateral elements has been taken from the NPARC Alliance Validation Archive [26] and corresponds to $y^+ = 10$ for the first grid point off the wall. The near-wall behaviour of the DG solutions is here presented in terms of skin-friction coefficient along the plate and velocity and non-dimensional ω profiles at $x/L = 0.923$. The results of k-ω model and linear and complete EARSM formulations, are compared to experimental data and in case the of u^+ and ω^+ to the law-of-the-wall profiles, Figs. 2–4. All the models result in an overall very good agreement with reference data. Comparing EARSM and standard k-ω model results, we observe that, rising the degree of polynomial approximation, EARSM provides more rapidly converging skin-friction coefficient distributions,

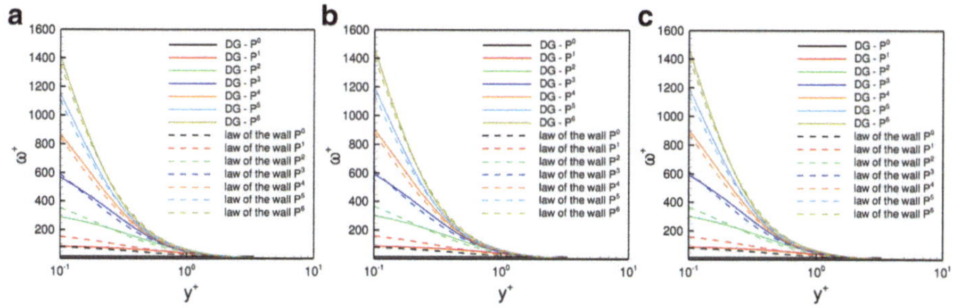

Fig. 4 Wieghardt flat plate: Non-dimensional specific dissipation-rate profile, ω^+. (**a**) k-ω. (**b**) EARSM1. (**c**) EARSM2

with a sort of transition visible even on the lowest order solution, see Fig. 2. No appreciable difference in the near-wall profiles of linear and non-linear EARSM results can be observed.

4.2 *VFE2 Medium-Radius Delta Wing*

The NASA 65° sweep delta wing has been proposed and investigated experimentally within the second international Vortex Flow Experiment (VFE-2) [29]. The farfield conditions of this test case are $M_\infty = 0.4$, $\alpha = 13.3°$ and $Re_{mac} = 3 \times 10^6$. The solution has been computed with the standard k-ω model and with EARSM1→3 up to $\mathbb{P}^3$ on a grid consisting of 13,816 20-node hexahedral elements. The high-order grid has been generated by means of an in-house agglomeration software starting from a linear finite volume grid [13]. In Fig. 5 the pressure coefficient distribution on the wing surface is compared with the experimental measurements [19,20,29]. The solution obtained employing the standard k-ω model shows a delayed onset of the vortex developing along the wing with respect to the experimental data. Using the EARSM model, in its linear and non-linear versions, the vortex moves towards the wing apex showing also a sharper definition of turbulence intensity, as depicted in Fig. 6. Figure 7 compares the computed and experimental pressure coefficient distributions at two wing sections normal the root chord c_r. Overall, EARSM results appear to improve standard k-ω predictions. In particular, the EARSM1 results are surprisingly good while EARSM2-3 appear slightly disappointing. This behaviour needs further investigation, but, as a first comment, we observe that the discrepancy with respect to experimental data is mainly due to an incorrect prediction of the primary vortex origin. This issue will require a closer numerical investigation of the flow behaviour around the nose of the wing.

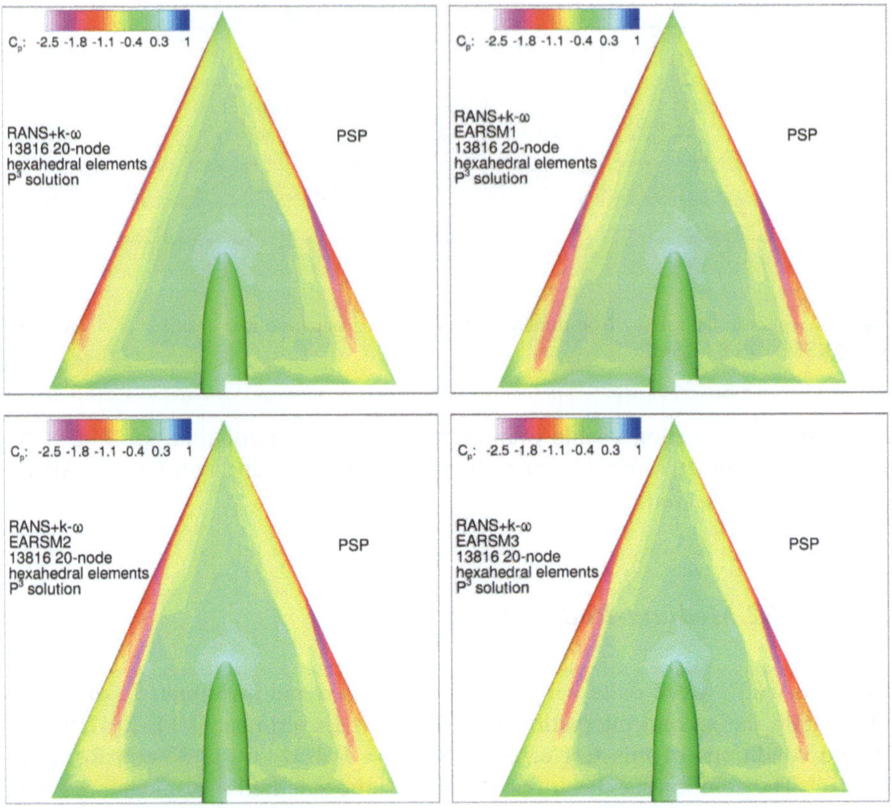

Fig. 5 VFE2: Pressure coefficient distribution, C_p. *Left side:* $\mathbb{P}^3$ results – *Right side:* Pressure sensitive paint, experimental data in [19, 20, 29]

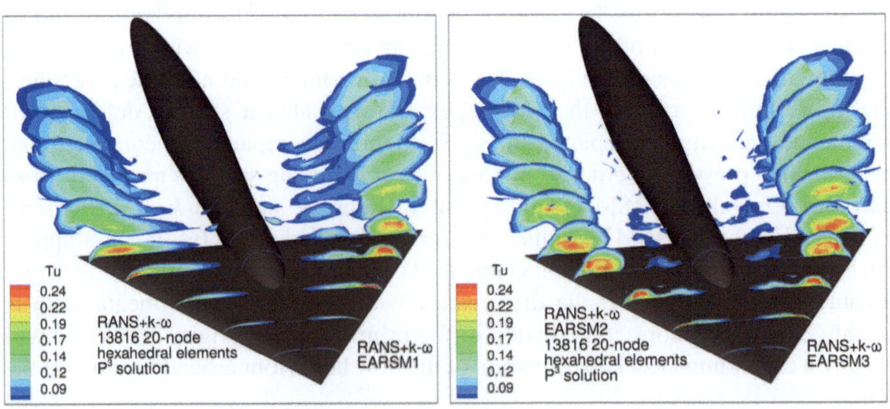

Fig. 6 VFE2: Turbulence intensity contours, $\mathbb{P}^3$ solutions

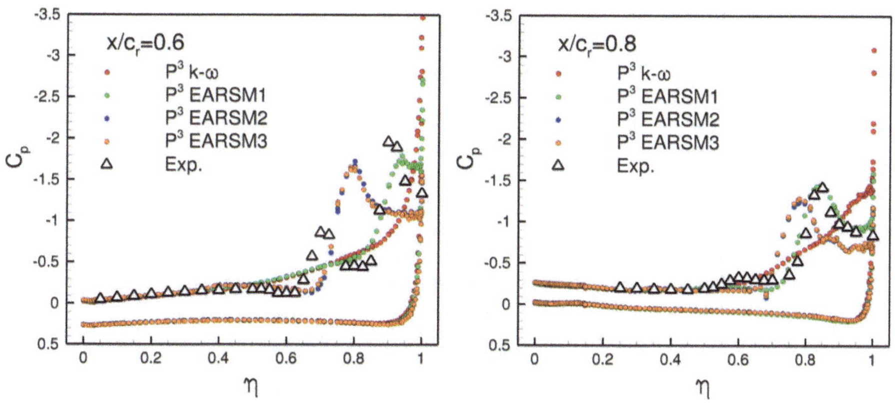

Fig. 7 VFE2: Pressure coefficient distribution at sections $x/c_r = 0.6$ and $x/c_r = 0.8$, experimental data in [19, 20, 29]

5 Conclusion

Fully implicit solution of RANS equations coupled with k-ω and EARSM model for external aerodynamics flow configurations has been reported in this work. The near-wall behavior provided by EARSM has been assessed by comparing to experimental data of a well documented flow over a flat plate. Very high-order solutions (up to seventh order) have been compared with the standard k-ω model showing that EARSM is able to provide more rapidly converging skin-friction coefficient distributions. First results for three-dimensional turbulent flows with EARSM1→3 have been presented for the VFE2 delta wing. The comparison of the computed high-order solutions with the experimental data demonstrates the effectiveness of EARSM in sharply modeling the vortices developing along the wing. Ongoing work deals with the implementation of the quartic terms of EARSM and the assessment of steady-state convergence properties of the solver on increasingly finer grids and using higher degree polynomial approximations. Moreover, a deep assessment of the predicting capabilities of EARSM compared to a standard linear eddy viscosity model will be performed on literature test cases. These topics will be addressed in a forthcoming paper.

Acknowledgements This work has been carried out within the EU FP7 IDIHOM project.

References

1. Project IDIHOM, Industrialisation of high-order methods a top-down approach, 2012.
2. D. N. Arnold, F. Brezzi, B. Cockburn, and D. Marini. Unified analysis of discontinuous Galerkin methods for elliptic problems. *SIAM J. Numer. Anal.*, 39(5):1749–1779, 2002.

3. S. Balay, K. Buschelman, W. D. Gropp, D. Kaushik, M. G. Knepley, L. C. McInnes, B. F. Smith, and H. Zhang. PETSc Web page, 2001. http://www.mcs.anl.gov/petsc.
4. F. Bassi, L. Botti, A. Colombo, A. Crivellini, N. Franchina, A. Ghidoni, and S. Rebay. Very high-order accurate discontinuous Galerkin computation of transonic turbulent flows on aeronautical configurations. In Norbert Kroll, Heribert Bieler, Herman Deconinck, Vincent Couaillier, Harmen van der Ven, and Kaare Sørensen, editors, *ADIGMA - A European Initiative on the Development of Adaptive Higher-Order Variational Methods for Aerospace Applications*, volume 113 of *Notes on Numerical Fluid Mechanics and Multidisciplinary Design*, pages 25–38. Springer Berlin / Heidelberg, 2010.
5. F. Bassi, L. Botti, A. Colombo, D.A. Di Pietro, and P. Tesini. On the flexibility of agglomeration based physical space discontinuous Galerkin discretizations. *Journal of Computational Physics*, 231(1):45–65, 2012.
6. F. Bassi, L. Botti, N. Colombo, A. Ghidoni, and S. Rebay. Discontinuous Galerkin for turbulent flows. In Z. J. Wang, editor, *Adaptive high-order methods in computational fluid dynamics*, volume 2 of *Advances in Computational Fluid Dynamics*. World Scientific, 2011.
7. F. Bassi, A. Crivellini, D. A. Di Pietro, and S. Rebay. An artificial compressibility flux for the discontinuous Galerkin solution of the incompressible Navier-Stokes equations. *J. Comput. Phys.*, 218:794–815, 2006.
8. F. Bassi, A. Crivellini, S. Rebay, and M. Savini. Discontinuous Galerkin solution of the Reynolds-averaged Navier-Stokes and k-ω turbulence model equations. *Comput. & Fluids*, 34:507–540, 2005.
9. F. Bassi and S. Rebay. A high order discontinuous Galerkin method for compressible turbulent flows. In *Discontinuous Galerkin Methods. Theory, Computation and Applications*, volume 11 of *Lecture Notes in Computational Science and Engeneering*, pages 77–88. Springer-Verlag, 2000. *First Internation Symposium on Discontinuous Galerkin Methods,* May 24–26, 1999, Newport, RI, USA.
10. F. Bassi, S. Rebay, G. Mariotti, S. Pedinotti, and M. Savini. A high-order accurate discontinuous finite element method for inviscid and viscous turbomachinery flows. In R. Decuypere and G. Dibelius, editors, *Proceedings of the 2nd European Conference on Turbomachinery Fluid Dynamics and Thermodynamics*, pages 99–108, Antwerpen, Belgium, March 5–7 1997. Technologisch Instituut.
11. L. Botti. Influence of reference-to-physical frame mappings on approximation properties of discontinuous piecewise polynomial spaces. *Journal of Scientific Computing*, 52(3):675–703, 2012.
12. F. Brezzi, G. Manzini, D. Marini, P. Pietra, and A. Russo. Discontinuous Galerkin approximations for elliptic problems. *Numer. Methods Partial Differential Equations*, 16:365–378, 2000.
13. S. Crippa. *Advances in vortical flow prediction methods for design of delta-winged aircraft*. PhD thesis, KTH, Aeronautical and Vehicle Engineering, 2008. QC 20100713.
14. D. A. Di Pietro and A. Ern. *Mathematical Aspects of Discontinuous Galerkin Methods*. Maths & Applications. Springer-Verlag, 2012.
15. V. Dolejší. Semi-implicit interior penalty discontinuous Galerkin methods for viscous compressible flows. *Commun. Comput. Phys.*, 4:231–274, 2008.
16. G. J. Gassner, F. Lörcher, and C.-D. Munz. A discontinuous Galerkin scheme based on a space-time expansion II. Viscous flow equations in multi dimensions. *J. Sci. Comput.*, 34:260–286, 2008.
17. T. B. Gatski and C. G. Speziale. On explicit algebraic stress models for complex turbulent flows. *Journal of Fluid Mechanics*, 254:59–78, 1993.
18. T. Jongen and T. B. Gatski. General explicit algebraic stress relations and best approximation for three-dimensional flows. *International Journal of Engineering Science*, 36(78):739–763, 1998.
19. R. Konrath, C. Klein, and A. Schrøder. PSP and PIV investigations on the VFE-2 configuration in sub- and transonic flow. *AIAA Paper*, (2008–379), 2008.

20. R. Konrath, C. Klein, and A. Schrøder. PSP and PIV investigations on the vfe-2 configuration in sub- and transonic flow. *Aerospace Science and Technology*, 24(1):22–31, 2013.
21. F. R. Menter, A. V. Garbaruk, and Y. Egorov. Explicit Algebraic Reynolds Stress Models for Anisotropic Wall-Bounded Flows. Versailles, July 6–9th 2009. EUCASS–3rd European Conference for Aero-Space Sciences.
22. G. Mompean, S. Gavrilakis, L. Machiels, and M. O. Deville. On predicting the turbulence-induced secondary flows using nonlinear k-epsilon models. *Physics of Fluids*, 8(7):1856–1868, 1996.
23. H. Naji, G. Mompeanr, and O. El Yahyaoui. Evaluation of explicit algebraic stress models using direct numerical simulations. *Journal of Turbulence*, page N38, 2004.
24. S. B. Pope. A more general effective-viscosity hypothesis. *Journal of Fluid Mechanics*, 72:331–340, 1975.
25. W. Rodi. A new algebraic relation for calculating the Reynolds stresses. *Z. Angew. Math. Mech*, 56:219–221, 1976.
26. J. W. Slater. NPARC alliance CFD verification and validation Web site, 2003. http://www.grc.nasa.gov/WWW/wind/valid/archive.
27. J. W. Slater, J. C. Dudek, and K. E. Tatum. The NPARC verification and validation archive. ASME Paper 2000-FED-11233, ASME, 2000.
28. P. Tesini. *An h-Multigrid Approach for High-Order Discontinuous Galerkin Methods*. PhD thesis, Università degli Studi di Bergamo, January 2008.
29. The second international Vortex Flow Experiment (VFE-2). http://vfe2.dlr.de.
30. S. Wallin and A. V. Johansson. An explicit algebraic Reynolds stress model for incompressible and compressible turbulent flows. *J. Fluid Mech.*, 403:89–132, 2000.
31. K. Wieghardt and W. Tillman. On the turbulent friction layer for rising pressure. Technical Memorandum 1314, NACA, 1951.
32. D. C. Wilcox. *Turbulence Modelling for CFD*. DCW industries Inc., La Cañada, CA 91011, USA, 1993.
33. D. A. Yoder and N. J. Georgiadis. Implementation and validation of the Chien k-ϵ turbulence model in the Wind Navier-Stokes code. AIAA Paper 99-0745, AIAA, 1999.

Investigation of Near-Wall Grid Spacing Effect in High-Order Discontinuous Galerkin RANS Computations of Turbomachinery Flows

F. Bassi, L. Botti, A. Colombo, A. Ghidoni, and S. Rebay

Abstract In the last decade, Discontinuous Galerkin (DG) methods have been the subject of extensive research effort because of their excellent performance in the high-order accurate discretization of advection-diffusion problems on general unstructured grids, and are nowadays finding use in several different applications. In this paper, the potential offered by a high-order accurate DG space discretization method with implicit time integration for the solution of the Reynolds-averaged Navier-Stokes equations coupled with the k-ω turbulence model is investigated in the numerical simulation of the turbulent flow through the well known T106A turbine cascade. The numerical results demonstrate that, by exploiting high order accurate DG schemes, it is possible to compute accurate simulations of this flow on grids with few elements.

1 Introduction

In the last two decades, CFD has been widely accepted as one of the main methods for evaluating the performance of new turbomachinery designs. Industrial CFD applications range from classical single- and multi-blade row simulations in steady and unsteady mode, to cavity flows, heat transfer and combustion chamber simulations. The accurate prediction of these flows requires a complete set of

F. Bassi · L. Botti · A. Colombo
University of Bergamo, Department of Industrial Engineering,
Viale Marconi 5, 24044 Dalmine, Bergamo, Italy
e-mail: francesco.bassi@unibg.it; lorenzo.botti@unibg.it; alessandro.colombo@unibg.it

A. Ghidoni (✉) · S. Rebay
University of Brescia, Department of Industrial and Mechanical Engineering, via Branze 38, 25123 Brescia, Italy
e-mail: antonio.ghidoni@ing.unibs.it; stefano.rebay@ing.unibs.it

M. Azaïez et al. (eds.), *Spectral and High Order Methods for Partial Differential Equations - ICOSAHOM 2012*, Lecture Notes in Computational Science and Engineering 95, DOI 10.1007/978-3-319-01601-6_9,

physical models and a high degree of numerical resolution due to the small scale of some of the flow features involved.

However, the numerical technology used in standard industrial codes is still mainly based on formally second-order accurate finite volume or finite element schemes, which can be often inadequate for these applications. This observation motivates the recent interest in higher-order accurate methods, which can cope with the complex flows encountered in turbomachinery analysis and design.

The discontinuous Galerkin (DG) method is one of the most promising techniques in this respect because of its robustness, accuracy and flexibility (see e.g. [1]).

At the moment standard industrial codes, because of their greater computational efficiency, can not be compared with DG methods, which show, however, a substantial room for improvement. For this reason the research effort has been recently devoted to devise more efficient computational strategies, both for the construction of DG space discretization operators and for the integration in time of the space discretized DG equations (see e.g. [2, 3, 5]).

The objective of this work is to investigate the effectiveness and limitations of a very high-order accurate DG space discretization in the numerical simulation of the compressible turbulent flow through the T106A turbine cascade. In particular, the aim is to show the feasibility of accurate simulations of such complex flows on grids with few elements (coarse meshes) by resorting to a sufficiently high-order accurate DG space discretization. The effect of near wall spacing grid on computation will be investigated by analyzing the behavior of the following quantities: the pressure coefficient and the skin friction on the blade, the velocity and the turbulence kinetic energy profiles in the boundary layer.

2 DG Space Discretization

RANS and turbulence model equations can be written in compact form as

$$\frac{\partial \mathbf{u}}{\partial t} + \nabla \cdot \mathbf{F}_c(\mathbf{u}) + \nabla \cdot \mathbf{F}_v(\mathbf{u},\nabla \mathbf{u}) + \mathbf{s}(\mathbf{u},\nabla \mathbf{u}) = \mathbf{0}, \tag{1}$$

where $\mathbf{u}$ and $\mathbf{s}$ are the vectors of the m conservative variables and source terms, and $\mathbf{F}_c, \mathbf{F}_v \in \mathbb{R}^m \otimes \mathbb{R}^d$ are defined as the arrays of the inviscid and viscous flux vectors. A weak formulation of the RANS equations is obtained multiplying each scalar conservation law in Eq. (1) by an arbitrary smooth test function $v_j \in \mathbf{v}$, $1 \le j \le m$, and integrating by parts, that is

$$\int_{\Omega} v_j \frac{\partial u_j}{\partial t}\,\mathrm{d}\mathbf{x} - \int_{\Omega} \nabla v_j \cdot \mathbf{F}_j(\mathbf{u}, \nabla \mathbf{u})\,\mathrm{d}\mathbf{x} + \int_{\partial\Omega} v_j \mathbf{F}_j(\mathbf{u}, \nabla \mathbf{u}) \cdot \mathbf{n}\,\mathrm{d}\sigma + \int_{\Omega} v_j \mathbf{s}_j(\mathbf{u}, \nabla \mathbf{u})\,\mathrm{d}\mathbf{x} = 0, \tag{2}$$

where $\mathbf{F}_j$ is the sum of the inviscid and viscous flux vectors, Ω the computational domain, $\partial\Omega$ the boundary of Ω, $\mathbf{n}$ the unit normal vector to the boundary in the outward direction.

Let Ω_h be an approximation of the domain Ω, and $\mathscr{T}_h = \{K\}$ a mesh of Ω_h, i.e. a collection of N "finite elements" K, and let $\mathscr{V}_h$ denote the space of piecewise polynomial functions on the element K, i.e.

$$\mathscr{V}_h = \{\mathbf{v}_h \in L^2(\Omega_h)^{d+2} : \mathbf{v}_h|_K \in \mathbb{P}^k, \forall K \in \mathscr{T}_h\},$$

where $\mathbb{P}^k(\mathrm{K})$ denotes the space of polynomials of degree at most k on the element K. The functions in $\mathscr{V}_h$ are in general discontinuous at element interfaces and the polynomial order k may in general be different from element to element. The solution $\mathbf{u}$ and the test function $\mathbf{v}$ are replaced with a finite element approximation $\mathbf{u}_h$ and a discrete test function $\mathbf{v}_h$ respectively, belonging to the space $\mathscr{V}_h$.

The discontinuous approximation of the numerical solution necessitates a specific treatment of the inviscid and viscous interface fluxes. In order to ensure conservation and correctly account for wave propagation the former is based on the Godunov flux computed with an exact Riemann solver, while for the latter the BR2 scheme is employed, proposed in [6] and theoretically analyzed in [7, 8].

Accounting for these aspects, the DG formulation of the compressible RANS and $\boldsymbol{k}$-$\boldsymbol{\omega}$ equations consists in seeking $\mathbf{u}_h \in \mathscr{V}_h$ such that

$$\begin{aligned}\sum_{K\in\mathscr{T}_h}\int_K v_{h,j}\frac{\partial u_{h,j}}{\partial t}\,\mathrm{d}\mathbf{x} - \sum_{K\in\mathscr{T}_h}\int_K \nabla_h v_{h,j}\cdot\mathbf{F}_j\left(\mathbf{u}_h,\nabla_h\mathbf{u}_h+\mathbf{R}\left([\![\mathbf{u}_h]\!]\right)\right)\,\mathrm{d}\mathbf{x}\\ +\sum_{F\in\mathscr{F}_h}\int_F [\![v_{h,j}]\!]\cdot\hat{\mathbf{F}}_j\left(\mathbf{u}_h^{\pm},\left(\nabla_h\mathbf{u}_h+\eta_F\mathbf{r}_F\left([\![\mathbf{u}_h]\!]\right)\right)^{\pm}\right)\,\mathrm{d}\sigma\\ +\sum_{K\in\mathscr{T}_h}\int_K v_{h,j}\mathbf{s}_j\left(\mathbf{u}_h,\nabla_h\mathbf{u}_h+\mathbf{R}\left([\![\mathbf{u}_h]\!]\right)\right)\,\mathrm{d}\mathbf{x} = 0 \qquad (3)\\ \forall\mathbf{v}_h\in\mathscr{V}_h,\end{aligned}$$

where K are the elements of the mesh, F the faces of the mesh, $[\![\ \ ]\!]$ the jump operator, $\mathbf{R}$ and $\mathbf{r}_F$ the global and local lifting operators, and η_F is a penalty parameter prescribed accordingly to [8].

The adopted turbulence model is the $\boldsymbol{k}$-$\boldsymbol{\omega}$ model proposed by Wilcox [9], which is here implemented in a non standard form since the variable $\tilde{\omega} = \log\omega$ instead of ω itself is used.

The variable $\tilde{\omega}$ in the source term and in the eddy viscosity equation is replaced by $\tilde{\omega}_r$, meaning that it must fulfill suitably defined "realizability" conditions, which set a lower bound on it. This limitation substantially improves the stability and robustness of the turbulent flow computations because there is numerical evidence that too small, though positive, values of $\omega = e^{\tilde{\omega}}$ can lead to sudden breakdown of the computations. The wall boundary condition for ω is prescribed following the approach proposed in [10].

The discrete problem corresponding to Eq. (3) can be written as

$$\mathbf{M}\frac{\mathrm{d}\mathbf{U}}{\mathrm{d}t} + \mathbf{R}(\mathbf{U}) = 0, \tag{4}$$

where **U** is the global vector of unknown degrees of freedom, **M** is the global block diagonal mass matrix, and **R** the residuals vector. The linear system is solved using the restarted GMRES algorithm preconditioned with the block Jacobi method (one block per process) or with additive Schwarz (two levels of overlapping) as available in the PETSc library [11].

3 Results

The purpose of this section is to demonstrate the performance of DG methods in the turbulent flow computation through a well known turbomachinery cascade. The investigation focuses on the effectiveness and limitations of the $k - \omega$ turbulence model (high-Reynolds version) in the prediction of the flow field, showing the feasibility of accurate simulations on very coarse grids by resorting to a sufficiently high-order accurate DG space discretization. In particular the effect of near wall spacing grid on the computations will be investigated by analyzing the behavior of the velocity and the turbulence kinetic energy inside the boundary layer, and of the skin friction, C_f, and the pressure coefficient, C_p, on the blade.

The test case chosen for the simulations is the T106A turbine cascade, which is a low-pressure turbine cascade designed by MTU Aero Engines. It has been extensively investigated in experimental and computational studies [4, 12, 13].

The computations are performed for a downstream isentropic Mach number $M_{2,is} = 0.59$, a Reynolds number based on the downstream isentropic conditions and on the blade chord $Re_{2,is} = 1.1 \times 10^6$, an inlet turbulence intensity $Tu_1 = 4.0\,\%$, and an inlet angles $\alpha_1 = 37.7°$.

The geometry has been represented with quadratic elements and three different family of grids, characterized by a different height of the elements adjacent to the wall, have been used, as summarized in Table 1. All the solutions have been computed through a sequence of $\mathbb{P}^{0\to6}$ approximations for the coarse grids, $\mathbb{P}^{0\to4}$ for the medium grids, and $\mathbb{P}^{0\to3}$ for the fine grid, starting from an uniform flow field at inlet conditions. The fine mesh has been used just to compute the reference solution, when experimental data are not available.

Figure 1 shows the coarse meshes for different values of the $y^+ \in \{2, 20, 40, 60\}$, while in Fig. 2 the Mach number and turbulence intensity contours computed on the coarse mesh ($y^+ = 2$) for a $\mathbb{P}^6$ solution approximation are depicted.

Figures 3 and 4 show a comparison of pressure coefficient distribution along the blade between experimental data and computed solution on coarse ($\mathbb{P}^6$ solution

Table 1 Main features of the meshes adopted for the simulations

Mesh	y^+	Elements
	2	1,271 quadrilateral elements
	20	965 quadrilateral elements
Coarse	40	863 quadrilateral elements
	60	812 quadrilateral elements
	100	761 quadrilateral elements
	2	4,626 quadrilateral elements
Medium	20	3,326 quadrilateral elements
	40	2,826 quadrilateral elements
Fine	2	9,882 quadrilateral elements

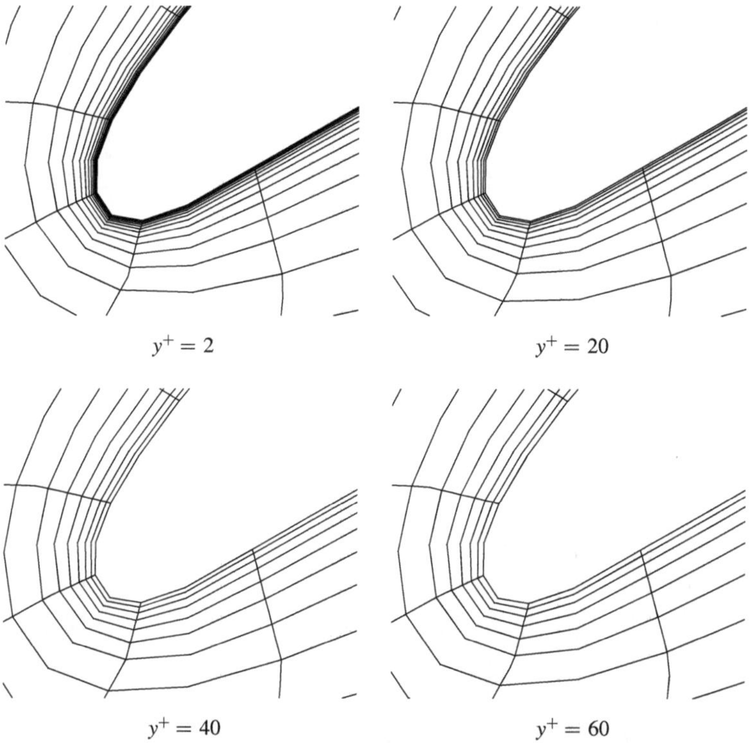

Fig. 1 Coarse meshes consisting of 1,271 (*top-left*), 965 (*top-right*), 863 (*bottom-left*) and 812 (*bottom-right*) quadrilateral elements

approximation) and on medium ($\mathbb{P}^4$ solution approximation) meshes. It can be seen that even for larger y^+ values, the chosen solution polynomial approximation allows to capture reasonably well the C_p distribution on the coarse and medium meshes. Only on the coarse meshes for $y^+ = 60$ and $y^+ = 100$ the curves are oscillating, probably denoting a rough resolution of the flow field.

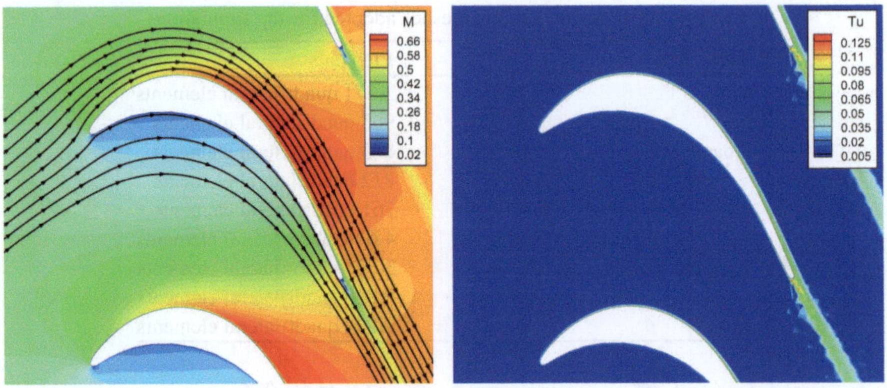

Fig. 2 Mach number and turbulence intensity contours computed on the coarse mesh, $y^+ = 2$, $\mathbb{P}^6$ polynomial approximation

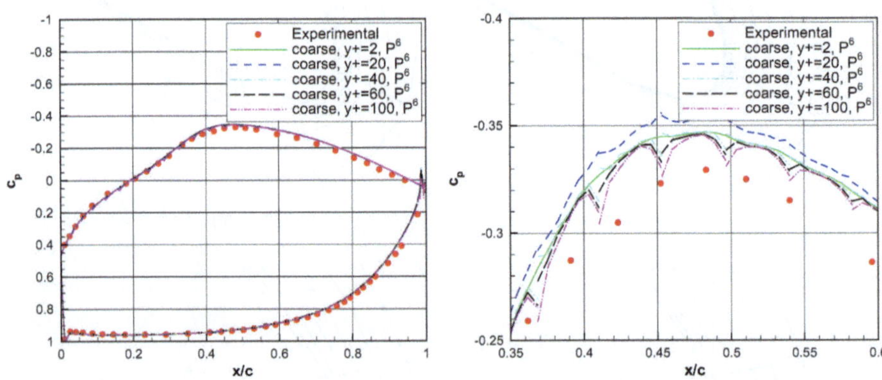

Fig. 3 Cp distribution along the blade on the coarse meshes, $\mathbb{P}^6$ solution approximation

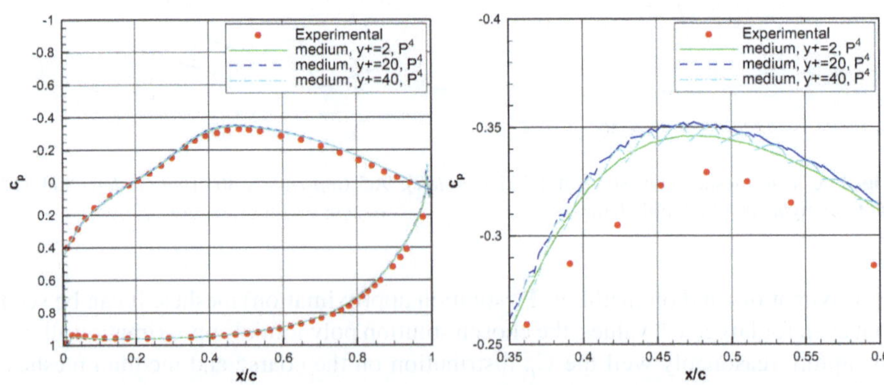

Fig. 4 Cp distribution along the blade on the medium meshes, $\mathbb{P}^4$ solution approximation

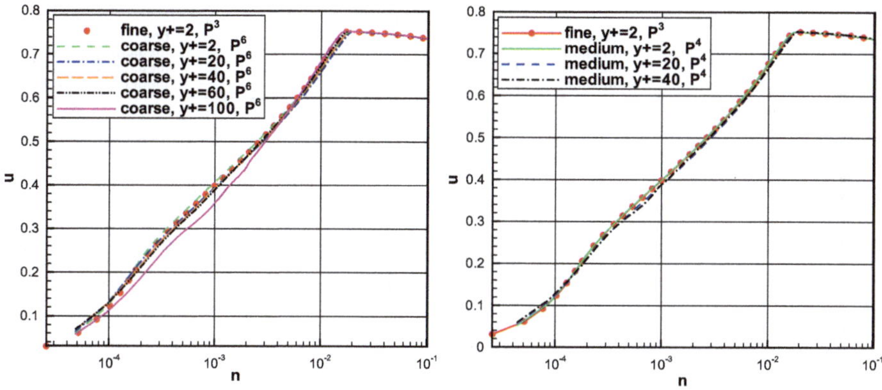

Fig. 5 Velocity profile in a boundary layer traverse ($x/c = 0.9$) on the coarse (*left*) and medium (*right*) meshes for a $\mathbb{P}^6$ and $\mathbb{P}^4$ solution approximation, respectively

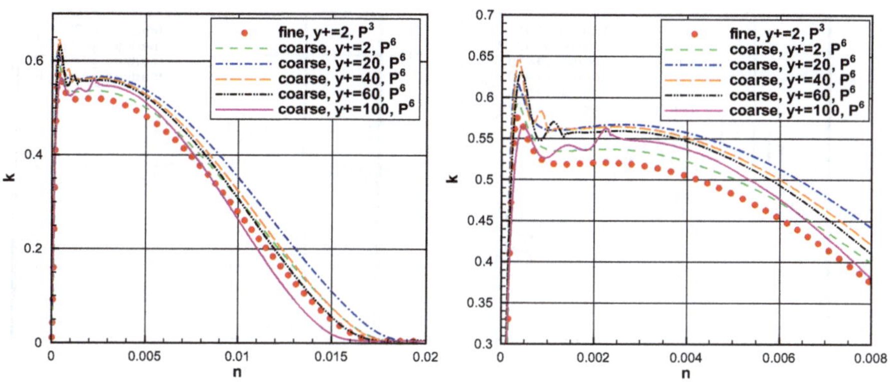

Fig. 6 Turbulence kinetic energy profile in a boundary layer traverse ($x/c = 0.9$) on the coarse meshes, $\mathbb{P}^6$ solution approximation

In Fig. 5 the velocity profile in a boundary layer traverse ($x/c = 0.9$, where c is the axial chord) on the suction side is shown, on the coarse meshes for a $\mathbb{P}^6$ solution approximation and on the medium meshes for a $\mathbb{P}^4$ solution approximation. It can be observed that only for the coarse mesh with $y^+ = 100$ the velocity profile differs significantly from the reference solution computed on the fine mesh, while for lower y^+ the curves are comparable.

Figures 6 and 7 show instead the turbulence kinetic energy profile in a boundary layer traverse ($x/c = 0.9$) on the coarse and medium meshes, for a $\mathbb{P}^6$ and $\mathbb{P}^4$ solution approximation, respectively. The solution computed with $y^+ = 2$ is superimposed with the reference solution, while increasing the y^+ the predicted peak of the turbulence kinetic energy profile becomes higher and some oscillations appear in the first cell, which are more evident increasing the y^+.

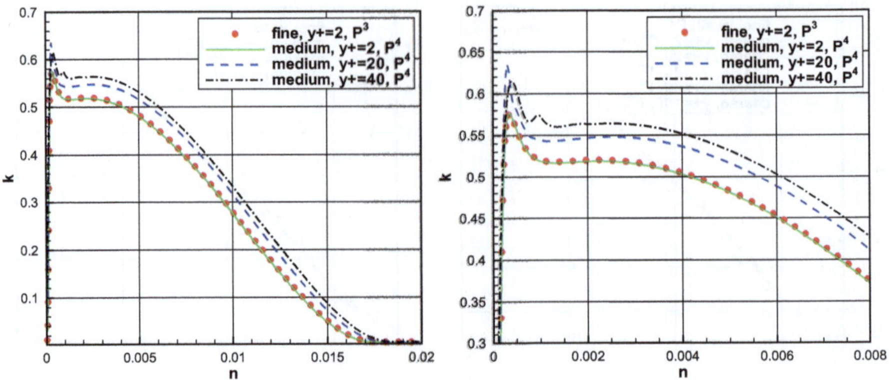

Fig. 7 Turbulence kinetic energy profile in a boundary layer traverse ($x/c = 0.9$) on the medium meshes, $\mathbb{P}^4$ solution approximation

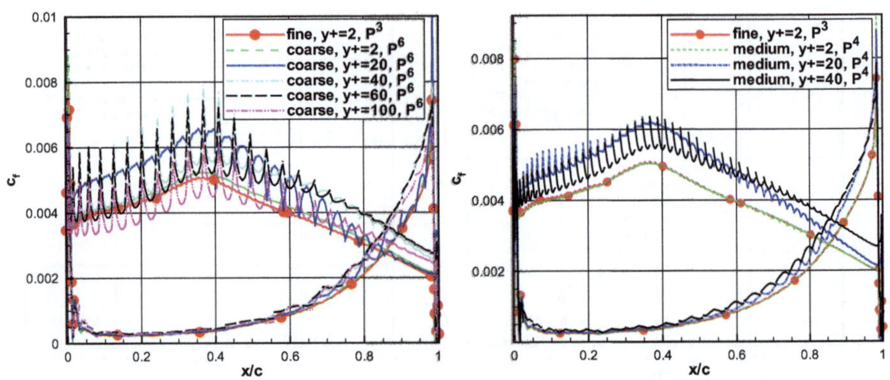

Fig. 8 Skin friction distribution along the blade on the coarse (*left*) and medium (*right*) meshes for a $\mathbb{P}^6$ and $\mathbb{P}^4$ solution approximation, respectively

Finally in Fig. 8 the skin friction distribution along the blade on the coarse and medium meshes for a $\mathbb{P}^6$ and $\mathbb{P}^4$ solution approximation, respectively, is represented. It can be noticed that the pressure side distribution is in good agreement with the reference solution for every y^+ on both meshes. On the suction side an oscillating behavior of the C_f can be observed for y^+ values different from 2, which is more evident for the coarse meshes.

4 Conclusions

An application of a DG method for the simulation of the subsonic turbulent flow through T106A low pressure turbine cascade has been presented. In particular the effect of the near wall grid spacing on the computation of the compressible turbulent flow by means of the $\boldsymbol{k}$-$\boldsymbol{\omega}$ turbulence model (high-Reynolds version) has been investigated.

The effect of high-order approximations on the solution accuracy has been thoroughly assessed by comparing a series of numerical results of increasing order of accuracy and obtained on different meshes characterized by different y^+, with the available experimental data and reference solutions obtained on a fine mesh.

The results show that, on the very coarse grids considered, the flow field can be accurately predicted by resorting to seventh and fifth order accurate approximations, while on these grids standard second order accuracy would clearly be unsuited to accurately simulate this flow.

However, in a future work further analysis is planned to investigate the accuracy of the method by comparison with more significant quantities for turbomachinery design, such as the efficiency and loss coefficients, to show if the oscillations present in the skin friction curves can affect negatively the prediction of these parameters.

Acknowledgements This work has been supported by the European Union, under the IDIHOM project on the development of innovative solution algorithms for aerodynamic simulations.

References

1. Bassi, F., Botti, L., Colombo, N., Ghidoni, A., and Rebay, S.: Discontinuous Galerkin for turbulent flows, In: Wang, Z.J. (eds.) Adaptive high-order methods in computational fluid dynamics, Vol. 2 of Advances in Computational Fluid Dynamics, World Scientific (2011)
2. Bassi, F., Ghidoni, A., Rebay, S., and Tesini, P.: High-order accurate p-multigrid discontinuous Galerkin solution of the Euler equations. Int. J. Numer. Meth. Fluids **8**, 847–865 (2009).
3. Bassi, F., Franchina, N., Ghidoni, A., and Rebay, S.: Spectral p-multigrid discontinuous Galerkin solution of the Navier-Stokes equations. Int. J. Numer. Meth. Fluids **11**, 1540–1558 (2011)
4. Bassi, F., Colombo, A., Ghidoni, A., and Rebay, S.: Simulation of the transitional flow in a low pressure gas turbine cascade with a high-order discontinuous Galerkin method. ASME Journal of Fluids Engineering **135**(7), (2013) doi:10.1115/1.4024107
5. Bassi, F., Franchina, N., Ghidoni, A., and Rebay, S.: A numerical investigation of a spectral-type nodal collocation discontinuous Galerkin approximation of the Euler and Navier-Stokes equations. Int. J. Numer. Meth. Fluids **71**, 1322–1339 (2012) doi: 10.1002/fld.3713
6. Bassi, F., Rebay, S., Mariotti, G., Pedinotti, S., and Savini, M.: A High-Order Accurate Discontinuous Finite Element Method for Inviscid and Viscous Turbomachinery Flows, In: Decuypere, R. and Dibelius, G. (eds.) 2nd European Conference on Turbomachinery Fluid Dynamics and Thermodynamics, 99–108 (1997)
7. Brezzi, F., Manzini, G., Marini, D., Pietra, P. and Russo, A.: Discontinuous Galerkin approximations for elliptic problems. Numer. Meth. for Part. Diff. Eq. **16**, 365–378 (2000)
8. Arnold, D.N., Brezzi, F., Cockburn, B., and Marini, D.: Unified analysis of discontinuous Galerkin methods for elliptic problems. SIAM J. Numer. Anal. **5**, 1749–1779 (2002)
9. Wilcox, D.C.: Turbulence Modelling for CFD, DCW industries Inc., (1993)
10. Bassi, F., Botti, L., Colombo, A., Crivellini, A., Franchina, N., Ghidoni, A., and Rebay, S.: Very High-Order Accurate Discontinuous Galerkin Computation of Transonic Turbulent Flows on Aeronautical Configurations, In: Kroll, N., Bieler, H., Deconinck, H., Couaillier, V., Van der Ven, H., and Sørensen, K. (eds.) ADIGMA - A European Initiative on the Development of

Adaptive Higher-Order Variational Methods for Aerospace Applications, Vol. 110 of Notes on Numerical Fluid Mechanics and Multidisciplinary Design, Springer Berlin Heidelberg, 25–38 (2010)
11. Balay, S., Buschelman, K., Eijkhout, V., Gropp, W.D., Kaushik, D. and Knepley, M.G., McInnes, L., Smith, B.S., and Zhang, H.: PETSc Users Manual, ANL-95/11 - Revision 3.0.0, Argonne National Laboratory, (2008)
12. Hoheisel, H.: Entwicklung neuer Entwurfskonzepte für zwei Turbinengitter, Teil III, Ergebnisse T106, Institut für Entwurfsaerodynamik, Braunschweig, (1981)
13. Marciniak, V. and Kugeler, E. and Franke, M.: Predicting transition on low-pressure turbine profiles, In: J. C. F. Pereira and A. Sequeira and J. M. C. Pereira and J. Janela and L. Borges (eds.) Proceedings of the V European Conference on Computational Fluid Dynamics ECCOMAS CFD 2010, (2010)

A Fourth-Order Compact Finite Volume Scheme for the Convection-Diffusion Equation

Christine Baur and Michael Schäfer

Abstract A fourth-order compact scheme for the convection-diffusion equation is presented. To adopt this approach to non-Cartesian grids a coordinate transformation is applied. The convection-diffusion equation is solved with a three-dimensional finite volume solver using boundary fitted, block-structured grids. The grid arrangement is collocated. The verification of the fourth-order method is done for analytical test cases. To show the influence of the boundary conditions some calculations with various conditions are performed. Furthermore, the grid dependance of solutions is studied. It is shown that the proposed approach constitutes an efficient high-order solution method for the convection-diffusion equation.

1 Introduction

Due to the requirement of highly accurate schemes for approximating PDE models for complex physical phenomena the compact method was introduced some decades ago in conjunction with classical finite difference methods [1, 7, 12]. In contrary to spectral methods the compact scheme is not limited to simple geometries and has a better fine-scale resolution than conventional finite difference approximations. Further, the higher accuracy is achieved without enlarging the computational discretization stencil size in comparison to explicit schemes.

In recent years the compact scheme became popular in the field of numerical simulation of turbulent flows and aeroacoustics [2, 8, 11, 14] because of the requirement of high accuracy.

C. Baur (✉) · M. Schäfer
Institute of Numerical Methods in Mechanical Engineering, Technische Universität Darmstadt, Dolivostr. 15, 64293 Darmstadt, Germany
e-mail: baur@fnb.tu-darmstadt.de; schaefer@fnb.tu-darmstadt.de

M. Azaïez et al. (eds.), *Spectral and High Order Methods for Partial Differential Equations - ICOSAHOM 2012*, Lecture Notes in Computational Science and Engineering 95, DOI 10.1007/978-3-319-01601-6_10,

In the context of the finite volume method some approaches were developed with application in the field of fluid dynamics (see e.g. [5, 15]). Kobayashi [9] proposed an approach by calculating the surface integrals directly by the cell averaged values. Thus, the surface and volume integrals are approximated by a simple midpoint rule. This is in contrast to explicit high-order finite volume schemes, where the fluxes are calculated at points and higher-order approximations of the surface and volume integrals have to be applied [13].

Some authors already dealt with the application of a compact scheme to non-Cartesian grids in the context of the finite volume scheme. While some authors [4, 10] calculate the fluxes directly in the physical space, others present a coordinate transformation [6, 16, 17].

In this paper the compact scheme proposed by Kobayashi [9] is studied on Cartesian grids and the formulation is extended to non-Cartesian grids. The set of linear equations resulting from the finite volume discretization of the convection-diffusion equation is solved with an ILU method. The discretization of the convective and diffusive fluxes leads to an additional tridiagonal system that is solved with the Thomas algorithm. To avoid oscillations in the solution the deferred correction approach is employed and the higher-order terms are treated explicitly.

2 Governing Equation

The steady three-dimensional convection-diffusion equation reads

$$\frac{\partial}{\partial x}\left(\varrho u\Phi - \Gamma_\Phi \frac{\partial \Phi}{\partial x}\right) + \frac{\partial}{\partial y}\left(\varrho v\Phi - \Gamma_\Phi \frac{\partial \Phi}{\partial y}\right) + \frac{\partial}{\partial z}\left(\varrho w\Phi - \Gamma_\Phi \frac{\partial \Phi}{\partial z}\right) = S_\Phi \tag{1}$$

where Φ is the unknown, Γ_Φ the constant diffusion coefficient, ϱ is the density, S_Φ is a source term. It is assumed that the velocity field (u, v, w) describes an incompressible flow. Dirichlet boundary conditions are prescribed on the whole boundary.

Applying the finite volume method, (1) becomes

$$\sum_c \int_{S_c} \left(\varrho u\Phi n_{cx} - \Gamma_\Phi \frac{\partial \Phi}{\partial x} n_{cx} + \varrho v\Phi n_{cy} - \Gamma_\Phi \frac{\partial \Phi}{\partial y} n_{cy} \right. \\ \left. + \varrho w\Phi n_{cz} - \Gamma_\Phi \frac{\partial \Phi}{\partial z} n_{cz}\right) \mathrm{d}S_c = \int_V S_\Phi \mathrm{d}V \tag{2}$$

where S_c are the surfaces of a control volume with $\cup S_c = V$ and n_{cx}, n_{xy}, n_{cz} the corresponding normal vectors in Cartesian coordinates.

For non-Cartesian grids, with the transformation $(x, y, z) \rightarrow (\xi, \eta, \zeta)$, the formulation is more complicated:

$$
\begin{aligned}
\sum_c \int_{S_c} \Bigg(\varrho U \Phi n_{c\xi} - \frac{\Gamma_\Phi}{J} \left(\frac{\partial \Phi}{\partial \xi} B^{11} + \frac{\partial \Phi}{\partial \eta} B^{21} + \frac{\partial \Phi}{\partial \zeta} B^{31} \right) n_{c\xi} \\
+ \varrho V \Phi n_{c\eta} - \frac{\Gamma_\Phi}{J} \left(\frac{\partial \Phi}{\partial \xi} B^{12} + \frac{\partial \Phi}{\partial \eta} B^{22} + \frac{\partial \Phi}{\partial \zeta} B^{32} \right) n_{c\eta} \\
+ \varrho W \Phi n_{c\zeta} - \frac{\Gamma_\Phi}{J} \left(\frac{\partial \Phi}{\partial \xi} B^{13} + \frac{\partial \Phi}{\partial \eta} B^{23} + \frac{\partial \Phi}{\partial \zeta} B^{33} \right) n_{c\zeta} \Bigg) \mathrm{d}S_c = \int_V J S_\Phi \mathrm{d}V
\end{aligned} \tag{3}
$$

$$U = u\beta^{11} + v\beta^{21} + w\beta^{31} \tag{4}$$

$$V = u\beta^{12} + v\beta^{22} + w\beta^{32} \tag{5}$$

$$W = u\beta^{13} + v\beta^{23} + w\beta^{33} \tag{6}$$

$$\mathrm{B}^{ik} = \beta_1^i \beta_1^k + \beta_2^i \beta_2^k + \beta_3^i \beta_3^k \tag{7}$$

where β^{ji} represents the matrix entries of the adjugate of the transformation matrix and J is the determinant of the Jacobian matrix. For a detailed description of the formulation and notation see, e.g., [3].

3 The Fourth-Order Compact Scheme

3.1 Calculation of Fluxes on Cartesian Grid

The basic idea of the compact finite volume scheme [9] is the calculation of cell averaged and face averaged fluxes instead of point values. The cell averaged values are

$$\overline{\Phi}_{i+\frac{1}{2},j+\frac{1}{2},k+\frac{1}{2}} = \frac{1}{\Delta x \Delta y \Delta z} \int_{x_i}^{x_{i+1}} \int_{y_j}^{y_{j+1}} \int_{z_k}^{z_{k+1}} \Phi(x,y,z) \mathrm{d}x\mathrm{d}y\mathrm{d}z \tag{8}$$

and the face averaged fluxes are

$$\overline{\Phi}_{i,j+\frac{1}{2},k+\frac{1}{2}} = \frac{1}{\Delta y \Delta z} \int_{y_j}^{y_{j+1}} \int_{z_k}^{z_{k+1}} \Phi(x_i,y,z) \mathrm{d}y\mathrm{d}z. \tag{9}$$

On Cartesian grids the approximation of convective fluxes is

$$
\begin{aligned}
a_1 \overline{\Phi}_{i-1,j+\frac{1}{2},k+\frac{1}{2}} + \overline{\Phi}_{i,j+\frac{1}{2},k+\frac{1}{2}} + a_2 \overline{\Phi}_{i+1,j+\frac{1}{2},k+\frac{1}{2}} \\
= b_1 \overline{\Phi}_{i+\frac{1}{2},j+\frac{1}{2},k+\frac{1}{2}} + b_2 \overline{\Phi}_{i-\frac{1}{2},j+\frac{1}{2},k+\frac{1}{2}}
\end{aligned} \tag{10}
$$

where

$$a_1 = \frac{(\Delta x_{i+1})^2}{(\Delta x_i + \Delta x_{i+1})^2}, \quad a_2 = \frac{(\Delta x_i)^2}{(\Delta x_i + \Delta x_{i+1})^2},$$

$$b_1 = \frac{2(\Delta x_i)^2 (\Delta x_i + 2\Delta x_{i+1})}{(\Delta x_i + \Delta x_{i+1})^3}, \quad b_2 = \frac{2(\Delta x_{i+1})^2 (2\Delta x_i + \Delta x_{i+1})}{(\Delta x_i + \Delta x_{i+1})^3}$$

and Δx_i the grid spacing. The coefficients are constant on uniform grids. In the other spatial directions the approximation is carried out analogously. The approximation of the diffusive fluxes is given by

$$a_3 \left(\overline{\frac{\partial \Phi}{\partial x}}\right)_{i-1,j+\frac{1}{2},k+\frac{1}{2}} + \left(\overline{\frac{\partial \Phi}{\partial x}}\right)_{i,j+\frac{1}{2},k+\frac{1}{2}} + a_4 \left(\overline{\frac{\partial \Phi}{\partial x}}\right)_{i+1,j+\frac{1}{2},k+\frac{1}{2}}$$
$$= b_3 \overline{\Phi}_{i+\frac{1}{2},j+\frac{1}{2},k+\frac{1}{2}} + b_4 \overline{\Phi}_{i-\frac{1}{2},j+\frac{1}{2},k+\frac{1}{2}} \quad (11)$$

where

$$a_3 = \frac{\Delta x_{i+1}}{\Delta x_i + \Delta x_{i+1}} - \frac{2(\Delta x_{i+1})^2}{(\Delta x_i)^2 + 3\Delta x_i \Delta x_{i+1} + (\Delta x_{i+1})^2},$$

$$a_4 = \frac{\Delta x_i}{\Delta x_i + \Delta x_{i+1}} - \frac{2(\Delta x_i)^2}{(\Delta x_i)^2 + 3\Delta x_i \Delta x_{i+1} + (\Delta x_{i+1})^2},$$

$$b_3 = -b_4 = \frac{12\Delta x_i \Delta x_{i+1}}{((\Delta x_i)^2 + 3\Delta x_i \Delta x_{i+1} + (\Delta x_{i+1})^2)(\Delta x_i + \Delta x_{i+1})}.$$

Again, on uniform grids the coefficients are constant and the formulation in the other spatial directions is analogous. Stencils of first, second and third order are used for the approximation at boundaries for diffusive fluxes. As Dirichlet conditions are prescribed at the boundary of the domain, no special boundary treatment is necessary for the approximation of convective fluxes.

The formulation for the second order approximation is exemplified for one boundary and is

$$\left(\overline{\frac{\partial \Phi}{\partial x}}\right)_{1,j+\frac{1}{2},k+\frac{1}{2}} + \frac{1}{2}\left(\overline{\frac{\partial \Phi}{\partial x}}\right)_{2,j+\frac{1}{2},k+\frac{1}{2}} = -\frac{3}{\Delta x_2}\overline{\Phi}_{1,j+\frac{1}{2},k+\frac{1}{2}} + \frac{3}{\Delta x_2}\overline{\Phi}_{\frac{3}{2},j+\frac{1}{2},k+\frac{1}{2}} \quad (12)$$

with the grid spacing Δx_2. The third order boundary approximation satisfies the condition

$$\left(\overline{\frac{\partial \Phi}{\partial x}}\right)_{1,j+\frac{1}{2},k+\frac{1}{2}} + a_5 \left(\overline{\frac{\partial \Phi}{\partial x}}\right)_{2,j+\frac{1}{2},k+\frac{1}{2}} = b_5 \overline{\Phi}_{1,j+\frac{1}{2},k+\frac{1}{2}} + b_6 \overline{\Phi}_{\frac{3}{2},j+\frac{1}{2},k+\frac{1}{2}}$$
$$+ b_7 \overline{\Phi}_{\frac{5}{2},j+\frac{1}{2},k+\frac{1}{2}}. \quad (13)$$

The coefficients of Eq. (13) are

$$a_5 = -\frac{3\Delta x_3}{\Delta x_2 - 2\Delta x_3} - 1$$
$$b_5 = -\frac{2}{\Delta x_2 + \Delta x_3} - \frac{3}{\Delta x_2} - \frac{1}{\Delta x_2 - 2\Delta x_3}$$
$$b_6 = -\frac{6\left(-2(\Delta x_2)^3 + 2\Delta x_2(\Delta x_3)^2 + (\Delta x_3)^3\right)}{\Delta x_2(\Delta x_2 + \Delta x_3)^2(\Delta x_2 - 2\Delta x_3)}$$
$$b_7 = -\frac{6(\Delta x_2)^2}{(\Delta x_2 + \Delta x_3)^2(\Delta x_2 - 2\Delta x_3)},$$

where Δx_2 and Δx_3 are the grid spacings. For the block boundary condition explicit formulations of second and fourth order are chosen. As the coefficients become very difficult only formulations on Cartesian grids are shown. The approximation of the convective fluxes at a block boundary is

$$\overline{\Phi}_{i,j+\frac{1}{2},k+\frac{1}{2}} = -\frac{1}{12}\overline{\Phi}_{i-\frac{3}{2},j+\frac{1}{2},k+\frac{1}{2}} + \frac{7}{12}\overline{\Phi}_{i-\frac{1}{2},j+\frac{1}{2},k+\frac{1}{2}} + \frac{7}{12}\overline{\Phi}_{i+\frac{1}{2},j+\frac{1}{2},k+\frac{1}{2}} - \frac{1}{12}\overline{\Phi}_{i+\frac{3}{2},j+\frac{1}{2},k+\frac{1}{2}}. \tag{14}$$

For the diffusive fluxes the formulation is given by

$$\left(\frac{\overline{\partial\Phi}}{\partial x}\right)_{i,j+\frac{1}{2},k+\frac{1}{2}} = \frac{1}{x_{i+\frac{1}{2}} - x_{i-\frac{1}{2}}}\left(\frac{1}{12}\overline{\Phi}_{i-\frac{3}{2},j+\frac{1}{2},k+\frac{1}{2}} - \frac{5}{4}\overline{\Phi}_{i-\frac{1}{2},j+\frac{1}{2},k+\frac{1}{2}} + \frac{5}{4}\overline{\Phi}_{i+\frac{1}{2},j+\frac{1}{2},k+\frac{1}{2}} - \frac{1}{12}\overline{\Phi}_{i+\frac{3}{2},j+\frac{1}{2},k+\frac{1}{2}}\right). \tag{15}$$

3.2 *Calculation of Fluxes on Non-Cartesian Grids*

The cell and face averaged fluxes are now interpreted as averaged fluxes of the transformed grid (see Eq. (3)). Because products and sums are averaged, the integral cannot be calculated directly and an additional approximation procedure is applied for nonlinear terms [15, 16]. The discretization of the Jacobian determinant and the cofactor matrix also plays an important role. Only a consistent interpolation yields the desired accuracy. We confine the formulation to the main terms to give an impression of the procedure. The approximation of the cofactor matrix and the Jacobi determinant are not shown here. A detailed description of the calculation of the cell and face averaged physical coordinates can be found in [6, 16, 17].

Analog to Eq. (10) the formulation of the convective fluxes is

$$\frac{1}{4}\overline{\Phi}_{i-1,j+\frac{1}{2},k+\frac{1}{2}} + \overline{\Phi}_{i,j+\frac{1}{2},k+\frac{1}{2}} + \frac{1}{4}\overline{\Phi}_{i+1,j+\frac{1}{2},k+\frac{1}{2}} = \frac{3}{4}\overline{\Phi}_{i+\frac{1}{2},j+\frac{1}{2},k+\frac{1}{2}} + \frac{3}{4}\overline{\Phi}_{i-\frac{1}{2},j+\frac{1}{2},k+\frac{1}{2}}. \tag{16}$$

For diffusive fluxes across a face the formulation is

$$\frac{1}{10}\left(\frac{\partial\overline{\Phi}}{\partial\xi}\right)_{i-1,j+\frac{1}{2},k+\frac{1}{2}} + \left(\frac{\partial\overline{\Phi}}{\partial\xi}\right)_{i,j+\frac{1}{2},k+\frac{1}{2}} + \frac{1}{10}\left(\frac{\partial\overline{\Phi}}{\partial\xi}\right)_{i+1,j+\frac{1}{2},k+\frac{1}{2}} = \frac{6}{5\Delta\xi}\left(\overline{\Phi}_{i+\frac{1}{2},j+\frac{1}{2},k+\frac{1}{2}} - \overline{\Phi}_{i-\frac{1}{2},j+\frac{1}{2},k+\frac{1}{2}}\right). \tag{17}$$

Additionally arise cross-diffusive terms, which have to be discretized in a different way

$$\left(\frac{\partial\overline{\Phi}}{\partial\eta}\right)_{i,j+\frac{1}{2},k+\frac{1}{2}} = \frac{1}{\Delta\eta\Delta\zeta}\int_{\Delta\zeta}\left(\Phi(\xi_i,\eta_{j+1},\zeta) - \Phi(\xi_i,\eta_j,\zeta)\right)\mathrm{d}\zeta \tag{18}$$

where the integrand can be evaluated in the same way as in Eq. (16).

3.3 Solution Procedure

Treating the calculation of the fluxes implicitly would significantly raise the complexity of solving the linear set of equations resulting from the finite volume discretization. Thus, the deferred correction approach is employed. While all higher-order terms are computed explicitly and added to the source term, the central difference scheme (CDS) part is treated implicitly:

$$\sum a_P\Phi_P - \sum_c a_c\Phi_c = S_\Phi - \left[F_{compact} + \sum a_P\Phi_P - \sum_c a_c\Phi_c\right]^{\text{old}}, \tag{19}$$

where P denotes the control volume the governing equation is approximated, c its direct neighbor control volumes and the coefficients a_P and a_c depend the used approximation. This keeps the memory requirement and computational effort small. More aspects of the deferred correction method are discussed in [3].

Terms discretized by the compact scheme lead to additional tridiagonal sets of equations, which are solved with the Thomas algorithm. The entire system of equations is solved with an ILU method (e.g. [20]).

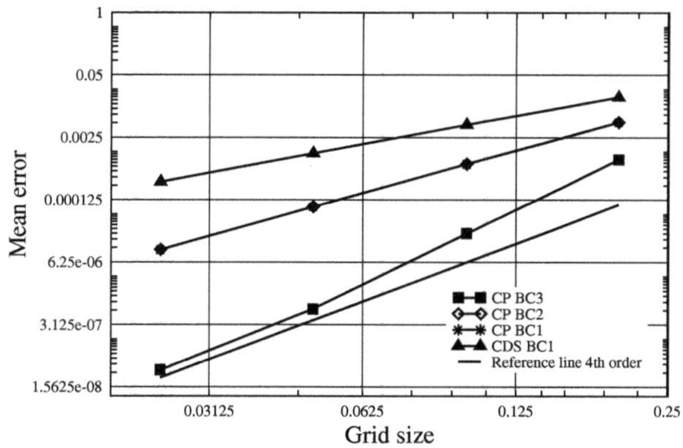

Fig. 1 Influence of boundary conditions of first, second and third order: CP BC3 – compact scheme with third order boundary condition, CP BC2 – compact scheme with second order boundary condition, CP BC1 – compact scheme with first order boundary condition, CDS BC1 – CDS with first order boundary condition

4 Verification

4.1 Grid Dependance Study on Uniform Grid

In recent years the method of manufactured solutions gained more popularity for code verification [18, 19, 21]. It enables the generating of analytical solutions for many relevant equations like incompressible Navier-Stokes equations by adapting the source term. Our test cases are manufactured by this method.

First, we study the dependance of the approximation order of the boundary condition on the global accuracy. We solve the convection-diffusion equation

$$\frac{\partial \Phi}{\partial x} + \frac{\partial \Phi}{\partial y} + \frac{\partial \Phi}{\partial z} - \frac{\partial^2 \Phi}{\partial x^2} - \frac{\partial^2 \Phi}{\partial y^2} - \frac{\partial^2 \Phi}{\partial z^2} = S_\Phi \tag{20}$$

with the analytical solution $\Phi = \sin(\pi x)\sin(\pi y)\sin(\pi z)$, $x, y, z \in [0, 1]^3$.

In Fig. 1 it is shown that the fourth order can only be achieved by the compact scheme in combination with a third-order boundary approximation (illustrated by squares) for diffusive fluxes on a uniform mesh. Any lower order treatment decreases the entire order. In [9] it is discussed that the global accuracy is affected by a third-order boundary approximation, but this could not be observed in our numerical tests. Using lower order boundary approximations (illustrated by stars and diamonds in the graphic) lowers the overall order to three. In comparison to the second-order scheme the compact scheme not only shows the expected higher order, but also smaller errors. The used mean error is the square root of the mean square error. To reach the same accuracy as the CDS solution on the finest grid the compact scheme requires more than a quarter less grid points in each direction.

Figure 2 shows the computational cost of the compact scheme in comparison to the second-order scheme. Although the compact scheme requires to solve additional systems of equations, which increases the computing time, the accuracy is much higher and the extra computational time negligible.

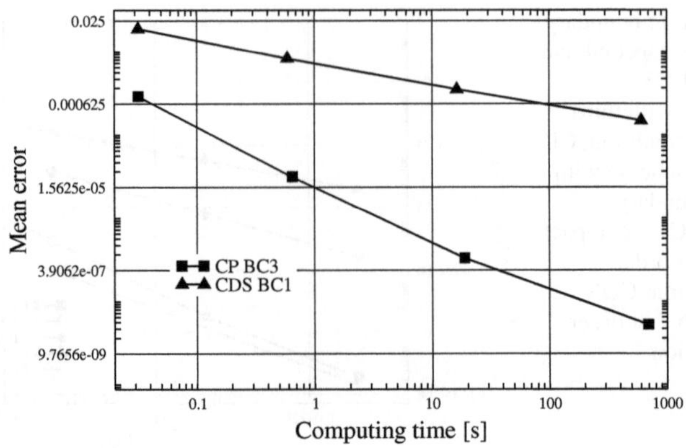

Fig. 2 Computational costs for second-order (CDS BC1) and compact fourth-order scheme (CP BC3)

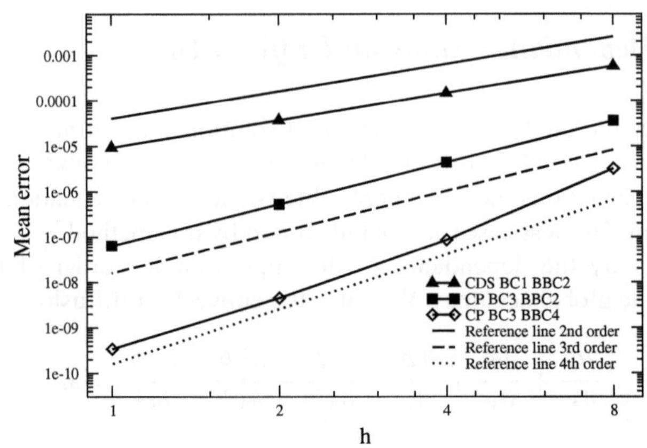

Fig. 3 Influence of block boundary condition: CDS with second order block boundary condition (CDS BC1 BBC2), compact scheme with second order block boundary condition (CP BC3 BBC2), compact scheme with fourth order block boundary condition (CP BC3 BBC4)

Another effect influencing the overall order is the block boundary condition. Figure 3 shows the influence of two different block boundary conditions for eight blocks solving the three-dimensional convection-diffusion equation. The analytical solution is $\Phi = \sin(\pi x)\sin(\pi y)\sin(\pi z),\ x, y, z \in [0, 1]^3$. The results are shown on four different grid levels, where h is the finest grid with 80 control volumes per direction. The domain is split into eight blocks of equal size. To achieve the fourth order diffusive fluxes are discretized by a fourth-order block boundary condition. The values of directly neighboring control volumes are not sufficient for such an approximation. Usually values of direct neighbors are stored in ghost cells and for the fourth-order accuracy one has to introduce a second ghost cell layer.

Table 1 Dependance of order on grid stretching for compact scheme in comparison to CDS

	Stretching factor							
	$S = 0.99$				$S = 0.9$			
	CDS		CP		CDS		CP	
Grid level h	Mean error	Order	Mean error	Order	Mean error	Order	Mean error	Order
4	$1.41 \cdot 10^{-2}$		$4.02 \cdot 10^{-4}$		$4.17 \cdot 10^{-2}$		$7.69 \cdot 10^{-3}$	
3	$8.56 \cdot 10^{-3}$	1.73	$1.17 \cdot 10^{-4}$	4.29	$2.73 \cdot 10^{-2}$	1.47	$2.92 \cdot 10^{-3}$	3.37
2	$4.24 \cdot 10^{-3}$	1.73	$1.93 \cdot 10^{-5}$	4.44	$1.48 \cdot 10^{-2}$	1.51	$6.09 \cdot 10^{-4}$	3.87
1	$1.32 \cdot 10^{-3}$	1.68	$7.81 \cdot 10^{-7}$	4.63	$5.01 \cdot 10^{-3}$	1.56	$3.83 \cdot 10^{-5}$	3.99

4.2 Grid Dependance Study on Stretched Grids

With the help of the three-dimensional convection-diffusion equation the influence of the grid stretching is studied. The stretching factor $0 < S < 1$ shows the concentration of grid points and is defined by the ratio of the length of two sequenced control volumes. The chosen analytical solution is $\Phi = \sin(\pi x)\sin(\pi y)\sin(\pi z)$. The number of grid points is constant with 20 per direction. That means, that the length of the geometry varies [21]. Table 1 lists the error norms for different stretching factors. On coarser grids the effect on error and order is the highest and the numerical order differs from the formal order. However, in comparison to the second-order scheme shown the compact scheme gives better results. The finer the grid, the closer is the calculated order to the formal order.

5 Conclusion

We have presented an efficient compact finite-volume scheme for the convection-diffusion equation. The fourth order could be achieved with a third-order boundary approximation for diffusive fluxes. Lower order boundary approximations reduce the entire order. The reduction of the order caused by third-order boundary approximations as stated in [9] could not be observed in our numerical simulations. Also crucial to maintain the desired order is the choice of the block boundary condition.

Despite of the need to solve additional sets of equations, the proposed compact scheme has much better accuracy at lower computational costs in comparison to second-order schemes.

For calculations on non-Cartesian grids a coordinate transformation approach was realized and applied to stretched grids. After these promising results the compact scheme should be applied to test cases with complex geometries.

In this paper only the steady convection-diffusion equation was studied. For extension to the unsteady equation a fourth-order time discretization scheme is necessary to obtain a fourth order of the whole discretization scheme. In literature an often used scheme is a fourth-order Runge-Kutta method [9, 15].

Due to the block-structured grids and the block by block approach, the solution procedure can be parallelized in a straightforward way.

References

1. Carpenter MH, Gottlieb D, Abarbanel S (1993) The stability of numerical boundary treatments for compact high-order finite-difference schemes. J Comput Phys 108:272–295
2. Ekaterinaris JA (1999) Implicit, high-resolution, compact schemes for gas dynamics and aeroacoustics. J Comput Phys 156:272–299
3. Ferziger JH, Perić M (2002) Computational methods for fluid dynamics. Springer, Berlin
4. Fosso A, Deniau H, Sicot F et al (2010) Curvilinear finite-volume schemes using high-order compact interpolation. J Comput Phys 229:5090–5122
5. Gaitonde D, Shang JS (1997) Optimized compact-difference-based finite-volume schemes for linear wave phenomena. J Comput Phys 138:617–643
6. Ghadimi M, Farshchi M (2012) Fourth order compact finite volume scheme on nonuniform grids with multi-blocking. Comput Fluids 56:1–16
7. Hirsh RS (1975) Higher order accurate difference solutions of fluid mechanics problems by a compact differencing technique. J Comput Phys 19:90–109
8. Hokpunna A, Manhart M (2010) Compact fourth-order finite volume method for numerical solutions of Navie-Stokes equations on staggered grids. J Comput Phys 229:7545–7570
9. Kobayashi MH (1999) On a Class of Padé Finite volume methods. J Comput Phys 156:137–180
10. Lacor C, Smirnov S, Baelmans M (2004) A finite volume formulation of compact central schemes on arbitrary structured grids. J Comput Phys 198:535–566
11. Lee C, Seo Y (2002) A new compact spectral scheme for turbulence simulations. J Comput Phys 183:438–469
12. Lele SK (1992) Compact finite difference schemes with spectral-like resolution. J Comput Phys 103:16–42
13. Lilek Ž, Perić M (1995) A fourth-order finite volume method with colocated variable arrangement. Comput Fluids 24:239–252
14. Nagarajan S, Lele SK, Ferziger JH (2003) A robust high-order compact method for large eddy simulation. J Comput Phys 191:392–419
15. Pereira JMC, Kobayashi MH, Pereira JCF (2001) A fourth-order-accurate finite volume compact method for the incompressible NavierStokes solutions. J Comput Phys 167:217–243
16. Piller M, Stalio E (2008) Compact finite volume schemes on boundary-fitted grids. J Comput Phys 227:4736–4762
17. Piller M, Stalio E (2004) Finite-volume compact schemes on staggered grids. J Comput Phys 197:299–340
18. Roache PJ (2002) Code verification by the method of manufactured solutions. J Fluid Eng 124:4–10
19. Salari K, Knupp P (2000) Code verification by the method of manufactured solutions. Sandia National Laboratories
20. Schäfer M (2006) Computational engineering: Introduction to numerical methods. Springer, Berlin
21. Veluri SP (2010) Code verification and numerical accuracy assessment for finite volume CFD codes. Aerospace Engineering, Virginia Polytechnic Institute and state University

On the Effect of Flux Functions in Discontinuous Galerkin Simulations of Underresolved Turbulence

Andrea D. Beck, Gregor J. Gassner, and Claus-Dieter Munz

Abstract In this work, the influence of the numerical flux functions in the context of an underresolved Discontinuous Galerkin discretization of turbulent compressible flows is investigated. We find that the impact of the choice of the numerical flux function strongly depends on the polynomial degree N, with a larger influence for lower order approximations. Overall, Discontinuous Galerkin discretizations are too dissipative compared to the reference DNS solution, with a lower error for the Roe flux function compared to the local Lax-Friedrichs flux function. This motivates further investigations into Discontinuous Galerkin-based implicit Large Eddy Simulation, an idea supported by results obtained with a low order approximation combined with a modified Roe flux function.

1 Introduction

Due to their geometric flexibility, excellent parallel scalability and high accuracy, high order Discontinuous Galerkin (DG) methods are attractive candidates for accurate and efficient resolution of complex multiscale phenomena like fluid transition and turbulence. For well-resolved, smooth problems, their low numerical error leads to a more faithful representation of the flow than for low order schemes. These favorable approximation properties also transfer to underresolved scenarios, where the concept of "order of convergence" loses its meaningfulness as Δh is large. Instead, the dissipation and dispersion properties over a large range of scales determine the overall quality of the approximation, and thus the numerical cost for a given error. As shown in [5], high order DG schemes can indeed recover more quality per degree of freedom for underresolved turbulence than low order variants.

A.D. Beck (✉) · G.J. Gassner · C.-D. Munz
Institute for Aerodynamics and Gasdynamics, University of Stuttgart, Stuttgart, Germany
e-mail: beck@iag.uni-stuttgart.de; gassner@iag.uni-stuttgart.de; munz@iag.uni-stuttgart.de

M. Azaïez et al. (eds.), *Spectral and High Order Methods for Partial Differential Equations - ICOSAHOM 2012*, Lecture Notes in Computational Science and Engineering 95, DOI 10.1007/978-3-319-01601-6_11,

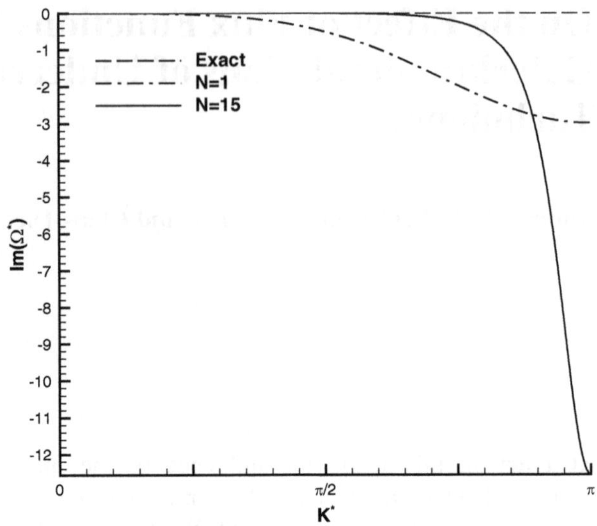

Fig. 1 Plot of the dissipation relation for low and high order discretization with N denoting the polynomial degree. $K^* = \frac{K}{N+1}$ is the normalized wavenumber and Ω^* is the corresponding modified normalized wavenumber

Supporting this statement, Fig. 1 highlights the dissipation error of the DG formulation for a high and low order variant. As obvious, the high order scheme has the advantageous property of introducing very little dissipation over a large range of well-resolved scales, and provides a very high numerical damping for the marginally resolved structures which can cause stability issues. These characteristics make the DG method a natural choice for simulations in underresolved scenarios, where the numerical dissipation is acting as a sink for the excess energy build-up in the higher modes, a method commonly referred to as Large Eddy Simulation with implicit (subgrid-) modelling (iLES) [11].
The upcoming interest in DG methods as a basis for iLES computations gives rise to the question of how the choice of the flux functions influences the solution quality in underresolved transitional and turbulent flows. While the influence of the flux functions for smooth problems as well as shock-dominated problems in the DG context has been investigated before (see e.g. [7–9, 12]), to the authors' knowledge this aspect has not been examined for turbulent situations.

The goal of this work to document the influence of the choice of the flux function of the DG discretization for underresolved turbulence.

This paper is organized as follows: Sect. 2 describes our analysis framework, with a special focus on the "analytical" DG formulation which has only the flux functions as a parameter to provide a sound evaluation ground for our investigations. In Sect. 3, we will highlight the influence of the choice of both the convective and viscous flux function on transitional flows, followed by a brief glance at optimization possibilities of the Riemann solver diffusion and an outlook in Sect. 4.

2 Analysis Framework

2.1 Physical Model

Since we are interested in the simulation of general fluid turbulence, we choose the three-dimensional compressible Navier-Stokes equations in conservative form as our physical model:

$$U_t + \nabla_x \cdot \mathbf{F}(U, \nabla U) = 0, \tag{1}$$

where U is the vector of conserved quantities and $\mathbf{F} = \mathbf{F}^C(U) - \mathbf{F}^V(U, \nabla U)$ their flux vector with convective and viscous contributions given by

$$U = \begin{pmatrix} \rho \\ \rho v_1 \\ \rho v_2 \\ \rho v_3 \\ \rho e \end{pmatrix}, \quad F_l^C(U) = \begin{pmatrix} \rho v_l \\ \rho v_1 v_l + \delta_{1l}\, p \\ \rho v_2 v_l + \delta_{2l}\, p \\ \rho v_3 v_l + \delta_{3l}\, p \\ \rho e v_l + p v_l \end{pmatrix}, \quad F_l^V(U, \nabla U) = \begin{pmatrix} 0 \\ \tau_{1l} \\ \tau_{2l} \\ \tau_{3l} \\ \tau_{lj} v_j - q_l \end{pmatrix}, \tag{2}$$

with $l = 1, 2, 3$ denoting the columns of the flux vectors. We use the established nomination of the physical quantities: ρ, $\mathbf{v} = (v_1, v_2, v_3)^T$, p, and e, denoting the density, the velocity vector, the pressure, and the specific total energy, respectively. The viscous stress tensor is given by

$$\underline{\tau} := \mu \left(\nabla \mathbf{v} + (\nabla \mathbf{v})^T - \frac{2}{3} (\nabla \cdot \mathbf{v}) \underline{I} \right), \tag{3}$$

and the heat flux expressed as a function of temperature T by $\mathbf{q} := (q_1, q_2, q_3)^T$ with

$$\mathbf{q} = -k \nabla T, \tag{4}$$

with the conductivity $k = \frac{c_p \mu}{Pr}$. The viscosity coefficient μ, the Prandtl number Pr and the adiabatic exponent $\kappa = \frac{c_p}{c_v}$ with the specific heats c_p, c_v depend on the fluid properties and are supposed to be constant in this work. The system is closed by the equation of state of a perfect gas:

$$p = \rho R T = (\kappa - 1)\rho(e - \frac{1}{2}\mathbf{v} \cdot \mathbf{v}), \quad e = \frac{1}{2}\mathbf{v} \cdot \mathbf{v} + c_v T, \tag{5}$$

with the specific gas constant $R = c_p - c_v$.

2.2 Discontinuous Galerkin Formulation

To derive a Discontinuous Galerkin formulation for systems of conservation equations such as (1), we first rewrite the hyperbolic-parabolic formulation as the following mixed first-order system

$$\begin{aligned} U_t + \nabla_x \cdot \mathbf{F}(U, W) &= U_t + \nabla_x \cdot \left(\mathbf{F}^C(U) - \mathbf{F}^V(U, W)\right) = 0, \\ W &= \nabla_x U. \end{aligned} \tag{6}$$

where we have introduced the gradients of the conservative variables $W = \nabla U$ as new unknowns. The first step of the actual DG discretization is then to subdivide the computational domain into grid cells Q and to choose a local polynomial ansatz of degree N

$$U(\mathbf{x}, t)\big|_Q \approx U_h(\mathbf{x}, t) = \sum_{j=1}^{\mathcal{N}(N)} \hat{U}_j(t)\, \phi_j(\mathbf{x}), \tag{7}$$

where $\mathcal{N}(N) = (N+1)^3$ is the number of coefficients for a given degree N, $\{\hat{U}_j(t)\}_{j=1}^{\mathcal{N}}$ are the time dependent nodal polynomial coefficients and $\{\phi_j(\mathbf{x})\}_{j=1}^{\mathcal{N}}$ a corresponding Lagrangean basis in the grid cell Q. For the interpolation points, we choose a tensor product of one-dimensional Gauss-Legendre points.

Inserting the trial function (7) into the Eq. (6), multiplying by a test function $\varphi(\mathbf{x}) \in \mathbb{P}_N(Q)$ and integrating over the grid cell Q yields the variational formulation of the system (6)

$$\begin{aligned} &\int_Q \left((U_h)_t + \nabla_x \cdot \mathbf{F}(U_h, W)\right) \varphi(\mathbf{x})\, d\mathbf{x} = 0, \\ &\int_Q W\, \varphi(\mathbf{x})\, d\mathbf{x} = \int_Q \nabla_x U_h\, \varphi(\mathbf{x})\, d\mathbf{x}. \end{aligned} \tag{8}$$

An integration by parts for the spatial derivatives is used to separate the boundary and the volume contribution and to couple the formulation across the elements

$$\begin{aligned} &\int_Q (U_h)_t\, \varphi(\mathbf{x})\, d\mathbf{x} = -\oint_{\partial Q} \left(\widehat{\left(\mathbf{F}^\mathrm{C} \cdot \mathbf{n}\right)} - \widehat{\left(\mathbf{F}^\mathrm{V} \cdot \mathbf{n}\right)}\right) \varphi\, dS + \int_Q \mathbf{F}(U_h, W)\, \nabla_x \varphi(\mathbf{x})\, d\mathbf{x}, \\ &\int_Q W\, \varphi(\mathbf{x})\, d\mathbf{x} = \oint_{\partial Q} \widehat{(\mathbf{n} * U_h)}\, \varphi\, dS - \int_Q U_h\, \nabla_x \varphi(\mathbf{x})\, d\mathbf{x}, \end{aligned} \tag{9}$$

where $\mathbf{n}$ denotes the outward pointing normal vector and $*$ the dyadic product. The separation of the boundary contribution allows us to introduce the numerical approximation of the flux traces $\widehat{(.)}$ at the grid cell interface ∂Q, which connects the local discontinuous approximations. For the convective flux traces $\widehat{\left(\mathbf{F}^{\mathbf{C}} \cdot \mathbf{n}\right)}$, we use Riemann-solver based flux functions further described in Sect. 2.3, while we rewrite the two viscous fluxes as

$$\widehat{(\mathbf{n} * U)} = \mathbf{n} * (\alpha U^{+} + (1-\alpha)\, U^{-}) \tag{10}$$

$$\widehat{\left(\mathbf{F}^{\mathbf{V}} \cdot \mathbf{n}\right)} = \mathbf{n} \cdot \left(\alpha\, \mathbf{F}^{\mathbf{V}^{-}} + (1-\alpha)\, \mathbf{F}^{\mathbf{v}^{+}})\right) \tag{11}$$

where the superscripts $\pm$ denote the left and right neighbor at the common interface, respectively, and the scalar $\alpha \in [0, 1]$ allows us to switch between different formulations for viscous fluxes. Details regarding the choice of α and the resulting formulations can be found below in Sect. 2.3.

It is important to note at this point that apart from the degree of the polynomial ansatz in (7) which gives a lower bound on the truncation error, the only approximation introduced in (9) is the choice of flux representations $\widehat{\mathbf{F}^{\mathbf{C}}}$ and $\widehat{\mathbf{F}^{\mathbf{V}}}$, so in that sense, (9) is the unique "analytical" weak form DG formulation for a given $\mathcal{N}$ when all inner products are evaluated exactly. In case of weakly compressible Navier-Stokes problems as investigated below, we found that $(2N+2)^3$ integration points are sufficient.

2.3 *Flux Functions*

For the Discontinuous Galerkin method presented in Sect. 2.2, the only remaining free parameters are the numerical flux functions. Therefore, their choice determines the consistency, accuracy and stability of the method. The task of the flux function is to couple two interfaces with non-unique solutions (and thus non-unique normal fluxes) by determining a suitable unique flux. One method of achieving this is the concept of defining this flux as an (approximate) solution to an initial value problem with constant states (a Riemann problem), an idea developed in the Finite Volume method community [10]. While an *exact* solution to the Riemann problem exists, this so-called *Godunov* flux function [6] is often too costly to evaluate. Instead, approximations of various degree of physical consistency exist, leading to a large class of flux functions for the convective and diffusive fluxes.

2.3.1 Euler Fluxes

For the DG method, the choice of the flux function for the Euler equations has been investigated by Qui and Qui et al. [8,9] for smooth and shock-dominated problems. They found that for smooth, well-resolved problems, the flux functions differ only marginally in terms of accuracy and thus the most computationally effective flux is a good candidate. For problems with discontinuities, the flux functions with higher physical approximation quality (more waves are considered) clearly outperform more dissipative variants. Kesserwani et al [7] also examined these aspects for the dam-break problem of the Shallow Water Equations with similar findings, while Wheatley et al. [12] performed similar numerical experiments with DG for the Magneto-Hydrodynamic (MHD) equations.

The focus of our investigation here differs from the aforementioned works in that we evaluate the flux functions for the situation of underresolved turbulence of the Navier-Stokes equations. For a finite Reynolds number, the underlying physics of this problem are smooth, but the introduction of a coarse numerical discretization introduces an "artifical roughness". In addition, the dam-break problems or shock dynamics investigated before are mostly governed by a few, very localized and strong non-viscous phenomena, while a turbulent field is characterized by a non-local range of fluctuations and viscous dissipation.

In our investigations, we focus on two representatives for the convective fluxes, namely the Local Lax-Friedrichs flux (LLF) and Roe's Approximate Riemann-solver. Our choice was governed by the difference in dissipation introduced by the flux and their widespread use among the DG community.

Lax-Friedrichs Flux Function

The Lax-Friedrichs (LF) flux and its local variant (LLF) are the simplest flux functions, disregarding all but the fastest wave and thus introducing the highest amount of numerical viscosity, see e.g. [10]. Due to its simplicity, robustness and computational efficiency, the LLF is widely used by the DG community. The convective numerical flux in (9) is approximated as

$$\widehat{\mathbf{F^C} \cdot \mathbf{n}} = -\frac{1}{2}\,\beta\,\lambda_{max}\left[U^+ - U^-\right] + \frac{1}{2}\left(\mathbf{F}_n^C(U^+) + \mathbf{F}_n^C(U^-)\right), \tag{12}$$

where $\mathbf{F}_n^C$ is the outward pointing normal flux component at an interface, β is a real number which allows control over the amount of numerical viscosity (with $\beta = 1$ being the classical LF definition), the superscripts $\pm$ denote the values from the neighbor and local element and λ_{max} corresponds to the maximum eigenvalue of the Euler flux matrix as

$$\lambda_{max} := \max_{U^+,U^-} (|\mathbf{v}| + c), \tag{13}$$

Here, c denotes the speed of sound waves computed as $c := \sqrt{\kappa RT}$. For the local LF variant considered in this work, the value of λ_{max} is computed from the local flow field.

Roe's Approximate Riemann Solver

Another favorite flux function in the DG community (see e.g. [11]) is the Approximate Riemann solver due to Roe, see e.g. [10], where the exact flux Jacobian $A = \frac{d\mathbf{F}}{dU}$ is replaced by a linearization $\tilde{A}$ about an average *Roe state*. The underlying system becomes linear with constant coefficients, i.e. instead of the exact Riemann problem, an approximation is generated, which is then solved *exactly*. The numerical flux is approximated as

$$\widehat{\mathbf{F^C} \cdot \mathbf{n}} = -\frac{1}{2}\beta \sum_{i=1}^{m} \tilde{\alpha_i} |\tilde{\lambda_i}| \mathbf{K}^{(\mathbf{i})} + \frac{1}{2}\left(\mathbf{F}_n^C(U^+) + \mathbf{F}_n^C(U^-)\right), \tag{14}$$

where the ˜ denotes the evaluation at the Roe state, m stands for number of eigenvalues $\lambda_i(U^+, U^-)$ of $\tilde{A}$, $\tilde{\alpha_i}(U^+, U^-)$ denote the wave strengths and $\mathbf{K}^{(\mathbf{i})}(U^+, U^-)$ are the corresponding right eigenvectors. There are two different approaches to finding the intermediate state and from there the wave strengths and eigenvectors which are detailed in [10], in our approach, we use the classical Roe formulation. Note that we have again introduced the parameter β as in (12), which allows control over the amount of numerical viscosity.

2.3.2 Viscous Fluxes

For the viscous fluxes, a large choice of formulations exists which like the convective fluxes lead to different stability and accuracy properties. An overview of the available options for the Laplace equation is e.g. given in [1]. Our implementation options shown in Sect. 2.2 (Eqs. (10) and (11)) allow a switch by selecting an α:

- $\alpha = \frac{1}{2}$ leads to the first variant of Bassi and Rebay (BR1) [2] by using the arithmetic mean for both fluxes. This flux is stable for parabolic problems, while it is known that it becomes unstable for purely elliptic cases.
- $\alpha = 0$ or $\alpha = 1$ lead to the local DG variant (LDG) by Cockburn and Shu [4].

3 Influence of the Numerical Fluxes

In this section, we report the influence of the choice of the flux function on the results of the numerical simulation of underresolved turbulence. As emphasized in Sect. 2, the only remaining parameter in our DG formulation apart from the

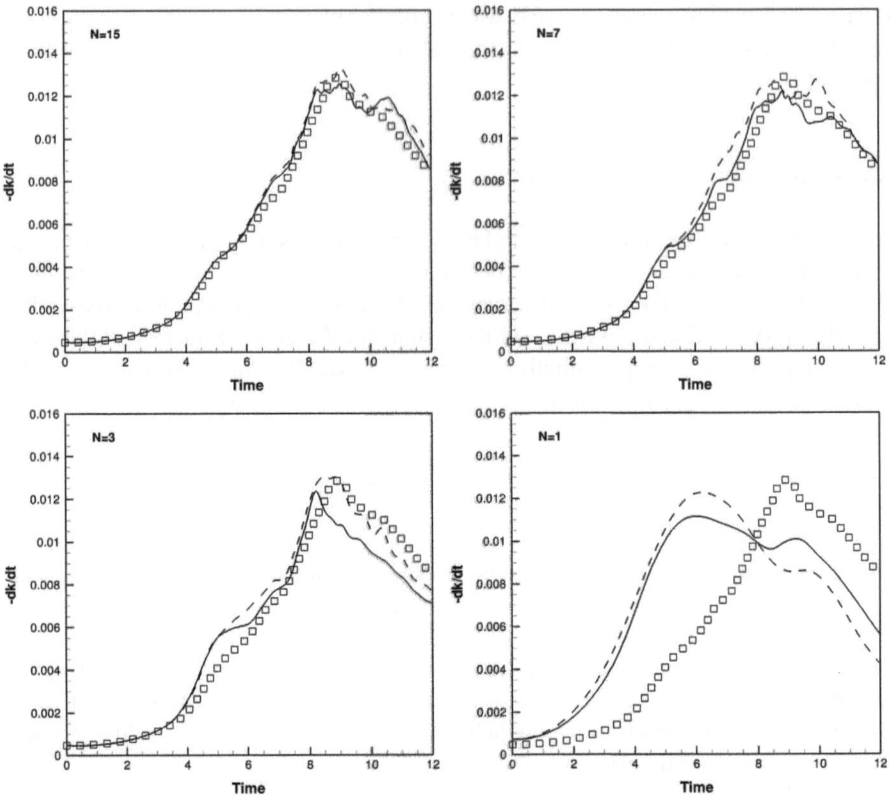

Fig. 2 Plot of the kinetic energy dissipation rate for the $Re = 1,600$ Taylor-Green vortex: comparison of convective numerical flux function influence for different polynomial degrees N with same total no. of DOF (64^3). *Square symbol* denotes the reference DNS solution [3], *dashed line* denotes the LLF flux result, the *solid line* denotes the Roe flux result

initial decision on the order of approximation are the numerical flux functions. We have performed underresolved computations with fixed total number of degrees of freedom (64^3 for all cases) of the well-established Taylor-Green vortex testcase, for a detailed description see [5].

For the weakly compressible Taylor-Green vortex flow, the dissipation rate of the kinetic energy is a suitable benchmark quantity for the evaluation of simulation fidelity. In Fig. 2, we show the comparison of this quantity for the two different convective fluxes described in Sect. 2.3 with decreasing polynomial approximation order but same total number of degrees of freedom. This means that for the cases $N = 15$, $N = 7$, $N = 3$ and $N = 1$ we use 4^3, 8^3, 16^3 and 32^3 grid cells, respectively. Thus, the nominal resolution (points per wavelength) is identical in all cases and is given by our choice of 64^3 DOF. We can clearly see that the polynomial degree has a strong influence. Although all computations are underresolved, the impact of the Riemann fluxes on the resulting kinetic energy dissipation is very low

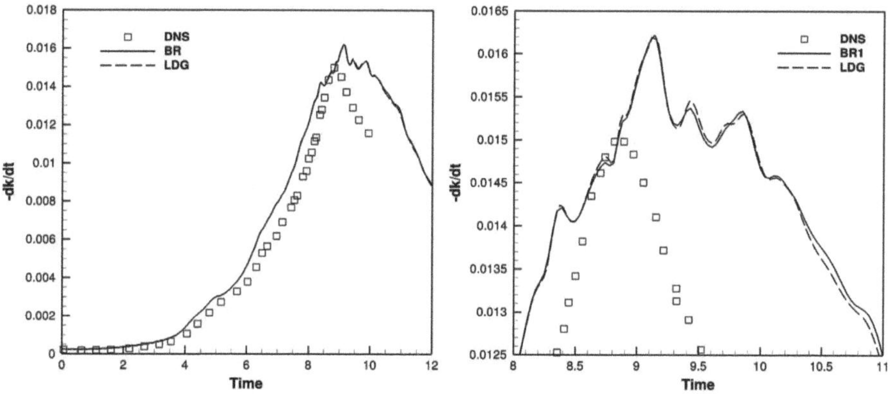

Fig. 3 Plot of the kinetic energy dissipation rate for the $Re = 3{,}000$ Taylor-Green vortex: comparison of LDG flux and BR1 flux (see Sect. 2.3.2) with DNS, 4^3 elements, polynomial degree $N = 15$. *right*: detailed view

for the case $N = 15$. This is in accordance with the dissipative behavior of the high order method, compare Fig. 1, where the onset of the dissipation is delayed to a higher wavenumber compared to the lower order variant. This implies that for the low order discretization, a larger fraction of the resolved scales is affected by the numerical dissipation. Consequently, we observe that the impact of the Riemann flux choice is very strong for the case $N = 1$. Comparing all results from Fig. 2, we see a clear progression from high order to low order in flux function impact. As the number of cell interfaces increases for a given overall resolution and the approximative strength of the local polynomial ansatz decreases, the influence of the flux function becomes more pronounced. Furthermore, it can be observed that for the $N = 7$ and $N = 1$ cases, the Roe flux is closer to the DNS result than the LLF formulation, while the situation is not so clear for the $N = 3$ case and thus warrants further investigations.

In contrast to the convective fluxes, the impact of the viscous flux function seems negligible in this advection-dominated case. For the low order variants, the numerical dissipation is totally governed by the dissipation mechanism of the Euler fluxes. Only for the case of very high polynomial degree, $N = 15$, where the influence of the Euler flux is low, we observe a measurable impact of the viscous flux choice. But as the results in Fig. 3 clearly show, the influence on the overall dissipation behavior even in this case is negligible .

4 Conclusion and Outlook

As shown in Sect. 3, the influence of the convective flux function on the solution quality can be substantial, especially for low order approximations. For the $N = 1$ computations in Fig. 2, the introduced numerical dissipation masks the flow

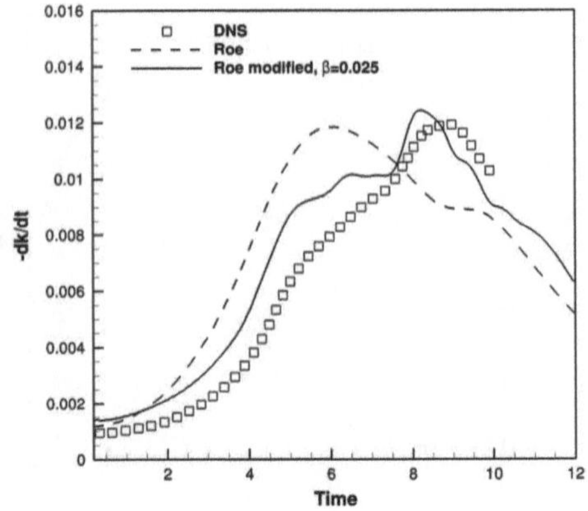

Fig. 4 Plot of the kinetic energy dissipation rate for the $Re = 800$ Taylor-Green vortex: comparison of classical Roe flux and modified Roe flux with DNS

dissipation and emulates a much lower Reynolds number, as evident by the shift of the dissipation rate maximum to about 5 s (see [3]). We have also observed that the Roe flux reproduces a better result than the LLF flux for low order approximations, and that the choice of the viscous flux function is not relevant in this advection-driven case.

Combining these aspects, we now investigate whether we can modify the Roe flux by tuning its dissipation parameter β in Eq. (14) and thereby improve the solution quality. Figure 4 shows the results for the $Re = 800$ simulation with the classical ($\beta = 1.0$) Roe flux and a modification ($\beta = 0.025$) with lowered dissipation. The overall agreement with the DNS results improves significantly for the modified version, and the physical structure of the dissipation rate reappears in the solution.

In conclusion, we have investigated the effects of the choice of the flux functions in Discontinuous Galerkin computations of transitional turbulence. We found a significant influence of the convective flux for underresolved low order simulations, and showed that a simple modification of the flux function dissipation significantly improves the solution quality. In future work, this observation is used as a basis for Discontinuous Galerkin based implicit Large Eddy Simulation of turbulent flows.

Acknowledgements The research presented in this paper was supported in parts by Deutsche Forschungsgemeinschaft (DFG) within the Schwerpunktprogramm 1276: MetStroem.

References

1. Arnold, D.N., Brezzi, F., Cockburn, B., Marini, L.D.: Unified analysis of discontinuous Galerkin methods for elliptic problems. SIAM J. Numer. Anal. **39**(5), 1749–1779 (2002)
2. Bassi, F., Rebay, S.: A high-order accurate discontinuous finite element method for the numerical solution of the compressible Navier-Stokes equations. J. Comput. Phys. **131**, 267–279 (1997)

3. Brachet, M.: Direct simulation of three-dimensional turbulence in the taylor–green vortex. Fluid Dynamics Research **8**(1–4), 1–8 (1991)
4. Cockburn, B., Shu, C.W.: The local discontinuous Galerkin method for time-dependent convection diffusion systems. SIAM Journal on Numerical Analysis **35**, 2440–2463 (1998)
5. Gassner, G., Beck, A.: On the accuracy of high-order discretizations for underresolved turbulence simulations. Theoretical and Computational Fluid Dynamics pp. 1–17 (2012)
6. Godunov, S.K.: A difference method for the numerical calculation of discontinuous solutions of hydrodynamic equations. Mat. Sbornik **47**, 271–306 (1959)
7. Kesserwani, G., Ghostine, R., Vazquez, J., Ghenaim, A., Mosé, R.: Riemann solvers with runge–kutta discontinuous galerkin schemes for the 1d shallow water equations. Journal of Hydraulic Engineering **134**(2), 243–255 (2008)
8. Qiu, J.: A numerical comparison of the Lax-Wendroff discontinuous Galerkin method based on different numerical fluxes. J. Sci. Comp. DOI: 10.1007/s10915-006-9109-5
9. Qiu, J., Khoo, B.C., Shu, C.W.: A numerical study for the performance of the Runge-Kutta discontinuous Galerkin method based on different numerical fluxes. J. Comput. Phys. **212**, 540–565 (2006)
10. Toro, E.: Riemann Solvers and Numerical Methods for Fluid Dynamics. Springer (1999)
11. Uranga, A., Persson, P.O., Drela, M., Peraire, J.: Implicit large eddy simulation of transition to turbulence at low reynolds numbers using a discontinuous galerkin method. International Journal for Numerical Methods in Engineering **87**(1–5), 232–261 (2011)
12. Wheatley, V., Kumar, H., Huguenot, P.: On the role of riemann solvers in discontinuous galerkin methods for magnetohydrodynamics. J. Comput. Phys. **229**(3), 660–680 (2010)

3. Brachet, M.E.: Direct simulation of three-dimensional turbulence in the Taylor-Green vortex. Fluid Dynamics Research 8(1-4), 1-8 (1991)
4. Cockburn, B., Shu, C.W.: The local discontinuous Galerkin method for time-dependent convection-diffusion systems. SIAM Journal on Numerical Analysis 35, 2440-2463 (1998)
5. Gassner, G., Beck, A.: On the accuracy of high-order discretizations for underresolved turbulence simulations. Theoretical and Computational Fluid Dynamics pp. 1-17 (2012)
6. Godunov, S.K.: A difference method for the numerical calculation of discontinuous solutions of hydrodynamic equations. Mat. Sbornik 47, 271-306 (1959)
7. Kesserwani, G., Liang, Q., Vazquez, J., Ghostine, R., Mosé, R.: Riemann solvers with Runge-Kutta discontinuous Galerkin schemes for the 1d shallow water equations. Journal of Hydraulic Engineering 134(2), 243-253 (2008)
8. Qiu, J.: A numerical comparison of the Lax-Wendroff discontinuous Galerkin method based on different numerical fluxes. J. Sci. Comp. DOI 10.1007/s10915-006-9109-5 (2006)
9. Qiu, J., Khoo, B.C., Shu, C.W.: A numerical study for the performance of the Runge-Kutta discontinuous Galerkin method based on different numerical fluxes. J. Comput. Phys. 212, 540-565 (2006)
10. Toro, E.: Riemann Solvers and Numerical Methods for Fluid Dynamics. Springer (1999)
11. Uranga, A., Persson, P.O., Drela, M., Peraire, J.: Implicit large eddy simulation of transition to turbulence at low Reynolds numbers using a discontinuous Galerkin method. International Journal for Numerical Methods in Engineering 87(1-5), 232-261 (2011)
12. Wheatley, V., Kumar, H., Huguenot, P.: On the role of Riemann solvers in discontinuous Galerkin methods for magnetohydrodynamics. J. Comput. Phys. 229(3), 660-680 (2010)

Generation of High-Order Polynomial Patches from Scattered Data

Karsten Bock and Jörg Stiller

Abstract Spectral element methods demand for curved grids when used with complex computational domains. This paper presents a method for the construction of high order polynomial surface patches. These patches are constructed from and approximate scattered surface data e.g. in form of scanning data triangulations. Accuracy and quality of the generated Bézier patches improve significantly by curving these fine grids before using them as "exact" surface representations. Iterative energy optimisation of the bounding curves during the construction process also shows positive effect on accuracy and quality of the patches. The combination of this methods results in patches of high polynomial order that feature a high quality surface approximation and almost smooth transitions between elements.

1 Introduction

Spectral element methods (SEM) are supposed to combine the superior accuracy of spectral methods and the geometrical versatility of finite element methods (FEM). Because of their high accuracy and convergence rates SEM allow for and demand much coarser grids then FEM or finite volume methods. To avoid that geometrical errors dominate the method, the geometrical approximation of the computational domain has to be done by curved elements because of the coarse grids used [1]. Often curved grids are derived from linear grids whereat invalid elements may occur. Different approaches to guarantee the validity of elements have been studied. The correction of invalid elements by face and edge swapping methods is possible [1]. Curvature based refinement of surface grids in combination with more robust prismatic elements for boundary regions can minimize the occurrence

K. Bock (✉) · J. Stiller
Institute of Fluid Mechanics (ISM), Technische Universität Dresden, 01062 Dresden, Germany
e-mail: Karsten.Bock@tu-dresden.de; Joerg.Stiller@tu-dresden.de

M. Azaïez et al. (eds.), *Spectral and High Order Methods for Partial Differential Equations - ICOSAHOM 2012*, Lecture Notes in Computational Science and Engineering 95, DOI 10.1007/978-3-319-01601-6_12,

Fig. 1 The geometry of the *rabbit aortic arch* built from scanning data and kindly provided by S. Sherwin, Imperial College London. The *tiny black frame* marks the section used for detail views

of invalid elements [7]. Another approach curves the initial straight sided mesh with a nonlinear elastic analogy [6].

This paper focuses on the construction of curved surface representations with high order polynomial patches. For many applications, including the field of biomedical flows, the definition of the computational domain is given exclusively by scanning data e.g. from computer tomography (CT), magnetic resonance imaging (MRI) or laser scanning. Therefore these patches need to be constructable from scattered data. As a starting point triangulations can be calculated from scanning data and then be used for the construction of coarse curved surface representations. Curvature based coarsening of these fine grids results in coarse linear grids featuring smaller elements in regions of high curvature. As an example Fig. 1 shows the fine grid computed from CT scanning data of a rabbit aortic arch cast [9] and a comparison of fine (385,023 triangles) and coarse (5,364 triangles) linear grids in detail views.

One method currently used to curve coarse linear grids are spherigon patches [11]. Though originally developed as a smoothing technique, they are used for interpolation here. Spherigon construction is carried out pointwise by blending of interpolants computed from data given in the vertices. These interpolants for a certain point are built as circular arcs that are orthogonal to the approximated normal of that point and the particular vertex normal. Depending on the blending function, G^0 and G^1 versions are available. Recently, this approach has been used for SEM flow simulations e.g. of vascular flows [3, 9]. In this paper we describe the construction of high order polynomial surface representations, namely Bézier curves and patches, from discrete surface definitions in from of scanning data triangulations. The effects of interpolating the fine triangulation and energy optimisation of curves during this process are investigated.

2 Surface Representation with Polynomial Patches

Triangular and quadrangular surface patches as well as the curves bordering them are expressed in Bernstein-Bézier form [2]. Using the Bernstein polynomials $B_i^n(t)$ of degree n the curve $\mathbf{c}(t)$ is written as

$$\mathbf{c}(t) = \sum_{i=0}^{n} \mathbf{b}_i \; B_i^n(t) \tag{1}$$

where $t \in [0, 1]$ is the parametric coordinate along the curve. Quadrangular Bézier surface patches are built in a tensor product form with two parametric coordinates $\xi, \eta \in [0, 1]$:

$$\mathbf{s}(\boldsymbol{\xi}) = \sum_{i=0}^{n} \sum_{j=0}^{n} \mathbf{b}_{ij} \; B_i^n(\xi) \, B_j^n(\eta) \; . \tag{2}$$

Barycentric coordinates $\boldsymbol{\tau}$ and the Bernstein polynomials $B_{ijk}^n(\boldsymbol{\tau})$ of degree n are used to express the triangular surface patch in Eq. (3) in Bernstein-Bézier form with the triple index ijk following the constraint $i + j + k = n$.

$$\mathbf{s}(\boldsymbol{\tau}) = \sum_{i,j,k} \mathbf{b}_{ijk} \; B_{ijk}^n(\boldsymbol{\tau}) \tag{3}$$

$\mathbf{b}_i$, $\mathbf{b}_{ij}$ and $\mathbf{b}_{ijk}$ are the control points defining the form of the curves and surface patches. A more detailed description of Bernstein polynomials, Bézier curves and surface patches is given in [2].

2.1 *Handling of Scattered Surface Data*

In order to fit polynomial patches to a given surface a projection method to the "exact" surface is provided. The normal projection $\mathscr{P}(\mathbf{p})$ of a point to an analytically defined surface can easily be achieved. This Section describes a normal projection method to surfaces defined by discrete data. A fine triangulation serves as the "exact" surface because it is the most accurate representation of the real surface available. Vertex normals of this fine grid are approximated with a method using area and angle weighting [5]. Optionally, Taubin smoothing [8] can be applied, if the input grid is too noisy.

The method used to normal project a given point $\mathbf{p}$ onto the fine surface grid is sketched in Fig. 2. Vertex coordinates and interpolated vertex normals of this fine grid are known. To project the point $\mathbf{p}$, each triangle $\mathbf{p}_1\mathbf{p}_2\mathbf{p}_3$ in the fine grid has to be checked, to see whether it contains a viable projection of this point. For this purpose

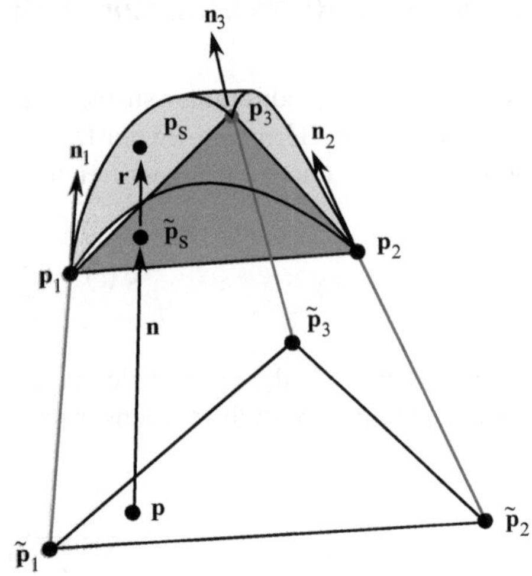

Fig. 2 Sketch of the method to project a point $\mathbf{p}$ onto a surface represented by scattered data (the *triangle* $\mathbf{p}_1\mathbf{p}_2\mathbf{p}_3$ is part of the fine surface grid)

an auxiliary triangle $\tilde{\mathbf{p}}_1\tilde{\mathbf{p}}_2\tilde{\mathbf{p}}_3$ is constructed in the plane parallel to the triangle $\mathbf{p}_1\mathbf{p}_2\mathbf{p}_3$ and containing the point $\mathbf{p}$. The vertices $\tilde{\mathbf{p}}_i$ are the projections of the vertices $\mathbf{p}_i$ in the according normal directions $\mathbf{n}_i$. If the point $\mathbf{p}$ is positioned inside this triangle $\tilde{\mathbf{p}}_1\tilde{\mathbf{p}}_2\tilde{\mathbf{p}}_3$, then its normal projection lies within the triangle $\mathbf{p}_1\mathbf{p}_2\mathbf{p}_3$ and its position is computed by mapping the barycentric coordinates from one triangle to the other. The projection direction of the point is equal to its Phong normal $\mathbf{n}$ [11] in the fine triangle $\mathbf{p}_1\mathbf{p}_2\mathbf{p}_3$, which is an interpolation of the triangles vertex normals. As the triangle $\tilde{\mathbf{p}}_1\tilde{\mathbf{p}}_2\tilde{\mathbf{p}}_3$ is generally curved, this projection method involves a linearisation error. Hence a subdivision scheme, in which the curved triangle is replaced by a finer linear triangulation, was introduced to minimize this error.

For every point to project there can be multiple viable normal projections on the fine surface grid. The projection, which requires the minimal projection distance, is then chosen as the projected point $\tilde{\mathbf{p}}_s$.

The fine grid is a linear approximation of the real surface and hence lacks high accuracy and smoothness. To improve the quality and accuracy of this surface representation, different interpolations computable from vertex data can be used [10, 11]. The results presented in this paper were obtained by applying spherigons with a quadratic blending function to curve fine grids. The position of the final projection $\mathbf{p}_s$ on the interpolated curved surface triangle can be calculated from $\tilde{\mathbf{p}}_s$ and the flat triangles $\mathbf{p}_1\mathbf{p}_2\mathbf{p}_3$ vertex data.

Especially for large grids it is very costly to check each triangle in the grid as described before. To minimize the computational costs, the space around the surface model is subdivided into voxel cells and the information which triangle intersects with which voxel is stored. If a point $\mathbf{p}$ is projected the voxel this point is positioned in can easily be computed. The search for possible projections starts with the triangles intersecting this voxel cell. Should no projection be found, the neighbouring voxel cells are checked until a projection is found or a stop criterion is met. As an example Fig. 3 shows the bounding box around a screw surface,

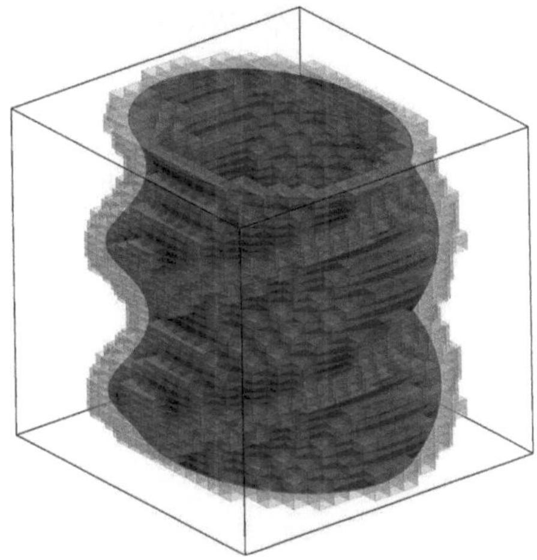

Fig. 3 Bounding box around the screw surface defining the space to be covered by the voxel grid. Only the voxels actually containing surface elements are displayed

which is covered by these voxel cells. Note that only voxels actually containing triangles of the fine grid are plotted.

2.2 Construction of Polynomial Patches

The construction of polynomial surface patches starts with establishing the boundary curves. Given a set of parameter values t_i, the sampling points $\mathbf{x}_i$ are constructed by projecting the linear interpolation between edge vertices $\mathbf{p}_0$ and $\mathbf{p}_1$ to the surface. We write this as

$$\mathbf{x}_i = \mathscr{P}\left[(1 - t_i)\,\mathbf{p}_0 + t_i \mathbf{p}_1\right] \quad , \tag{4}$$

where $\mathscr{P}(\mathbf{p})$ represents the normal projection of a point $\mathbf{p}$ onto the surface. The control points $\mathbf{b}_0$ and $\mathbf{b}_n$ of the curve are set to match the edge vertices $\mathbf{p}_0$ and $\mathbf{p}_1$. The remaining, inner control points $\mathbf{b}_i$ are computed by minimisation of the distance functional

$$\mathscr{E}_d = \sum_{i=1}^{m} |\mathbf{c}(t_i) - \mathbf{x}_i|^2 \to \min , \tag{5}$$

which is evaluated using m sampling points.

Given the curves, the boundary coefficients of the patches are defined. The inner control points follow from a minimisation of a quadratic distance functional similar to Eq. (5). The required sampling points are first defined in parameter space, then their position is approximated using BBG or Coons patches [2] interpolating the boundary curves. Finally they are projected to the exact or fine surface representation.

2.3 Iterative Optimization of Bézier Curves

Energy minimisation of splines and curves is a known method used to improve the quality of curves on surfaces [4]. Minimizing the energy functional in Eq. (6) leads to geodesic curves, which are the shortest curves connecting the vertices on the surface. The strain energy of a curve is defined by Eq. (7). Both of these energy functionals are readily computed from Bézier curve control points.

$$\mathscr{E}_1 = \int_0^1 \dot{\mathbf{c}}^2(t)\,\mathrm{d}t \tag{6}$$

$$\mathscr{E}_2 = \int_0^1 \ddot{\mathbf{c}}^2(t)\,\mathrm{d}t \tag{7}$$

The minimisation of these energy functionals for a Bézier curve on a surface is done by iterative minimization of the functional J in Eq. (8) while the curve is confined to the surface. The starting curve $\mathbf{c}^0(t)$ is constructed as described in Sect. 2.2. The energy functionals $\mathscr{E}_1^0$ and $\mathscr{E}_2^0$ and the distance functional $\mathscr{E}_d^0$ of this starting curve are evaluated by Eqs. (6), (7) and (5) and will be used for normalisation within the functional

$$J = w_x \frac{\mathscr{E}_d}{\mathscr{E}_d^0} + w_e \left[(1-\alpha)\frac{\mathscr{E}_1}{\mathscr{E}_1^0} + \alpha \frac{\mathscr{E}_2}{\mathscr{E}_2^0} \right] \rightarrow \min \quad . \tag{8}$$

w_x, w_e and α name weighting factors within the functional. A higher value of weight w_e in comparison to w_x allows for higher deviations from the surface but also for faster decreasing values of the energy functionals during one iteration step.

At the beginning of the actual iteration process sampling points are distributed on the starting curve with a set of parameter coordinates t_i. After projecting these points to the surface, the sampling points $\mathbf{x_i} = \mathscr{P}[\mathbf{c}(t_i)]$ can be used to calculate the distance functional $\mathscr{E}_d$ during the minimisation of the normalized functional J. The minimisation results in a curve that approximates the surface and has reduced in the energies $\mathscr{E}_1$ and $\mathscr{E}_2$. New sampling points are then created on this curve and again projected to the surface. After a minimisation of the sampled distance error functional $\mathscr{E}_d$ follows an energy reduced curve lying on the surface. These steps are repeated until a maximum number of iterations is reached or a stop criterion is reached. The reduction rates of J, $\mathscr{E}_d$, $\mathscr{E}_1$ and $\mathscr{E}_2$ during the iteration steps are monitored and used as a stop criterion, which terminates the iteration when a certain threshold δ_{min} is reached.

Figure 4 illustrates this process for a long curve on the screw surface. The full line representing the curve after the iteration is clearly minimized in energies in comparison with the dashed starting curve and lies on the screw surface.

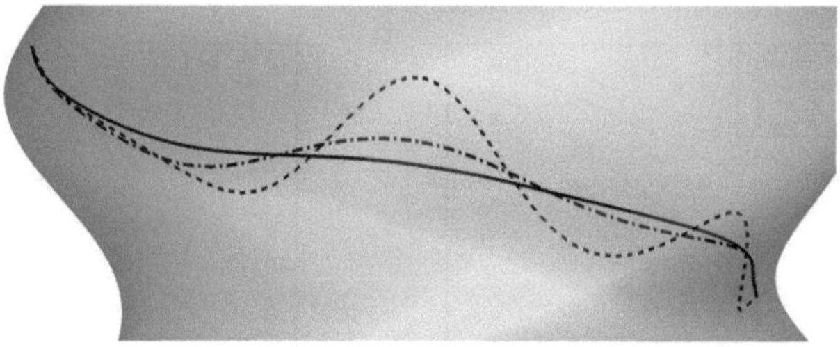

Fig. 4 Example of the curve iteration process. The *dashed line* represents the *starting curve*, the *dash-dotted line* shows the result after 10 iterations and the *full line* the result of 50 iterations (Polynomial degree $n = 10$; iteration weights and parameters: $w_x = w_e = 0.5$, $\alpha = 0.8$, $\delta_{min} = 0.01$)

3 Numerical Results

We performed studies comparing different triangular surface patches of exact and discrete surface definitions. A screw surface shown e.g. in Fig. 4 serves as the example for exact surface definitions. The triangulated scanning data of the rabbit aortic arch illustrated in Fig. 1 was used as example for discrete surface representations.

The error of triangular Bézier patches of different polynomial orders and of spherigon patches to an exactly defined screw surface are plotted in Fig. 5a in maximum error norm. The grid size h and the error ε_∞ are normalized with the minimum curvature radius r_c of the screw surface. The projection during the Bézier patch construction was performed onto the exact surface. The Coarse grid vertex coordinates and normals necessary to compute spherigon patches and were also calculated by projection on the exact definition of the screw surface.

Bézier triangles with increasing polynomial order n achieve decreasing errors compared to spherigons. Therefore it is possible to maintain a high accuracy in surface representation even when using coarse surface grids. Spherigons in the form used here are G^1 continuous over the edges of adjacent patches but not directly expressed as polynomials. In Fig. 5b the normal vector deviation at the common edge of neighbouring patches is shown in maximum error norm. The plot indicates, that other than spherigons Bézier patches are not able to produce a smooth surface representation. However, the normal deviations decrease rapidly with higher polynomial orders n of the Bézier patches.

The rabbit aortic arch pictured in Fig. 1 is used as an example for the construction of polynomial surface patches from scattered scanning data. A comparison of the results of different surface patch construction methods is given in Fig. 6 where a high curvature region of this grid is selected because such a region is particularly hard to represent by coarse grids. The top left figure shows spherigon patches producing a smooth but bumpy surface because of the enforcement of smooth transitions between surface patches. These spherigons were constructed from coarse grid vertex coordinates and normals that were improved by projection to a curved fine grid

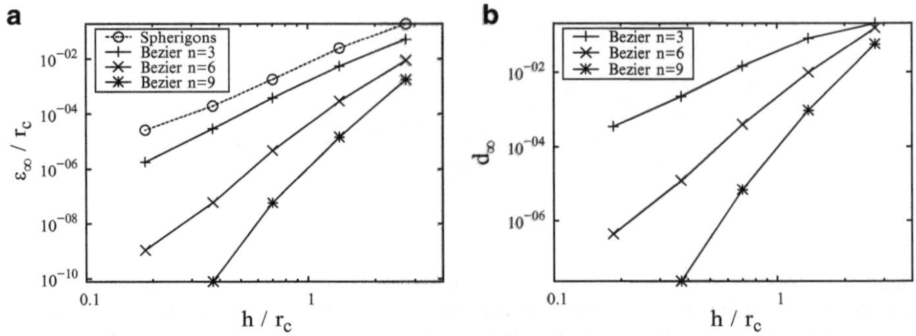

Fig. 5 (**a**) Error to exact screw surface in maximum error norm (normalisation with minimal curvature radius $r_c \approx 0.117$ is used). (**b**) Maximum difference between normal vectors of adjacent edges

surface representation, which was interpolated with G^0 spherigons. All Bézier triangles pictured in Fig. 1 are of polynomial order $n = 10$. At such high polynomial order the coarse curved surface patches represent the fine grid surface model very well. However, this results in a very bumpy surface imitating the linear fine grid when using a projection to this grid when constructing Bézier triangles, as the top right figure shows. This surface representation is not of the desired quality. The bottom surface plots display Bézier triangles constructed with projection to a fine grid interpolated by G^0 spherigons. In the bottom left picture the Bézier triangles can not retain smooth patch transitions but overall render a much less bumpy surface. They were built by the method described in Sect. 2.2 with projection to the interpolated fine grid. The patches shown in the bottom right plot are Bézier triangles using iterative energy minimised Bézier curves as a basis. These patches provide much smoother transitions and maintain a good overall visual quality of the surface. It becomes clear, that the quality of curved edge representations has a great impact on the quality of the resulting Bézier patches.

These observations are backed by the error data listed in Table 1, where only Bézier patches with projection to an interpolated fine grid are listed. We omitted patches calculated with projection to a linear fine grid because they were not able to produce high quality surface representations for high polynomial orders. The positional errors in maximum norm ε_∞ and L_2 norm ε_2 are calculated from surface normal distances to the most exact surface representation of the rabbit aortic arch available, the G^0 spherigon interpolated fine grid. The data point out a significant reduction of positional errors of the surface patches as well as edges when using iterative optimised Bézier curves as a basis of Bézier triangle construction. The ε_2 error of the Bézier triangles is magnitudes lower than that of the spherigons. The maximum error ε_∞ reaches slightly higher values but is of the same order of magnitude. While Bézier patches built from not iterated Bézier curves exhibit high normal deviations in maximum and L_2 norm d_∞ and d_2, patches formed from iterative optimised curves feature lower deviations and thus smoother patch transitions at critical areas as shown in Fig. 6.

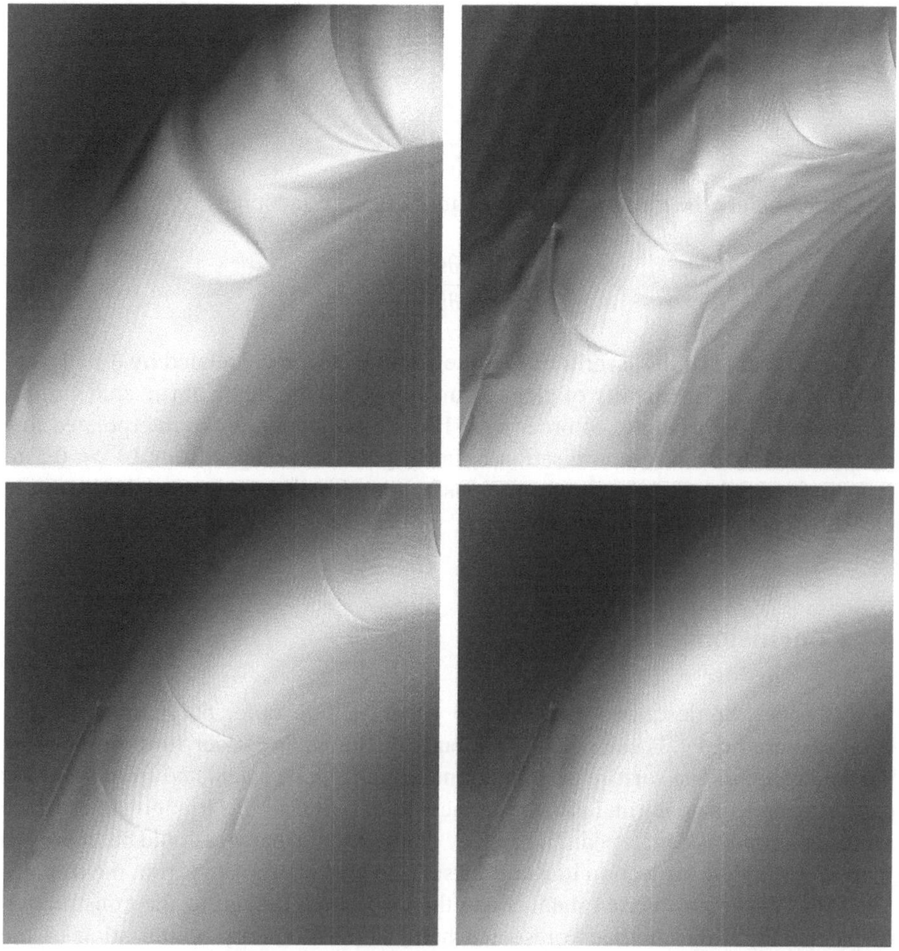

Fig. 6 Comparison of different curved surface patches modelling a high curvature region of the rabbit aortic arch surface. *Top left*: Spherigon patches. *Top right*: Bézier patches with projection to linear fine grid. *Bottom left*: Bézier patches approximating an interpolated triangulation. *Bottom right*: Bézier patches constructed from iterative optimised Bézier curves with projection to interpolated fine grid (Bézier patches of degree $n = 10$; iteration parameters used: $w_x = w_e = 0.5$, $\alpha = 0.8$, $\delta_{min} = 0.03$)

Table 1 Error data of edges and surface patches to spherigon interpolated fine grid as the most exact surface representation available

	Edges				Patches	
	ε_∞	ε_2	d_∞	d_2	ε_∞	ε_2
Spherigons	–	–	–	–	2.26e−02	1.21e−01
Original curves	7.13e−02	1.28e−02	1.11	1.42	7.13e−02	1.64e−03
Optimised curves	2.71e−02	7.68e−03	3.26e−01	9.05e−01	2.71e−02	1.62e−03

Surface patch mapping $\mathbf{s}(\boldsymbol{\xi})$ and $\mathbf{s}(\boldsymbol{\tau})$ quality can be assessed by the distortion measure

$$I^S = \frac{J^S_{min}}{J^S_{max}} \quad . \tag{9}$$

J^S_{min} and J^S_{max} are the minimum and maximum surface Jacobians

$$J^S = \left| \frac{\partial \mathbf{s}}{\partial \xi} \times \frac{\partial \mathbf{s}}{\partial \eta} \right| \tag{10}$$

in one element [1], where for triangles the parameters are calculated by $\xi = 1 - \tau_1$ and $\eta = 1 - \tau_2$. The quality of element mappings for the rabbit aortic arch surface built from Bézier triangles of order $n = 10$ with projection to the interpolated fine grid and iterative curve optimisation is fairly good and ranges from $I^S \approx 0.5$ to 1 apart from very high curvature regions at the bifurcations where the distortion measure reaches its minimum value of $I^S \approx 0.23$

4 Conclusion

In this paper we described a method to generate high order polynomial patches from scattered data. These patches are built in Bernstein-Bézier form on basis of the data given in fine surface triangulations, which can be calculated from scanning data. A approximate normal projection method of points to this triangulated surface representation was developed. The quality of polynomial patches could be improved by performing the projection to a curved surface grid interpolated from the triangulated fine linear grid vertex data. Since the curves are crucial to the construction of patches with the methods presented here, iterative energy optimisation of the curves representing the element edges leads to error reduction and increasing quality of surface patches as well as the bordering curves. A similar approach for the construction of patches could be tested in the future to further improve element quality, accuracy and smoothness of the curved grid. The polynomial patches showed high accuracy in modelling the geometry and offer some advantages over spherigons, which are one method currently in use to generate curved surfaces for SEM methods. Especially in regions of high curvature the Bézier patches were able to provide a much higher surface quality. Although they do not guarantee smooth transitions between surface patches that spherigons feature, they generate an visually smooth surface with rapidly decreasing normal deviations d_n at element borders especially when using patches of high order.

References

1. Dey, S., O'Bara, R.M., Shephard, M.S.: Towards curvilinear meshing in 3d: the case of quadratic simplices. Computer-Aided Design **33**(3), 199–209 (2001)
2. Farin, G.: Curves and Surfaces for CAGD - A Practical Guide, 5. edn. Academic Press (2002)
3. Grinberg, L., Anor, T., Madsen, J.R., Yakhot, A., Karniadakis, G.E.: Large-scale simulation of the human arterial tree. Clinical and Experimental Pharmacology and Physiology **36**, 194–205 (2009)
4. Hofer, M., Pottmann, H.: Energy-minimizing splines in manifolds. ACM Trans. Graph. **23**(3), 284–293 (2004)
5. Max, N.: Weights for computing vertex normals from facet normals. Journal of Graphics, GPU, and Game Tools **4**(2), 1–6 (1999)
6. Persson, P.O., Peraire, J.: Curved mesh generation and mesh refinement using lagrangian solid mechanics. In: Proc. of the 47th AIAA Aerospace Sciences Meeting and Exhibit (2009)
7. Sherwin, S.J., Peiró, J.: Mesh generation in curvilinear domains using high-order elements. Int. J. Num. Meth. Engrg. **53**(1), 207–223 (2002)
8. Taubin, G.: A signal processing approach to fair surface design. In: Proceedings of the 22nd annual conference on Computer graphics and interactive techniques, SIGGRAPH '95, pp. 351–358. ACM (1995)
9. Vincent, P.E., Plata, A.M., Hunt, A.A.E., Weinberg, P.D., Sherwin, S.J.: Blood flow in the rabbit aortic arch and descending thoracic aorta. Journal of The Royal Society Interface (2011)
10. Vlachos, A., Peters, J., Boyd, C., Mitchell, J.L.: Curved pn triangles. In: Proceedings of the 2001 symposium on Interactive 3D graphics, I3D '01, pp. 159–166. ACM (2001)
11. Volino, P., Thalmann, N.M.: The spherigon: A spimple polygon patch for smoothing quickly your polygonal meshes. Proceedings of Computer Animation Ca'98 pp. 72–79 (1998)

Towards a High Order Fourier-SEM Solver of Fluid Models in Tokamaks

A. Bonnement, S. Minjeaud, and R. Pasquetti

Abstract We investigate a fluid modeling approach to describe the plasma behavior in tokamaks. For the numerical approximation, we use a high order method based on Fourier expansions in the toroidal direction and the spectral element method (SEM) in the poloidal plane. We first focus on anisotropic diffusion, because in tokamaks diffusion strongly dominates along the magnetic field lines, and provide some comparisons with finite element results. Then we give preliminary results for a plasma two fluid (ions and electrons) numerical model.

1 Introduction

The production of energy by fusion of light nuclei like Deuterium and Tritium may be achieved by Magnetic Confinement Fusion. This is done in annular apparatus called tokamaks, where the reacting material is under the form of plasma (ionized gas at very high temperature). A strong magnetic field is used to confine the plasma, in order to overcome the pressure gradient and curvature effects. The ITER device is presently under construction in Cadarache (France) [17].

Simulating the plasma behavior is extremely difficult, e.g. due to the various space and time scales which should be considered. In the core of the plasma kinetic (or gyrokinetic) approaches, based on the resolution of a six-dimensional (or five-dimensional) Boltzmann-like equation, are usually preferred. In the edge region of the plasma, where the geometry is more complex and the temperature less high, fluid approaches may be relevant (this notion is certainly not shared by all tokamak physicists). Especially, they can be of interest beyond or close to the separatrix,

A. Bonnement · S. Minjeaud · R. Pasquetti (✉)
Lab. J.A. Dieudonné, UMR CNRS 7351 & project INRIA CASTOR, University of Nice-Sophia Antipolis, Parc Valrose, 06108 Nice Cedex 2, France
e-mail: richard.pasquetti@unice.fr

M. Azaïez et al. (eds.), *Spectral and High Order Methods for Partial Differential Equations - ICOSAHOM 2012*, Lecture Notes in Computational Science and Engineering 95, DOI 10.1007/978-3-319-01601-6_13,

which separates the core region, where the magnetic surfaces are closed, and the scrape off layer (SOL region), where the magnetic surfaces are open, see e.g. [16].

On the basis of a two fluid modeling, our goal is to develop a three-dimensional Fourier-SEM code to describe the turbulence transport phenomena in the SOL region. The fluid model is based on the usual conservation equations of mass, momentum and energy expressed for both ions and electrons and on the assumption of quasi-neutrality of the plasma. The so-called divertor configuration, which will be used for ITER, is considered. In a given poloidal plane, it is characterized by the presence of an X-point in the magnetic lines. Such a configuration is out of reach of codes that make use of Fourier expansions in the poloidal angle, coupled to finite differences or finite elements in the radial one, see e.g. [1].

2 Governing Equations

The governing equations express the conservation of density, momentum and energy of each species $s = \{i, e\}$, with i for ion and e for electron. Moreover, we assume that the fluctuations of the magnetic field $\boldsymbol{B}$ are negligible and consequently that the electric field derives from an electric potential U (from Faraday's law). With:

- n_s, m_s, e_s for the volume fraction, mass and electric charge, respectively,
- $\boldsymbol{u}_s$, p_s, Π_s, ε_s, $\boldsymbol{\varphi}_s$ for velocity, pressure, deviatoric part of the pressure tensor, internal energy and associated flux density, respectively,
- $\boldsymbol{R}_s$ for the friction forces due to ion-electron collisions and Q_s for the energy exchange due to collisions, one obtains:

$$\begin{aligned}
&\partial_t n_s + \nabla \cdot (n_s \boldsymbol{u}_s) = 0, \\
&\partial_t (n_s m_s \boldsymbol{u}_s) + \nabla \cdot (m_s n_s \boldsymbol{u}_s \boldsymbol{u}_s + p_s I + \Pi_s) = n_s e_s (-\nabla U + \boldsymbol{u}_s \wedge \boldsymbol{B}) + \boldsymbol{R}_s, \\
&\partial_t \varepsilon_s + \nabla \cdot (\varepsilon_s \boldsymbol{u}_s + \boldsymbol{\varphi}_s) = -p_s \nabla \cdot \boldsymbol{u}_s - \Pi_s : \nabla \boldsymbol{u}_s + Q_s\,.
\end{aligned} \tag{1}$$

These equations are completed with the perfect gas law for each species:

$$p_s = n_s T_s \quad , \quad \varepsilon_s = p_s/(\gamma - 1), \tag{2}$$

where the temperature T_s has here the dimension of an energy and $\gamma = 5/3$. The system is closed using the Braginskii closure [4] which provides the expressions:

- $\Pi_s \equiv \Pi_s(\boldsymbol{u}_s)$,
- $\boldsymbol{R}_s \equiv \boldsymbol{R}_s(T_e, n_e, \boldsymbol{j})$, where $\boldsymbol{j} = \sum_s n_s e_s \boldsymbol{u}_s$ is the current density,
- $\boldsymbol{\varphi}_s \equiv \boldsymbol{\varphi}_s(T_s, p_s, \boldsymbol{j})$,
- $Q_s \equiv Q_s(T_e, T_i, \boldsymbol{j})$.

An additional reasonable assumption is the electroneutrality of the plasma, which means:

$$\sum_s n_s e_s = 0 \rightarrow \begin{cases} n_e = Z n_i, \\ \nabla \cdot \sum_s n_s e_s \boldsymbol{u}_s \equiv \nabla \cdot \boldsymbol{j} = 0, \end{cases} \tag{3}$$

where $Z = -e_i/e_e$. We thus observe that the current $\boldsymbol{j}$ is divergence free.

When taking into account this additional constraint, we obtain a system of 10 non-linear and coupled Partial Differential Equations (PDE) for the variables $n(= n_e)$, U, $\boldsymbol{u}_s$ and ε_s. Such a problem appears extremely difficult because being:

- Steep, as a result of (i) $m_e \ll m_i$ and (ii) $\boldsymbol{B}$ strong;
- Multiscale in space: The Larmor radius, associated to the spiral motion of the ions and electrons around the magnetic lines, is much smaller than the size of the ITER device;
- Multiscale in time: The cyclotron period is much smaller than the turbulence time scale which itself is much smaller than the discharge time (duration of an experiment);
- Strongly anisotropic: Diffusion is indeed very dominant along the magnetic field lines. This difficulty is addressed in the next Section.

It should be noted that the considered PDE system does not make use of the so-called drift velocity assumption. We refer to [14] and to the works carried out in the frame of the ESPOIR ANR project for this kind of approaches, see e.g. [9, 15]. One can also note that simpler fluid modelings, often based on the MHD equations, have been and are still investigated, see e.g. [1, 5, 13].

3 Anisotropic Diffusion

The Braginskii closure yields expressions of anisotropic form, e.g. for the energy flux density (subscripts i or e are omitted in this section):

$$\varphi = -\chi_{\parallel} \nabla_{\parallel} T - \chi_{\perp} \nabla_{\perp} T + \chi_{\wedge} (\boldsymbol{b} \wedge \nabla T), \tag{4}$$

where $\boldsymbol{b} = \boldsymbol{B}/|\boldsymbol{B}|$, $\nabla_{\parallel} T = (\boldsymbol{b} \cdot \nabla T)\boldsymbol{b}$ and $\nabla_{\perp} T = \nabla T - \nabla_{\parallel} T$. Such expressions are strongly anisotropic. Indeed one has:

$$\frac{\chi_{\perp}}{\chi_{\parallel}} \sim \frac{1}{(\omega_c \tau)^2}, \quad \frac{\chi_{\wedge}}{\chi_{\parallel}} \sim \frac{1}{\omega_c \tau},$$

where, see e.g. [11], $\omega_c = |\boldsymbol{B}|e/m$ is the cyclotron frequency and $\tau \propto m^{1/2}(kT)^{3/2}/(ne^4)$ is the collision time (with k, Boltzmann constant). The resulting values of the product $\omega_c \tau$ for the plasma core and plasma edge regions are given in Table 1.

Table 1 Typical values of the temperature, density and $\omega_c\tau$ product for the ions and electrons in the core and edge regions of the plasma

	Plasma core	Plasma edge
Temperature (K)	1.16 10^8	5.8 10^5
Density (m^{-3})	10^{20}	10^{19}
$\omega_c\tau$ for electrons	3.39 10^7	1.2 10^5
$\omega_c\tau$ for ions	1.12 10^6	3.96 10^3

Because we plan to use an unstructured mesh, a priori not aligned to the magnetic field lines, our implementation of Eq. (4) is based on a tensorial form of the diffusion coefficient:

$$\begin{aligned}\varphi &= \chi_\parallel(\boldsymbol{b}\cdot\nabla T)\boldsymbol{b} + \chi_\perp(\nabla T - (\boldsymbol{b}\cdot\nabla T)\boldsymbol{b}) + \chi_\wedge(\boldsymbol{b}\wedge\nabla T)\\ &= (\chi_\parallel - \chi_\perp)(\boldsymbol{b}\cdot\nabla T)\boldsymbol{b} + \chi_\perp\nabla T + \chi_\wedge(\boldsymbol{b}\wedge\nabla T)\\ &= K\nabla T \end{aligned} \tag{5}$$

where $K = (\chi_\parallel - \chi_\perp)\,\boldsymbol{bb} + \chi_\perp\mathbb{I} + \chi_\wedge B$, with an easily identifiable antisymmetric matrix B, such that $\chi_\wedge(\boldsymbol{b}\wedge\nabla T) = \chi_\wedge B\nabla T$.

The validity of our Fourier – SEM approach has first been checked on the anisotropic diffusion problem $\partial_t T = K\nabla T$. In time we use a standard fourth Runge-Kutta (RK4) scheme and in space Fourier expansions in the toroidal direction together with spectral elements in the poloidal plane. Using the Galerkin Fourier method allows us to substitute a set of two-dimensional problems to the initial 3D one. These 2D problems are then solved by using the SEM, see e.g. [7, 12].

We consider a test problem of the CEMM (Center for Extended MHD Modeling, Princeton), see e.g. [8], in a toroidal geometry of square poloidal cross-section. In the cylindrical coordinate system (R, ϕ, z), the initial condition $T_0 = T(t = 0)$ is a "pulse" of Gaussian shape located at $(R = R_1, \phi = 0, z = 0)$ and of standard deviation δ:

$$T_0 = \exp(-((R - R_1)^2 + (R_1\phi)^2 + z^2)/\delta^2). \tag{6}$$

The magnetic field $\boldsymbol{B}$ is defined in the toroidal coordinate system (r, θ, ϕ) :

$$\boldsymbol{B} = \frac{1}{R}(\boldsymbol{e}_\phi - \frac{1}{R_0 q_0}\frac{r}{1 + (r/a)^2}\boldsymbol{e}_\theta), \tag{7}$$

with: R_0, q_0: radius of the torus, safety factor (tilting parameter of the magnetic lines); a: radius of the torus section; $\boldsymbol{e}_\phi, \boldsymbol{e}_\theta$: unit vectors versus ϕ and θ directions. Then, the magnetic lines make spirals on closed tubular surfaces $r = const$.

We first present a test case assuming $\chi_\parallel = 1$ and $\chi_\perp = \chi_\wedge = 0$, that is $K = \boldsymbol{BB}/B^2$. Such an input is of course not physical but here our goal is only to check the capability of the algorithm in the most extreme case. Figure 1 (top) shows isotherms at the initial and final time of the computations, whereas Fig. 1 (bottom) shows the ϕ-averaged solutions at two intermediate times. The computation has

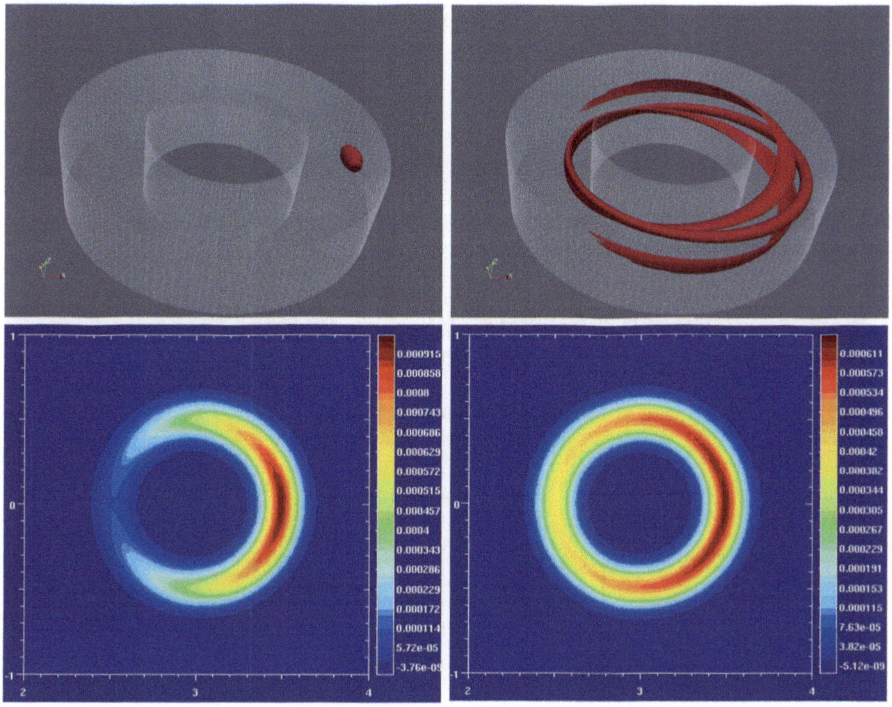

Fig. 1 *Top*: Isotherms at the initial and final time, $t = 0$ and $t = 148.44$. *Bottom*: ϕ-averaged solutions at $t = 11.72$ and $t = 31.25$

been done with 64 toroidal Fourier modes, a polynomial approximation degree $N = 4$ in each element and 9,409 grid-points in the poloidal plane, so that the total number of grid-points is 1,204,352. The mesh is simply aligned along the horizontal (r) and vertical (z) directions. The time-step was taken equal to $7.8125\ 10^{-4}$. As can be observed, despite the fact that the mesh is not aligned on the magnetic field lines, the anisotropic diffusion phenomenon is well described.

We now investigate the influence of the $\chi_\wedge(\boldsymbol{b} \wedge \nabla T)$ term. In fact one has:

$$\nabla \cdot \chi_\wedge(\boldsymbol{b} \wedge \nabla T) = (\nabla \chi_\wedge \wedge \boldsymbol{b} + \chi_\wedge(\nabla \wedge \boldsymbol{b})) \cdot \nabla T, \tag{8}$$

which means that this "diffusion term" behaves like a transport term with velocity $\boldsymbol{u}_\wedge = \nabla \chi_\wedge \wedge \boldsymbol{b} + \chi_\wedge \nabla \wedge \boldsymbol{b}$.

Simulation results are provided in Fig. 2. The $\boldsymbol{b}$ vector being essentially parallel to $\boldsymbol{e}_\theta$, because $\nabla \wedge \boldsymbol{e}_\theta = \boldsymbol{e}_z/R$ one observes a transport phenomenon in the vertical direction which sense depends on the sign of $\chi_\wedge$.

More quantitative tests have been carried out in two-dimension, with circular magnetic lines and diffusivity such that $\chi_\parallel = 1$ and $\chi_\perp = \chi_\wedge = 0$. Using as initial condition a temperature distribution only depending on the radius, one expects that the solution does not evolve in time. Two kinds of radius dependencies have been

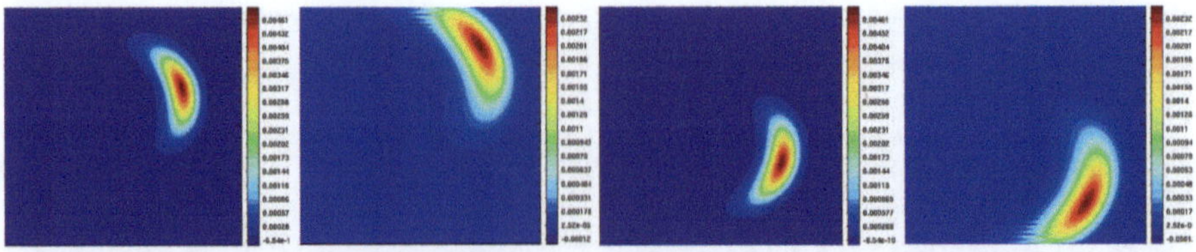

Fig. 2 CEMM test, $\chi_{\parallel} = 1$, $\chi_{\perp} = 0$ and $\chi_{\wedge} = 1$ (the two visualizations at *left*) or $\chi_{\wedge} = -1$ (at *right*). ϕ-averaged solutions at two different times

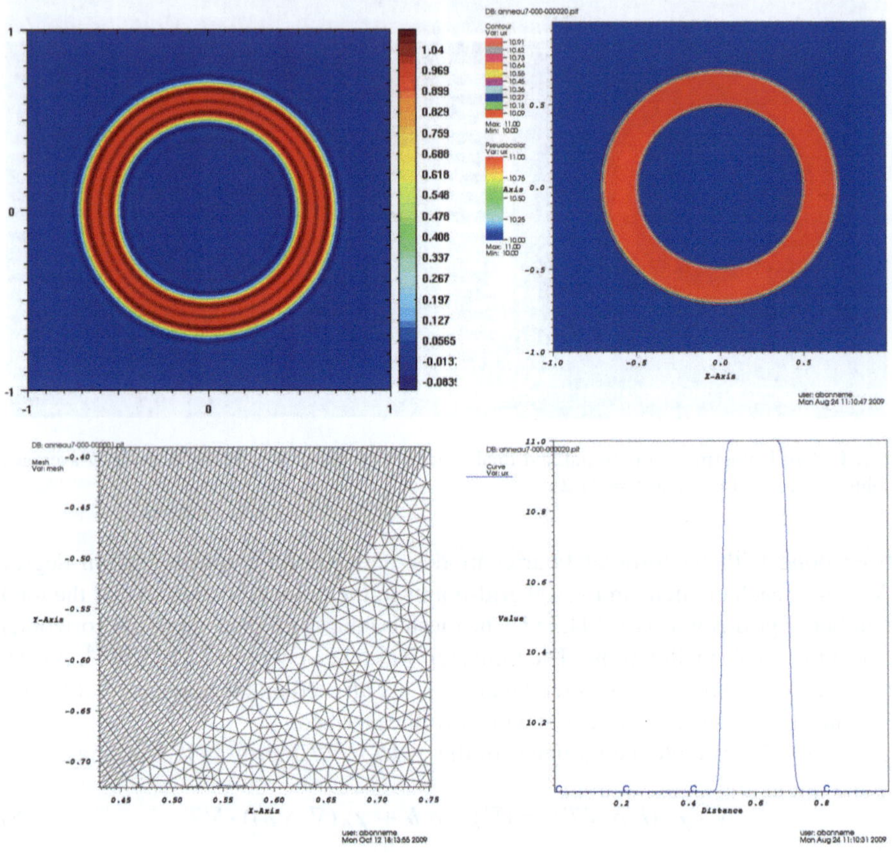

Fig. 3 The characteristic function of a ring is used as initial condition. *Top*: SEM ($N = 4$) and P_1-FEM solutions at $t = 2$. *Bottom*: Details of the FEM-mesh, which inner part is aligned on the circular magnetic lines, and profile of the FEM solution

tested, smooth (Gaussian) or steep (characteristic function). Nice results have been obtained in both cases, except of course of the expected Gibbs phenomenon in the stiff case, see Fig. 3 (top-left). Comparisons have been made with the standard P_1-FEM approach, based on the FluidBox / Plato software [18], for which it turned out to be necessary to use a mesh aligned on circles to obtain satisfactory results in the steep case. Figure 3 (bottom-left) shows a zoom of such a mesh, which allows to compute the steep problem without any oscillations of the solution, see Fig. 3 (right). Details are provided in [3].

4 Towards the Full Two-Fluid Braginskii Model

Combining the conservation equations introduced in Sect. 2, one obtains the momentum – current $\boldsymbol{q}$ - $\boldsymbol{j}$ system, see e.g. [6], which is equivalent to the ion – electron momentum $\boldsymbol{q}_i - \boldsymbol{q}_e$ system that directly results from the Eqs. (1) to (3). With $\rho = \sum_s n_s m_s$, $\boldsymbol{q} = \sum_s \boldsymbol{q}_s$ and when taking into account that $\sum_s \boldsymbol{R}_s = 0$ one obtains:

$$
\begin{aligned}
&\partial_t \rho + \nabla \cdot \boldsymbol{q} = 0 \\
&\partial_t \boldsymbol{q} + \nabla \cdot \sum_s (\boldsymbol{q}_s \boldsymbol{u}_s + p_s I + \Pi_s) = \boldsymbol{j} \wedge \boldsymbol{B} \\
&\partial_t \boldsymbol{j} + \nabla \cdot \sum_s w_s (\boldsymbol{q}_s \boldsymbol{u}_s + p_s I + \Pi_s) = -c_\rho \rho \nabla U + (c_q \boldsymbol{q} + c_j \boldsymbol{j}) \wedge \boldsymbol{B} + \sum_s w_s \boldsymbol{R}_s \\
&\nabla \cdot \boldsymbol{j} = 0 \\
&\partial_t \varepsilon_s + \nabla \cdot (\varepsilon_s \boldsymbol{u}_s + \boldsymbol{\varphi}_s) = -p_s \nabla \cdot \boldsymbol{u}_s - \Pi_s : \nabla \boldsymbol{u}_s + Q_s
\end{aligned}
\tag{9}
$$

where c_ρ, c_q, c_j and $w_s = e_s/m_s$ are given coefficients. This system must be completed by the state laws, for the ions and electrons, and by the Braginskii closure. The present formulation clearly points out that one has to solve for a compressible dynamics, to get ρ and $\boldsymbol{q}$, and an incompressible one, to get U and $\boldsymbol{j}$. Thus, the potential U appears as the Lagrange multiplier which allows the current density $\boldsymbol{j}$ to be solenoidal.

Looking at the set of PDEs (9), when taking into account that $m_e \ll m_i$ and if assuming (i) that $T_i = T_e = T$, so that with $n = \sum_s n_s$, $p = nT$, and (ii) that the viscous stresses are negligible, it turns out to be relevant to check the capability of the Fourier-SEM approach on the Euler system. The Euler system may however yield discontinuous solutions and so a stabilization technique is required. To this end, we have implemented the entropy viscosity technique, that relies on the idea of introducing a non-linear viscous term, which amplitude is controlled by a viscosity coefficient proportional to the absolute value of the entropy residual and bounded from above by a $O(h)$ viscosity (h is the grid-size) [10]. An example of result is presented in Fig. 4 (top) for an axisymmetric Euler computation in a domain showing the limiter (rather than the divertor) configuration. This is e.g. the case of the Tore-Supra device in Cadarache. One observes that the wave front is rather well described and, as required by the Bohm boundary condition, that the Mach number equals one at the plates. When taking into account the viscous terms and then using the usual closure of Newtonian fluids, one obtains the Navier-Stokes system. Figure 4 (bottom) shows the influence of viscosity on the previous simulation

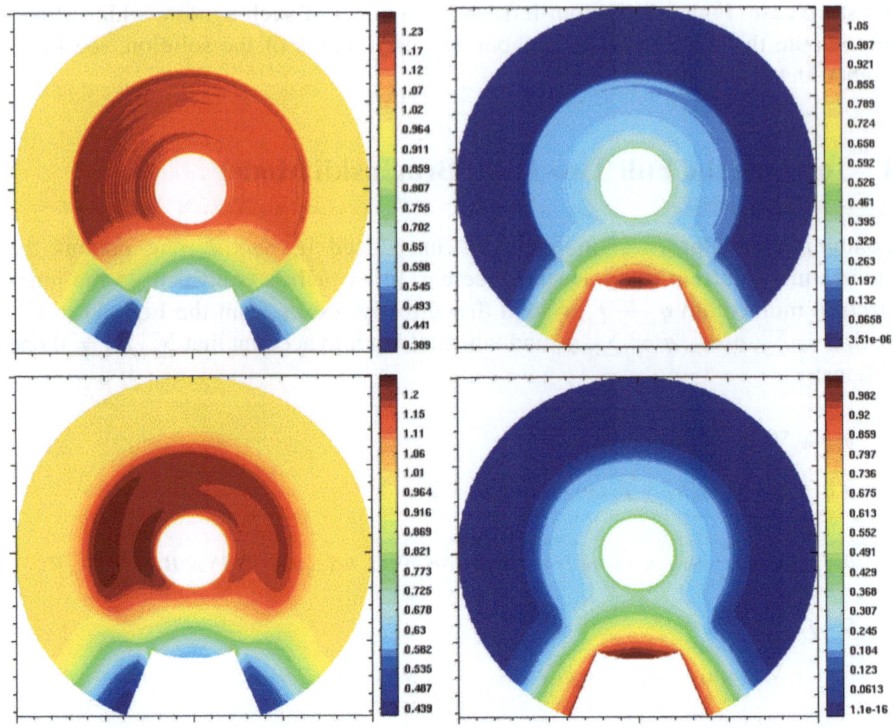

Fig. 4 Euler (*top*) and Navier-Stokes (*bottom*) results, density (at *left*) and Mach number (at *right*). SEM ($N = 4$)-EV approximation. RK4 scheme. Initial condition: Fluid at rest, constant density and pressure. Boundary conditions: Inflow imposed at the inner boundary; Free-slip at the outer one. The "Bohm boundary condition" $M \geq 1$ is imposed at the plates

result. Considering such a simplified single fluid approach with Euler, Navier-Stokes and also Braginskii-like closure was investigated in [3], using in space a finite element / finite volume approximation.

To solve the full $\boldsymbol{q}$ - $\boldsymbol{j}$ system or equivalently the $\boldsymbol{q}_i - \boldsymbol{q}_e$ one, we use in time a third order RK3 IMEX scheme [2], in such a way that the flux terms are treated explicitly whereas the (.) $\wedge$ $\boldsymbol{B}$ terms are treated implicitly. Note however that no additional cost is required because such terms do not involve space derivatives. Finally, we use a projection method to compute U such that $\nabla \cdot \boldsymbol{j} = 0$. This requires to solve the elliptic equation:

$$\nabla \cdot (\rho \nabla \delta U) = \nabla \cdot \boldsymbol{j}^\star, \quad \partial_n \delta U|_\Gamma = 0 \tag{10}$$

with $\boldsymbol{j}^\star$ the provisional current obtained by solving the IMEX scheme, discarding the divergence free constraint, and δU a potential increment.

Axisymmetric computations have been done using the geometry of the JET tokamak in Culham [16], considering only the edge region. At the initial time we use the data provided by the resolution of a Grad-Shrafranov equilibrium, using the code JOREK [5], i.e. the ion density, the total pressure $p = p_i + p_e$ and the

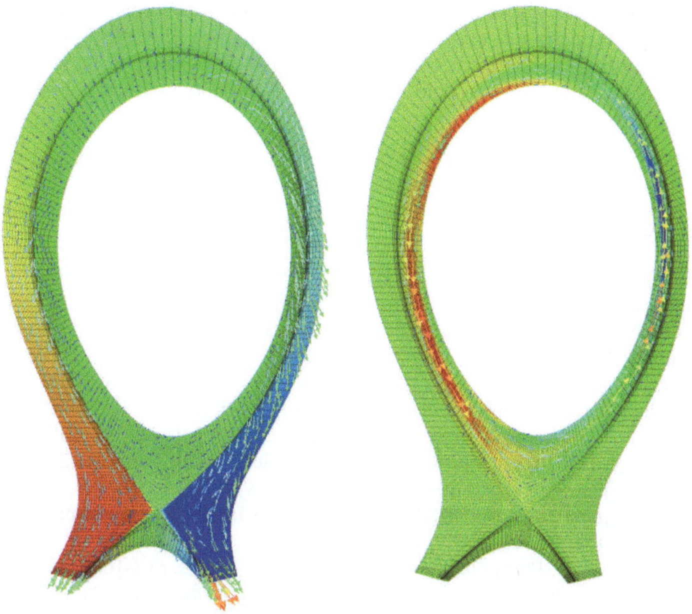

Fig. 5 Ion velocity (at *left*) and current density (at *right*). The vectors display the poloidal component and the color the toroidal one. SEM ($N = 2$) – RK3 IMEX scheme. The "Bohm boundary condition" $M \geq 1$ is imposed at the plates

magnetic potential. From that one can derive the magnetic field and the ion and electron internal energies. The initial current density is assumed toroidal, so that $\boldsymbol{j} \cdot \nabla p = 0$ and $\nabla \cdot \boldsymbol{j} = 0$, and computed in such a way the $\boldsymbol{j} \wedge \boldsymbol{B}$ term compensates at best the pressure gradient. The initial ion velocity is set equal to 0 inside the separatrix. In the SOL, it is taken co-linear to the magnetic field and at the plates such that $\boldsymbol{u}_i = \pm c\, \boldsymbol{b}$, where c is the sound velocity. Free-slip conditions are used everywhere except at the plates where we use the Bohm boundary condition $M \geq 1$, with M for the ion Mach number. The mesh is the one provided by the JOREK code. It is aligned on the magnetic surfaces and is essentially structured, except at the X point where eight quadrangular elements use it as a vertex. This is well supported by the SEM approximation, which is designed to support non-structured meshes. Steep gradients however occur, especially because at the initial time the ion and electron velocities are not continuous at the separatrix and moreover show four different values about the X-point. The computations have been done with $N = 2$. Increasing this value of the polynomial approximation turns out to be difficult with the appearance of negative values of the pressure at the plates. This seems strongly due to the JOREK mesh, that includes, especially at the plates, elements of very high aspect ratio. Figure 5 shows snapshots of the ion velocity field and of the current density for an Euler closure of the governing equations.

Acknowledgements This work is carried out within (i) the FR-FCM project "Two fluid numerical modeling of edge plasma in tokamaks" and (ii) the ANR project ESPOIR, in collaboration with IRFM (Ph. Ghendrih, P. Tamain), lab. M2P2 (G. Ciraolo, F. Schwander, E. Serre) and the INRIA CASTOR project (L. Combe, H. Guillard, B. Nkonga). We also thank J.L. Guermond (Texas A & M University) for useful discussions.

References

1. D.V. Anderson, W.A. Cooper, R. Gruber, S. Merazzi, U. Schwenn, TERPSICHORE: A three-dimensional ideal MHD stability program, Scientific Computing on Supercomputers II, Devreese and Van Camp, eds, Plenum Press, NY, 1990.
2. U.M. Ascher, S.J. Ruuth, R.J. Spiteri, Implicit-explicit Runge-Kutta methods for time-dependent partial differential equations, Appl. Numer. Math., 25:151–167, 1997.
3. A. Bonnement, Modélisation numérique par approximation fluide du plasma de bord des tokamaks (projet ITER), PHD thesis, University of Nice-Sophia Antipolis, 2012.
4. S.I. Braginskii, Transport processes in a plasma. Review of Plasma Physics, 1: 205–311, 1965.
5. O. Czarny, G.T.A. Huysmans, MHD stability in x-point geometry : simulation of ELMS, Nuclear fusion, 47:659–666, 2007.
6. J.L. Delcroix, A. Bers, Physique des plasmas (II), InterEditions, / CNRS Editions, 1994.
7. M.O. Deville, P.F. Fischer, E.H. Mund, *High-order methods for incompressible flows*, Cambridge University Press, 2002.
8. P.F. Fischer, Anisotropic diffusion in a toroidal geometry, Journal of Physics: Conference Series, 16:446, 2005.
9. Ph. Ghendrih et al., Impact on Divertor Operation of the Pattern of Edge and SOL Flows Induced by Particle Sources and Sinks, IAEA paper, submitted.
10. J.L. Guermond, R. Pasquetti, B. Popov, Entropy viscosity method for non-linear conservation laws, J. of Comput. Phys., 230 (11), 4248–4267, 2011.
11. J.D. Huba, *NRL Plasma formulary*, The office of naval research, Washington DC, 2009.
12. G.E. Karniadakis, S.J. Sherwin, *Spectral hp element methods for CFD*, Oxford Univ. Press, London, 1999.
13. D. Reiter, M. Baelmans, P. Borner, The EIRENE and B2-EIRENE codes, Fusion Science and Technology, 47, 172–186, 2005.
14. P. Tamain, Etude des flux de matière dans le plasma de bord des tokamaks: alimentation, transport et turbulence, PHD thesis, Aix-Marseille University, 2007.
15. P. Tamain et al., Tokam-3d: A fluid code for tranport and turbulence in the edge plasma of tokamaks, J. of Comput. Phys., 229(2), 361–378, 2010.
16. http://www.edfa.org
17. http://www.iter.org
18. http://www-sop.inria.fr/pumas/plato.php

Whitney Forms, from Manifolds to Fields

Alain Bossavit and Francesca Rapetti

Abstract Theses notes present the duality of Whitney forms as a tool to describe manifolds by chains and fields by cochains. Relying on this duality, we can construct compatible discretization methods for PDEs, that are methods which respect the nature of the fields involved in the equations (the degrees of freedom have a physical meaning) as well as the geometric and topological structure of the continuous model. We briefly recall how it is possible to define high-order Whitney forms just refining the chains that describe the manifolds.

1 Introduction

Variational methods have an ancient and well known history. The solution to some field problem often happens to be the one, among a family of a priori "eligible" fields, that minimizes some energy-related quantity, or at least makes this quantity stationary with respect to small variations. By restricting the search of this optimum to a well-chosen finite subfamily of eligible fields, one obtains the desired finite system of equations, the solution of which provides a near-optimum. This powerful heuristics, or Rayleigh-Ritz method, applicable to most areas of physics, leads in a quite natural way to finite element (FE) methods. The finite dimensional subspace of the infinite dimensional one is constructed from a triangulation of the

A. Bossavit
Laboratoire de Génie Electrique de Paris – CNRS & SUPELEC, 11 rue Joliot Curie, Plateau de Moulon, 91192 Gif sur Yvette, France
e-mail: bossavit@lgep.supelec.fr

F. Rapetti (✉)
Laboratoire de Mathématiques "J.A. Dieudonné", UMR 7351 CNRS & Université de Nice Sophia-Antipolis, Parc Valrose, 06108 Nice Cedex 02, France
e-mail: frapetti@unice.fr; Francesca.Rapetti@unice.fr

M. Azaïez et al. (eds.), *Spectral and High Order Methods for Partial Differential Equations - ICOSAHOM 2012*, Lecture Notes in Computational Science and Engineering 95, DOI 10.1007/978-3-319-01601-6_14,

given domain and polynomial spaces on each simplex, pieced together by a certain assembling process. With Maxwell equations, FEs should be able to represent a field from a finite number of degrees of freedom (dofs) but the nature of dofs, fluxes, circulations, etc., associate them to geometric mesh elements other than nodes. Since the beginning of the 1980s, the magnetomotive force (mmf) or edge circulations along branches seemed indeed preferable to describe the magnetic field, H in an eddy-current computation (see first numerics in [2]). The problem was to be able to interpolate from edge mmfs: Knowing the circulation of H along edges of a tetrahedral finite element mesh, by which interpolating formula to express the magnetic field inside each tetrahedron? The paper [9] was decisive in proposing such a formula: $H(x) = a \times x + b$, where x is the position, and a, b, two ordinary vectors, tetrahedron-dependent. However, the analytic form of these shape functions was a puzzle (why precisely $a \times x + b$?) and remained so for several years. A key piece was provided in [8] where a connection was made with a less well-known concept from differential geometry, namely Whitney forms. This was the beginning of a long work of reformulation [3] and the basis for several further developments. In Sect. 2 we detail the three aspects: fields as differential forms, Whitney forms on a tetrahedron and their recursive formula, $a \times x + b$ and $\alpha x + b$ as proxies (i.e., representing vectors) for Whitney 1- and 2-forms. Section 3 is dedicated to high-order forms. This contribution is part of a collection on differential forms and their applications. To avoid repetitions, we introduce the notation which is strictly necessary to the understanding of this work. For further details, we refer to [1, 3], and references therein.

2 From Proxies to p-Forms and Whitney Finite Element Bases

In physics, a field is a function, defined over the entire space or over a portion of it, which associates to each point x a value related to the physical quantity under exam. The concept of field is very useful to describe the perturbations of the spacetime properties due to the presence of a source. However, the human eye cannot see the electromagnetic field itself but only its effects on suitable objects.

2.1 Fields as Differential Forms

The presence of an electric field E in a given domain Ω is detected through its force acting on an electric charge q positioned in Ω. Indeed, let $E(x)$ be the electric field, that we assume well defined, at a point $x \in \Omega$. The vector $E(x)$ represents the force exerted by the field over a unit electric charge q placed at x. In first approximation, this force is known via the virtual work $E(x) \cdot v$ involved in moving q from x to $x+v$, where v stands for the virtual displacement. Here, we assume that q is small enough not to perturb E and that v is small enough to have $E(x) \approx E(x+v)$. The mapping $v \rightarrow E \cdot v$ being linear and continuous on the space V of three-dimensional real vectors, it is an element e of V^*, the dual of V. The elements of V^* are called *covectors* and the duality pairing between a vector v and a covector e is denoted $\langle v; e\rangle$. With this new notation, we have $v \cdot E = \langle v; e\rangle$, which indicates that the electric field is totally described by the application $x \rightarrow e(x)$, that is a field of covectors. It is named *differential* 1*-form* of polynomial degree 1 [5]. While comparing the two sides of the identity $v \cdot E = \langle v; e\rangle$, we may remark the two possible mathematical descriptions of the electric field. The one on the left, which uses the vector E in combination with the metric structure defined on the space by the dot product "·", and the one on the right which uses the 1-form e, a mathematical object which is metric-independent. We say that the vector E is the proxy for the entity e.

To understand the importance of the 1-form e in the description of the electric field, let us consider the electromotive force along an oriented curve c. The curve c, which starts in x_0 and ends in x_n, is replaced by $p_n^t c$ (the transposition will be soon clarified), an oriented polygonal line with vertices $x_j \in c$, $j = 0, \ldots, n$, and sides $\{x_{i-1}, x_i\}$, $i = 1, \ldots, n$. Calling v_i the vector from x_{i-1}, to x_i, we compute $\sum_{i=1}^n \langle v_i; e(x_{i-1})\rangle$. By increasing n to infinity, if $p_n^t c$ is well defined, we have

$$\lim_{n\to\infty} \langle p_n^t c; e\rangle = \lim_{n\to\infty} \sum_{i=1}^{n} \langle v_i; e(x_{i-1})\rangle = \langle c; e\rangle := \int_c e.$$

This limit number is the electromotive force along c and it is what a voltmeter would measure if it were connected to the extremities x_0 and x_n of c. Since c can be any smooth oriented curve between x_0 and x_n, the 1-form e contains all the information which is physically important about the electric field. The curve c is the probe, the 1-form is the electric field and $\langle c; e\rangle$ is the measurement of the electromotive force.

The value $\langle c; e\rangle$ is classically computed by evaluating the *circulation* of the proxy field E, namely, the line integral $\int_c \tau(x) \cdot E(x)dx$, where $\tau(x)$ is the unit tangent vector at a point $x \in c$, oriented in the forward direction (from x_0 to x_n) along c and dx the elementary measure of lengths. Comparing the two expressions for the electromotive force, $\langle v; e\rangle = \int_c \tau(x) \cdot E(x)dx$, one remarks that the metric of the space induced by the dot product "·" is not essential on the left side, thus it does not participate to the physical description of the electric field. Whereas with proxies on the right, if the metric changes for another one associated to a new scalar product $\bullet$, then we should modify the proxy for the electric field, say $E^\bullet$, define a new unit tangent vector $\tau^\bullet$ to the curve c and a new elementary measure $d^\bullet x$ of lengths, in order to preserve the equality $\int_c \tau(x) \cdot E(x)dx = \int_c \tau^\bullet(x) \bullet E^\bullet(x)d^\bullet x$ between the two expressions. Indeed, the electromotive force of E along c has a unique value which is $\langle c; e\rangle$. Thus, fields are not vectors, and the 1-form e should be used to represent the electric field as a physical entity.

Similar considerations apply to the magnetic induction. The presence of a magnetic field B in a given domain Ω is detected through the current flowing in a closed loop in motion across Ω or through the displacement of the wheel of a magnetic compass set in Ω. One fully knows it when, for any oriented surface S, the induction flux $\langle S; b\rangle$ embraced by S is known. Faraday's law then connects the rate of variation of this flux with the electromotive force $\langle \partial S; e\rangle$ along the boundary ∂S:

$$d_t\langle S; b\rangle + \langle \partial S; e\rangle = 0.$$

Note that here S must not change in time, so that one has $d_t\langle S; b\rangle = \langle S; \partial_t b\rangle$. Orienting the surface means providing it with a gyratory sense. Moreover, the orientations of S and of its boundary ∂S should match, that is, the forward direction along ∂S should agree with the gyratory sense assigned to S. To understand $\langle S; b\rangle$, where b is a 2-form, one should imagine S replaced by a collection of small patches S_i, each with a vertex at $x_i \in S$ and compute $p_n^t S = \sum_{i=1}^{n} \pm\langle S_i; b(x_{i-1})\rangle$, where the $\pm$ sign depends on whether the patch S_i has or not the same orientation of S. As soon as we increase n to infinity, if $p_n^t S$ is well defined, we have

$$\lim_{n\to\infty} \langle p_n^t S; b\rangle = \lim_{n\to\infty} \sum_{i=1}^{n} \langle S_i; b(x_i)\rangle = \langle S; b\rangle := \int_S b.$$

The behavior of B as proxy of the mapping b from surfaces to reals (fluxes) is more complex than that of E. Not only has B to change if the metric changes in order to keep the flux $\langle S; b\rangle$ invariant, but with unchanged metric the sign of B depends on orientation conventions. But the metric has not disappeared from Maxwell equations when replacing proxies by forms, it is hidden in the constitutive relations among fields. These relations written in terms of forms make sense if the physical parameters are interpreted as (Hodge) operators (see additional details in [3, 7]).

2.2 Whitney Forms in a Finite Element Context

We now focus on Whitney 1-forms which allow us to describe e in a finite element context. Given the domain Ω, we consider a mesh m over $\overline{\Omega}$ by d-simplices and label n, a, f, t the nodes, edges, etc., each with its own orientation and belonging to the set $\mathcal{N}$, $\mathcal{A}$, $\mathcal{F}$, $\mathcal{T}$, respectively. We denote by w^n, w^a, etc., the corresponding Whitney forms of degree 1 associated to $n, a, \ldots$ that we introduce in a moment. In a finite element framework, the 1-form e is approximated by $\sum_{a\in\mathcal{A}} e_a w^a$ which we shall denote by $p_{\mathrm{m}} e$, where p_{m} is the interpolation operator of a field onto the Whitney forms. The mapping $e \to \sum_{a\in\mathcal{A}} e_a w^a$ is the composition of the de Rham map $e \to \mathbf{e} = (e_a)_{a\in\mathcal{A}}$ and of the Whitney map $\mathbf{e} \to \sum_{a\in\mathcal{A}} e_a w^a$. Suppose that we replace $c \in \Omega$ by a p-chain $p_{\mathrm{m}}^t c = \sum_{a\in\mathcal{A}} \langle c; w^a\rangle a$, where p_{m}^t is the operator mapping a p-manifold c in its "finite" representation (weighted sum of

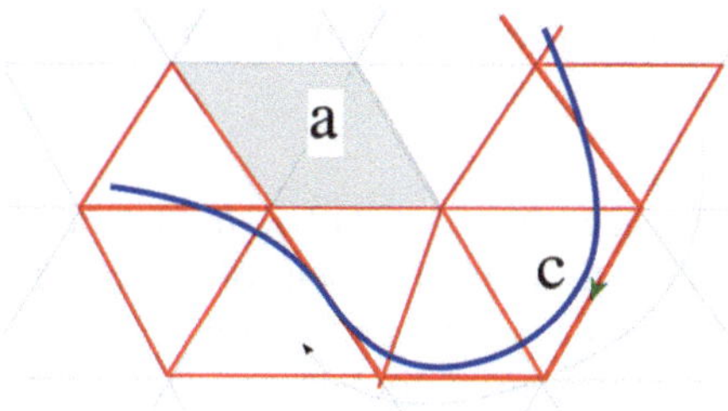

Fig. 1 Let e_a be the electromotive force of e along a and $\langle c; w^a \rangle = \int_c w^a$ the weight of the edge a in the representation of c. If $c \sim p^t_{\mathrm{m}} c = \sum_{a \in \mathscr{A}} \langle c; w^a \rangle a$ then $\int_c e \sim \sum_{a \in \mathscr{A}} \langle c; w^a \rangle \int_a e$ by linearity, from which we get $\int_c e \sim \sum_{a \in \mathscr{A}} \langle c; w^a \rangle e_a = \langle c; \sum_{a \in \mathscr{A}} e_a w^a \rangle$ and thus $e \sim p_{\mathrm{m}} e = \sum_{a \in \mathscr{A}} e_a w^a$

p-simplices in m) and let us interpret the scalars e_a as the elementary values $\int_a e$ (circulations, here). Then a natural approximation of $\int_c e$ is obtained by substituting $p^t_{\mathrm{m}} c$ for c. Hence an approximate knowledge of the field e, i.e., of all its measurable attributes, from the array $\mathbf{e} = \{e_a : a \in \mathscr{A}\}$. The problem is then: "how best to represent c by a chain ?". Solving it yields, by duality, a *definition* of Whitney forms [13]: w^a, for instance, is, like the field e itself, a covector, a map from lines c to real numbers $\langle c; w^a \rangle$. Note that, with this convention,

$$\begin{aligned}\langle p^t_{\mathrm{m}} c; e \rangle &= \langle \textstyle\sum_{a \in \mathscr{A}} \int_c w^a a; e \rangle = \sum_{a \in \mathscr{A}} \int_c w^a \langle a; e \rangle \\ &= \textstyle\sum_{a \in \mathscr{A}} \langle c; w^a e_a \rangle = \langle c; \sum_{a \in \mathscr{A}} w^a e_a \rangle = \langle c; p_{\mathrm{m}} e \rangle.\end{aligned}$$

This formula states that the evaluation of the *exact* electric field e along the *approximated* curve $p^t_{\mathrm{m}} c$ is equal to the evaluation of the *approximated* electric field $p_{\mathrm{m}} e$ along the exact curve c. We may remark how the same w^a appears in both expressions $p^t_{\mathrm{m}} c$ and $p_{\mathrm{m}} e$, thus the strong connection between geometry and physical variables associated with that geometry. For the moment, we just say that w^a is the Whitney form of polynomial degree 1 associated to a.

In Fig. 1, we represent the curve c by a 1-chain, a weighted average of mesh edges. Variations in the thickness of edges suggest differences in the weights $\langle c; w^a \rangle$ which are assigned in the 1-chain $p^t_{\mathrm{m}} c$ to the different mesh edges a. To edges as a, whose domain of influence (gray area) does not intersect c, we assign weight zero. A weight can be positive or negative depending on the orientation of the mesh edge with respect to that of c. How these weights should be assigned is the central point in the construction of Whitney's forms. Note how this justifies the " p^t_{m} " notation. Whitney's forms, on the one hand, reconstruct a cochain from a vector of dofs (this is p_{m}) and, on the other hand, represent a p-manifold by a p-chain (this is p^t_{m}).

2.3 Recursive Formula for Whitney Forms on a Simplicial Mesh

In the following, we use the same symbol n to denote a node as well as its position vector with respect to a given system of coordinates. To each node n of the mesh m in $\bar{\Omega}$, we attribute a function whose value at point x is 0, if none of the tetrahedra with a vertex in n contains x, otherwise, it is the barycentric coordinate at x with respect to n, denoted by $\lambda_n(x)$, in the affine basis provided by the vertices of the tetrahedron to which x belongs. Note that by construction, $\lambda_n(x) \geq 0$ and $\sum_{n\in\mathcal{N}} \lambda_n(x) = 1$ for all $x \in \overline{\Omega}$. The λ_ns themselves are often called "hat functions". Any point x of the meshed domain can be represented by $p^t_{\mathrm{m}}(x) = \sum_{n\in\mathcal{N}} \lambda_n(x)n = \sum_{n\in\mathcal{N}}\langle x;\lambda_n\rangle n$, where λ_n is the only piecewise affine (by restriction to each tetrahedron) function that takes value 1 at node n and 0 at all other nodes. Note that $x \equiv p^t_{\mathrm{m}}x$, for any point $x \in \bar{\Omega}$. The definition of $p_{\mathrm{m}}u = \sum_{n\in\mathcal{N}} u_n w^n$ for any scalar field u defined on Ω is obtained by transposition:

$$\begin{aligned}\langle p^t_{\mathrm{m}}x; u\rangle &= \langle \textstyle\sum_{n\in\mathcal{N}} \lambda_n(x)n; u\rangle = \sum_{n\in\mathcal{N}} \lambda_n(x)\langle n; u\rangle \\ &= \textstyle\sum_{n\in\mathcal{N}} w^n(x)u_n = \langle x; \sum_{n\in\mathcal{N}} w^n u_n\rangle = \langle x; p_{\mathrm{m}}u\rangle.\end{aligned}$$

Again Whitney forms have a double feature: the 0-forms $w^n(\equiv \lambda_n)$ allow to represent a generic point as a linear combination of the mesh nodes, as well as reconstruct a scalar field from its nodal values at the mesh nodes. This way of reasoning is repeated for any p-form, $0 < p \leq d$, where d is the dimension of the ambient space Ω, to prove the explicit (recursive) formula for Whitney p-forms of polynomial degree 1 firstly stated in [4] (see [10] for a related algorithm to compute the weights). To this purpose, we need to introduce the incidence matrices of a simplicial complex talking about the *inner orientation* of p-simplices, which is intrinsic and does not depend on the simplex being embedded in a larger space. But, giving a crossing direction for a surface is outer orientation. Outer orienting a line is the same as giving a way to turn around it. Outer orientation is involved in the definition of twisted forms (as the electric current density j, see [3] for a complete presentation).

We attribute an orientation to nodes, just assigning $+1$ to each $n \in \mathcal{N}$. Note that $\partial n = 0$ for each n. An edge a (resp. a face f, a volume v) is by definition an ordered couple (resp. triplet, quadruplet) of vertices, not merely a collection. We will make the convention that the edge $a = \{m, n\}$ is oriented from m to n. All edges of the mesh are oriented, and the opposite edge $\{n, m\}$ is not supposed to belong to $\mathcal{A}$ if a does. We may define the so-called incidence numbers $G^{mn}_m = -1$, $G^{mn}_n = 1$, and $G^{mn}_k = 0$ for nodes k other than m and n. These numbers form a rectangular matrix G with $\mathcal{N}$ and $\mathcal{A}$ as column set and row set, which describes how edges connect to nodes. The expression $\partial a = \sum_{k\in\mathcal{N}} \mathbf{G}^k_a k$ expresses the boundary of the edge a as a formal linear combination of nodes (such a thing is called 1-chain). Here $\partial a = n - m$. Faces also are oriented and we shall adopt a similar convention to give the list of nodes: a face $f = \{l, m, n\}$ has the three vertices l, m, n and an inner orientation inducing the orientation $l \to m \to n \to l$ on its boundary ∂S. We regard even permutations of nodes, $\{m, n, l\}$ and $\{n, l, m\}$, as being the same face, and odd permutations as defining the oppositely oriented face, which is not supposed to belong to $\mathscr{F}$ if f does. In terms of incidence numbers, $\mathbf{R}_a^f$ is $+1$ if a runs along the boundary of f, -1 otherwise, and 0 if a is not one of the edges of f. Hence a matrix $\mathbf{R}$, indexed over $\mathscr{A}$ and $\mathscr{F}$, which describes how edges bound faces. The expression $\partial f = \sum_{k \in \mathscr{A}} \mathbf{R}^a_f a$ expresses the boundary of the face f as a formal linear combination of edges (such a thing is called 2-chain). For the given f, we have $\partial f = lm + mn - ln$ if edges $lm, mn, ln \in \mathscr{A}$. A matrix $\mathbf{D}$, indexed over $\mathscr{F}$ and $\mathscr{V}$, is similarly defined to describe how faces bound volumes: $\mathbf{D}^v{}_f = \pm 1$ if face f bounds tetrahedron v, the sign depending on whether the orientations of f and of the boundary of v match or not. In the adopted convention, if $v = \{k, l, m, n\}$, the vectors kl, km, and kn, in this order, define a positive frame. Note that $\{l, m, n, k\}$ has the opposite orientation, so it does not belong to $\mathscr{V}$ if v does.

Proposition 1. *For any integer $0 < p \le d$, where d is the ambient dimension in Ω, the Whitney p-form w^s associated to the p-simplex s of a mesh m in $\bar{\Omega}$ satisfies*

$$w^s(x) = \sum_{\sigma \in \{(p-1)-\text{simplices}\}} \partial^s_\sigma \lambda_{s-\sigma}(x) \mathrm{d}w^\sigma, \qquad x \in \Omega \tag{1}$$

where ∂^s_σ is the incidence matrix entry linking σ to s, d is the exterior derivative operator from $(p-1)$-forms to p-forms (which is related to the boundary operator ∂ from p-chains to $(p-1)$-chains by the Stokes theorem: $\langle S; \mathrm{d}w \rangle = \langle \partial S; w \rangle$, for all p-chains S and $(p-1)$-forms w) and w^σ is the $(p-1)$-form associated to σ.

Proof. The proof relies on a reasoning by recurrence on the dimension p of manifolds, $0 \le p \le d$. More precisely, we know how a $(p-1)$-manifold can be represented by a $(p-1)$-chain, and we wish to use this information to represent a p-manifold by a p-chain. We take $d = 3$ for the proof. For a compact formula, we denote the node l by $f - a$ (with $\lambda_{f-a} = \lambda_l$) and k by $v - f$ (with $\lambda_{v-f} = \lambda_k$).

Let xy be the oriented segment going from point x to point y entirely contained in the simplicial mesh of $\bar{\Omega}$. We know that for one point y we have $p^t_{\mathrm{m}}(y) = \sum_{n \in \mathscr{N}} \langle y; w^n \rangle n$, and we expand $p^t_{\mathrm{m}}(xy)$ by linearity: $p^t_{\mathrm{m}}(xy) = \sum_{n \in \mathscr{N}} \langle y; w^n \rangle p^t_{\mathrm{m}}(xn)$. Figure 2 suggests the only possible way to find out $p^t_{\mathrm{m}}(xn)$: take the "average" of all mesh edges with one extremity in n and weights given by the barycentric weights of x with respect to the other extremity of these edges. Here $\partial^s_\sigma = \mathbf{G}^a_n$ and we may write $p^t_{\mathrm{m}}(xn) = \sum_{a \in \mathscr{A}} \mathbf{G}^a_n \lambda_{a-n}(x) a$. Then

$$p^t_{\mathrm{m}}(xy) = \sum_{a \in \mathscr{A}} \langle xy; w^a \rangle a = \sum_{n \in \mathscr{N}, a \in \mathscr{A}} \mathbf{G}^a_n \lambda_{a-n}(x) \langle y; w^n \rangle a.$$

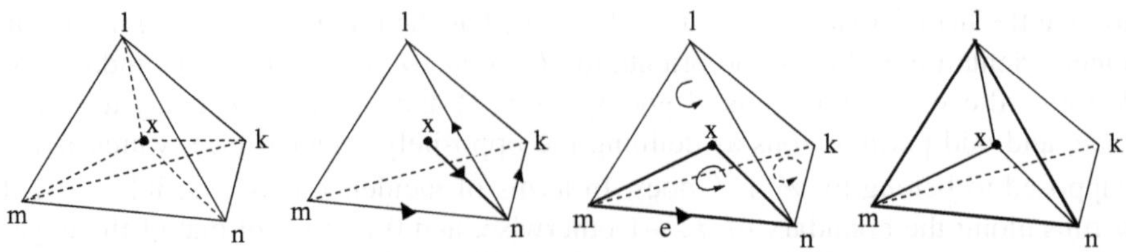

Fig. 2 Let $d = 3$. This figure explains how a p-manifold can be represented by a p-chain. The 0-chain $p^t_{\mathrm{m}}x$ is $\sum_{i=m,n,k,l} \lambda_i(x)i$. The 1-chain $p^t_{\mathrm{m}}(xn)$ associated with the oriented segment xn is $\lambda_m(x)mn - \lambda_k(x)nk - \lambda_l(x)nl$. The minus sign in front of $\lambda_l(x)$ (resp. $\lambda_k(x)$) is due to the fact that nl (resp. nk) starts in n, and ends in l (resp. k). Edges as ml or lk which do not have n as vertex make no contribution to the 1-chain for xn. The 2-chain $p^t_{\mathrm{m}}(xmn)$ associated with the oriented face xmn is $\lambda_l(x)mnl + \lambda_k(x)mnk$. Faces as mkl or nkl, which do not have mn as edge, make no contribution to $p^t_{\mathrm{m}}(xmn)$. The 3-chain $p^t_{\mathrm{m}}(xmnl)$ associated with the oriented volume $xmnl$ is $\lambda_k(x)mnlk$

Hence $\langle xy; w^a\rangle = \sum_{n\in\mathscr{N}} \mathbf{G}^a_n \lambda_{a-n}(x)\langle y; w^n\rangle$. Subtracting $\langle xx; w^a\rangle = 0$ from $\langle xy; w^a\rangle$, with $\langle xx; w^a\rangle = \sum_{n\in\mathscr{N}} \mathbf{G}^a_n \lambda_{a-n}(x)\langle x; w^n\rangle$, and being d the dual of ∂, we get

$$\begin{aligned}\langle xy; w^a\rangle &= \textstyle\sum_{n\in\mathscr{N}} \mathbf{G}^e_n \lambda_{a-n}(x)\langle y - x; w^n\rangle \\ &= \textstyle\sum_{n\in\mathscr{N}} \mathbf{G}^a_n \lambda_{a-n}(x)\langle \partial(xy); w^n\rangle \\ &= \textstyle\sum_{n\in\mathscr{N}} \mathbf{G}^a_n \lambda_{a-n}(x)\langle xy; \mathrm{d}w^n\rangle = \langle xy; \sum_{n\in\mathscr{N}} \mathbf{G}^a_n \lambda_{a-n}(x)\mathrm{d}w^n\rangle.\end{aligned}$$

Let xyz be the oriented face entirely contained in the cluster of tetrahedra in $\bar{\Omega}$. We know that $p^t_{\mathrm{m}}(yz) = \sum_{a\in\mathscr{A}} \langle yz; w^a\rangle a$, and we figure out $p^t_{\mathrm{m}}(xyz)$ by linearity: $p^t_{\mathrm{m}}(xyz) = \sum_{a\in\mathscr{A}} \langle yz; w^a\rangle\, p^t_{\mathrm{m}}(xa)$, with xa the face of vertices x and those of the edge a. Figure 2 suggests the only reasonable way to find out $p^t_{\mathrm{m}}(xa)$: take the "average" of all mesh faces with one edge on a and weights given by the barycentric weights of x with respect to the other vertices of these faces that are not ending points of a. Here $\partial^s_\sigma = \mathbf{R}^f_a$ and $p^t_{\mathrm{m}}(xa) = \sum_{f\in\mathscr{F}} \mathbf{R}^f_a \lambda_{f-a}(x) f$. Then

$$p^t_{\mathrm{m}}(xyz) = \sum_{a\in\mathscr{A}, f\in\mathscr{F}} \mathbf{R}^f_a \lambda_{f-a}(x)\langle yz; w^a\rangle f = \sum_{f\in\mathscr{F}} \langle xyz; w^f\rangle f.$$

Hence $\langle xyz; w^f\rangle = \sum_{a\in\mathscr{A}} \mathbf{R}^f_a \lambda_{f-a}(x)\langle yz; w^a\rangle$. Adding $\langle xzx; w^f\rangle = 0$ and $\langle xxy; w^f\rangle = 0$ to $\langle xyz; w^f\rangle$, we have

$$\begin{aligned}\langle xyz; w^f\rangle &= \textstyle\sum_{a\in\mathscr{A}} \mathbf{R}^f_a \lambda_{f-a}(x)\langle yz + zx + xy; w^a\rangle \\ &= \textstyle\sum_{a\in\mathscr{A}} \mathbf{R}^f_a \lambda_{f-a}(x)\langle \partial(xyz); w^a\rangle \\ &= \textstyle\sum_{a\in\mathscr{A}} \mathbf{R}^f_a \lambda_{f-a}(x)\langle xyz; \mathrm{d}w^a\rangle = \langle xyz; \sum_{a\in\mathscr{A}} \mathbf{R}^f_a \lambda_{f-a}(x)\mathrm{d}w^a\rangle.\end{aligned}$$

Finally, let $xyzr$ be the oriented volume entirely contained in the cluster of tetrahedra in $\bar{\Omega}$. We know that $p^t_{\mathrm{m}}(yzr) = \sum_{f\in\mathscr{F}} \langle yzr; w^f\rangle f$, and we express

$p^t_{\mathrm{m}}(xyzr)$ by linearity: $p^t_{\mathrm{m}}(xyzr) = \sum_{f\in\mathcal{F}} \langle yzr; w^f\rangle p^t_{\mathrm{m}}(xf)$, with xf the volume of vertices x and those of the face f. For $p = 3$, we have $\partial^s_\sigma = \mathbf{D}^v_f$ and as suggested in Fig. 2, $p^t_{\mathrm{m}}(xf) = \sum_{v\in\mathcal{V}} \mathbf{D}^v_f \lambda_{v-f}(x)v$. Thus

$$p^t_{\mathrm{m}}(xyzr) = \sum_{v\in\mathcal{V}} \langle xyzr; w^v\rangle v = \sum_{f\in\mathcal{F},v\in\mathcal{V}} \mathbf{D}^v_f \lambda_{v-f}(x)\langle yzr; w^f\rangle v.$$

Hence $\langle xyzr; w^v\rangle = \sum_{f\in\mathcal{F}} \mathbf{D}^v_f \lambda_{v-f}(x)\langle yzr; w^f\rangle$. Adding $\langle xxzy; w^v\rangle = 0$, $\langle xxyr; w^v\rangle = 0$ and $\langle xxrz; w^f\rangle = 0$ to $\langle xyzr; w^v\rangle$ we obtain

$$\begin{aligned}\langle xyzr; w^v\rangle &= \textstyle\sum_{f\in\mathcal{F}} \mathbf{D}^v_f \lambda_{v-f}(x)\langle yzr + xzy + xyr + xrz; w^f\rangle \\ &= \textstyle\sum_{f\in\mathcal{F}} \mathbf{D}^v_f \lambda_{v-f}(x)\langle \partial(xyzr); w^f\rangle \\ &= \textstyle\sum_{f\in\mathcal{F}} \mathbf{D}^v_f \lambda_{v-f}(x)\langle xyzr; \mathrm{d}w^f\rangle = \langle xyzr; \sum_{f\in\mathcal{F}} \mathbf{D}^v_f \lambda_{v-f}(x)\mathrm{d}w^f\rangle. \diamond\end{aligned}$$

2.4 *Proxies for Whitney Forms in the Finite Element Context*

The connection of Whitney forms with the lowest order Nédélec elements (see [9], Definitions 2 and 4 with $d = 3$ and $k = 1$) is revisited in the following lemma.

Proposition 2. *Let $t = \{m, n, k, l\}$ be a tetrahedron. The proxy $\mathbf{w}^{mn}$ associated to the 1-form w^{mn} reads as $a \times x + b$, where $a, b \in R^d$. The proxy $\mathbf{w}^{klm}$ associated to the 2-form w^{klm} reads as $\alpha x + b$ where $\alpha \in R$ and $b \in R^d$.*

Proof. The expression of the 1-form associated to mn as stated in Proposition 1 is $w^{mn} = w^m \mathrm{d}w^n - w^n \mathrm{d}w^m$. It is sufficient to replace the exterior derivative d by the gradient operator ∇ to have the proxy of w^{mn}, namely $\mathbf{w}^{mn} = w^m\nabla w^n - w^n\nabla w^m$. Let $|t|$ denote the volume of t and kl the vector starting in k and ending in l. We can write that $\mathbf{w}^{mn} = (kl \times kx)/(6|t|)$ since $kl \times kx \cdot nm = 6|t|$ for a point x lying on the edge mn and $\mathbf{w}^{mn} \cdot t_{mn} = 1$, t_{mn} is the unit tangent vector to the edge mn. Let o be the origin in the Cartesian coordinates, we obtain

$$\mathbf{w}^{mn} = [kl \times (ox + ok)]/(6|t|) = kl/(6|t|) \times x + ok/(6|t|) = a \times x + b.$$

The expression of the 2-form associated to klm as stated in Proposition 1 is

$$w^{klm} = w^m \mathrm{d}(w^k \mathrm{d}w^l - w^l \mathrm{d}w^k) + w^k \mathrm{d}(w^l \mathrm{d}w^m - w^m \mathrm{d}w^l) + w^l \mathrm{d}(w^m \mathrm{d}w^k - w^k \mathrm{d}w^m).$$

To obtain the expression of $\mathbf{w}^{klm}$, we need to involve additional properties, that are: (i) $\mathrm{d} \circ \mathrm{d} = 0$ (related to the fact that $\partial \circ \partial = 0$); (ii) $\mathrm{d}(\kappa w) = \mathrm{d}\kappa \wedge w + \kappa \mathrm{d}w$, where κ is a scalar field, w is a form and $\wedge$ the exterior product between forms (see [5] for more details); (iii) ${}^1u \wedge {}^1v = {}^2(\mathbf{u} \times \mathbf{v})$, where ${}^1u$ denotes a 1-form, ${}^2u$ a 2-form and $\times$ is the cross product between vectors.

Using (i) and (ii), we have, for instance, that $\mathrm{d}(w^l \mathrm{d}w^m) = w^l \mathrm{dd}w^m + \mathrm{d}w^l \wedge \mathrm{d}w^m = \mathrm{d}w^l \wedge \mathrm{d}w^m$. We thus obtain $w^f = 2(w^n \mathrm{d}w^l \wedge \mathrm{d}w^m + w^l \mathrm{d}w^m \wedge \mathrm{d}w^n + w^m \mathrm{d}w^n \wedge \mathrm{d}w^l)$. Using (iii), we have the proxy associated to w^f that reads $\mathbf{w}^f = 2(w^n \nabla w^l \times \nabla w^m + w^l \nabla w^m \times \mathrm{d}w^n + w^m \nabla w^n \times \nabla w^l)$. Let h be the height of node n above the plane of $f = \{k,l,m\}$. Then $|t| = h\,|\{k,l,m\}|/3 = |\{k,l,m\}|/(3|\nabla w^n|)$. Thus, we may write $\nabla w^m \times \nabla w^n = [(kn \times kl) \times \nabla w^n]/(6|t|)$ and $\mathbf{w}^f = 2(w^l kl + w^m km + w^n kn)/(6|t|)$. If we choose k as the origin in the Cartesian coordinates from $x = \sum_i w^i(x) i$ we have $kx = \sum_i w^i(x) ki$. Thus $\mathbf{w}^f(x) = kx/(3|t|) = [ox + ok]/(3|t|) = \alpha x + b$. $\diamond$

3 High-Order Whitney Forms

Three key heuristic points underlie the construction of high-order Whitney p-forms: (1) High order forms should satisfy a certain "partition of unity" property (which stems for consistency); (2) They should pair up with integration domains of dimension p (which we shall build from "small" p-simplices, appropriate homothetic images of the mesh p-simplices); (3) The spaces they span should constitute an exact sequence. On each tetrahedron, higher order p-forms are here obtained as product of Whitney p-forms of degree 1 by suitable homogeneous monomials in the barycentric coordinate functions of the simplex (we invite the interested reader to find more details in [12]). The construction we propose can be summarized in the following two steps.

Let v be a tetrahedron of the mesh m in $\bar{\Omega}$ and $\mathscr{I}(d+1,k)$ the set of multi-indices $\mathbf{k}$ with non-negative integer $d+1$ components k_i and weight $k = \sum_i k_i$. To each multi-integer $\mathbf{k} \in \mathscr{I}(d+1,k)$ corresponds a map, denoted by $\tilde{\mathbf{k}}$, from v into itself. Let $\tilde{k}_i$ denote the affine function that maps [0, 1] onto $[k_i/(k+1), (1+k_i)/(k+1)]$. If $\lambda_i(x)$, $0 \le i \le d$, are the barycentric coordinates of point $x \in v$ with respect to the vertex n_i of v, its image $\tilde{\mathbf{k}}(x)$ has barycentric coordinates $\tilde{k}_i(\lambda_i(x)) = (\lambda_i(x) + k_i)/(k+1)$. We call small p-simplices of v, $0 \le p \le d$, the images $\tilde{\mathbf{k}}(S)$ for all (big) p-simplices $S \in v$ and all $\mathbf{k} \in \mathscr{I}(d+1,k)$, and denote them by $s = \{\mathbf{k}, S\}$.

Whitney p-forms of higher degree in each volume v are more "numerous" therefore they should be associated to a finer geometric structure in v. Indeed, they are associated to the geometric collection in v defined by the $\tilde{\mathbf{k}}$ map for all possible multi-indices $\mathbf{k} \in \mathscr{I}(d+1,k)$ as follows. Let $\lambda^{\mathbf{k}}$ denote the homogeneous polynomial $\Pi_{i=0,\dots,d} \lambda_i^{k_i}$ of degree k. Whitney p-forms of polynomial degree $k+1$ in a volume v are the $w^s = \lambda^{\mathbf{k}} w^S$, where s is a pair $\{\mathbf{k}, S\}$, made of a multi-index $\mathbf{k} \in \mathscr{I}(d+1,k)$ and a (big) p-simplex $S \in v$, and w^S the Whitney p-form of polynomial degree 1 associated to S (see Proposition 1). The space of Whitney p-forms of polynomial degree $N = k+1$ in v is $W^p_{k+1}(v) = span\{w^s : s = \{\mathbf{k}, S\}, \mathbf{k} \in \mathscr{I}(d+1,k), S \in \mathscr{S}^p(v)\}$. This construction is completely in the spirit of what has been presented in the previous section: Whitney forms are best viewed as a device to represent manifolds by simplicial chains. The representation gets better and better as the simplices get smaller and smaller (we talk about h-refinement with the big simplices, N-refinement with the small ones) and, by duality, we improve the

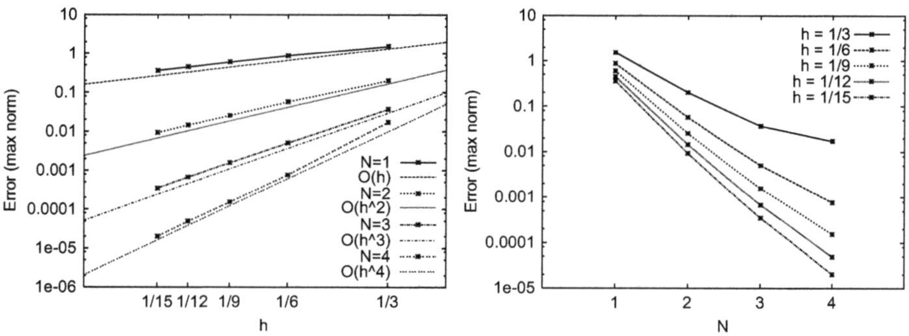

Fig. 3 Algebraic h-convergence (log-plot, *left*) and spectral N-convergence (semi log-plot, *right*) error rates for Whitney 1-forms [11] when applied to the model problem $[I + \nabla \times (\nabla \times)]u = f$ in $\Omega = [1/2, 3/2] \times [1/4, 3/4]$, with homogeneous Dirichlet boundary conditions on $\partial\Omega$ and internal source consistent with $u = (2\pi \sin(\pi x)\cos(2\pi y), -\pi \cos(\pi x)\sin(2\pi y))^t$ as exact solution; h is the maximal diameter of the volumes in $\mathfrak{m}$ and N is the polynomial degree of the forms

approximation of the differential form associated to the manifold, as shown in Fig. 3. Note that the forms $\lambda^{\mathbf{k}} w^S$ are not all linearly independent and in [6] we explain how it is possible to work with redundant bases without making a pre-selection on the basis functions which would break the symmetry of the construction (see [9] for the dimension of the spaces).

References

1. Arnold, D.N., Falk, R., Winther, R.: Finite element exterior calculus, homological techniques, and applications. Acta Numerica, 1–155 (2006)
2. Bossavit, A., Vérité, J.C.: A Mixed FEM-BIEM Method to Solve Eddy-Current Problems. IEEE Trans. on Magn. **18**, 431–435 (1982)
3. Bossavit, A.: Computational Electromagnetism. Academic Press, New York (1998)
4. Bossavit, A.: Generating Whitney forms of polynomial degree one and higher. IEEE Trans. on Magn. **38**, 341–344 (2002)
5. Burke, W.L.: Applied differential geometry. Cambridge Univ. Press, Cambridge U.K. (1985)
6. Christiansen, S.H., Rapetti, F.: On high order finite element spaces of differential forms. Preprint (2013)
7. Gerritsma, M., Hiemstra, R., Kreeft, J., Palha, A., Rebelo, P., Toshniwal, D.: The geometric basis of mimetic spectral approximations. Icosahom 2012 procs, to appear (2012)
8. Kotiuga, P.R.: Hodge Decompositions and Computational Electromagnetics. PhD Thesis, Dept. of Electrical Engineering, McGill University, Montréal (1984)
9. Nédélec, J.-C.: Mixed finite elements in R^3. Numerische Mathematik **35**, 315–341 (1980)
10. Rapetti, F.: Weight computation for simplicial Whitney forms of degree one. C. R. Acad. Sci. Paris Ser. I **341**(8), 519–523 (2005)
11. Rapetti, F.: High order edge elements on simplicial meshes. Meth. Math. en Anal. Num. **41**(6), 1001–1020 (2007)
12. Rapetti, F., Bossavit, A.: Whitney forms of higher degree. SIAM J. Numer. Anal. **47**(3), 2369–2386 (2009)
13. Whitney, H.: *Geometric integration theory*. Princeton Univ. Press (1957).

Fig. 3 Algebraic h convergence (log plot, *left*) and spectral N convergence (semi-log plot, *right*) of error rates for Whitney 1-forms [12] when applied to the model problem $[I + \nabla \times (\nabla \times)]u = f$ in $\Omega = [1/2, 3/2] \times [1/4, 3/4]$, with homogeneous Dirichlet boundary conditions on $\partial\Omega$ and internal source consistent with $u = (2\pi \sin(\pi x)\cos(2\pi y), -\pi \cos(\pi x)\sin(2\pi y))$ as exact solution. h is the maximal diameter of the simplices in m and N is the polynomial degree of the forms

approximation of the differential forms associated to the manifold, as shown in Fig. 3. Note that the forms $\lambda^k w^l$ are not all linearly independent and in [6] we explain how it is possible to work with redundant bases without making a pre-selection on the basis functions, which would break the symmetry of the construction. See [9] for the construction of the spaces.

References

1. Arnold, D.N., Falk, R., Winther, R.: Finite element exterior calculus, homological techniques, and applications. Acta Numerica, 1–155 (2006)
2. Bossavit, A., Vérité, J.C.: A Mixed FEM-BIEM Method to Solve Eddy-Current Problems. IEEE Trans. on Magn. 18, 431–435 (1982)
3. Bossavit, A.: Computational Electromagnetism. Academic Press, New York (1998)
4. Bossavit, A.: Generating Whitney forms of polynomial degree one and higher. IEEE Trans. on Magn. 38, 341–344 (2002)
5. Burke, W.L.: Applied differential geometry. Cambridge Univ. Press, Cambridge U.K. (1985)
6. Christiansen, S.H., Rapetti, F.: On high order finite element spaces of differential forms. Preprint (2013)
7. Gerritsma, M., Hiemstra, R., Kreeft, J., Palha, A., Rebelo, P., Toshniwal, D.: The geometric basis of numerical spectral approximations. Icosahom 2012 procs., to appear (2013)
8. Kotiuga, P.R.: Hodge Decompositions and Computational Electromagnetics. PhD Thesis, Dept. of Electrical Engineering, McGill University, Montreal (1984)
9. Nédélec, J.-C.: Mixed finite elements in R^3. Numerische Mathematik 35, 315–341 (1980)
10. Rapetti, F.: Weight computation for simplicial Whitney forms of degree one. C. R. Acad. Sci. Paris Ser. I 341(8), 519–523 (2005)
11. Rapetti, F.: High order edge elements on simplicial meshes. Meth. Math. en Anal. Num. 41(6), 1001–1020 (2007)
12. Rapetti, F., Bossavit, A.: Whitney forms of higher degree. SIAM J. Numer. Anal. 47(3), 2369–2386 (2009)
13. Whitney, H.: Geometric Integration Theory. Princeton Univ. Press (1957)

Exponential Convergence of the hp Version of Isogeometric Analysis in 1D

Annalisa Buffa, Giancarlo Sangalli, and Christoph Schwab

Abstract We establish exponential convergence of the hp-version of isogeometric analysis for second order elliptic problems in one spacial dimension. Specifically, we construct, for functions which are piecewise analytic with a finite number of algebraic singularities at a-priori known locations in the closure of the open domain Ω of interest, a sequence $(\Pi_\sigma^\ell)_{\ell\geq 0}$ of interpolation operators which achieve exponential convergence. We focus on localized splines of reduced regularity so that the interpolation operators $(\Pi_\sigma^\ell)_{\ell\geq 0}$ are Hermite type projectors onto spaces of piecewise polynomials of degree $p \sim \ell$ whose differentiability increases linearly with p. As a consequence, the degree of conformity grows with N, so that asymptotically, the interpoland functions belong to $C^k(\Omega)$ for any fixed, finite k. Extensions to two- and to three-dimensional problems by tensorization are possible.

1 Introduction

Isogeometric Analysis (IGA) is an innovative technique for the discretization of partial differential equations which has been proposed by T.J.R. Hughes et al. in 2005 in [6]. IGA is gaining a growing interest in different communities: mechanical engineering, numerical analysis and geometric modeling. In its simplest

---

A. Buffa
IMATI "E. Magenes", CNR, Via Ferrata 1, 27100 Pavia, Italy
e-mail: annalisa@imati.cnr.it

G. Sangalli
Dipartimento di Matematica "F. Casorati", Via Ferrata 1, 27100 Pavia, Italy
e-mail: giancarlo.sangalli@unipv.it

C. Schwab (✉)
SAM, ETH Zürich, HG G57.1, ETH Zentrum, CH 8092 Zürich, Switzerland
e-mail: schwab@math.ethz.ch; christoph.schwab@sam.math.ethz.ch

M. Azaïez et al. (eds.), *Spectral and High Order Methods for Partial Differential Equations - ICOSAHOM 2012*, Lecture Notes in Computational Science and Engineering 95, DOI 10.1007/978-3-319-01601-6_15,

formulation, IGA consists in solving a PDE with a Galerkin technique projecting onto the space of splines. In this paper, we consider IGA in conjunction with the hp- paradigm, i.e., simultaneous h and p refinements. We prove exponential convergence of IGA for a model elliptic PDE with piecewise analytic solutions.

Exponential convergence of piecewise polynomial approximations with a fixed degree of conformity for analytic functions with a point singularity was shown first for free-knot, variable degree spline approximation in [4,8] and the references there. Inspired by these results, the hp-version of the finite element method (FEM) for the numerical solution of elliptic problems was proposed in the mid 1980s by I. Babuška and B.Q. Guo (see [1] and the references there). Exponential convergence rates $\exp(-b\sqrt{N})$ with respect to the number of degrees of freedom N for the hp version of the standard, C^0-conforming FEM in one dimension were shown by Babuška and Gui in [5] for the model singular solution $u(x) = x^\alpha - x \in H_0^1(\Omega)$ in $\Omega = (0, 1)$. This result required σ-geometric meshes with *any subdivision ratio* $\sigma \in (0, 1)$ (in particular, for $\sigma = 1/2$ geometric element sequences Ω_i are obtained by successive element bisection towards $x = 0$) while the constant b in the convergence estimate $\exp(-b\sqrt{N})$ depends on the singularity exponent α as well as on σ. In one space dimension, the results were further refined and optimal values of σ as well as estimates on the actual value of b are known. Among all $\sigma \in (0, 1)$, the optimal value was shown in [5, 8] to be $\sigma_{\mathrm{opt}} = (\sqrt{2} - 1)^2 \approx 0.17$, see, in particular, [5, Theorem 3.2], provided that the geometric mesh refinement is combined with nonuniform polynomial degrees $p_i \geq 1$ in Ω_i which are *s-linear*, i.e., $p_i \sim si$, with the optimal slope s being $s_{\mathrm{opt}} = 2(\alpha - 1/2)$. In this case, the finite element error converges as $\exp(-b\sqrt{N})$ where $b = 1.76\ldots \times \sqrt{(\alpha - 1/2)}$. For the bisected geometric mesh where $\sigma = 1/2$ and for linear polynomial degree distributions with slope $s_{\mathrm{opt}} = 0.39\ldots \times (\alpha - 1/2)$, one has $b = 1.5632\ldots \times \sqrt{(\alpha - 1/2)}$, whereas for $\sigma = 1/2$ and uniform polynomial degree, $b = 1.1054\ldots \times \sqrt{(\alpha - 1/2)}$; see [5, Table 1]. It was left open in [4,5,8] if the convergence rate $\exp(-b\sqrt{N})$ is optimal.

In the present paper, we investigate the rate of convergence of the hp version of isogeometric analysis when local splines are used. Indeed, we consider the space of splines, defined on an open knot vector on [0, 1], of degree p and of conformity $\lfloor \frac{p-1}{2} \rfloor$ which is proportional to the polynomial degree p.

In this case, the Hermite-type interpolant and the related analytic convergence estimates proposed in [2] are available and are used here to establish exponential convergence of the hp-version of isogeometric FEM for solutions which are piecewise analytic. Indeed, we prove that for piecewise analytic functions in one space dimension. We prove that for all functions in a countably normed space equipped with a family of weighted Sobolev norms which contain in particular the singular functions $u(x) = x^\alpha - x \in H_0^1(\Omega)$, the interpolation error on reduced spline spaces defined on families $\{\mathcal{G}_\sigma^M\}_{M\geq 1}$ of geometric knot meshes is exponentially decreasing at the rate $\exp(-b\sqrt{N})$. We should note that in terms of the number of degrees of freedom, the approximations we consider are linear, i.e. non-adaptive.

Our numerical examples show that exponential convergence rate $\exp(-b\sqrt{N})$, with respect to the number of degrees of freedom N, is attained. We also show that the constant b is considerably larger than in the case of standard hp-finite elements (with constant p), which enforce merely interelement continuity, but not smoothness across interelement boundaries.

2 Model Problem

In the bounded interval $\Omega = (0, 1)$, we consider the model Dirichlet problem

$$-(a(x)u')' + c(x)u = f \quad \text{in} \quad \Omega\,, \quad u(0) = 0,\ u(1) = 0\,. \tag{1}$$

We shall consider the Finite Element discretization of (1) based on the (standard) variational formulation. To this end, we introduce the space $V = H_0^1(\Omega)$. Then the variational formulation of (1) reads: find

$$u \in V: \quad a(u, v) = (f, v) \quad \forall v \in V\,, \tag{2}$$

where the bilinear form $a(\cdot,\cdot)$ is given by

$$a(w, v) = \int_0^1 (a(x)w'v' + c(x)wv)dx$$

and where $(\cdot,\cdot)$ denotes the $L^2(\Omega)$ innerproduct. Assuming that $a, c \in Ł^\infty(\Omega)$ and positivity of $a(x)$, i.e.,

$$\operatorname{ess\,inf}\{a(x) : x \in \Omega\} \geq \underline{a} > 0\,, \qquad \operatorname{ess\,inf}\{c(x) : x \in \Omega\} \geq 0\,, \tag{3}$$

there hold continuity and coercivity, i.e. exists $C(\underline{a}) > 0$ such that, for every $v, w \in V$ holds

$$a(v, v) \geq C(\underline{a})\|v\|^2_{H^1(\Omega)}\,, \quad |a(v, w)| \leq \max\{\|a\|_{L^\infty(\Omega)}, \|c\|_{L^\infty(\Omega)}\}\|v\|_{H^1(\Omega)}\|w\|_{H^1(\Omega)}\,, \tag{4}$$

and by the Lax-Milgram Lemma, for every $f \in (H^1(\Omega))^*$ exists a unique solution of (2).

Let $\{V_\ell\}_{\ell=0}^\infty$ denote a sequence of subspaces $V_\ell \subset V$ of finite dimensions $N_\ell = \dim V_\ell$. Below, we shall be interested in particular in the case when the subspaces are nested, i.e. when

$$V_0 \subset V_1 \subset \ldots \subset V_\ell \subset \ldots \subset V$$

and that V_ℓ are dense, i.e.

$$\forall u \in V : \quad \lim_{\ell\to\infty} \inf_{v_\ell \in V_\ell} \|u - v_\ell\|_{H^1(\Omega)} = 0 . \tag{5}$$

By (4), for every $\ell \geq 0$, there exists a unique Finite Element solution of the Galerkin approximation of (2) find

$$u_\ell \in V_\ell : \quad a(u_\ell, v) = (f, v) \quad \forall v \in V_\ell \tag{6}$$

which is quasioptimal, i.e. there holds

$$\|u - u_\ell\|_{H^1(\Omega)} \leq C \inf_{v_\ell \in V_\ell} \|u - v_\ell\|_{H^1(\Omega)} . \tag{7}$$

We next quantify, for particular classes of data f in (1), the regularity of solutions. Subsequently, we shall exhibit choices of FE spaces V_ℓ for which high (exponential) rates of convergence can be achieved.

3 Regularity

We shall be in particular interested in *piecewise analytic* solutions u of (1) which exhibit singularities at $x = 0$ (multiple, but finitely many, singularities in $\overline{\Omega}$ could be also considered and everything that follows will apply to this case with straightforward modifications).

To quantify the analytic regularity, for $x \in \Omega = (0, 1)$, we consider the weight function $\Phi_\beta(x) = x^\beta$, $\beta \in \mathbb{R}$. For integer $\ell \geq 0$ and for $k = \ell, \ell+1, \ldots$, we define the weighted seminorms

$$|u|_{H^{k,\ell}_\beta(\Omega)} = \|\Phi_{\beta+k-\ell} D^k u\|_{L^2(\Omega)} \tag{8}$$

and the weighted norms $\|u\|_{H^{k,\ell}_\beta(\Omega)}$ by

$$\|u\|^2_{H^{k,\ell}_\beta(\Omega)} = \begin{cases} \|u\|^2_{H^{\ell-1}(\Omega)} + \sum_{k=\ell}^{m} |u|^2_{H^{k,\ell}_\beta(\Omega)} & \text{if} \quad \ell > 0 , \\ \sum_{k=\ell}^{m} |u|^2_{H^{k,\ell}_\beta(\Omega)} & \text{if} \quad \ell = 0 . \end{cases} \tag{9}$$

We shall be interested in classes of functions u which are analytic in $(0, 1]$ with a point singularity at $x = 0$ as follows.

Definition 1. We say that $u \in \mathscr{B}^\ell_\beta(\Omega)$ if $u \in \bigcap_{m\geq\ell} H^{m,\ell}_\beta(\Omega)$ and if there exist constants $C_u > 0$, $d_u \geq 1$ such that

$$|u|_{H^{k,\ell}_\beta(\Omega)} \le C_u d_u^{k-\ell}(k-\ell)! \quad k = \ell, \ell+1, \dots . \tag{10}$$

Functions u in the set $\mathscr{B}^\ell_\beta(\Omega)$ are analytic in $(0,1]$ with possibly an algebraic singularity at $x = 0$. It follows directly from the definition that for $0 < \beta < 1$ and for $\ell \ge 1$ it holds that $\mathscr{B}^\ell_\beta(\Omega) \subset H^{\ell-1}(\Omega)$.

The spaces $\mathscr{B}^\ell_\beta(\Omega)$ (and closely related spaces $\mathfrak{C}^2_\beta(\Omega)$) were introduced in [1], in plane polygonal domains with curved boundaries.

For the model problem (1), piecewise analyticity of the right hand side f implies corresponding smoothness of the solution u. We have the following precise regularity result.

Theorem 1. *If the coefficient functions a and c in* (1) *are analytic in $\overline{\Omega}$ and if* (3) *holds, then for every $f \in \mathscr{B}^0_\beta(\Omega)$ for some $0 < \beta < 1$, the unique weak solution $u \in H^1_0(\Omega)$ of* (1) *belongs to $\mathscr{B}^2_\beta(\Omega)$.*

Proof. The proof of the $\mathscr{B}^2_\beta(\Omega)$ regularity follows from the integral representation of the exact solution u of (1), and from the assumed analyticity of the coefficient functions $a(x)$ and $c(x)$ in $\overline{\Omega}$.

4 *hp*-IsoGeo FEM

By the quasioptimality (7), the error in the FE approximations of the solution u is bounded by the best approximation of u in the $H^1(\Omega)$ norm.

We shall be interested in establishing *exponential* convergence rates for approximations of functions $u \in \mathscr{B}^\ell_\beta(\Omega)$ from spaces of piecewise polynomials in Ω, expressed in terms of the number of degrees of freedom, i.e. of their dimension N. As indicated in the introduction, particular attention will be paid to *smoothest hp-approximations*, i.e. to Finite Element spaces with substantial extra regularity beyond the (minimal) C^0-interelement regularity which is necessary for conformity $V_\ell \subset V$.

We start by introducing notation for meshes, polynomial degrees and interelement conformity. We denote by $\{\Omega_j : j = 1, \dots, M\}$ a partition of Ω into open, nonempty intervals such that $\overline{\Omega} = \bigcup_{j=1}^M \overline{\Omega}_j$. We denote $\Omega_j = (x_{j-1}, x_j)$, with the endpoints given by $0 = x_0 < x_1 < x_2 < \dots < x_M = 1$. We denote by $h_j = |\Omega_j| = x_j - x_{j-1}$. On Ω_j, we consider spaces of polynomial functions of degree at most $p_j \ge 1$, denoted by $\mathbb{P}_{p_j}(\Omega_j)$. We collect the polynomial degrees p_i in a *degree vector* $\mathbf{p} = \{p_j\}_{j=1}^M$. At the node $x_j = \overline{\Omega_j} \cap \overline{\Omega_{j+1}}$, $j = 1, 2, \dots, M-1$, we enforce *interelement compatibility* of orders $0 \le k_j \le p_j \wedge p_{j-1}$ by the condition

$$[\![u^{(m)}]\!](x_j) = 0 \quad m = 0, 1, 2, \dots, k_j - 1 . \tag{11}$$

We combine also the interelement compatibilities into the *conformity vector* $\mathbf{k} = \{k_j\}_{j=1}^{M-1}$ and all elements Ω_j into the mesh $\mathscr{T} = \{\Omega_j\}_{j=1}^{M}$. Then, we denote

$$\begin{aligned} S_{\mathbf{k}}^{\mathbf{p}}(\Omega;\mathscr{T}) := \big\{ v \in L^2(\Omega) : v|_{\Omega_j} \in \mathbb{P}_{p_j}(\Omega_j) \,,\ [\![v^{(m)}]\!](x_j) = 0 \\ \text{for } j = 1,2,\ldots,M-1,\ m = 0,1,\ldots,k_j - 1 \big\} \,. \end{aligned} \tag{12}$$

The number of degrees of freedom in the space $S_{\mathbf{k}}^{\mathbf{p}}(\Omega;\mathscr{T})$ is easily seen to be

$$N = \dim(S_{\mathbf{k}}^{\mathbf{p}}(\Omega;\mathscr{T})) = \sum_{j=1}^{M}(p_j + 1) - \sum_{j=1}^{M-1} k_j \,.$$

If $p_j = p \geq 1$ and if $k_j = k \geq 0$ for all j, we also write $S_k^p(\Omega;\mathscr{T})$ in place of $S_{\mathbf{k}}^{\mathbf{p}}(\Omega;\mathscr{T})$. Then

$$N = \dim(S_k^p(\Omega;\mathscr{T})) = M(p+1) - (M-1)k \,. \tag{13}$$

In the context of approximation of functions $u \in \mathscr{B}_\beta^\ell(\Omega)$, we shall use so-called *geometric meshes* $\mathscr{T}$. We say that a mesh $\mathscr{T}$ is a *geometric mesh on* Ω *with* $M > 1$ *layers* if $x_j = \sigma^{M-j}$ for $j = 1,2,\ldots,M$ for a *geometric grading factor* $\sigma \in (0,1)$. We denote such meshes by $\mathscr{T} = \mathscr{G}_\sigma^M$ and note that $h_1 = x_1 = \sigma^{M-1}$ and that for $j \geq 2$ it holds that $h_j = x_j - x_{j-1} = \lambda x_{j-1}$ where $\lambda = (1-\sigma)/\sigma = \sigma^{-1} - 1$ is independent of j. We observe that $h_j = \lambda x_{j-1}$ for $j = 2,3,\ldots,M$.

5 Basic Local Interpolation Operators

We obtain convergence rate estimates by constructing global *hp* interpolation operators with high conformity **k**. As usual in FE analysis, these operators will be built from local, i.e. elemental interpolation operators which are constructed and analyzed on the reference element $\Lambda = (-1,1)$, and then transported to the physical elements $\Omega_j \in \mathscr{G}_\sigma^M$ by an affine mapping. We will give two constructions: the first one is based on a spectral-like elemental approximation proposed in [BBRS10], whereas the second will be based on a classical, nodal interpolation operator.

For $u \in \mathscr{B}_\beta^\ell(\Omega)$, we construct a family of spectral *hp*-interpolants based on a construction which was introduced in [BBRS10]. These interpolants are based on L^2 projections on spaces of discontinuous polynomials of a certain derivative of the function to be interpolated, and by subsequent enforcement of interelement conformity of order k_i. This interpolant is a generalization of the one proposed in [7, Chap. 3] for the analysis of C^0-conforming *hp*-FEM. We require the following result which is Corollary 2 in [BBRS10].

Proposition 1. *Let $\Lambda = (-1,1)$, $p,k,s \geq 0$ integers, with polynomial degree $p \geq \max\{0, 2k-1\}$, and with $\kappa := p-k+1$. Then there exists a quasi-interpolation operator $\hat{\pi}_k^p$ such that, for any function $\hat{u} : \Lambda \mapsto \mathbb{R}$ such that $\hat{u}^{(k)} \in H^s(\Lambda)$, for every $0 \leq s \leq \kappa$ holds the interpolation error bound*

$$\left\|\hat{u}^{(j)} - \left(\hat{\pi}_k^p \hat{u}\right)^{(j)}\right\|_{L^2(\Lambda)}^2 \leq \frac{(\kappa-s)!}{(\kappa+s)!}\frac{(\kappa-(k-j))!}{(\kappa+(k-j))!}\left\|\hat{u}^{(k)}\right\|_{H^s(\Lambda)}^2, \quad j = 0,1,\ldots,k\,. \tag{14}$$

Moreover, the interpolating polynomial $\hat{\pi}_k^p \hat{u}$ satisfies

$$\left(\hat{\pi}_k^p \hat{u}\right)^{(j)}(\pm 1) = \hat{u}^{(j)}(\pm 1)\,, \quad j = 0,1,\ldots,k-1 \tag{15}$$

(with (15) *understood to be void in the case $k = 0$).*

We specialize this general result to maximal smoothness. To avoid fractional bounds for indices, we substitute in (14)

$$p = 2q-1\,, \quad q \geq 1\,.$$

Then $1 \leq k$ and $2k-1 \leq 2q-1$. Therefore, we may choose in (14) $k = q \geq 1$. This implies that $\kappa = q$ and that $0 \leq s \leq p$. This implies, upon scaling (14) to a generic interval $J = (a,b) \subset \Omega = (0,1)$ of length $h = b-a > 0$ that there holds, for every $0 \leq s \leq q$, $0 \leq j \leq q$, $q \geq 1$, the interpolation error bound

$$\left\|u^{(j)} - \left(\pi_q^{2q-1} u\right)^{(j)}\right\|_{L^2(J)}^2 \leq \left(\frac{h}{2}\right)^{2(q+s-j)} \frac{(q-s)!}{(q+s)!}\frac{i!}{(2q-j)!}\left\|u^{(q+s)}\right\|_{L^2(J)}^2\,. \tag{16}$$

By (15), the interpolant $\hat{\pi}_q^{2q-1}$ ensures interelement continuity of, roughly speaking, the first $k-1 = q-1 = O(p/2)$ many derivatives of the piecewise polynomial, interpolating function. As increasing conformity of the interpolating function reduces the dimension N of the subspaces $S_{\mathbf{k}}^{\mathbf{p}}(\Omega, \mathscr{T})$, it is readily verified from (13), that conformity of order $O(p/2)$ will imply

$$\dim(S_q^{2q-1}(\Omega, \mathscr{G}_\sigma^M)) = O((M+1)p/2)$$

as $M, p \to \infty$. If, in particular, $M = O(p)$ (as will be the case in hp-FEM), we find $\dim(S_q^{2q-1}(\Omega, \mathscr{G}_\sigma^M)) = O(p^2)$. From (13) it is thus evident that an asymptotic complexity reduction of hp-FEM is only possible for $k_i \geq p_i - \bar{k}$, i.e. for *subspaces $S_k^p(\Omega, \mathscr{G}_\sigma^M)$ whose conformity orders k_i equal, up to an absolute gap $\bar{k}$, with p_i*. If this gap is proportional to (any power of) p_i, $\dim(S_k^p(\Omega, \mathscr{G}_\sigma^M))$ will scale polynomially in p for $M = O(p)$ elements in the mesh $\mathscr{G}_\sigma^M$.

6 Exponential Convergence

We now "assemble" scaled versions of the elementwise quasi-interpolation projectors π_k^p into corresponding global interpolators Π_k^p and prove exponential convergence of these projectors for functions $u \in \mathscr{B}_\beta^2(\Omega)$, from within the hp-FE space $S_q^p(\Omega; \mathscr{G}_\sigma^p)$ where $p = 2q - 1 \geq 1$ under the *provision that the geometric grading factor* $0 < \sigma < 1$ *is sufficiently large*, depending on the constant d_u in (10), following the analysis in [7, Chap. 3]. From these results, exponential convergence of the hp-FEM will follow via (7).

We start by considering a generic element $J = (a, b) \in \mathscr{G}_\sigma^p$ not abutting at the singular support $x = 0$ of u. If $u \in \mathscr{B}_\beta^2(\Omega)$, we have for any such J and any integer $s \geq 0$ that

$$|u|^2_{H_\beta^{s+1,\ell}(\Omega)} \geq \|u^{(s+1)}\Phi_{\beta+s+1-\ell}\|^2_{L^2(J)} \geq a^{2(\beta+s+1-\ell)}\|u^{(s+1)}\|^2_{L^2(J)} \,.$$

This implies that there holds with $\ell = 2$

$$\|u^{(s+1)}\|^2_{L^2(J)} \leq a^{-2(\beta+s+1-\ell)}|u|^2_{H_\beta^{s+1,\ell}(\Omega)} \,. \tag{17}$$

We replace now s with $q + s - 1$, $j = 0, 1$ and $\ell = 2 > j$ to obtain

$$\|u^{(q+s)}\|^2_{L^2(J)} \leq a^{-2(\beta+q+s-\ell)}|u|^2_{H_\beta^{q+s,\ell}(\Omega)} \,.$$

Using this bound for $J = \Omega_i \in \mathscr{G}_\sigma^p, 2 \leq i \leq p$, we get from (16) the bound

$$\left\|u^{(j)} - (\pi_q^{2q-1}u)^{(j)}\right\|^2_{L^2(\Omega_i)} \leq \left(\frac{h_i}{2}\right)^{2(q+s-j)} \frac{(q-s)!}{(q+s)!(2q-j)!}\|u^{(q+s)}\|^2_{L^2(\Omega_i)} \,,$$

for $j = 0, 1$. Using here (17), we find with (10) the bound

$$\begin{aligned}
&\left\|u^{(j)} - (\pi_q^{2q-1}u)^{(j)}\right\|^2_{L^2(\Omega_i)} \\
&\leq \left(\frac{h_i}{2}\right)^{2(q+s-j)} \frac{(q-s)!}{(q+s)!(2q-j)!} x_{i-1}^{-2(\beta+q+s-\ell)}|u|^2_{H_\beta^{q+s,\ell}(\Omega)} \\
&\leq C\frac{(q-s)!((q+s-\ell)!)^2}{(q+s)!(2q-j)!} x_{i-1}^{-2(\beta+j-\ell)} \left(\frac{\lambda d_u}{2}\right)^{2(q+s-j)} \,.
\end{aligned} \tag{18}$$

Since for $i = 2, 3, \ldots, M$ it holds $x_i = \sigma^{M-i}$, we can write for $j = 0, 1$ and for $\ell = 2$

$$
\begin{aligned}
&\left\| u^{(j)} - (\pi_q^{2q-1} u)^{(j)} \right\|^2_{L^2(\Omega_i)} \\
&\leq C \tfrac{(q-s)!((q+s-\ell)!)^2}{(q+s)!(2q-j)!} \sigma^{2(M-i+1)(\ell-\beta-j)} \left(\tfrac{\lambda d_u}{2}\right)^{2(q+s-j)} \\
&\leq C \underbrace{\frac{(q-s)!((q+s-\ell)!)^2}{(q+s)!(2q-j)!}}_{*} \sigma^{2M(1-\beta)} \left(\frac{\lambda d_u}{2}\right)^{2(q+s-j)} \sigma^{2(1-i)(1-\beta)} .
\end{aligned}
$$

Remark 1. Note that so far, in all error bounds the differentiation order s is an integer. In what follows, we shall use also norms of fractional order s for which the corresponding error bounds can be obtained by classical interpolation arguments.

We now analyze the factorial expression $*$.

Lemma 1. *There is a constant $C > 0$ such that for every $q \geq 1$ and for the choice $s = \alpha q$ for some $0 < \alpha < 1$ there holds the estimate*

$$
| * | \leq C G(\alpha)^q , \quad \textit{where} \quad G(\alpha) := \frac{1}{4}(1+\alpha)^{1+\alpha}(1-\alpha)^{1-\alpha} . \tag{19}
$$

Here, the factorial expressions in $*$ are continued to noninteger arguments obtained by choosing $s = \alpha q$ using the Gamma function.

Proof. Throughout the proof, $C > 0$ and $\lesssim$ denotes a constant and a bound, respectively, which are independent of s, p and of n. We start by recalling the *Stirling inequalities*

$$
\forall n \in \mathbb{N} : \quad \frac{n! e^n}{e\sqrt{n}} \leq n^n \leq \frac{n! e^n}{\sqrt{2\pi n}}
$$

which imply $n! \geq n^n e^{-n} \sqrt{2\pi n} \geq c n^{n+1/2} e^{-n}$, and $n! \leq n^n e^{-n} e\sqrt{n} \leq c n^{n+1/2} e^{-n}$. We then estimate $(*)$ in the case $j = 1$ as follows:

$$
\begin{aligned}
|(*)| &\leq C 2q \frac{(q-s)^{q-s+1/2} e^{-(q-s)} \left[(q+s-2)^{(q+s-2)+1/2} e^{-q-s+2}\right]^2}{(q+s)^{q+s+1/2} e^{-q-s} (2q)^{2q+1/2} e^{-2q}} \\
&\leq C q \frac{(q-s)^{q-s} (q+s)^{2(q+s)-s}}{(q+s)^{q+s} (2q)^{2q+1/2}} \\
&\leq C q^{1/2} (q+s)^{-3} \frac{(q-s)^{q-s} (q+s)^{q+s}}{(2q)^{2q}} .
\end{aligned}
$$

In this bound, we now *choose* $s = \alpha q$ for some $0 < \alpha < 1$ to be selected.
Then

$$
q^{1/2}(q+s)^{-3} = q^{1/2} q^{-s} (1+\alpha)^{-3} \leq C(\alpha) < \infty .
$$

Therefore

$$
\begin{aligned}
|(*)| &\le C\frac{(q-s)^{q-s}(q+s)^{q+s}}{(2q)^{2q}} = C\frac{(q^2-s^2)^q(q+s)^s(q-s)^{-s}}{(2q)^{2q}} \\
&= C\frac{q^{2q}(1-\alpha^2)^q q^{\alpha q}(1+\alpha)^{\alpha q} q^{-\alpha q}(1-\alpha)^{-\alpha q}}{(2q)^{2q}} \\
&= 2^{-2q}(1-\alpha^2)^q\left(\frac{1+\alpha}{1-\alpha}\right)^{\alpha q} \\
&= \left[\frac{1-\alpha^2}{4}\left(\frac{1+\alpha}{1-\alpha}\right)^{\alpha}\right]^q = \left[\frac{1}{4}(1+\alpha)^{1+\alpha}(1-\alpha)^{1-\alpha}\right]^q = [G(\alpha)]^q\,.
\end{aligned}
$$

We observe that the function $G(\cdot)$ defined in (19) satisfies for $\alpha \in [0,1]$

$$
1/4 \le G(\alpha) \le 1\,, \quad G(0) = \frac{1}{4}\,, \quad \text{and} \quad \lim_{\alpha\to 1^-} G(\alpha) = 1\,.
$$

Inserting the bound obtained in Lemma 1 into (18), we find

$$
\begin{aligned}
&\left\|(u-\Pi_q^{2q-1}u)^{(j)}\right\|^2_{L^2(x_1,1)} \\
&= \sum_{i=2}^{M}\left\|u^{(j)} - (\pi_q^{2q-1}u)^{(j)}\right\|^2_{L^2(\Omega_i)} \\
&\le C(\alpha,d_u)G(\alpha)q\left(\frac{\lambda d_u}{2}\right)^{2q(1+\alpha)}\left(\frac{\lambda d_u}{2}\right)^{-2j}\sum_{i=2}^{M}\sigma^{2(M+1-i)(1-\beta)} \\
&\le C(\alpha,\sigma,d_u)\left[G(\alpha)\left(\frac{\lambda d_u}{2}\right)^{2(1+\alpha)}\right]^q .
\end{aligned}
$$

For $0 < \sigma < 1$, we have

$$
\lambda d_u = \left(\frac{1}{\sigma} - 1\right) d_u < 2
$$

if

$$
1 > \sigma > (1 + 2/d_u)^{-1} > 0\,. \tag{20}
$$

Choosing $\alpha > 0$ sufficiently small, we find for such σ with (20) that

$$
\left[G(\alpha)\left(\frac{\lambda(\sigma)d_u}{2}\right)^{2(1+\alpha)}\right] \le F(\sigma,d_u) < 1\,.
$$

We have proved

Theorem 2. *Assume that* $u \in \mathscr{B}^2_\beta(\Omega)$ *for some* $0 < \beta \leq 1$, $\Omega = (0,1)$, *and that* (10) *holds with some* $C_u, d_u > 0$. *Then, for any* $0 < \sigma < 1$ *satisfies* (20), *there exist* $b, C > 0$ *such that there holds for all* $p \geq 1$

$$\inf_{v \in S^{\mathbf{p}}_q(\Omega, \mathscr{G}^p_\sigma)} \|u - v\|_{H^1(\Omega)} \leq C \exp(-b(\sigma,\beta)p) \tag{21}$$

Here, $\mathbf{p} = (q+1, p, \ldots, p)$ *with* $p = 2q - 1$ *and* $q \in \mathbb{N}$. *By* (13), *in terms of the number* N *of degrees of freedom, we have the estimate*

$$\inf_{v \in S^{\mathbf{p}}_q(\Omega, \mathscr{G}^p_\sigma)} \|u - v\|_{H^1(\Omega)} \leq C' \exp(-b'(\sigma,\beta)\sqrt{N}) \tag{22}$$

with possibly different constants $b', C' > 0$.

Proof. The assertion follows from the previous bounds on the elementwise interpolation error in $(x_1, 1)$, *for either of the global interpolation operators*, i.e. for $\Pi^{\mathbf{p}}_{\mathbf{k}}$ and a Hardy-Type estimate (e.g. (3.3.68) in [7]) in $\Omega_1 = (0, x_1) = (0, \sigma^p)$.

7 Numerical Results

In this section we present numerical results in one space dimension with the aim of demonstrating the validity of the convergence theorems presented in the previous section, and also to add some insight on the numerical behavior of the *hp*-isogeometric method.

Instead of dealing directly with the interpolation operator, we show numerical errors of the Galerkin projection u_h of u onto spline spaces, obtained solving the simple Poisson problem

$$-u'' = f \quad \text{in} \quad \Omega = (0,1)\,, \quad u(0) = 0,\, u(1) = 0, \tag{23}$$

with right hand side f is selected in order to have exact solution $u = x^{0.6} - x$.

We first set $\sigma = 1/2$ and plot in Fig. 1 the corresponding L^2-error for the three cases $u_h \in S^{2q-1}_1(\Omega, \mathscr{G}^{2q-1}_\sigma)$, $u_h \in S^{2q-1}_q(\Omega, \mathscr{G}^{2q-1}_\sigma)$, $u_h \in S^{2q-1}_{2q-1}(\Omega, \mathscr{G}^{2q-1}_\sigma)$, that is, for C^0, C^{q-1} (the maximum regularity allowed by our interpolant) and C^{2q-2} continuity respectively. We clearly see exponential convergence $\|u - u_h\|_{L^2} = C\exp(-b\sqrt{N})$ in all cases, with higher b for higher regularity.

Furthermore, we want to investigate numerically the sharpness of condition (20). It is very easy to see that for our choice $u(x) = x^{-0.6} - x$, the corresponding d_u is equal to 1 and thus, (20) prescribes $\sigma > 1/3$. We present in Fig. 2 the error plot for $u_h \in S^{2q-1}_q(\Omega, \mathscr{G}^{2q-1}_\sigma)$ and $\sigma = 0.1$. The plot shows exponential convergence, if we exclude the last computed value, where u_h is strongly affected by round-off error.

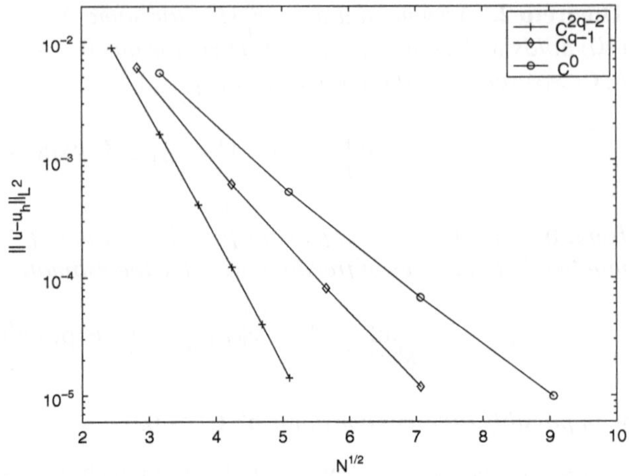

Fig. 1 L^2-error of the Galerkin projection versus degrees-of-freedom number N: comparison between splines and standard finite element approximation; as a reference, the line $\exp(-\sqrt{N})$ is shown

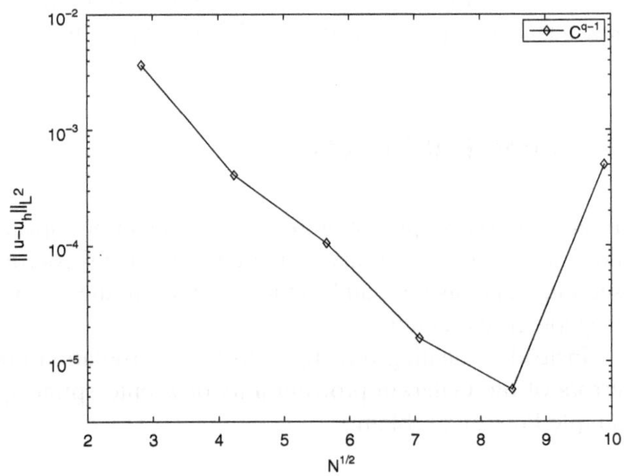

Fig. 2 L^2-error of the Galerkin projection versus degrees-of-freedom number N for degree $2q-1$ and C^{q-1} continuity; $\sigma = 0.1$. Round-off error pollutes the last computed entry

Indeed, for this last plot entry we have $q = 6$, that is degree 13, minimum mesh-size $0.1^{12} = 10^{-12}$, and the condition number of the linear system gets $> 10^{17}$.

In conclusion, there is numerical evidence that the mesh condition (20) is not necessary to approximate a singular solution in the space $S_{2q-1}^{2q-1}(\Omega, \mathcal{G}_\sigma^{2q-1})$. However, we also stress that the B-spline basis is not suitable for a strong geometric grading and high degree, since it induces a severe ill-conditioning of the linear system of equations.

We observe that each of the proposed hp-FEM discretizations converge with exponential rate $\exp(-b\sqrt{N})$. The constant $b > 0$ in the exponential rate depends on σ and on the conformity in the hp-space. In particular, the hp-IGA will afford substantially larger values of b in the exponential convergence bound, leading to several orders of magnitude error reduction at a given budget of N degrees of freedom over standard, C^0-conforming hp-FEM. The increased efficiency is at the expense of a rather large condition number of the (small) linear systems of equations which, presumably, is due to the use of spline-bases in the current implementation of hp-IGA. These work savings by enhanced conformity afforded by the hp-version of isogeometric analysis are expected to be even more pronounced in two and three spatial dimensions, and the unfavourable conditioning of the stiffness matrices is expected to be alleviated by the use of an orthonomal basis.

Acknowledgements A. Buffa and G. Sangalli were partially supported by the European Research Council through the FP7 Ideas Starting Grant n. 205004, by the Italian MIUR through the FIRB Grant RBFR08CZ0S and by the European Commission through the FP7 Factories of the Future project *TERRIFIC*. C. Schwab was supported by the European Research Council through the FP7 Ideas Advanced Grant n. 247277. This support is gratefully acknowledged.

References

1. I. Babuška and B.Q. Guo, The h-p version of the finite element method for domains with curved boundaries, SIAM J. Numer. Anal., **25**(1988) 837–861.
2. L. Beirao da Veiga, A. Buffa, J. Rivas, G. Sangalli, Some estimates for h-p-k-refinement in Isogeometric Analysis. Numer. Mathem. **118** (2011), no. 2, 271–305.
3. L. Beirao da Veiga, D. Cho, and G. Sangalli. Anisotropic NURBS approximation in isogeometric analysis. Comput. Meth. Appl. Mech. Eng., **209** (2012), 1–11.
4. W. Dahmen and K. Scherer, Best approximation by piecewise polynomials with variable knots and degrees, Journ. Approx. Theory **26** (1979) pp. 1–13.
5. W. Gui and I. Babuška, The h, p and h-p versions of the finite element method in 1 dimension. II. The error analysis of the h- and h-p versions, Numerische Mathematik **49**(1986) 613–657.
6. T. J. R. Hughes, J. A. Cottrell, and Y. Bazilevs, Isogeometric analysis: CAD, finite elements, NURBS, exact geometry and mesh refinement, Comput. Methods Appl. Mech. Engrg., 194 (2005), pp. 4135–4195.
7. C. Schwab, p- and hp-FEM – Theory and Application to Solid and Fluid Mechanics", Oxford University Press, Oxford, 1998.
8. K. Scherer, On optimal global error bounds obtained by scaled local error estimates, Numerische Mathematik, **36** (1981) pp. 257–277.

High-Order Locally Implicit Time Integration Strategies in a Discontinuous Galerkin Method for Maxwell's Equations

S. Descombes, S. Lanteri, and L. Moya

Abstract The starting point for the present study is a second-order locally implicit time integration method for a nondissipative discontinuous Galerkin (DG) discretisation of Maxwell's equations. The system is split into the explicit and the implicit parts based on the geometry of the mesh: locally refined regions are treated implicitly while the rest of the domain is treated explicitly. When combined with an explicit time integration method one of the main drawbacks of the DG time-domain (DGTD) method is the restriction on the time step when high-order elements are used. If the region of refinement is small relative to the computational domain, the implicit-explicit (IMEX) method allows to overcome this efficiency issue without needing to solve a linear system at each time step for the entire size of the problem.

The topic of this study is to propose higher order time integration techniques based on a second-order locally implicit method to fully exploit the attractive features of the IMEX approach combined with a DG discretisation which allows to easily increase the spatial convergence order.

S. Descombes
J.A. Dieudonné Mathematics Laboratory, CNRS UMR 7351,
University Nice Sophia Antipolis,
06108 Nice Cedex, France

Nachos project-team, Inria Sophia Antipolis – Méditerranée, 2004 Route des Lucioles, BP 93,
06902 Sophia Antipolis Cedex, Nice Cedex, France
e-mail: stephane.descombes@unice.fr

S. Lanteri · L. Moya (✉)
Nachos project-team, Inria Sophia Antipolis – Méditerranée, 2004 Route des Lucioles, BP 93,
06902 Sophia Antipolis Cedex, Nice Cedex, France
e-mail: stephane.lanteri@inria.fr,ludovic.moya@inria.fr

M. Azaïez et al. (eds.), *Spectral and High Order Methods for Partial Differential Equations - ICOSAHOM 2012*, Lecture Notes in Computational Science and Engineering 95, DOI 10.1007/978-3-319-01601-6_16,

1 Introduction

We consider the time-domain Maxwell equations

$$\partial_t \mathbf{D} = \mathrm{curl}\,(\mathbf{H}) - \mathbf{J}^f, \quad \partial_t \mathbf{B} = -\mathrm{curl}\,(\mathbf{E}), \tag{1}$$

$$\mathrm{div}\,(\mathbf{D}) = \rho^f, \quad \mathrm{div}\,(\mathbf{B}) = 0, \tag{2}$$

where $\mathbf{E}$ and $\mathbf{D}$ are the electric field and the electric flux density, $\mathbf{H}$ and $\mathbf{B}$ the magnetic field and the magnetic flux density, $\mathbf{J}^f$ the free current density and ρ^f the free charge density. For many application we can assume that the underlying medium is isotropic, linear and time-invariant. We then have the relations

$$\mathbf{D} = \varepsilon \mathbf{E} \quad \text{and} \quad \mathbf{B} = \mu \mathbf{H}, \tag{3}$$

where ε and μ are the dielectric permittivity and magnetic permeability of the medium, respectively. With the constitutive relation (3), Eqs. (2) are just the consistency conditions for (1). Indeed if we take the divergence of (1), make use (2) and (3), the resulting equation represents the charge conservation law, i.e. $\partial_t \rho^f + \mathrm{div}\left(\mathbf{J}^f\right) = 0$. Thus, as long as initial conditions satisfy (2) and the electromagnetic field evolves according to (1), the solution at any time will also satisfy (2). Consequently we can only consider the Eqs. (1) in which the constitutive relations (3) are included i.e.

$$\begin{cases} \varepsilon \partial_t \mathbf{E} &= \mathrm{curl}\,(\mathbf{H}) - \mathbf{J}^f, \\ \mu \partial_t \mathbf{H} &= -\mathrm{curl}\,(\mathbf{E}). \end{cases} \tag{4}$$

The free current density $\mathbf{J}^f$ includes the conduction current density and the source current density denoted by $\mathbf{J}^c$ and $\mathbf{J}^s$, respectively. The relation between an electric field and the conduction current density which is generated at any point of the conducting material is given by Ohm's law $\mathbf{J}^c = \sigma \mathbf{E}$, where σ is the conductivity of the medium. The source current density $\mathbf{J}^s$ may is associated to external sources or generators and is often called driven or impressed current. In this study we will consider non-conductive materials ($\sigma = 0$) such that the time-domain Maxwell equations write as form

$$\begin{cases} \varepsilon \partial_t \mathbf{E} &= \mathrm{curl}\,(\mathbf{H}) - \mathbf{J}^s, \\ \mu \partial_t \mathbf{H} &= -\mathrm{curl}\,(\mathbf{E}), \end{cases} \tag{5}$$

The starting point of the present study is the nondissipative DG formulation presented in [5]. By applying this nodal DG approach based on a discontinuous piecewise polynomial space for the approximation of the electromagnetic field within an element of the mesh, we obtain the global semi-discrete Maxwell system

$$\begin{cases} M^{\varepsilon}\partial_t E = SH - j^s, \\ M^{\mu}\partial_t H = -S^T E, \end{cases} \tag{6}$$

where the matrices M^{ϵ}, M^{μ} are the DG mass matrices which contain the values of the dielectric permittivity and magnetic permeability coefficients. The matrix S emanates from the discretization of the curl operator and j^s represents the discretized source current density. Similarly to [1, 7, 10] we introduce the Cholesky factorization of the mass matrices

$$M^{\epsilon} = L_{M^{\epsilon}} L_{M^{\epsilon}}^T \text{ and } M^{\mu} = L_{M^{\mu}} L_{M^{\mu}}^T, \tag{7}$$

where $L_{M^{\epsilon}}$ and $L_{M^{\mu}}$ are triangular matrices. Then by introducing the change of variables $\tilde{E} = L_{M^{\epsilon}}^T E$ and $\tilde{H} = L_{M^{\mu}}^T H$ in (6), we write

$$\begin{cases} \partial_t \tilde{E} = \tilde{S}\tilde{H} - \tilde{j}^s, \\ \partial_t \tilde{H} = -\tilde{S}^T \tilde{E}, \end{cases} \tag{8}$$

where

$$\tilde{S} = L_{M^{\epsilon}}^{-1}\, S\, \left(L_{M^{\mu}}^{-1}\right)^T, \text{ and } \tilde{j}^s = L_{M^{\epsilon}}^{-1}\, j^s. \tag{9}$$

For convenience of presentation in the following we will omit "$\sim$" in (8) and (9).

2 The Second-Order Locally Implicit Method from [10]

A popular time integration method for the semi-discrete Maxwell system (8) is the second order Leap-Frog scheme that we write in the three-stage form, emanating from Verlet's method

$$\Phi_{\Delta t}^{LF2} : \begin{cases} \dfrac{H^{n+1/2} - H^n}{\Delta t} = -\dfrac{1}{2} S^T E^n, \\ \dfrac{E^{n+1} - E^n}{\Delta t} = S H^{n+1/2} - \dfrac{1}{2}\left(j^s(t_{n+1}) + j^s(t_n)\right), \\ \dfrac{H^{n+1} - H^{n+1/2}}{\Delta t} = -\dfrac{1}{2} S^T E^{n+1}, \end{cases} \tag{10}$$

where $\Delta t = t_{n+1} - t_n$ denotes the time step and upper indices refer to time levels. This method has consistency order two, is explicit in S, conditionally stable with a critical step size determined by Botchev et al. [1]

$$\Delta t \leq 2 \times \rho\left(SS^T\right)^{-\frac{1}{2}}, \tag{11}$$

where ρ denotes the spectral radius, this inequality being strict for zero conduction ($D = 0$). Hence DG applied with its attractive feature of local grid refinement may lead to an unduly time step size restriction. An alternative to (10) is the second order, unconditionally stable Crank-Nicolson method that we write in the three-stage form

$$\Phi_{\Delta t}^{CN2}: \begin{cases} \dfrac{H^{n+1/2} - H^n}{\Delta t} &= -\dfrac{1}{2} S^T E^n, \\ \dfrac{E^{n+1} - E^n}{\Delta t} &= \dfrac{1}{2} S \left(H^{n+1} + H^n\right) - \dfrac{1}{2} \left(j^s(t_{n+1}) + j^s(t_n)\right), \\ \dfrac{H^{n+1} - H^{n+1/2}}{\Delta t} &= -\dfrac{1}{2} S^T E^{n+1}, \end{cases} \tag{12}$$

which only differs from (10) in the middle stage in the time level for H. For consistency and stability of this implicit method we refer to [12]. The expense for the implicit computation is too large to consider (12) as an attractive alternative to (10), especially in 3D (see e.g. [12]). If the region of refinement is small relative to the computational domain, the unduly time step restriction of (10) and the overhead of (12) can be overcome by blending the two methods yielding locally implicit approaches where only variables associated to the smallest grid elements are implicitly treated. The IMEX time integration method from [10] is a blend of (10) and (12) applied to the semi-discrete Maxwell system (8)

$$\Phi_{\Delta t}^{IMEX2}: \begin{cases} \dfrac{H^{n+1/2} - H^n}{\Delta t} = -\dfrac{1}{2} S^T E^n, \\ \dfrac{E^{n+1} - E^n}{\Delta t} = S_0 H^{n+1/2} + \dfrac{1}{2} S_1 \left(H^{n+1} + H^n\right) - \dfrac{1}{2} \left(j^s(t_{n+1}) + j^s(t_n)\right), \\ \dfrac{H^{n+1} - H^{n+1/2}}{\Delta t} = -\dfrac{1}{2} S^T E^{n+1}, \end{cases}$$

where $S = S_0 + S_1$ is a matrix splitting. This method is symmetric and implicit in S_1, explicit in S_0. For $S_0 = 0$ we recover (12) and for $S_1 = 0$ the method (10). The matrix splitting adopted in [10] is defined as $S_1 = SS_H$, where S_H is a diagonal matrix of dimension the length of H with

$$(S_H)_{jj} = \begin{cases} 0, & \text{component } H_j \text{ of } H \text{ to be treated explicitly,} \\ 1, & \text{component } H_j \text{ of } H \text{ to be treated implicitly.} \end{cases}$$

The locally implicit time integration method can be written as

$$\Phi_{\Delta t}^{IMEX2}: \begin{cases} H^{n+1/2} &= H^n - \dfrac{\Delta t}{2} S^T E^n, \\ \mathcal{M} E^{n+1} &= b_{n+1}, \\ H^{n+1} &= H^{n+1/2} - \dfrac{\Delta t}{2} S^T E^{n+1}, \end{cases} \tag{13}$$

where $\mathcal{M} = I + \frac{\Delta t^2}{4} S_1 S^T$ and

$$b_{n+1} = E^n + \Delta t\ S_0 H^{n+1/2} + \frac{\Delta t}{2} S_1 \left(H^{n+1/2} + H^n\right) - \frac{\Delta t}{2} \left(j^s\left(t_{n+1}\right) + j^s\left(t_n\right)\right).$$

Noticing that $S_1 S^T = S S_H S^T = S S_H S_H S^T = S_1 S_1^T$, then the matrix splitting $S_1 S^T$ is symmetric which facilitates the resolution of the linear system of the second stage of the IMEX method (13). The matrix $\mathcal{M}$ is then given by

$$\mathcal{M} = I + \frac{\Delta t^2}{4} S_1 S_1^T. \tag{14}$$

For $S_1 = S$ we recover the matrix of the linear system of the fully implicit method (12), [12]. With the adopted splitting the matrix $\mathcal{M}$ will be significantly more sparse than without splitting, enabling us to solve the linear system at lower costs. Similarly to the explicit and implicit methods (10) and (12) for $n \geq 1$ the third stage derivative computation can be copied to the first stage at the next time step. The IMEX method (13) is conditionally stable with a critical step size determined by (see [3])

$$\Delta t < 2 \times \rho\left(S_0 S_0^T\right)^{-\frac{1}{2}}. \tag{15}$$

This condition is similar to the stability condition of the explicit scheme (11) with S_0 instead of S, allowing to let the definition of Δt be restricted to the subset of the coarse grid elements. Thus in the presence of a local refinement, the purpose of the IMEX method is achieved since the most severe stability constraints on explicit time integration methods can be overcome. Finally it is proven in [10] that the component splitting is not detrimental to the second-order ODE convergence of the method (13), under stable simultaneous space-time grid refinement towards the exact underlying PDE solution. This property is an attractive feature of the locally implicit method (13). Indeed for IMEX approaches, the component splitting can introduce order reduction which makes use of a high-order DG spatial discretization less appealing due to the errors introduced by the lower temporal order, see e.g. [7].

3 High-Order Time-Integration Methods

3.1 Symmetric Composition of Symmetric Methods

High-order composition methods have been extensively studied for geometric composition, see e.g. [6, 9]. It is possible to construct arbitrary high-order composition methods. We are interested in a composition which at most of order 4, following the construction presented in [11]. Let $\Phi_{\Delta t}^{CO4}$ defined by

$$\Phi_{\Delta t}^{CO4} = \Phi_{\alpha_s \Delta t}^{IMEX2} \circ \cdots \circ \Phi_{\alpha_1 \Delta t}^{IMEX2}, \tag{16}$$

where $\alpha_1 + \cdots + \alpha_s = 1$ and $\alpha_1^3 + \cdots + \alpha_s^3 = 0$. Denote $(E^{\alpha_0}, H^{\alpha_0}) = (E^n, H^n)$ and $(E^{\alpha_s}, H^{\alpha_s}) = (E^{n+1}, H^{n+1})$, the composition scheme can be written as

$$\text{for } k = 1 : s \begin{cases} \dfrac{H^{\beta_k} - H^{\alpha_{k-1}}}{\alpha_k \Delta t} &= -\dfrac{1}{2} S^T E^{\alpha_{k-1}}, \\ \dfrac{E^{\alpha_k} - E^{\alpha_{k-1}}}{\alpha_k \Delta t} &= S_0 H^{\beta_k} + \dfrac{1}{2} S_1 \left(H^{\alpha_k} + H^{\alpha_{k-1}}\right) - \dfrac{1}{2}\left(j^s(t_{\alpha_k}) + j^s(t_{\alpha_{k-1}})\right), \\ \dfrac{H^{\alpha_k} - H^{\beta_k}}{\alpha_k \Delta t} &= -\dfrac{1}{2} S^T E^{\alpha_k}, \end{cases}$$

with time levels $t_{\alpha_0} = t_n$, $t_{\beta_k} = t_{\alpha_{k-1}} + \alpha_k \Delta t/2$, $t_{\alpha_k} = t_{\alpha_{k-1}} + \alpha_k \Delta t$, spanning the interval $[t_n, t_{n+1}]$, for $k = 1, \cdots, s$. The composition method reads

$$\text{for } k = 1 : s \begin{cases} H^{\beta_k} &= H^{\alpha_{k-1}} - \dfrac{\alpha_k \Delta t}{2} S^T E^{\alpha_{k-1}}, \\ \mathscr{M}_k E^{\alpha_k} &= b_k, \\ H^{\alpha_k} &= H^{\beta_k} - \dfrac{\alpha_k \Delta t}{2} S^T E^{\alpha_k}, \end{cases} \tag{17}$$

where $\mathscr{M}_k = I - \dfrac{(\alpha_k \Delta t)^2}{4} S_1 S^T$ and

$$b_k = E^{\alpha_{k-1}} + \alpha_k \Delta t \, S_0 H^{\beta_k} + \frac{\alpha_k \Delta t}{2} S_1 \left(H^{\beta_k} + H^{\alpha_{k-1}}\right) - \frac{\alpha_k \Delta t}{2}\left(j^s(t_{\alpha_k}) + j^s(t_{\alpha_{k-1}})\right).$$

Two fourth-order compositions of interest are those obtained for $s = 3$ and $s = 5$ with coefficient sets [6, 9]

$$\alpha_1 = \alpha_3 = \frac{1}{2 - 2^{1/3}}, \quad \alpha_2 = -\frac{2^{1/3}}{2 - 2^{1/3}} \quad \text{and}$$

$$\alpha_1 = \alpha_2 = \alpha_4 = \alpha_5 = \frac{1}{4 - 4^{1/3}}, \quad \alpha_3 = -\frac{4^{1/3}}{4 - 4^{1/3}},$$

For $s = 3$ and 5 the composition methods are three and five times more expensive than the base method (13).

3.2 *Richardson Extrapolation*

In this study we also examine the extension of the second-order locally implicit method to fourth-order through the Richardson extrapolation technique for symmetric methods, see e.g. [6]. Denote $\Phi_{\Delta t}^{REX4}$ the Richardson extrapolation of the IMEX

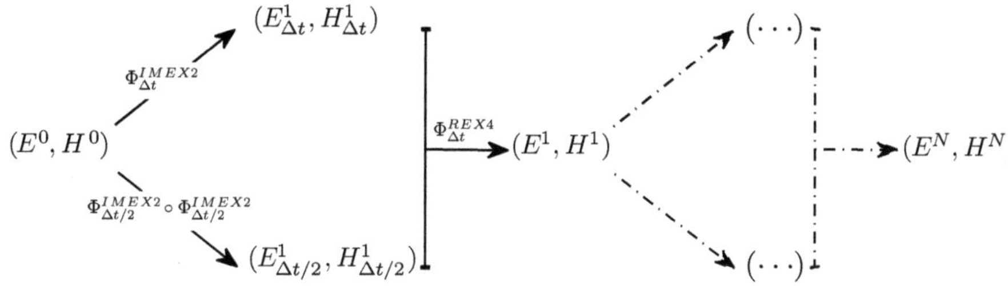

Fig. 1 Schematic diagrams of active Richardson extrapolation (final time $T = N\Delta t$)

method (13), we read

$$\Phi^{REX4}_{\Delta t} = \frac{4}{3}\Phi^{IMEX2}_{\Delta t/2} \circ \Phi^{IMEX2}_{\Delta t/2} - \frac{1}{3}\Phi^{IMEX2}_{\Delta t}. \tag{18}$$

Richardson extrapolation can be implemented in two different ways: active or passive Richardson extrapolation. We denote the approximate electromagnetic field at time t_n by $(E^n_{\Delta t}, H^n_{\Delta t})$ when we apply the IMEX method (13) with time step Δt, by $(E^n_{\Delta t/2}, H^n_{\Delta t/2})$ when we apply the composition of the IMEX method (13) with time step $\Delta t/2$; and by (E^n, H^n) when we apply the Richardson extrapolation. The two implementations of the Richardson extrapolation are depicted in Figs. 1–2. We observe that in the active form the value of (E^n, H^n) is used to calculate $(E^{n+1}_{\Delta t}, H^{n+1}_{\Delta t})$ and $(E^{n+1}_{\Delta t/2}, H^{n+1}_{\Delta t/2})$ while in the passive form the value of the approximation (E^n, H^n) is never used in the further computations. The passive Richardson extrapolation has the same stability properties as the second-order method while the active Richardson extrapolation leads to a new time integration method which might not share the good stability properties of the base method, it may cause instability of the computational process. For example the computation of the fully implicit method (12) together with the active Richardson extrapolation will in general be unstable, see e.g. [4]. This also happens for the IMEX method (13) which is a blend of the methods (10)–(12). Thus, to extend the IMEX method to fourth-order we will only consider the passive Richardson extrapolation which requires only three times more computation.

4 Numerical Results for a Two-Dimensional Test Problem

The 2D Transverse Magnetic model (TM) for components $\tilde{H}^{\tilde{x}}$, $\tilde{H}^{\tilde{y}}$ and $\tilde{E}^{\tilde{z}}$ is given as

$$\begin{cases} \mu(\tilde{\mathbf{x}})\,\partial_{\tilde{t}}\tilde{H}^{\tilde{x}}(\tilde{\mathbf{x}},\tilde{t}) = -\partial_{\tilde{y}}\tilde{E}^{\tilde{z}}(\tilde{\mathbf{x}},\tilde{t}), \\ \mu(\tilde{\mathbf{x}})\,\partial_{\tilde{t}}\tilde{H}^{\tilde{y}}(\tilde{\mathbf{x}},\tilde{t}) = \partial_{\tilde{x}}\tilde{E}^{\tilde{z}}(\tilde{\mathbf{x}},\tilde{t}), \\ \epsilon(\tilde{\mathbf{x}})\,\partial_{\tilde{t}}\tilde{E}^{\tilde{z}}(\tilde{\mathbf{x}},\tilde{t}) = \partial_{\tilde{x}}\tilde{H}^{\tilde{y}}(\tilde{\mathbf{x}},\tilde{t}) - \partial_{\tilde{y}}\tilde{H}^{\tilde{x}}(\tilde{\mathbf{x}},\tilde{t}) - \tilde{J}^{s}(\tilde{\mathbf{x}},\tilde{t}). \end{cases} \tag{19}$$

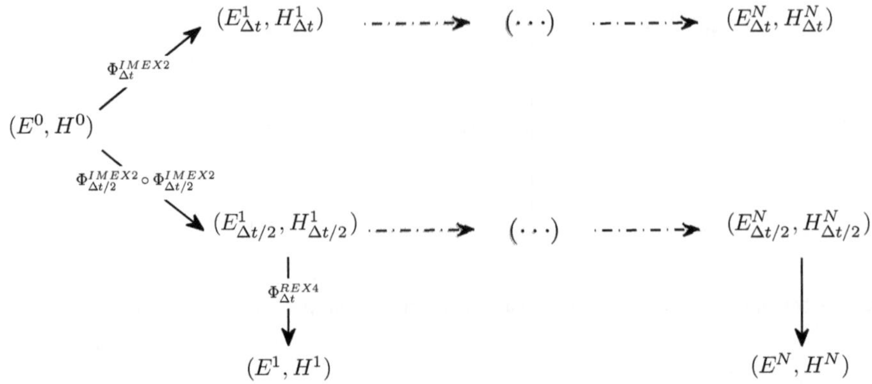

Fig. 2 Schematic diagrams passive Richardson extrapolation (final time $T = N\Delta t$)

The magnetic permeability $\mu(\tilde{\mathbf{x}})$, and the electric permittivity $\varepsilon(\tilde{\mathbf{x}})$, reflect the material coefficients. In the following, we model a metallic air-filled cavity, then $\Omega = [0, 1]^2$, $\mu = \mu_0$ and $\varepsilon = \varepsilon_0$ are constant vacuum values. Then we introduce the normalized space, time variables and physical fields through the relations

$$\mathbf{x} = \tilde{\mathbf{x}},\ t = c_0\,\tilde{t},\ E = \tilde{E},\ H = Z_0\tilde{H} \text{ and } J^s = Z_0\tilde{J}^s, \tag{20}$$

where $c_0 = 1/\sqrt{\varepsilon_0\mu_0}$ is the speed of light in vacuum ($c_0 \simeq 3 \times 10^8 \text{m}\cdot\text{s}^{-1}$) and $Z_0 = \sqrt{\mu_0/\varepsilon_0}$ is the free space intrinsic impedance. The normalized time variables are now expressed in meter (m) and the electric and magnetic fields in volt per meter ($\text{V}\cdot\text{m}^{-1}$). Substituting the normalized space, time variables and fields (20) into (19) we write

$$\begin{cases} \partial_t H^x(\mathbf{x}, t) = -\partial_y E^z(\mathbf{x}, t), \\ \partial_t H^y(\mathbf{x}, t) = \partial_x E^z(\mathbf{x}, t), \\ \partial_t E^z(\mathbf{x}, t) \ = \partial_x H^y(\mathbf{x}, t) - \partial_y H^x(\mathbf{x}, t) - J^s(\mathbf{x}, t). \end{cases} \tag{21}$$

The Eqs. (21) are space discretized using a DG method formulated on triangular meshes. In the preliminary implementation of this DG method, the approximation of the electromagnetic field components within a triangle τ_i relies on a nodal $\mathbb{P}_k$ interpolation method. The a priori convergence analysis for the error in $C^0([0, T], L^2(\Omega))$ and the adopted DG method, formulated on simplicial meshes and based on a centered numerical flux for the approximation of the boundary integral term at the interface between neighboring elements, shows that the convergence rate is $\mathcal{O}(h^k)$ for a k-th interpolation order [5]. The convergence result is slightly weaker than available results for upwind fluxes [2, 8], nevertheless this setting allows to obtain the conservation of a discrete form of the electromagnetic energy. A triangle τ_i is characterized by its height h_i. Denote Ω_h the computational domain and Ω_h^{exp} the set of triangles that belong to the explicit region. The critical time step Δt_c used in the numerical tests is given by

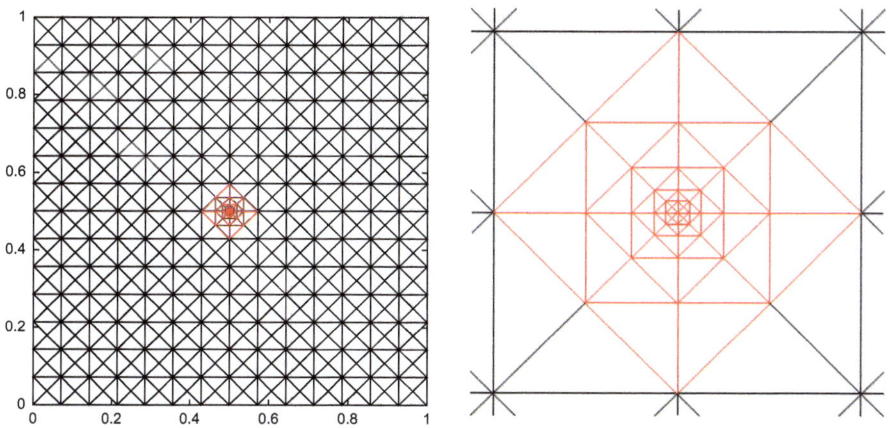

Fig. 3 Example of a mesh used in numerical tests (# elements: 848, implicit treatment: *red region*)

$$\Delta t_c = CFL \times \min\{h_i, \quad \tau_i \in \Omega_h^{exp}\}. \tag{22}$$

The values of the CFL number, corresponds to the numerical stability, i.e. the limit beyond which we observe a growth of the discrete energy.

We consider the propagation of an eigenmode in a unitary perfectly electrically conducting cavity. In this problem there is no source term i.e. $J^s = 0$ in (21) and the exact solution is given by

$$\begin{cases} H^x(\mathbf{x}, t) = -(k\pi/\omega)\sin(l\pi x)\cos(k\pi y)\sin(\omega t), \\ H^y(\mathbf{x}, t) = (l\pi/\omega)\cos(l\pi x)\sin(k\pi y)\sin(\omega t), \\ E^z(\mathbf{x}, t) = \sin(l\pi x)\sin(k\pi y)\cos(\omega t), \end{cases} \tag{23}$$

where the resonance frequencies, ω, are given as $\omega = \pi\sqrt{k^2 + l^2}$. For numerical tests we put $k = l = 1$ and we initialize the electromagnetic field with the exact analytical solution at time $t = 0$. We impose a metallic boundary condition such that the tangential component of the electric field vanishes on the boundaries, i.e. $n \times E^z = 0$ on $\partial\Omega$, where n denotes the unit outward normal to $\partial\Omega$. The total simulation time is set to $T = 5$. We investigate the space-time convergence order of the composition method (16) for $s = 3$ and 5, and the passive Richardson extrapolation (18) based on the second-order IMEX method (13). We consider a sequence of 6 successively refined triangular meshes, see Fig. 3 and Table 1 for an example and the characteristics of the different meshes. The critical time step is determined by the smallest height in the region treated explicitly; for the structured meshes and the implicit regions used in numerical tests it is equal to h^{max}, since all refined triangles belong to the implicit region. To estimate the order of convergence we measure the maximal L^2-norm of the error and we plot this error as a function of the square root of the number of degrees of freedom (DOF), in logarithmic scale.

Table 1 Data of the six successively refined triangular meshes (the total number of DOF is indicated for a DGTD-$\mathbb{P}_4$ method)

# elements	# DOF	h^{min}	h^{max}
208	3,120	0.00736	0.16667
464	6,960	0.00442	0.10000
848	12,720	0.00316	0.07143
2,368	35,520	0.00184	0.04167
4,688	70,320	0.00130	0.02941
7,808	117,120	0.00100	0.02273

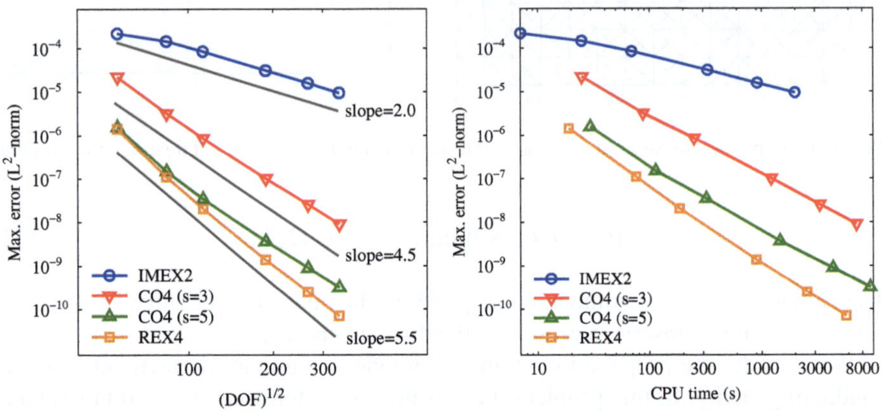

Fig. 4 Convergence and maximum error (L^2-norm) in function of final CPU time for the locally implicit DGTD-$\mathbb{P}_4$ methods (*left – right*, respectively)

We use the DGTD-$\mathbb{P}_4$ method so that the spatial error is not detrimental to the temporal convergence orders. Figure 4 shows orders of convergence about 4.5 for composition methods and about 5.5 for the Richardson extrapolation. We also plot the error as a function of the CPU time. For a given error or a given CPU time we observe the high efficiency of the fourth-order time integration methods compared to the second-order method. Finally for this problem Richardson extrapolation is the most accurate method.

These promising results should not prevent us to be cautious; we can not conclude that the fourth-order will be preserved whichever the problem considered. Indeed, the source term and the presence of a damping term which models conduction may be the cause of a reduction order, see for e.g. [1, 11]. Consequently in a near future, a theoretical convergence analysis will be conducted. Nevertheless, even if a reduction order was observed, the accuracy of the high-order locally implicit DG methods proposed in this study will certainly be very interesting.

References

1. Botchev, M.A., Verwer, J.G.: Numerical Integration of Damped Maxwell Equations. SIAM J. SCI. Comput. 31(2), 1322–1346 (2009)
2. Cockburn, B., Karniadakis, G.E., Shu, C.-W.: Discontinuous Galerkin methods. Theory, computation and applications. Springer-Verlag, Berlin (2000)
3. Descombes, S., Lanteri, S., Moya, L.: Locally implicit time integration strategies in a discontinuous Galerkin method for Maxwell's equations. J. Sci. Comput. 56(1), 190–218 (2013).
4. Faragó, I., Havasi, Á., Zlatev, Z.: Efficient implementation of stable Richardson extrapolation algorithms. Comput. Math. Appl. 60(8), 2309–2325 (2010)
5. Fezoui, L., Lanteri, S., Lohrengel, S., Piperno, S.: Convergence and stability of a discontinuous Galerkin time-domain method for the 3D heterogeneous Maxwell equations on unstructured meshes. ESAIM: M2AN 39(6), 1149–1176 (2005)
6. Hairer, E.,Lubich, C., Wanner, G.: Geometric Numerical Integration. Second edition, Springer - Verlag, Berlin (2002)
7. Moya, L.: Temporal convergence of a locally implicit discontinuous Galerkin method for Maxwell's equations. ESAIM: M2AN 46(5), 1225–1246 (2012)
8. Sármány, D., Botchev, M.A., Van der Vegt, J.J.W.: Dispersion and dissipation error in high-order Runge-Kutta discontinuous Galerkin discretisations of the Maxwell equations. J. Sci. Comput. 33(1), 47–74 (2007)
9. Suzuki, M.: Fractal decomposition of exponential operators with applications to many-body theories and Monte-Carlo simulations. Phys. Lett. A 146, 319–323 (1990)
10. Verwer, J.G.: Component splitting for semi-discrete Maxwell equations. BIT Numer. Math. 51(2), 427–445 (2010)
11. Verwer, J.G.: Composition methods, Maxwell's equations and source term. SIAM J. Numer. Anal. 50(2), 439–457 (2012)
12. Verwer, J.G., Botchev, M.A.: Unconditionaly stable integration of Maxwell's equations. Linear Algebra and its Applications 431(3–4),300–317 (2009)

High-Order ADI Schemes for Convection-Diffusion Equations with Mixed Derivative Terms

B. Düring, M. Fournié, and A. Rigal

Abstract We present new high-order Alternating Direction Implicit (ADI) schemes for the numerical solution of initial-boundary value problems for convection-diffusion equations with cross derivative terms. Our approach is based on the unconditionally stable ADI scheme proposed by Hundsdorfer (Appl Numer Math 42:213–233, 2002). Different numerical discretizations which lead to schemes which are fourth-order accurate in space and second-order accurate in time are discussed.

1 Introduction

We consider the multi-dimensional convection-diffusion equation

$$u_t = \operatorname{div}(D\nabla u) + c \cdot \nabla u \tag{1}$$

on a rectangular domain $\Omega \subset \mathbb{R}^2$, supplemented with initial and boundary conditions. In (1),

$$c = \begin{pmatrix} c_1 \\ c_2 \end{pmatrix}, \quad D = \begin{pmatrix} d_{11} & d_{12} \\ d_{21} & d_{22} \end{pmatrix},$$

B. Düring
Department of Mathematics, University of Sussex, Pevensey 2, Brighton, BN1 9QH, UK
e-mail: b.during@sussex.ac.uk

M. Fournié (✉) · A. Rigal
Institut de Mathématiques de Toulouse, Equipe 'Mathématiques pour l'Industrie et la Physique', CNRS, Unité Mixte 5219, Universités de Toulouse, 118, route de Narbonne, 31062 Toulouse Cedex, France
e-mail: michel.fournie@math.univ-toulouse.fr; alain.rigal@math.univ-toulouse.fr

M. Azaïez et al. (eds.), *Spectral and High Order Methods for Partial Differential Equations - ICOSAHOM 2012*, Lecture Notes in Computational Science and Engineering 95, DOI 10.1007/978-3-319-01601-6_17,

are a given nonzero convection vector and a given, fully populated (non-diagonal), and positive definite diffusion matrix, respectively. Thus, both mixed derivative and convection terms are present in (1).

After rearranging, problem (1) may be formulated as

$$\frac{\partial u(x,y,t)}{\partial t} = \underbrace{(d_{12}+d_{21})\frac{\partial^2 u}{\partial x \partial y}}_{=:F_0(u)} + \underbrace{(c_1\frac{\partial u}{\partial x} + d_{11}\frac{\partial^2 u}{\partial x^2})}_{=:F_1(u)} + \underbrace{(c_2\frac{\partial u}{\partial y} + d_{22}\frac{\partial^2 u}{\partial y^2})}_{=:F_2(u)}. \quad (2)$$

This type of convection-diffusion equations with mixed derivatives arise frequently in many applications, e.g. in financial mathematics for option pricing in stochastic volatility models or in numerical mathematics when coordinate transformations are applied. Such transformations are particularly useful to allow working on simple (rectangular) domains or on uniform grids (to have better accuracy). Thus, this approach allows to consider complex domains or to define non-uniform meshes to take into account the stiffness behavior of the solution in some part of the domain.

In the mathematical literature, there exist a number of numerical approaches to approximate solutions to (1), e.g. finite difference schemes, spectral methods, finite volume and finite element methods. Here, we consider (1) on a rectangular domain $\Omega \subset \mathbb{R}^2$. In this situation a finite difference approach seems most straight-forward.

The Alternating Direction Implicit (ADI) method introduced by Peaceman and Rachford [1], Douglas [4,5], Fairweather and Mitchell [7] is a very powerful method that is especially useful for solving parabolic equations on rectangular domains. Beam and Warming [2], however, have shown that no simple ADI scheme involving only discrete solutions at time levels n and $n+1$ can be second-order accurate in time in the presence of mixed derivatives ($F_0 \neq 0$ in (2)). To overcome this limitation and construct an unconditionally stable ADI scheme of second order in time, a number of results have been given by Hundsdorfer [11,12] and more recently by in 't Hout and Welfert [10]. These schemes are second-order accurate in time and space.

High-Order Compact (HOC) schemes (see, e.g. [8, 14]) employ a nine-point computational stencil using the eight neighbouring points of the reference grid point only and show good numerical properties. Several papers consider the application of HOC schemes (fourth order accurate in space) for two-dimensional convection-diffusion problems *with mixed derivatives* [3,6] but *without ADI* splitting. Moreover, the HOC approach introduces a high algebraic complexity in the derivation of the scheme.

We are interested in obtaining efficient, *high-order* ADI schemes, i.e. schemes which have a consistency order equal to two in time and to four in space, which are unconditionally stable and robust (no oscillations). We combine the second-order ADI splitting scheme presented in [10, 12] with different high-order schemes to approximate F_0, F_1, F_2 in (2). We note that some results on coupling *HOC with ADI* have been presented in [13], however, *without mixed derivative terms* present in the equation.

Up to the knowledge of the authors there are currently no results for ADI-HOC in the presence of mixed derivative terms. In this preparatory work we validate the coupling of ADI and HOC by numerical experiments.

2 Splitting in Time

In time, we consider the following splitting scheme presented in [10, 12]. We consider (2), and we look for a (semi-discrete) approximation $U^n \approx u(t_n)$ with $t_n = n\Delta_t$ for a time step Δ_t. The scheme used corresponds to

$$\begin{cases} Y^0 = U^{n-1} + \Delta_t F(U^{n-1}), \\ Y^1 = Y^0 + \theta\Delta_t(F_1(Y^1) - F_1(U^{n-1})), \\ Y^2 = Y^1 + \theta\Delta_t(F_2(Y^2) - F_2(U^{n-1})), \\ \tilde{Y}^0 = Y^0 + \sigma\Delta_t(F(Y^2) - F(U^{n-1})), \\ \tilde{Y}^1 = \tilde{Y}^0 + \theta\Delta_t(F_1(\tilde{Y}^1) - F_1(Y^2)), \\ \tilde{Y}^2 = \tilde{Y}^1 + \theta\Delta_t(F_2(\tilde{Y}^2) - F_2(Y^2)), \\ U^n = \tilde{Y}^2, \end{cases} \tag{3}$$

with constant parameters θ and σ, and $F = F_0 + F_1 + F_2$. To ensure second-order consistency in time we choose $\sigma = 1/2$. The parameter θ is arbitrary and typically fixed to $\theta = 1/2$. The choice of θ is discussed in [12]. Larger θ gives stronger damping of implicit terms and lower values return better accuracy (some numerical results for $\theta = 1/2 + \sqrt{3}/6$ are given in Sect. 4).

We note that F_0 is treated explicitly, whereas F_1, F_2 (unidirectional contributions in F) are treated implicitly. In the following section, we discuss different high-order (fourth order) strategies for the discretization in space.

3 High-Order Approximation in Space

For the discretization in space, we replace the rectangular domain $\Omega = [L_1, R_1] \times [L_2, R_2] \subset \mathbb{R}^2$ with $R_1 > L_1$, $R_2 > L_2$ by a uniform grid $Z = \{x_i \in [L_1, R_1] : x_i = L_1 + (i-1)\Delta_x, i = 1, \ldots, N\} \times \{y_j \in [L_2, R_2] : y_j = L_2 + (j-1)\Delta_y, j = 1, \ldots, M\}$ consisting of $N \times M$ grid points, with space steps $\Delta_x = (R_1 - L_1)/(N-1)$ and $\Delta_y = (R_2 - L_2)/(M-1)$. Let $u_{i,j}$ denote the approximate solution in (x_i, y_j) at some fixed time (we omit the superscript n to simplify the notation).

We present different fourth-order schemes to approximate F_0, F_1, F_2 in (3). The first one uses five nodes in each direction and the second one is compact. Both schemes are considered with boundary conditions of either periodic or Dirichlet type.

3.1 Fourth-Order Scheme Using Five Nodes

We denote by δ_{x0}, δ_{x+} and δ_{x-}, the standard central, forward and backward finite difference operators, respectively. The second-order central difference operator is denoted by δ_x^2,

$$\delta_x^2 u_{i,j} = \delta_{x+}\delta_{x-}u_{i,j} = \frac{u_{i+1,j} - 2u_{i,j} + u_{i-1,j}}{\Delta_x^2}.$$

The difference operators in the y-direction, δ_{y0}, δ_{y+}, δ_{y-} and δ_y^2, are defined analogously. Then it is possible to define fourth-order approximations based on,

$$\begin{aligned}
(u_x)_{i,j} &\approx \left(1 - \frac{\Delta_x^2}{6}\delta_x^2\right)\delta_{x0}u_{i,j} = \frac{-u_{i+2,j} + 8u_{i+1,j} - 8u_{i-1,j} + u_{i-2,j}}{12\Delta_x},\\
(u_y)_{i,j} &\approx \left(1 - \frac{\Delta_y^2}{6}\delta_y^2\right)\delta_{y0}u_{i,j} = \frac{-u_{i,j+2} + 8u_{i,j+1} - 8u_{i,j-1} + u_{i,j-2}}{12\Delta_y},\\
(u_{xx})_{i,j} &\approx \left(1 - \frac{\Delta_x^2}{12}\delta_x^2\right)\delta_x^2 u_{i,j} = \frac{-u_{i+2,j} + 16u_{i+1,j} - 30u_{i,j} + 16u_{i-1,j} - u_{i-2,j}}{12\Delta_x^2},\\
(u_{yy})_{i,j} &\approx \left(1 - \frac{\Delta_y^2}{12}\delta_y^2\right)\delta_y^2 u_{i,j} = \frac{-u_{i,j+2} + 16u_{i,j+1} - 30u_{i,j} + 16u_{i,j-1} - u_{i,j-2}}{12\Delta_y^2},\\
(u_{xy})_{i,j} &\approx \left(1 - \frac{\Delta_x^2}{6}\delta_x^2\right)\delta_{x0}\left(1 - \frac{\Delta_y^2}{6}\delta_y^2\right)\delta_{y0}u_{i,j}\\
&= \frac{1}{144\Delta_x\Delta_y}\Big[64(u_{i+1,j+1} - u_{i-1,j+1} + u_{i-1,j-1} - u_{i+1,j-1})\\
&\qquad +8(-u_{i+2,j+1} - u_{i+1,j+2} + u_{i-1,j+1} + u_{i-2,j+1}\\
&\qquad\quad -u_{i-2,j-1} - u_{i-1,j-2} + u_{i+1,j-2} + u_{i+2,j-1})\\
&\qquad +(u_{i+2,j+2} - u_{i-2,j+2} + u_{i-2,j-2} - u_{i+2,j-2})\Big].
\end{aligned} \tag{4}$$

For each differential operators appearing in F_0, F_1 and F_2, we use these five-points fourth-order difference formulae.

Combining this spatial discretization with the time splitting (3), we obtain a high-order, five-points ADI scheme denoted HO5. Its order of consistency is two in time and four in space.

3.2 Fourth-Order Compact Scheme

We start by deriving a fourth-order HOC scheme for

$$F_1(u) = d_{11}\frac{\partial^2 u}{\partial x^2} + c_1\frac{\partial u}{\partial x} = g, \tag{5}$$

with some arbitrary right hand side g. We employ central difference operators to approximate the derivatives in (5) using

$$\frac{\partial u}{\partial x}(x_i, y_j) = \delta_{x0}u_{i,j} - \frac{\Delta_x^2}{6}\frac{\partial^3 u}{\partial x^3}(x_i, y_j) + \mathcal{O}(\Delta_x^4), \tag{6}$$

$$\frac{\partial^2 u}{\partial x^2}(x_i, y_j) = \delta_x^2 u_{i,j} - \frac{\Delta_x^2}{12}\frac{\partial^4 u}{\partial x^4}(x_i, y_j) + \mathcal{O}(\Delta_x^4). \tag{7}$$

By differentiating (5), we can compute the following auxiliary relations for the derivatives appearing in (6), (7) (in the following, for the sake of brevity we omit the argument (x_i, y_j) of the continuous functions)

$$\frac{\partial^3 u}{\partial x^3} = \frac{1}{d_{11}}\frac{\partial g}{\partial x} - \frac{c_1}{d_{11}}\frac{\partial^2 u}{\partial x^2}, \tag{8}$$

$$\frac{\partial^4 u}{\partial x^4} = \frac{1}{d_{11}}\frac{\partial^2 g}{\partial x^2} - \frac{c_1}{d_{11}}\frac{\partial^3 u}{\partial x^3} = \frac{1}{d_{11}}\frac{\partial^2 g}{\partial x^2} - \frac{c_1}{d_{11}}\left(\frac{1}{d_{11}}\frac{\partial g}{\partial x} - \frac{c_1}{d_{11}}\frac{\partial^2 u}{\partial x^2}\right). \tag{9}$$

Hence, using (8) and (9) in (6) and (7), respectively, Eq. (5) can be approximated by

$$d_{11}\delta_x^2 u_{i,j} + c_1\delta_{x0}u_{i,j} = g_{i,j} + \frac{\Delta_x^2}{12}\left(\frac{c_1}{d_{11}}\frac{\partial g}{\partial x} + \frac{\partial^2 g}{\partial x^2} - \frac{c_1^2}{d_{11}}\frac{\partial^2 u}{\partial x^2}\right) + \mathcal{O}(\Delta_x^4). \tag{10}$$

We note that all derivatives on the right hand side of (10) can be approximated on a compact stencil using second-order central difference operators. This yields a high-order compact scheme of fourth order for (5) which is given by

$$d_{11}\delta_x^2 u_{i,j} + c_1\delta_{x0}u_{i,j} + \frac{\Delta_x^2}{12}\frac{c_1^2}{d_{11}}\delta_x^2 u_{i,j} = g_{i,j} + \frac{\Delta_x^2}{12}\left(\frac{c_1}{d_{11}}\delta_{x0}g_{i,j} + \delta_x^2 g_{i,j}\right). \tag{11}$$

In a similar fashion we can discretize the operator $F_2(u) = g$ by a high-order compact scheme of fourth order given by

$$d_{22}\delta_y^2 u_{i,j} + c_2\delta_{y0}u_{i,j} + \frac{\Delta_y^2}{12}\frac{c_2^2}{d_{22}}\delta_y^2 u_{i,j} = g_{i,j} + \frac{\Delta_y^2}{12}\left(\frac{c_2}{d_{22}}\delta_{y0}g_{i,j} + \delta_y^2 g_{i,j}\right). \tag{12}$$

Defining vectors $U = (u_{1,1}, \ldots, u_{N,M})$ and $G = (g_{1,1}, \ldots, g_{N,M})$, we can state these schemes (11) and (12) in matrix form $A_x U = B_x G$ (for $F_1(u) = g$) and $A_y U = B_y G$ (for $F_2(u) = g$), respectively. We apply these HOC schemes to find the unidirectional contributions Y^1, $\tilde{Y}^1$, and Y^2, $\tilde{Y}^2$ in (3), respectively. For example, to compute

$$Y^1 = Y^0 + \frac{\Delta_t}{2}(F_1(Y^1) - F_1(U^{n-1}))$$

in the second step of (3) (which is equivalent to $F_1(Y^1 - U^{n-1}) = -\frac{2}{\Delta_t}(Y^0 - Y^1)$), we use $A_x(Y^1 - U^{n-1}) = B_x(-\frac{2}{\Delta_t}(Y^0 - Y^1))$ that can be rewrite into

$$\left(B_x - \frac{\Delta_t}{2}A_x\right)Y^1 = B_xY^0 - \frac{\Delta_t}{2}A_xU^{n-1}.$$

Note that the matrix $(B_x - (\Delta_t/2)\,A_x)$ appears twice in (3), in steps 2 and 5. Similarly, $(B_y - (\Delta_t/2)\,A_y)$ appears in steps 3 and 6 of (3). Hence, using LU-factorisation, only two matrix inversions are necessary in each time step of (3). Moreover, for the case of constant coefficients, these matrices can be LU-factorized before iterating in time to obtain an even more efficient algorithm.

To compute Y^0 and $\tilde{Y}^0$ in steps 1 and 4 of (3) which require evaluation of F_0 (mixed term) we use an explicit approximation using the five-points fourth-order formulae (4).

Combining this spatial discretization with the time splitting (3), we obtain a high-order compact ADI scheme denoted HOC. Its order of consistency is two in time and four in space.

4 Numerical Experiments

We present numerical experiments on a square domain $\Omega = [0,1] \times [0,1]$ for two types of boundary conditions, periodic and Dirichlet type. The initial condition is given at time $T_0 = 0$ and the solution is computed at the final time $T_f = 0.1$ with different meshes $\Delta_x = \Delta_y = h$ and different time steps Δ_t. In our numerical tests we focus on the errors with respect to time and to space.

In the first part, we consider the periodic boundary value problem considered in [10]. We implement the scheme detailed in [10] based on second-order finite difference approximations (referred to as CDS below) and compare its behaviour to our new schemes HO5 (Sect. 3.1) and HOC (Sect. 3.2). In the second part, we consider Dirichlet boundary conditions and restrict our study to the more interesting HOC scheme. In that part, we extend the splitting scheme to a convection-diffusion equation with source term.

4.1 *Periodic Boundary Conditions*

The problem given in [10] is formulated on the domain $\Omega = [0,1] \times [0,1]$. The solution u satisfies (1) where

$$c = -\begin{pmatrix}2\\3\end{pmatrix}, \quad D = 0.025\begin{pmatrix}1 & 2\\2 & 4\end{pmatrix},$$

Table 1 Numerical convergence rates in time for $\theta = \frac{1}{2}$

	l_2-error convergence rate			l_∞-error convergence rate		
Scheme	$h = 0.1$	$h = 0.025$	$h = 0.00625$	$h = 0.1$	$h = 0.025$	$h = 0.00625$
CDS	2.2002	2.1975	2.1969	2.1973	2.1958	2.1956
HO5	2.1999	2.1973	2.1969	2.1992	2.1953	2.1955
HOC	2.2002	2.1973	2.1969	2.2007	2.1953	2.1955

Table 2 Numerical convergence rates in space of l_2-error for fixed μ as Δ_x, $\Delta_t \to 0$ and $\theta = \frac{1}{2}$

Scheme	$\mu = 0.4$	$\mu = 0.2$	$\mu = 0.1$	$\mu = 0.05$
CDS	1.7828	1.7909	1.7821	1.7845
HO5	2.2291	2.5188	2.8153	3.0672
HOC	2.2685	2.5191	2.8152	3.0671

with periodic boundary conditions and initial condition $u(x, y, T_0) = e^{-4(\sin^2(\pi x)+\cos^2(\pi y))}$. We employ the splitting (3) with $\sigma = 1/2$ and $\theta = 1/2$.

We first present a numerical study to compute the order of convergence in time of the schemes CDS, HO5 and HOC. Asymptotically, we expect the error ε to converge as

$$\varepsilon = C\Delta_t^m$$

at some rate m with C representing a constant. This implies

$$\log(\varepsilon) = \log(C) + m\log(\Delta_t).$$

Hence, the double-logarithmic plot ε against Δ_t should be asymptotic to a straight line with slope m that corresponds to the order of convergence in time of the scheme. We denote by ε_2 and ε_∞ the errors in the l_2-norm and l_∞-norm, respectively. We refer to Table 1 for the order of convergence in time computed for different fixed mesh widths $h \in \{0.1, 0.0.025, 0.00625\}$ and time steps $\Delta_t \in [T_f/30, T_f/90]$. The solution computed for $\Delta_t = T_f/100$ is considered as reference solution to compute the errors. The global errors for the splitting behave like $C(\Delta_t)^2$. We also observe that the constant C only depends weakly on the spatial mesh widths h.

In the following, we study the spatial convergence. The double-logarithmic plots ε_2 and ε_∞ against h give the rates of convergence. Contrary to the time convergence, the order now depends on the parabolic mesh ratio $\mu = \Delta_t/\Delta_x^2$, so the numerical tests are performed for a set of different constant values of μ. For simulations, μ is fixed at constant values $\mu \in \{0.4, 0.2, 0.1, 0.005\}$ while $\Delta_x = \Delta_y = h \to 0$ (Δ_t is then given by $\Delta_t = \mu h^2$). The results for the l_2-error are given in Table 2 and for the l_∞-error in Table 3. The solution computed for $h = 0.00625$ is used as reference solution to compute the errors.

Remark. The choice of the parameter θ is discussed in [12]. However, for the convergence rates, θ seems to have little influence. For example, for the scheme HO5 with $\theta = 1/2 + \sqrt{3}/6$ we obtain very similar results as shown in Table 4.

Table 3 Numerical convergence rates in space of l_∞-error for fixed μ as $\Delta_x, \Delta_t \to 0$ and $\theta = \frac{1}{2}$

Scheme	$\mu = 0.4$	$\mu = 0.2$	$\mu = 0.1$	$\mu = 0.05$
CDS	1.7170	1.7125	1.7040	1.7038
HO5	2.2931	2.6166	2.9182	3.1584
HOC	2.3175	2.6176	2.9184	3.1584

Table 4 Numerical convergence rates in space for HO5 for fixed μ as $\Delta_x, \Delta_t \to 0$ and $\theta = \frac{1}{2} + \frac{\sqrt{3}}{6}$

	$\mu = 0.4$	$\mu = 0.2$	$\mu = 0.1$	$\mu = 0.05$
l_2 rate	2.2310	2.5186	2.8152	3.0671
l_∞ rate	2.2938	2.6164	2.9181	3.1584

4.2 Dirichlet Boundary Conditions

In this section we only consider the HOC scheme which presents more interesting properties than the other schemes. Indeed, compared to CDS, its accuracy is larger and compared to HO5, no specific treatment at the boundaries is required for the unidirectional terms F_1, F_2, the compact scheme is optimal in this respect. A particular treatment is necessary when ghost points appear in the explicit approximation of the mixed term F_0. To preserve the global performance, the accuracy of the approximation near the boundary conditions has to be sufficiently high. We have used a sixth-order approximation in one direction (although lower order may also be used [9]). For example, for $u_{0,j}$ on the boundary, at a ghost point $u_{-1,j}$ we impose

$$u_{-1,j} = 5u_{0,j} - 10u_{1,j} + 10u_{2,j} - 5u_{3,j} + u_{4,j}.$$

For the numerical tests, we consider the problem

$$u_t = \operatorname{div}(D\nabla u) + c \cdot \nabla u + S$$

on the domain $\Omega = [0,1] \times [0,1]$ where

$$c = -\begin{pmatrix} 2 \\ 3 \end{pmatrix}, \quad D = 0.025 \begin{pmatrix} 1 & 2 \\ 2 & 4 \end{pmatrix},$$

and the source term S is determined in such a way that the solution is equal to $u(x,y,t) = -\frac{1}{t+1}\sin(\pi x)\sin(\pi y)$. The Dirichlet boundary condition and initial condition are deduced from the solution. To incorporate the source term S in the splitting (3), F needs to be replaced by $F + S$. More specifically, $F(U^{n-1})$ is replaced by $F(U^{n-1}) + S(t^{n-1})$ and $F(Y^2)$ by $F(Y^2) + S(t^n)$. We perform the same numerical experiments as in the previous section. The final time is fixed to $T_f = 0.1$ and the errors are computed with respect to a reference solution computed on a fine grid in space ($\Delta_x = \Delta_y = 0.00625$). Different meshes in space are considered

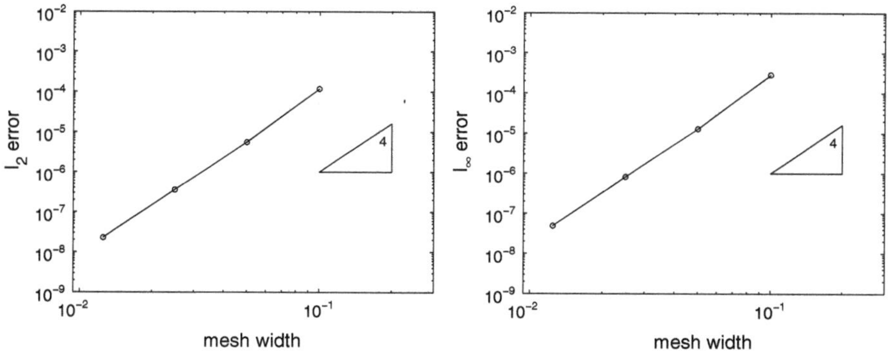

Fig. 1 Numerical convergence rate in space for HOC ($\theta = \frac{1}{2}$) and $\mu = 0.4$

Table 5 Numerical convergence rates of l_2-error and l_∞-error for HOC ($\theta = \frac{1}{2}$) for different constant values of μ (dirichlet boundary conditions)

	$\mu = 0.4$	$\mu = 0.2$	$\mu = 0.1$	$\mu = 0.05$
l_2 rate	4.0971	4.1875	4.2129	4.2196
l_∞ rate	4.1530	4.2372	4.2717	4.2806

for $\Delta_x = \Delta_y = h$ and $h \in \{0.1, 0.05, 0.025, 0.0125\}$. For $\mu = 0.4$ the double-logarithmic plots ε_2 and ε_∞ against h are given in Fig. 1.

The results of several numerical tests are reported in Table 5 for fixed parabolic mesh ratio $\mu = \Delta_t/\Delta_x^2$ while Δ_x, $\Delta_t \rightarrow 0$. In all situations, the new HOC scheme shows a good performance with fourth-order convergence rates in space, independent of the parabolic mesh ratio μ.

5 Conclusion

We have presented new high-order Alternating Direction Implicit (ADI) schemes for the numerical solution of initial-boundary value problems for convection-diffusion equations with mixed derivative terms. Using the unconditionally stable ADI scheme from [12] we have proposed different spatial discretizations which lead to schemes which are fourth-order accurate in space and second-order accurate in time.

We have performed a numerical convergence analysis with periodic and Dirichlet boundary conditions where high-order convergence is observed. In some cases, the order depends on the parabolic mesh ratio. More detailed discussions of these schemes including this dependence and a stability analysis will be presented in a forthcoming paper.

References

1. D.W. Peaceman and H.H. Rachford Jr., The numerical solution of parabolic and elliptic differential equations, *J. Soc. Ind. Appl. Math.*, **3**, 28–41, (1959).
2. R.M. Beam and R.F. Warming, Alternating Direction Implicit methods for parabolic equations with a mixed derivative, *Siam J. Sci. Stat. Comput.*, **1**(1), (1980).
3. M. Fournié and S. Karaa, Iterative methods and high-order difference schemes for 2D elliptic problems with mixed derivative, *J. Appl. Math. & Computing*, **22** (3), 349–363, (2006).
4. J. Douglas, Alternating direction methods for three space variables, *Numer. Math.*, **4**, 41–63, (1962).
5. J. Douglas and J. E. Gunn, A general formulation of alternating direction methods. I. Parabolic and hyperbolic problems, *Numer. Math.*, **6**, 428–453, (1964).
6. B. Düring and M. Fournié, High-order compact finite difference scheme for option pricing in stochastic volatility models. *J. Comput. Appl. Math.* **236**(17), 4462–4473, (2012).
7. G. Fairweather and A. R. Mitchell, A new computational procedure for A.D.I. methods, *SIAM J. Numer. Anal.*, **4**, 163–170, (1967).
8. M.M. Gupta, R.P. Manohar and J.W. Stephenson, A single cell high-order scheme for the convection-diffusion equation with variable coefficients, *Int. J. Numer. Methods Fluids*, **4**, 641–651, (1984).
9. B. Gustafsson, The convergence rate for difference approximation to general mixed initial-boundary value problems, *SIAM J. Numer. Anal.* **18**(2), 179–190, (1981).
10. K.J. in 't Hout and B.D. Welfert, Stability of ADI schemes applied to convection-diffusion equations with mixed derivative terms, *Appl. Num. Math.*, **57**, 19–35, (2007).
11. W. Hundsdorfer and J.G Verwer, Numerical solution of time-dependent advection-diffusion-reaction equations, Springer Series in Computational Mathematics, **33**, Springer-Verlag, Berlin, (2003).
12. W. Hundsdorfer, Accuracy and stability of splitting with stabilizing corrections, *Appl. Num. Math.*, **42**, 213–233, (2002).
13. S. Karaa and J. Zhang, High-order ADI method for solving unsteady convection-diffusion problems, *J. Comput. Phys.*, **198**(1), 1–9, (2004).
14. A. Rigal, Schémas compacts d'ordre élevé: application aux problèmes bidimensionnels de diffusion-convection instationnaire I, *C.R. Acad. Sci. Paris. Sr. I Math.*, **328**, 535–538, (1999).

A Numerical Study of Averaging Error Indicators in p-FEM

Philipp Dörsek and J. Markus Melenk

Abstract We consider the averaging error indicator in the context of the p-FEM. We explain how a proof of reliability and efficiency might look, and why the error indicator will behave differently than for low order methods. Using two model problems, one with nonsmooth, the other one with smooth solution, we identify appropriate spaces for the averaged fluxes in order to obtain reasonable reliability and efficiency bounds on the averaging error indicator for p-FEM. In particular, averaging over two neighbouring elements using global polynomials of the same polynomial degree as the finite element solution leads to reliability and efficiency up to a factor of order $O(p)$.

1 Introduction

The averaging error indicator, also called gradient recovery, superconvergent patch recovery, or Zienkiewicz-Zhu error indicator, going back originally to [17], is a widely used method for gauging errors in finite element methods and steering adaptive mesh refinements. Its main advantage is that it is very simple to compute, requiring only a local averaging of the numerical fluxes. A mathematical analysis in the low order context was performed in [1–3, 6, 11, 14–16]. In [7], the proof of reliability was reduced to the existence of approximation operators with certain additional orthogonality properties, and such approximation operators were then constructed for arbitrary, but fixed polynomial degree. It is also stated in [7, p. 991]

P. Dörsek (✉)
ETH Zürich, Rämistrasse 101, CH-8092 Zürich, Switzerland
e-mail: philipp.doersek@gmail.com

J.M. Melenk
Vienna University of Technology, Wiedner Hauptstraße 8–10, A-1040 Vienna, Austria
e-mail: melenk@tuwien.ac.at

M. Azaïez et al. (eds.), *Spectral and High Order Methods for Partial Differential Equations - ICOSAHOM 2012*, Lecture Notes in Computational Science and Engineering 95, DOI 10.1007/978-3-319-01601-6_18,

that the numerical behaviour observed in an hp-adaptive strategy "suggests that those constants depend only moderately on p", where the constants referred to are the reliability and efficiency constants of the averaging error indicator.

It is therefore our aim in this paper to analyse whether the proof for reliability and efficiency in [7] can be carried over to the p-FEM. A counting argument on the degrees of freedom shows quickly that the usual good efficiency estimate (efficiency with constant 1 up to a term of higher order) cannot be expected in the high order setting at least for algebraic rates of convergence, as this would require too many degrees of freedom in the approximation space for the averaged fluxes. Hence, we perform numerical computations for two model problems, one with nonsmooth, the other with smooth solution. Our results suggest that increasing the polynomial degree by one, as is commonly done in the low order context, leads to reasonable results if the averaging is performed over four quadrilateral elements. However, in this case, we observe the p-gap, similarly as in the residual error indicator due to [8, 12], which can be removed using equilibration techniques, see [9].

2 The Averaging Error Indicator

Let $\Omega \subset \mathbb{R}^d$, $d = 2, 3$, be a bounded polygonal domain, and $f \in \mathrm{H}^{-1}(\Omega)$, where $\mathrm{H}^{-1}(\Omega) = (\mathrm{H}^1_0(\Omega))^*$ are the usual Sobolev spaces. We denote the L^2 norm by $\|u\|_0 := \left(\int_\Omega u^2 \mathrm{d}x\right)^{1/2}$ and the H^1 seminorm by $|u|_1 := \left(\int_\Omega |\nabla u|^2 \mathrm{d}x\right)^{1/2}$, where $|\cdot|$ is the Euclidean norm. Consider for simplicity the Poisson problem with homogeneous Dirichlet boundary conditions,

$$-\Delta u = f \quad \text{in } \Omega, \quad u = 0 \quad \text{on } \partial\Omega; \tag{1}$$

the analysis of more general boundary conditions is also possible. Defining $V := \mathrm{H}^1_0(\Omega)$, its weak formulation reads: find $u \in V$ such that

$$a(u, v) = \ell(v) \quad \text{for all } v \in V, \tag{2}$$

with

$$a(w, v) := \int_\Omega \nabla w \cdot \nabla v \mathrm{d}x \quad \text{and} \quad \ell(v) := \int_\Omega f v \mathrm{d}x. \tag{3}$$

We approximate u from the conforming hp-finite element space $V_N \subset V$, i.e., with a triangulation $\mathcal{T}_N$ of Ω into quadrilaterals and a vector $(p_{N,T})_{T \in \mathcal{T}_N}$ of polynomial degrees, we consider

$$V_N := \{v \in V : v|_T \in \mathbb{Q}^{p_{N,T}}\}, \tag{4}$$

where $\mathbb{Q}^k$ is the usual space of tensor product polynomials of degree k in every component. Then, $u_N \in V_N$ is defined through

$$a(u_N, v_N) = \ell(v_N) \quad \text{for all } v_N \in V_N. \tag{5}$$

Let $\Sigma_N \subset \mathrm{H}(\nabla\cdot, \Omega) := \{\tau \in (\mathrm{L}^2(\Omega))^2 : \nabla \cdot \tau \in \mathrm{L}^2(\Omega)\}$, then the global error indicator is defined by

$$\eta_N := \inf_{\tau_N \in \Sigma_N} \|\tau_N - \nabla u_N\|_0. \tag{6}$$

Let $\sigma_N \in \Sigma_N$ denote the uniquely determined argument where the above infimum is attained. If Σ_N is finite-dimensional, it is clear that this quantity can be calculated by solving a system of linear equations.

Proposition 1 (Reliability). *Let $I_N : V \to V_N$ be a linear operator with*

$$|I_N v|_1 \le C_N |v|_1 \quad \textit{for all } v \in V. \tag{7}$$

Assume that $\nabla u \in \mathrm{H}(\nabla\cdot, \Omega)$. Then, the error indicator η defined in (6) *satisfies*

$$|u - u_N|_1 \le (1 + C_N)\eta_N + \sup_{v \in V\setminus\{0\}} \frac{\int_\Omega (f + \nabla \cdot \sigma_N)(v - I_N v)\mathrm{d}x}{|v|_1}. \tag{8}$$

Proof. As $\sigma_N \in \Sigma_N \subset \mathrm{H}(\nabla\cdot, \Omega)$, the Galerkin orthogonality yields

$$\begin{aligned}
|u - u_N|_1 &= \sup_{v \in V\setminus\{0\}} \frac{a(u - u_N, v)}{|v|_1} = \sup_{v \in V\setminus\{0\}} \frac{a(u - u_N, v - I_N v)}{|v|_1} \\
&= \sup_{v \in V\setminus\{0\}} \frac{\int_\Omega (f + \nabla \cdot \sigma_N)(v - I_N v)\mathrm{d}x}{|v|_1} \\
&\quad + \frac{\int_\Omega (\sigma_N - \nabla u_N) \cdot \nabla(v - I_N v)\mathrm{d}x}{|v|_1} \\
&\le \sup_{v \in V\setminus\{0\}} \frac{\int_\Omega (f + \nabla \cdot \sigma_N)(v - I_N v)\mathrm{d}x}{|v|_1} + (1 + C_N)\|\sigma_N - \nabla u_N\|_0.
\end{aligned} \tag{9}$$

This proves the claimed estimate. □

Remark 1. The above result suggests to look for a linear operator $I_N : V \to V_N$ such that, ideally, its norm in V is bounded independently of N and, additionally, it has the orthogonality property

$$\int_\Omega w_N (v - I_N v)\mathrm{d}x = 0 \quad \text{for all } v \in V \text{ and } w_N \in W_N, \tag{10}$$

where W_N is a sufficiently large discrete space satisfying $\nabla \cdot \Sigma_N \subset W_N$. In this case, we observe

$$\int_{\Omega} \nabla \cdot \sigma_N (v - I_N v) \mathrm{d}x = 0 \quad \text{for all } v \in V \tag{11}$$

and hence

$$|u - u_N|_1 \leq C\eta_N + C\gamma_N \inf_{f_N \in W_N} \|f - f_N\|_0, \tag{12}$$

i.e., reliability with a generic constant. Here, γ_N is defined by

$$\gamma_N := \sup_{v \in V \setminus \{0\}} \frac{\|v - I_N v\|_0}{|v|_1} \tag{13}$$

and usually behaves like $\gamma_N \sim h_N p_N^{-1}$ on quasi-uniform meshes and polynomial degree distributions, i.e., the last term in (12) is of higher order compared to $|u-u_N|_1$ if W_N is large enough.

Remark 2. If the polynomial degree is fixed and the mesh is refined, an operator I_N as required above is constructed in [7]. Their construction, however, does not generalise directly to the p-version.

In order to obtain an operator I_N for the p-version, a first step would be to let I_N be the L^2-projection operator onto $\mathbb{Q}^{p_N}$, global polynomials of degree p_N, if we assume that $W_N = \mathbb{Q}^{p_N}$ consists of global polynomials, as well. This assumption makes sense in a pure p-version context on a reasonably coarse mesh. If we ignore the issue of boundary conditions, e.g., by considering a pure Neumann problem, [10, Theorem 2.4] yields that on a quasi-uniform mesh with uniform polynomial degree,

$$\|I_N v\|_1 \leq C(p_N + 1)^{1/2} \|v\|_1 \quad \text{for all } v \in \mathrm{H}^1(\Omega); \tag{14}$$

see also [13, Theorem 1.3] for a corresponding result for triangular and tetrahedral meshes. In this case, we obtain

$$\int_{\Omega} w_N (v - I_N v) \mathrm{d}x = 0 \quad \text{for all } v \in V \text{ and } w_N \in W_N. \tag{15}$$

Choosing $\Sigma_N := \mathbb{Q}^{(p_N+1)\times p_N} \times \mathbb{Q}^{p_N \times (p_N+1)}$, we observe $\nabla \cdot \Sigma_N \subset W_N$, and hence Proposition 1 yields

$$|u - u_N|_1 \leq C(p_N + 1)^{1/2} \eta_N + C p_N^{-1} \inf_{f_N \in W_N} \|f - f_N\|_0. \tag{16}$$

Proposition 2 (Efficiency). *The error indicator* η *defined in* (6) *satisfies*

$$\eta_N \leq |u - u_N|_1 + \inf_{\tau_N \in \Sigma_N} \|\tau_N - \nabla u\|_0. \tag{17}$$

Proof. We see that

$$\|\sigma_N - \nabla u_N\|_0 \leq \inf_{\tau_N \in \Sigma_N} \|\tau_N - \nabla u_N\|_0 \leq |u - u_N|_1 + \inf_{\tau_N \in \Sigma_N} \|\tau_N - \nabla u\|_0, \tag{18}$$

from which the claim follows. □

Remark 3. In order to ensure efficiency of the error indicator, the gradient L^2 projection error

$$\xi_N := \inf_{\tau_N \in \Sigma_N} \|\tau_N - \nabla u\|_0 \tag{19}$$

needs to be small. In the h-version context, [7] shows that ξ_N is indeed of higher order if local averaging over edge patches is done using polynomials of degree p_N. It is unclear whether this is possible when averaging globally, see [7, Remark 4.3].

For the p-version, we cannot hope that ξ_N is of higher order: if u_N is approximated using polynomials of degree p, then, in order that ξ_N is of higher order, we need that Σ_N consists of polynomials of degree $p^{1+\alpha}$ for some $\alpha > 0$. But this is not possible if we simultaneously want to ensure existence of an operator I_N as outlined in Remark 1, as in this case $\dim \Sigma_N$ grows faster than $\dim V_N$, which is incompatible with $I_N\colon V \to V_N$ being orthogonal to $W_N \supset \nabla \cdot \Sigma_N$. However, the following argument lets us hope for efficiency, at least if the convergence is only algebraic and we are prepared to accept a p-gap. Let us restrict ourselves for ease of exposition to Ω being a square and the right-hand side being smooth; general polygonal domains can be treated in a similar fashion. Then, [4, Theorem 2.7 and 2.10] yield the sharp convergence bounds

$$c(1 + p_N)^{-4} \leq |u - u_N|_1 \leq C(1 + p_N)^{-4}. \tag{20}$$

Similarly, as the gradient of a singularity function is again a singularity function, we obtain, assuming $\mathbb{Q}^{p_N} \subset \Sigma_N$, from [4, Theorem 2.7] that

$$\inf_{\tau_N \in \Sigma_N} \|\tau_N - \nabla u\|_0 \leq C(1 + p_N)^{-3}. \tag{21}$$

Together with (17), this implies

$$\eta_N \leq C(1 + p_N)|u - u_N|_1. \tag{22}$$

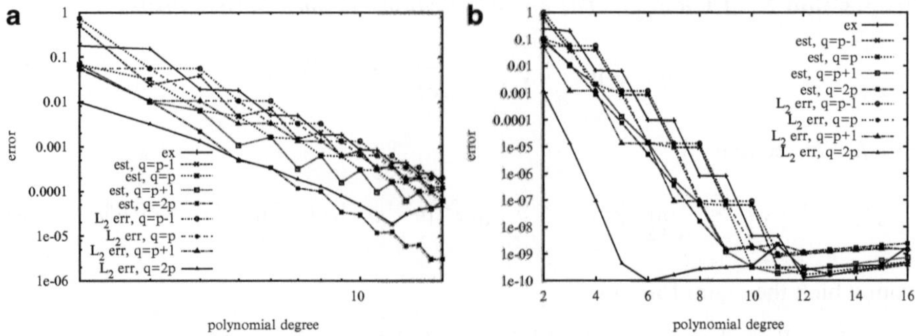

Fig. 1 Errors and error indicators, two elements. (**a**) Algebraic rate, two elements. (**b**) Exponential rate, two elements

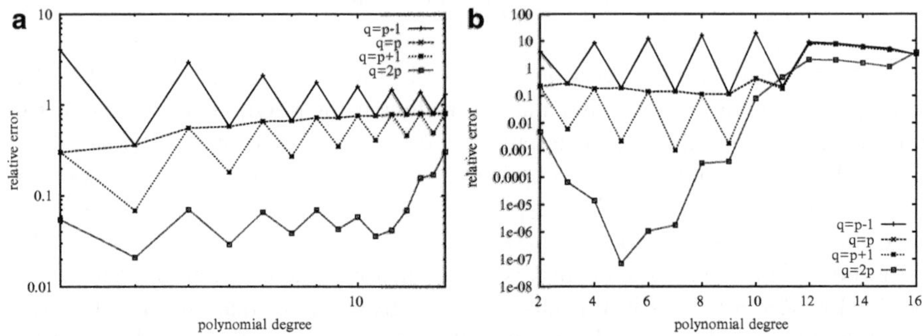

Fig. 2 Gradient L^2 projection error ξ_N relative to Galerkin error, two elements. (**a**) Algebraic rate. (**b**) Exponential rate

3 Numerical Examples

For our numerical computations, we consider the square domain $\Omega = (0, \pi)^2$ and solve the homogeneous Dirichlet problem

$$-\Delta u = f \quad \text{in } \Omega, \qquad u = 0 \quad \text{on } \partial\Omega, \tag{23}$$

with the two right-hand sides $f = 1$ and $f(x, y) = 2\sin(x)\sin(y)$. In the first case, the solution is known in terms of a Fourier series, and it is in $H^{3-\varepsilon}(\Omega)$ for all $\varepsilon > 0$. As the singularities are in the corners of the domain and can therefore be described using the corresponding singularity functions, the rate of convergence is known to be p^{-4}, see [5, Sect. 4.2], and this is confirmed in Figs. 1 and 4. In the second case, the solution $u(x, y) = \sin(x)\sin(y)$ is analytic, hence the convergence is exponential, and this is also confirmed in Figs. 1 and 4.

We consider two triangulations, one with two quadrilateral elements, $\mathcal{T}_2 = \{(0, \pi/2) \times (0, \pi), (\pi/2, \pi) \times (0, \pi)\}$, and the second with four elements,

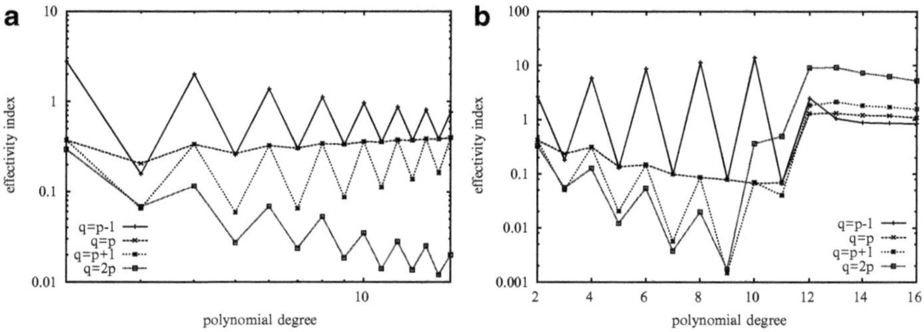

Fig. 3 Effectivity indices ζ_N, two elements. (**a**) Algebraic rate. (**b**) Exponential rate

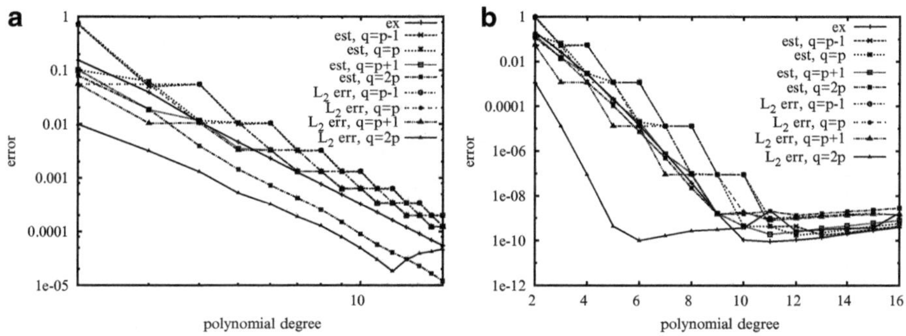

Fig. 4 Errors and error indicators, four elements. (**a**) Algebraic rate, four elements. (**b**) Exponential rate, four elements

$\mathcal{T}_4 = \{(0,\pi/2)\times(0,\pi/2), (\pi/2,\pi)\times(0,\pi/2), (0,\pi/2)\times(\pi/2,\pi), (\pi/2,\pi)\times(\pi/2,\pi)\}$. The finite element space is

$$V_N^{(\ell)} := \{v \in V : v|_T \in \mathbb{Q}^{p_N} \text{ for } T \in \mathcal{T}_\ell\}, \quad \ell = 2, 4, \tag{24}$$

and the approximation space for the averaged fluxes is chosen to be global polynomials. More precisely, we set $\Sigma_N := \mathbb{Q}^{q_N+1,q_N} \times \mathbb{Q}^{q_N,q_N+1}$ with $\mathbb{Q}^{q_1,q_2}$ the space of tensor product polynomials of degree q_1 in the first and q_2 in the second component, i.e., we average the numerical flux over two or four elements using Raviart-Thomas elements.

Given p_N, we consider for q_N the values $p_N - 1$, p_N, $p_N + 1$ and $2p_N$. This choice of values can be explained as follows: for $q_N = p_N - 1$, it is reasonable to expect that we can prove reliability by following the strategy laid out in Remark 1. Additionally, the computation of the error indicator is cheapest with this choice. A direct generalisation of the *h*-version error indicator from [7] leads to $q_N = p_N$. The choices $q_N = p_N + 1$ and $q_N = 2p_N$, finally, should yield better efficiency. Finally, more general choices such as $q_N = p_N^\beta$ are not suitable as explained in

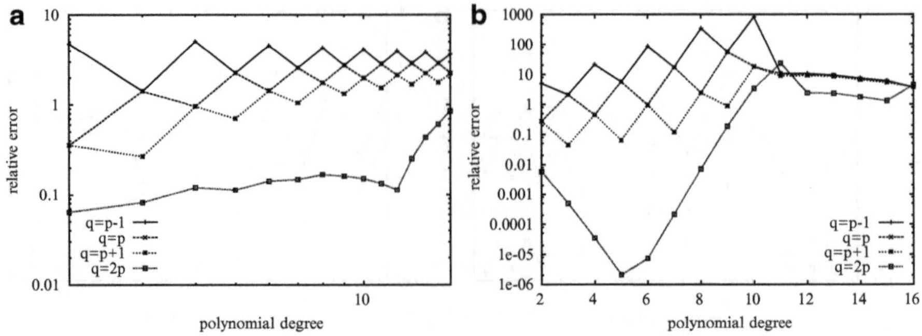

Fig. 5 Gradient L^2 projection error ξ_N relative to Galerkin error, four elements. (**a**) Algebraic rate. (**b**) Exponential rate

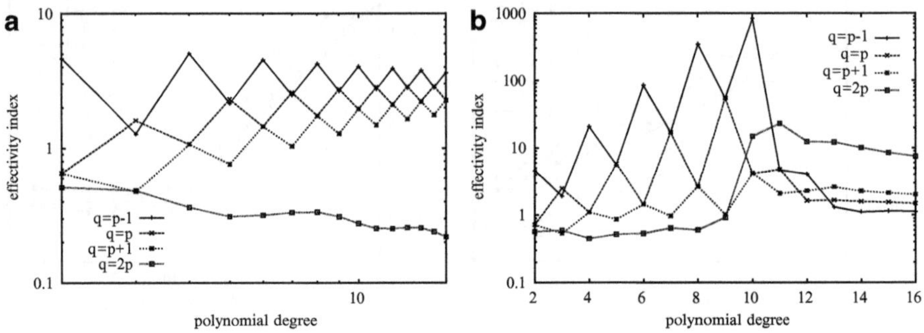

Fig. 6 Effectivity indices ζ_N, four elements. (**a**) Algebraic rate. (**b**) Exponential rate

Remark 3. We therefore believe that our experiments cover the relevant choices for the approximation space Σ_N.

A good choice for q_N should ensure that the effectivity indices do not decay too quickly in p, and that the gradient L^2 projection error in (17) is at least not more important than the error. Hence, we plot the gradient L^2 projection error ξ_N of ∇u from Σ_N defined in (19) relative to the Galerkin error $\|u - u_N\|_1$,

$$\frac{\xi_N}{|u - u_N|_1}; \tag{25}$$

and the effectivity indices ζ_N defined by

$$\zeta_N := \frac{\eta_N}{|u - u_N|_1}. \tag{26}$$

obtained in our numerical experiments.

Our numerical results show that the gradient L^2 projection error ξ_N is of higher order relative to the Galerkin error only for exponentially decaying error, and even then only for $q_N = 2p_N$. For two elements, the choices $q_N = p_N$ and $q_N = p_N + 1$ at least lead to ξ_N being not larger than the Galerkin error. When averaging over four elements, even that is only achieved using $q_N = 2p_N$.

Let us now turn to the effectivity indices ζ_N. For two elements, the most reasonable choice is given by $q_N = p_N$; it leads to a reliable and efficient error indicator in the nonsmooth model problem with effectivity indices varying between 0.2 and 0.4, and only to a moderate loss of reliability (of the order $O(p)$) in the smooth model problem. Setting $q_N = p_N + 1$ is adequate in the nonsmooth model problem, but the loss of reliability in the smooth model problem is pronounced. For four elements, both choices $q_N = p_N$ and $q_N = p_N + 1$ are reliable in both model problems, but lead to a loss of efficiency (of the order $O(p^{1.35})$ and $O(p^{0.85})$, respectively). The choice $q_N = 2p_N$, finally, leads to a slight loss in reliability in the nonsmooth model problem (of the order $O(p^{0.35})$), and is reliable and efficient in the smooth problem.

4 Conclusions

In contrast to low order finite elements, the use of the averaging error indicator in p-FEM leads to certain difficulties. The standard methods of proof cannot be used to obtain reliability and efficiency in the same sense as for the low order case. As explained in Remark 3, the gradient L^2 projection error present in the efficiency estimate cannot be made to be of higher order relative to the Galerkin error.

Averaging the numerical fluxes over two neighbouring quadrilaterals using Raviart-Thomas elements of degree q, reasonable results (reliability up to a factor of the order $O(p)$ and efficiency, i.e., a p-gap) in two model problems are obtained if q is set equal to the local approximation order. This choice is practically the most relevant, as this corresponds to what is known to work in h-FEM and can therefore be expected to be used in hp-FEM. When averaging over four elements, we observe the p-gap when setting $q = p$ or $q = p + 1$. In this case, however, the gradient L^2 projection error in the efficiency estimate even dominates the Galerkin error, which might be of concern theoretically. Finally, averaging over four elements and setting $q = 2p$ leads to an efficient estimator that is reliable up to $O(p^{0.35})$.

Acknowledgements The first author gratefully acknowledges support by the ETH Foundation.

References

1. Ainsworth, M., Craig, A.: A posteriori error estimators in the finite element method. Numer. Math. **60**(4), 429–463 (1992). DOI 10.1007/BF01385730. URL http://dx.doi.org/10.1007/BF01385730

2. Ainsworth, M., Oden, J.T.: A posteriori error estimation in finite element analysis. Pure and Applied Mathematics (New York). Wiley-Interscience [John Wiley & Sons], New York (2000). DOI 10.1002/9781118032824. URL http://dx.doi.org/10.1002/9781118032824
3. Ainsworth, M., Zhu, J.Z., Craig, A.W., Zienkiewicz, O.C.: Analysis of the Zienkiewicz-Zhu a posteriori error estimator in the finite element method. Internat. J. Numer. Methods Engrg. **28**(9), 2161–2174 (1989). DOI 10.1002/nme.1620280912
4. Babuška, I., Guo, B.: Optimal estimates for lower and upper bounds of approximation errors in the p-version of the finite element method in two dimensions. Numer. Math. **85**(2), 219–255 (2000). DOI 10.1007/PL00005387. URL http://dx.doi.org/10.1007/PL00005387
5. Babuška, I., Suri, M.: The p and h-p versions of the finite element method, basic principles and properties. SIAM Rev. **36**(4), 578–632 (1994). DOI 10.1137/1036141. URL http://dx.doi.org/10.1137/1036141
6. Babuška, I.M., Rodríguez, R.: The problem of the selection of an a posteriori error indicator based on smoothening techniques. Internat. J. Numer. Methods Engrg. **36**(4), 539–567 (1993). DOI 10.1002/nme.1620360402
7. Bartels, S., Carstensen, C.: Each averaging technique yields reliable a posteriori error control in FEM on unstructured grids. II. Higher order FEM. Math. Comp. **71**(239), 971–994 (electronic) (2002). DOI 10.1090/S0025-5718-02-01412-6
8. Bernardi, C.: Indicateurs d'erreur en h-N version des éléments spectraux. RAIRO Modél. Math. Anal. Numér. **30**(1), 1–38 (1996)
9. Braess, D., Pillwein, V., Schöberl, J.: Equilibrated residual error estimates are p-robust. Comput. Methods Appl. Mech. Engrg. **198**(13–14), 1189–1197 (2009). DOI 10.1016/j.cma.2008.12.010
10. Canuto, C., Quarteroni, A.: Approximation results for orthogonal polynomials in Sobolev spaces. Math. Comp. **38**(157), 67–86 (1982). DOI 10.2307/2007465. URL http://dx.doi.org/10.2307/2007465
11. Carstensen, C., Verfürth, R.: Edge residuals dominate a posteriori error estimates for low order finite element methods. SIAM J. Numer. Anal. **36**(5), 1571–1587 (1999). DOI 10.1137/S003614299732334X
12. Melenk, J.M., Wohlmuth, B.I.: On residual-based a posteriori error estimation in hp-FEM. Adv. Comput. Math. **15**(1–4), 311–331 (2002) (2001). DOI 10.1023/A:1014268310921
13. Melenk, J.M., Wurzer, T.: On the stability of the polynomial L^2-projection on triangles and tetrahedra. Tech. rep., Institute for Analysis and Scientific Computing, Vienna University of Technology (2012)
14. Rank, E., Zienkiewicz, O.: A simple error estimator in the finite element method. Commun. Appl. Numer. Methods **3**, 243–249 (1987). DOI 10.1002/cnm.1630030311
15. Rodríguez, R.: Some remarks on Zienkiewicz-Zhu estimator. Numer. Methods Partial Differential Equations **10**(5), 625–635 (1994). DOI 10.1002/num.1690100509
16. Zienkiewicz, O., Li, X.K., Nakazawa, S.: Iterative solution of mixed problems and the stress recovery procedures. Commun. Appl. Numer. Methods **1**, 3–9 (1985). DOI 10.1002/cnm.1630010103
17. Zienkiewicz, O.C., Zhu, J.Z.: A simple error estimator and adaptive procedure for practical engineering analysis. Internat. J. Numer. Methods Engrg. **24**(2), 337–357 (1987). DOI 10.1002/nme.1620240206

Coupling of an Exact Transparent Boundary Condition with a DG Method for the Solution of the Time-Harmonic Maxwell Equations

M. El Bouajaji, N. Gmati, S. Lanteri, and J. Salhi

1 Introduction

The numerical simulation of electromagnetic wave propagation in open domains involving scattering objects or/and inhomogeneous regions naturally raises the question of the appropriate treatment of the artificial truncation of the infinite propagation domain. The main issue is to find a suitable way to efficiently model the far-field propagation region, by limiting the extent of the volume discretization. Two main approaches can be considered. The first class leads to *approximate methods* in the sense that an inherent error remains present even without any discretization error. In this class of methods, one finds the absorbing boundary conditions [2], the method of perfectly matched layers initiated by Bérenger [4] and adapted and used later in many works, and the method of unbounded elements [1]. In the second class, the only error comes from the discretization, leading to *exact methods* which are generally based on a coupling of a volume discretization method (e.g. a finite element method) with a boundary element method (BEM) [5] or with an integral representation [3–10]. The approach that we consider here belongs to the latter class, however it differs from existing solutions in two aspects: first, a high order discontinuous Galerkin (DG) method is used for the volume discretization [7]; second, the finite element/integral representation coupling strategy is adopted

M. El Bouajaji (✉) · S. Lanteri
NACHOS project-team, INRIA Sophia Antipolis – Méditerranée research center, F-06902 Sophia Antipolis Cedex, France
e-mail: Mohamed.El_bouajaji@inria.fr; Stephane.Lanteri@inria.fr

N. Gmati · J. Salhi
ENIT – LAMSIN, BP 37, 1002 Tunis, Tunisia
e-mail: nabil.gmati@ipein.rnu.tn; jamil.salhi@yahoo.fr

M. Azaïez et al. (eds.), *Spectral and High Order Methods for Partial Differential Equations - ICOSAHOM 2012*, Lecture Notes in Computational Science and Engineering 95, DOI 10.1007/978-3-319-01601-6_19,

([3–11]) and adapted to the DG discretization framework. The resulting numerical methodology is illustrated here by considering 2D test problems for time-harmonic electromagnetic wave propagation in homogeneous and inhomogeneous media.

2 Presentation of the Problem

Let Ω_i be a bounded domain in $\mathbb{R}^d$ $(d = 2, 3)$ with a regular boundary Γ representing a scattering objet. Let Ω_e be the unbounded complementary domain of Ω_i. We consider here the scattering problem of a time-harmonic electromagnetic wave by the obstacle Ω_i. We are interested in solving the time-harmonic Maxwell'equations

$$i\omega\varepsilon\mathbf{E} - \text{curl}\,\mathbf{H} = 0, \qquad i\omega\mu\mathbf{H} + \text{curl}\,\mathbf{E} = 0, \tag{1}$$

where $\mathbf{E}$ and $\mathbf{H}$ denote the electric and magnetic field, ε is the electric permittivity and μ the magnetic permeability. The positive real parameter ω is the pulsation of the harmonic wave. In the case where the obstacle is impenetrable, we impose on $\Gamma := \Gamma_m$ a perfect electric conductor (PEC) condition $\mathbf{n} \times \mathbf{E} = 0$, where $\mathbf{n}$ denotes the unit outward normal, and we consider a radiation condition at infinity

$$\lim_{r\to\infty} r\left((\mathbf{E} - \mathbf{E}^{inc}) - Z_0(\mathbf{H} - \mathbf{H}^{inc}) \times \mathbf{n}\right) = 0, \tag{2}$$

where $(\mathbf{E}^{inc}, \mathbf{H}^{inc})$ is an incident electromagnetic field and Z_0 the impedance of the vacuum. This boundary value problem has a unique solution which belongs to the space (see [13], [12]) $X_{loc}(\Omega_e, \Gamma_m) := \{\mathbf{v} \in \mathbf{H}_{loc}(\Omega_e, \Gamma_m) : \mathbf{n} \times \mathbf{v} \in [L^2(\Gamma_m)]^3 \text{ and } (\mathbf{n} \times \mathbf{v}) \cdot \mathbf{n} = 0\}$. This problem can be solved by using an integral representation of $\mathbf{E}$ and $\mathbf{H}$. The solution, denoted $(\mathbf{E}_{RI}, \mathbf{H}_{RI})$, is given by the Stratton-Chu formulae [6]

$$\begin{aligned}
\mathbf{E}_{RI}(\mathbf{x}) &= \mathbf{E}^{inc} + \text{curl}_{\mathbf{x}} \int_{\Gamma} \mathbf{J}(\mathbf{y})G(\mathbf{x},\mathbf{y})d\mathbf{y} - \frac{1}{i\omega\mu}\text{curl}_{\mathbf{x}}\text{curl}_{\mathbf{x}} \int_{\Gamma} \mathbf{M}(\mathbf{y})G(\mathbf{x},\mathbf{y})d\mathbf{y}, \\
\mathbf{H}_{RI}(\mathbf{x}) &= \mathbf{H}^{inc} + \text{curl}_{\mathbf{x}} \int_{\Gamma} \mathbf{M}(\mathbf{y})G(\mathbf{x},\mathbf{y})d\mathbf{y} + \frac{1}{i\omega\varepsilon}\text{curl}_{\mathbf{x}}\text{curl}_{\mathbf{x}} \int_{\Gamma} \mathbf{J}(\mathbf{y})G(\mathbf{x},\mathbf{y})d\mathbf{y},
\end{aligned} \tag{3}$$

where $\mathbf{J}(y) = \mathbf{n}(\mathbf{y}) \times \mathbf{E}(\mathbf{y})$, $\mathbf{M}(y) = \mathbf{n}(\mathbf{y}) \times \mathbf{H}(\mathbf{y})$, $\mathbf{y} \in \Gamma_m$, $\mathbf{x} \in \Omega_e$ and G is the Green function (with $k = \omega\sqrt{\varepsilon\mu}$)

$$G(\mathbf{x},\mathbf{y}) = \frac{\exp(ik|\mathbf{x}-\mathbf{y}|)}{4\pi|\mathbf{x}-\mathbf{y}|} \text{ (3D case)}, \quad G(\mathbf{x},\mathbf{y}) = \frac{\mathrm{i}}{4}H_0^{(1)}(k|\mathbf{x}-\mathbf{y}|) \text{ (2D case)},$$

where $H_0^{(1)}$ is the Hankel function of the first kind and of order 0. We use (3) to build an exact transparent condition (TC) coupled to a DG method for the discretization of Ω_i. Then, it is necessary to truncate the exterior domain by an artificial boundary Γ_a, and (1) are solved in a bounded domain Ω. On Γ_a, we impose an impedance condition $\mathscr{B}_{\mathbf{n}}(\mathbf{E},\mathbf{H}) = \mathscr{B}_{\mathbf{n}}(\mathbf{g}_1,\mathbf{g}_2)$ with $\mathscr{B}_{\mathbf{n}}(\mathbf{E},\mathbf{H}) = \mathbf{n}\times\frac{\mathbf{E}}{Z} - \mathbf{n}\times(\mathbf{H}\times\mathbf{n})$. This condition is equivalent to imposing Dirichlet conditions on the incoming characteristics if we consider the hyperbolic nature of the underlying time-dependent system. Setting $(\mathbf{g}_1,\mathbf{g}_2) = (\mathbf{E}^{inc},\mathbf{H}^{inc})$, we obtain the well-known Silver-Müller absorbing condition (SMC). It is an approximation of (2) at finite distance that produces a truncation error. To reduce this error, the artificial boundary Γ_a must be placed sufficiently far from the scatterer. Then we get a larger computational domain Ω thus increasing the computing cost. In order to reduce the size of Ω, we introduce an exact TC at finite distance by taking $(\mathbf{g}_1,\mathbf{g}_2) = (\mathbf{E}_{RI},\mathbf{H}_{RI})$, and we solve the boundary value problem

$$\begin{cases} i\omega G_0\mathbf{W} + G_x\partial_x\mathbf{W} + G_y\partial_y\mathbf{W} + G_z\partial_z\mathbf{W} = 0 \text{ in } \Omega, \\ (M_{\Gamma_a} - G_{\mathbf{n}})(\mathbf{W} - \mathbf{W}^{\mathrm{inc}}) = 0 \text{ on } \Gamma_a \text{ (if we use a SMC condition)}, \\ (M_{\Gamma_a} - G_{\mathbf{n}})(\mathbf{W} - \mathbf{W}_{RI}) = 0 \text{ on } \Gamma_a \text{ (if we use a TC condition)}, \\ (M_{\Gamma_m} - G_{\mathbf{n}})\mathbf{W} = 0 \text{ on } \Gamma_m, \end{cases} \tag{4}$$

where $\mathbf{W}_{RI}(\mathbf{x}) = \mathbf{W}^{\mathrm{inc}}(\mathbf{x}) + \int_{\Gamma_m} M_{RI}(\mathbf{x},\ \mathbf{y})\mathbf{W}(\mathbf{y})dy$, $\mathbf{W} = (\mathbf{E},\mathbf{H})^T$, and

$$G_0 = \begin{pmatrix} \varepsilon\,\mathrm{I}_3 & 0_3 \\ 0_3 & \mu\,\mathrm{I}_3 \end{pmatrix},\quad G_l = \begin{pmatrix} 0_3 & N_{\mathbf{e}^l} \\ N_{\mathbf{e}^l}^T & 0_3 \end{pmatrix},\quad N_{\mathbf{v}} = \begin{pmatrix} 0 & v_z & -v_y \\ -v_z & 0 & v_x \\ v_y & -v_x & 0 \end{pmatrix},$$

with $l \in \{x,y,z\}$ where $(\mathbf{e}^{\mathbf{x}},\mathbf{e}^{\mathbf{y}},\mathbf{e}^{\mathbf{z}})$ is the canonical basis of $\mathbb{R}^3$; I_3 and 0_3 are the identity and null matrices, both of dimension 3×3. The real part of G_0 is symmetric positive definite and its imaginary part, which appears in the case of conductive materials, is symmetric negative. In the following we denote by $G_{\mathbf{n}}$ the sum $G_xn_x + G_yn_y + G_zn_z$ and by $G_{\mathbf{n}}^+$ and $G_{\mathbf{n}}^-$ its positive and negative parts. We recall that if $T\Lambda T^{-1}$ is the eigen-decomposition of $G_{\mathbf{n}}$, then $G_{\mathbf{n}}^{\pm} = T\Lambda^{\pm}T^{-1}$ where Λ^+ (respectively Λ^-) only gathers the positive (respectively negative) eigenvalues. We also define $|G_{\mathbf{n}}| = G_{\mathbf{n}}^+ - G_{\mathbf{n}}^-$. We have that M_{Γ_a} is given by $M_{\Gamma_a} = |G_{\mathbf{n}}|$.

3 Discretization by a Discontinuous Galerkin Method

Let Ω_h be a discretization of Ω into simplicial elements K and define $\mathbb{P}_p(K)$ as the space of vectors with polynomial components of order at most p over K. Let $V_h = \{\mathbf{U} \in [L^2(\Omega)]^3 \ /\ \forall K \in \Omega_h,\ \mathbf{U}_{|K} \in \mathbb{P}_p(K)\}$. The DG discretization of system (4) leads to the formulation of the following discrete problem for $\mathbf{W}_h$ in $V_h \times V_h$

$$
\begin{aligned}
&\int_{\Omega_h} (\mathrm{i}\omega G_0 \mathbf{W}_h)^T \overline{\mathbf{V}} dv + \sum_{K\in\Omega_h} \int_K \left(\sum_{l\in\{x,y,z\}} G_l \partial_l(\mathbf{W}_h) \right)^T \overline{\mathbf{V}} dv \\
&+ \sum_{F\in\Gamma^m\cup\Gamma^a} \int_F \left(\frac{1}{2}(M_{F,K} - I_{FK}G_{\mathbf{n}_F})\mathbf{W}_h \right)^T \overline{\mathbf{V}} ds \\
&-\alpha \sum_{F\in\Gamma^a} \int_F \left(\frac{1}{2}(M_{F,K} - I_{FK}G_{\mathbf{n}_F}) \int_{\Gamma_m} M_{RI}(\mathbf{x},\mathbf{y})\mathbf{W}_h(\mathbf{y})d\mathbf{y} \right)^T \overline{\mathbf{V}} ds \qquad (5) \\
&- \sum_{F\in\Gamma^0} \int_F (G_{\mathbf{n}_F}[\![\mathbf{W}_h]\!])^T \{\overline{\mathbf{V}}\} ds + \sum_{F\in\Gamma^0} \int_F (S_F[\![\mathbf{W}_h]\!])^T [\![\overline{\mathbf{V}}]\!] ds \\
&= \sum_{F\in\Gamma^a} \int_F \left(\frac{1}{2}(M_{F,K} - I_{FK}G_{\mathbf{n}_F})\mathbf{W}^{\mathrm{inc}} \right)^T \overline{\mathbf{V}} ds, \;\; \forall \mathbf{V} \in V_h \times V_h, \mathbf{x} \in F,
\end{aligned}
$$

where $\alpha = 0$ if we use a SMC condition, $\alpha = 1$ if we use a TC condition; Γ^0, Γ^a and Γ^m respectively denote the sets of interior faces, boundary faces on Γ_a and boundary faces on Γ_m. The unit normal on the oriented face F is $\mathbf{n}_F$ and I_{FK} stands for the incidence matrix between oriented faces whose entries are equal to 0 if $F \notin K$, 1 if $F \in K$ and their orientations match, and -1 if $F \in K$ and their orientations do not match. For $F = \partial K \cap \partial \tilde{K}$, we also define $[\![\mathbf{V}]\!] = I_{FK}\mathbf{V}_{|K} + I_{F\tilde{K}}\mathbf{V}_{|\tilde{K}}$ and $\{\mathbf{V}\} = \frac{1}{2}(\mathbf{V}_{|K} + \mathbf{V}_{|\tilde{K}})$. Finally, the matrix S_F, which is Hermitian positive definite, permits the penalization of the jump of a field, and the matrix $M_{F,K}$, insures the asymptotic consistency with the boundary conditions of the continuous problem. Problem (5) is often interpreted in terms of local problems in each element K of Ω_h coupled by the introduction of an element boundary term called numerical flux (see also [8]). We usually consider two options for the definitions for S_F and $M_{F,K}$

- **Centered numerical flux** (see [9] for the time-domain equivalent)

$$S_F = 0 \text{ and } M_{F,K} = |G_{\mathbf{n}_F}| \text{ if } F \in \Gamma^a. \qquad (6)$$

- **Upwind numerical flux** (see [8, 14])

$$S_F = \frac{1}{2}\begin{pmatrix} N_{\mathbf{n}_F}N_{\mathbf{n}_F}^T & 0_3 \\ 0_3 & N_{\mathbf{n}_F}^T N_{\mathbf{n}_F} \end{pmatrix} \text{ and } M_{F,K} = |G_{\mathbf{n}_F}| \text{ if } F \in \Gamma^a. \qquad (7)$$

The above formulation of the DG scheme actually applies to a homogeneous medium. In the heterogeneous case the DG scheme can be written formally as in (5) by using $Z^K = \sqrt{\frac{\mu^K}{\varepsilon^K}} = \frac{1}{Y^K}$, $Z^F = \frac{Z^K + Z^{\tilde{K}}}{2}$ and $Y^F = \frac{Y^K + Y^{\tilde{K}}}{2}$, and modifying S_F as

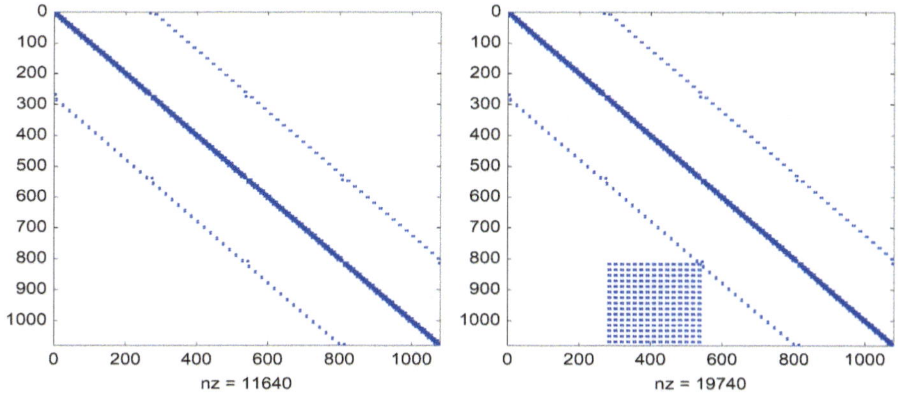

Fig. 1 Sparsity pattern of the matrix with TC (*right*) and SMC (*left*) conditions

$$S_F = \frac{1}{2}\begin{pmatrix} \frac{1}{Z^F} N_{\mathbf{n}_F} N_{\mathbf{n}_F}^T & 0_3 \\ 0_3 & \frac{1}{Y^F} N_{\mathbf{n}_F}^T N_{\mathbf{n}_F} \end{pmatrix}, \tag{8}$$

and by using for the average a weighted average $\{\cdot\}_F$ for each face F

$$\{\mathbf{V}\}_F = \frac{1}{2}\left(\begin{pmatrix} \frac{Z^{\tilde{K}}}{Z^F} & 0_3 \\ 0_3 & \frac{Y^{\tilde{K}}}{Y^F} \end{pmatrix}\mathbf{V}_{|K} + \begin{pmatrix} \frac{Z^K}{Z^F} & 0_3 \\ 0_3 & \frac{Y^K}{Y^F} \end{pmatrix}\mathbf{V}_{|\tilde{K}}\right). \tag{9}$$

Within each mesh element K the electromagnetic field $(\mathbf{E}, \mathbf{H})^T$ is approximated as $(\mathbf{E}_h)_{|K} = \sum_{i=1}^{n_{pK}} \mathbf{E}_i^K \varphi_i^K$ and $(\mathbf{H}_h)_{|K} = \sum_{i=1}^{n_{pK}} \mathbf{H}_i^K \varphi_i^K$, and $M_{RI}(\mathbf{x}, \mathbf{y})$ is evaluated as $M_{RI}(\mathbf{x}, \mathbf{y}) = \sum_{l=1}^{n_{pK}} \sum_{i=m}^{n_{pF}} M_{RI}(\mathbf{x}_l, \mathbf{y}_m)\varphi_l^K(\mathbf{x})\varphi_m^F(\mathbf{y})$, where $\mathbf{x} \in K$, $\mathbf{y} \in F$, $\mathbf{x}_l$ and $\mathbf{y}_m$ are respectively the degrees of freedom (d.o.f) on K and F, $\mathbf{E}_i^K$ and $\mathbf{H}_i^K$ are the vectors of local d.o.f corresponding to the basis expansion $\{\varphi_i^K\}_{i=1,\cdots,n_{pK}}$ of $\mathbb{P}_p(K)$. In the present study, we adopt the classical Lagrange nodal basis functions defined on a simplex. The resulting method is denoted as DGTH-$\mathbb{P}_p$. The discretization (5) leads to a large sparse complex linear system of equations $(\mathscr{A} + \alpha\mathscr{C})\,\mathbf{W}_h = b$. The matrix $\mathscr{A}$ is sparse and it comes from the volume discretization while $\mathscr{C}$ contains a dense block which is due to the discretization of the non-local transparent condition. Figure 1 shows typical sparsity pattern of these matrices. The matrix $(\mathscr{A} + \mathscr{C})$ is ill-conditioned. The system can be preconditioned by $\mathscr{A}^{-1}$ and the system $(I + \mathscr{A}^{-1}\mathscr{C})\mathbf{X} = \mathscr{A}^{-1}b$ is solved by a Krylov method.

4 Numerical Results

We present numerical results for a preliminary implementation in the 2D case and for that purpose we consider the transverse magnetic (TM) Maxwell equations

$$\mathrm{i}\omega\mu H_x + \frac{\partial E_z}{\partial y} = 0, \;\; \mathrm{i}\omega\mu H_y - \frac{\partial E_z}{\partial x} = 0, \;\; \mathrm{i}\omega\varepsilon E_z - \frac{\partial H_y}{\partial x} + \frac{\partial H_x}{\partial y} = 0. \tag{10}$$

We have $\mathbf{W}_{RI}(\mathbf{x}) = \mathbf{W}^{\mathrm{inc}}(\mathbf{x}) + \int_{\Gamma_m} M_{RI}(\mathbf{x}, \mathbf{y})\mathbf{W}(\mathbf{y})d\mathbf{y}$ with $\mathbf{W} = (H_x, H_y, E_z)$ and

$$M_{RI}=\begin{pmatrix} n_2(\mathbf{y})\dfrac{\partial}{\partial_{y_2}} & -n_1(\mathbf{y})\dfrac{\partial}{\partial_{y_2}} & \dfrac{1}{\mathrm{i}\omega\mu}[n_2(\mathbf{y})\dfrac{\partial^2}{\partial^2_{y_1}} - n_1(\mathbf{y})\dfrac{\partial^2}{\partial_{y_2}\partial_{y_1}} + n_2(\mathbf{y})k^2] \\ -n_2(\mathbf{y})\dfrac{\partial}{\partial_{y_1}} & n_1(\mathbf{y})\dfrac{\partial}{\partial_{y_1}} & \dfrac{-1}{\mathrm{i}\omega\mu}[n_1(\mathbf{y})\dfrac{\partial^2}{\partial^2_{y_2}} - n_2(\mathbf{y})\dfrac{\partial^2}{\partial_{y_1}\partial_{y_2}} + n_1(\mathbf{y})k^2] \\ \dfrac{1}{\mathrm{i}\omega\mu}n_2(\mathbf{y})k^2 & -\dfrac{1}{\mathrm{i}\omega\mu}n_1(\mathbf{y})k^2 & n_2(\mathbf{y})\dfrac{\partial}{\partial_{y_2}} + n_1(\mathbf{y})\dfrac{\partial}{\partial_{y_1}} \end{pmatrix} G.$$

where $\mathbf{x} = (x_1, x_2) \in \Gamma_a$, $\mathbf{y} = (y_1, y_2) \in \Gamma_m$, $\mathbf{n}(\mathbf{y}) = (n_1(\mathbf{y}), n_2(\mathbf{y}))$ denotes the unit outward normal to Γ_m at $\mathbf{y}$ and G is the Green function.

4.1 Scattering of a Plane Wave by a PEC Cylinder

We consider the test problem of the scattering of a plane wave by a perfectly conducting (PEC) cylinder of radius $r_0 = 0.5\ m$, see Fig. 5 (left figure). The boundary Γ_a is a circle of radius r_2; we set $\mathbf{W}^{inc} = (H_x^{inc}, H_y^{inc}, E_z^{inc}) = (\frac{1}{Z_0}, \frac{-1}{Z_0}, 1)e^{-ikx}$, with $k = 10\pi/3$. We assess here the influence of the position of Γ_a while moving the latter closer to the obstacle Γ_m by varying r_2. We summarize in Table 1 the L^2 errors as a function of r_2 for the DGTH-$\mathbb{P}_1$ and DGTH-$\mathbb{P}_2$ methods with SMC and TC conditions. We observe that the error deteriorates with the SMC while the TC allows to move Γ_a closer to Γ_m without altering the accuracy. A comparison of the contour lines of the real part of E_z in the case where $r_2 = 0.8$ m and using the DGTH-$\mathbb{P}_2$ method is shown on Fig. 2. We clearly see on this figure the influence of the exact TC.

4.2 Scattering of a Plane Wave by a PEC Square

The second test problem is similar to the previous one but this time the scatterer is non-smooth (PEC square). The boundary Γ_a is a circle of radius r. We assess

Table 1 Scattering of a plane wave by a PEC cylinder: performances results of the DGTH-$\mathbb{P}_1$ and DGTH-$\mathbb{P}_2$ methods as a function of r_2

r_2	2.0	1.4	1.2	1.0	0.8
DGTH-$\mathbb{P}_1$ method, frequency F = 500 MHz					
With TC	1.3×10^{-1}	1.0×10^{-1}	9.9×10^{-2}	8.6×10^{-2}	6.9×10^{-2}
With SMC	1.5×10^{-1}	1.3×10^{-1}	1.4×10^{-1}	1.2×10^{-1}	1.4×10^{-1}
DGTH-$\mathbb{P}_2$ method, frequency F = 500 MHz					
With TC	8.7×10^{-3}	7.3×10^{-3}	6.6×10^{-3}	5.8×10^{-3}	5.0×10^{-3}
With SMC	5.6×10^{-3}	7.4×10^{-2}	8.8×10^{-2}	1.0×10^{-1}	1.2×10^{-1}

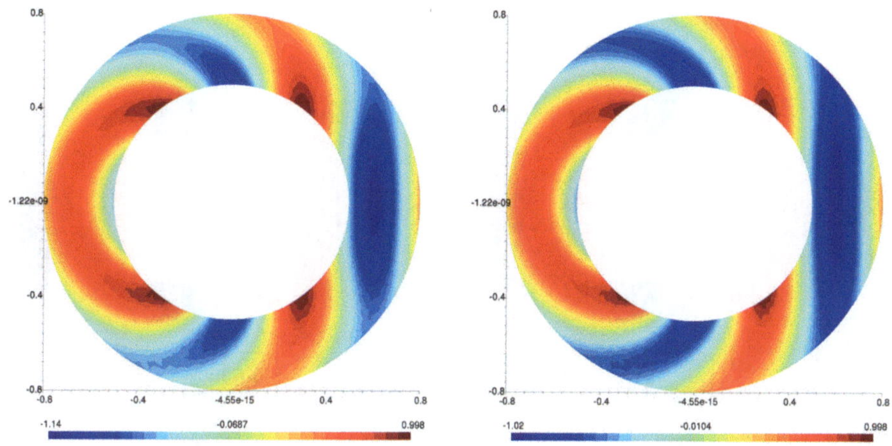

Fig. 2 Scattering of a plane wave by a PEC square cylinder: contour lines of the real part of E_z with DGTH-$\mathbb{P}_2$ method for SMC condition (*left*) and TC condition (*right*)

here the influence of the position of Γ_a on the convergence of the BiCGStab Krylov method for solving the coupled system. We summarize in Table 2 some performance results (the number of iterations needed for convergence, i.e to reach a relative residual of 10^{-8}, and the CPU time) as a function of r for the DGTH-$\mathbb{P}_1$ and DGTH-$\mathbb{P}_2$ methods. We observe that the number of iteration increases when r decreases, but the CPU time decreases too. This result shows that for a given error level, the coupling with the TC is a more efficient strategy than when a SMC is adopted. A comparison of the contour lines of the real part of E_z in the case where $r = 0.8$ m and using the DGTH-$\mathbb{P}_2$ method is shown on Fig. 2. We again clearly see on this figure the influence of the TC (Fig. 3).

4.3 Scattering of a Plane Wave by a Penetrable Cylinder

The test problem that we consider now is the scattering of a plane wave by a dielectric cylinder of radius $r_0 = 0.5$ m. The boundary Γ_a is a circle of radius r_2. The

Table 2 Scattering of a plane wave by a PEC square: performance results of the DGTH-$\mathbb{P}_1$ and DGTH-$\mathbb{P}_2$ methods as a function of r_2

r	2.0	1.4	1.2	1.0	0.8
DGTH-$\mathbb{P}_1$ method, frequency F = 500 MHz					
Iteration	6	7	10	14	20
CPU time (s)	35.2	12.6	10.3	9.8	11.9
DGTH-$\mathbb{P}_2$ method, frequency F = 500 MHz					
Iteration	5	6	7	9	15
CPU time (s)	22.1	10.3	7.7	8.8	10.2

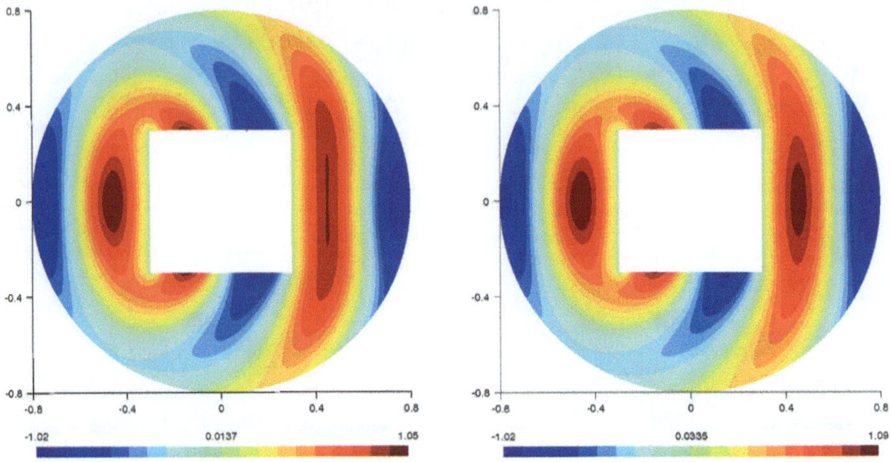

Fig. 3 Scattering of a plane wave by a dielectric cylinder: contour lines of the real part of E_z with the DGTH-$\mathbb{P}_2$ method for SMC condition (*left*) and TC condition (*right*)

Table 3 Scattering of a plane wave by a dielectric cylinder: performances results of the DGTH-$\mathbb{P}_1$ and DGTH-$\mathbb{P}_2$ methods as a function of r_2

r_2	2.0	1.4	1.2	1.0	0.8
DGTH-$\mathbb{P}_1$ method, frequency F = 500 MHz					
With TC	5.2×10^{-1}	1.6×10^{-1}	1.1×10^{-1}	9.1×10^{-2}	7.2×10^{-2}
With SMC	5.3×10^{-1}	2.1×10^{-1}	1.4×10^{-1}	2.2×10^{-1}	2.5×10^{-1}
DGTH-$\mathbb{P}_2$ method, frequency F = 500 MHz					
With TC	4.2×10^{-2}	1.6×10^{-2}	1.3×10^{-2}	1.2×10^{-2}	1.0×10^{-2}
With SMC	9.1×10^{-2}	1.6×10^{-1}	1.5×10^{-1}	2.0×10^{-1}	2.8×10^{-1}

relative permittivity of the inner cylinder is $\varepsilon_2 = 2.25$ while the vacuum ($\varepsilon_1 = 1$) is assumed for the rest of the domain. We summarize in Table 3 the L^2 errors as a function of r_2 for the DGTH-$\mathbb{P}_1$ and DGTH-$\mathbb{P}_2$ methods with SMC and TC conditions. A comparison of the contour lines of the real part of E_z in the case where $r_2 = 0.8$ m and with DGTH-$\mathbb{P}_2$ method is shown on Fig. 4. As for the previous test problem, the benefit of using the TC condition is clearly demonstrated.

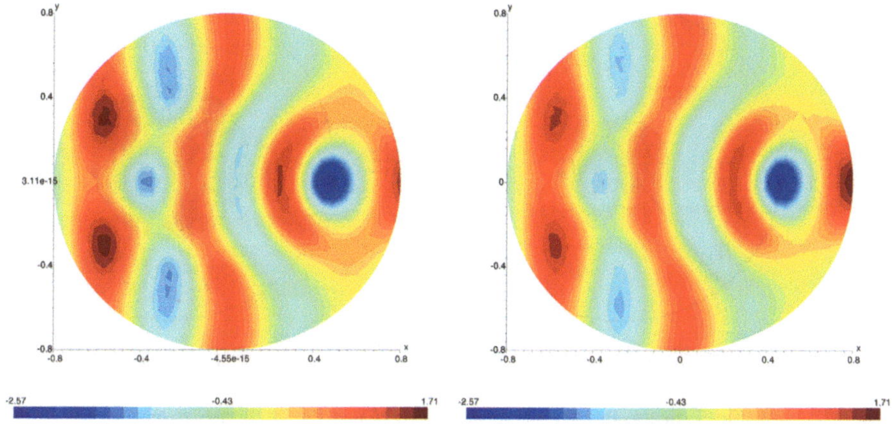

Fig. 4 Scattering of a plane wave by a dielectric cylinder: contour lines of the real part of E_z with the DGTH-$\mathbb{P}_2$ method for SMC condition (*left*) and TC condition (*right*)

Table 4 Scattering of a plane wave by a coated PEC cylinder: performances results of the DGTH-$\mathbb{P}_2$ method using the centered and upwind fluxes

Mesh (# elements)	Method	Error $L^2(\mathbf{E}, \mathbf{H})$
Centered flux		
870	DGTH-$\mathbb{P}_2$ + TC	9.0×10^{-2}
-	DGTH-$\mathbb{P}_2$ + SMC	8.4×10^{-1}
3540	DGTH-$\mathbb{P}_2$ + TC	1.1×10^{-2}
-	DGTH-$\mathbb{P}_2$ + SMC	3.2×10^{-1}
14280	DGTH-$\mathbb{P}_2$ + TC	1.6×10^{-3}
-	DGTH-$\mathbb{P}_2$ + SMC	3.1×10^{-1}
Upwind flux		
870	DGTH-$\mathbb{P}_2$ + TC	8.2×10^{-2}
-	DGTH-$\mathbb{P}_2$ + SMC	4.1×10^{-1}
3540	DGTH-$\mathbb{P}_2$ + TC	1.1×10^{-2}
-	DGTH-$\mathbb{P}_2$ + SMC	3.2×10^{-1}
14280	DGTH-$\mathbb{P}_2$ + TC	1.5×10^{-3}
-	DGTH-$\mathbb{P}_2$ + SMC	3.1×10^{-1}

4.4 *Scattering of a Plane Wave by a Coated PEC Cylinder*

We finally consider the problem of the scattering of a plane wave by a conducting cylinder (of radius $r_0 = 0.5$ m) coated by a dielectric layer (of radius $r_1 = 0.8$ m) characterized by $\varepsilon = 2.25$, and $k = 2\pi$ (see Fig. 5, right figure). We summarize in Table 4 the L^2 errors as a function of mesh refinement for the DGTH-$\mathbb{P}_2$ method using the centered and upwind fluxes. We observe a deterioration of the convergence when using the SMC condition. Convergence order are given in Table 5.

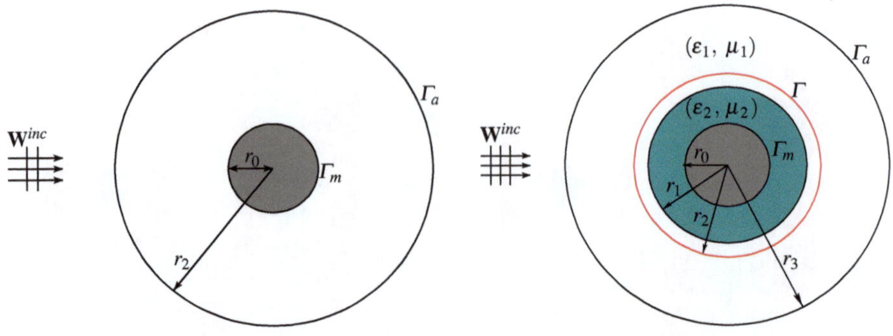

Fig. 5 Scattering of a plane wave by a PEC cylinder (*left*) or a coated PEC cylinder (*right*)

Table 5 Scattering of a plane wave by a coated PEC cylinder: numerical convergence of the DGTH-$\mathbb{P}_1$, DGTH-$\mathbb{P}_2$ and DGTH-$\mathbb{P}_3$ methods using the centered and upwind fluxes and the TC condition

Numerical flux type	DGTH-$\mathbb{P}_1$	DGTH-$\mathbb{P}_2$	DGTH-$\mathbb{P}_3$
Centered flux	1.2	2.8	3.7
Upwind flux	2	2.9	4.1

5 Conclusion

We have studied the coupling of a DG method for the discretization of the time-harmonic Maxwell equations with an exact transparent condition for the numerical modeling of electromagnetic wave propagation problems in open domains. Preliminary numerical results in the 2D case have shown that the accuracy of the proposed solution strategy is better than that obtained for transparent boundary conditions of sommerfeld type. The additional cost due to the dense block introduced by the non-local exact TC is largely compensated by the ability to choose the truncation boundary fairly close to the boundary of the obstacle.

References

1. R.J. Astley. Infinite elements for wave problems: a review of current formulations and an assessment of accuracy. *Int. J. Numer. Meth. Engng.*, **49**, 951–976, (2000).
2. X. Antoine, H. Barucq and A. Bendali. Bayliss-Turkel like radiation conditions on surfaces of arbitrary shape. *J. Math. Anal. Appl.*, **229**, 184–211, (1999).
3. F. Ben Belgacem, M. Fournié, N. Gmati et F. Jelassi. *On the Schwarz algorithms for elliptic exterior boundary value problems. ESAIM-M2AN, Vol 39, N.4, 2005.*
4. J.-P. Berenger. A perfectly matched layer for the absorption of electromagnetic waves. *J. Comput. Phys.*, **114**, 185–200, (1994).
5. Y. Boubendir, A. Bendali and M. Fares. Coupling of a non-overlapping domain decomposition method for a nodal finite element method with a boundary element method. *Int. J. Numer. Meth. Engng.*, **73**, 1624–1650, (2008).

6. D. Colton and R. Kress. Integral Equations in Scattering Theory. *Pure and Applied Mathematics, John Wiley and Sons*, New York, (1983).
7. M. El Bouajaji and S. Lanteri. High order discontinuous Galerkin method for the solution of 2D time-harmonic Maxwell's equations. *Appl. Math. Comput.*, to appear, (2012).
8. A. Ern and J.-L. Guermond. Discontinuous Galerkin methods for Friedrichs systems I. General theory. *SIAM J. Numer. Anal.*, **44**, 753–778, 2, (2006)
9. L. Fezoui and S. Lanteri and S. Lohrengel and S. Piperno. Convergence and stability of a discontinuous Galerkin time-domain method for the 3D heterogeneous Maxwell equations on unstructured meshes. *ESAIM: Math. Model. Num. Anal.*, **39**, 6, 1149–1176, (2005)
10. C. Hazard and M. Lenoir. On the solution of time-harmonic scattering problems for Maxwell's equations. *SIAM J. Numer. Anal.*, **27**, 1597–1630, (1996)
11. J. Liu and J.-M. Jin. A novel hybridization of higher order finite element and boundary integral methods for electromagnetic scattering and radiation problems. *IEEE Trans. Antennas and Propagation*, **49**, 1794–1806, (2001).
12. P. Monk. Finite Element Methods for Maxwell's Equations, Numerical Mathematics and Scientific Computation. *Oxford Science Publication*, UK, (2003).
13. J.-C. Nédélec. Acoustic and Electromagnetic Equations: Integral Representations of Harmonic Problems. *Appl. Appl. Math. Sci 144*, Springer, Berlin, (2001).
14. S. Piperno. L^2-stability of the upwind first order finite volume scheme for the Maxwell equations in two and three dimensions on arbitrary unstructured meshes. *M2AN: Math. Model. Numer. Anal.*, **34**, 139–158, 1, (2000)

A New Proof for Existence of $\mathcal{H}$-Matrix Approximants to the Inverse of FEM Matrices: The Dirichlet Problem for the Laplacian

Markus Faustmann, Jens M. Melenk, and Dirk Praetorius

Abstract We study the question of approximability of the inverse of the FEM stiffness matrix for the Laplace problem with Dirichlet boundary conditions by blockwise low rank matrices such as those given by the $\mathcal{H}$-matrix format introduced in Hackbusch (Introd $\mathcal{H}$-Matrices Comput 62(2):89–108, 1999). We show that exponential convergence in the local block rank r can be achieved. Unlike prior works Bebendorf and Hackbusch (Numer Math 95(1):1–28, 2003) and Börm (Numer Math 115(2):165–193, 2010) our analysis avoids any a priori coupling $r = \mathcal{O}(|\log^{\alpha} h|)$ of r and the mesh width h. Moreover, the techniques developed can be used to analyze other boundary conditions as well.

1 Introduction

The format of $\mathcal{H}$-matrices was introduced in [8] as blockwise low-rank matrices that permit storage, application, and even a full (approximate) arithmetic with log-linear complexity. This data-sparse format is well suited to represent exactly sparse matrices arising from discretizations of differential operators and to represent at high accuracy matrices stemming from discretizations of many integral operators, for example, those appearing in boundary integral equation methods.

The inverse of the finite element (FEM) stiffness matrix corresponding to the Dirichlet problem for elliptic operators with bounded coefficients can be approximated in the format of $\mathcal{H}$-matrices with an error that decays exponentially in the block rank employed. This was first observed numerically in [7]. Using properties of the continuous Green's function, [2] proves this exponential decay in

M. Faustmann (✉) · J.M. Melenk · D. Praetorius
Institut für Analysis und Scientific Computing, Technische Universität Wien, A–1040 Wien, Austria
e-mail: markus.faustmann@tuwien.ac.at

M. Azaïez et al. (eds.), *Spectral and High Order Methods for Partial Differential Equations - ICOSAHOM 2012*, Lecture Notes in Computational Science and Engineering 95, DOI 10.1007/978-3-319-01601-6_20,

the block rank up to the discretization error. The work [3] improves on the result [2] in several ways, in particular, by proving a corresponding approximation result in the framework of $\mathcal{H}^2$-matrices. Whereas the analysis of [2,3] is based on the solution operator on the continuous level (e.g., by studying the Green's function), the present approach works on the discrete level. The exponential approximability in the block rank shown here is therefore not limited by the discretization error. Moreover, in [2, 3] the block rank r and the mesh width h are coupled by $r \sim |\log^\alpha h|$, which is not needed in our case, since we prove an error estimate that is explicit in both r and h. We mention that the approach taken here generalizes quite naturally to other boundary conditions such as Neumann boundary conditions, which were not studied in [2, 3] and have not been treated yet. The reader is referred to the forthcoming work [5], where the case of higher order Galerkin discretizations is discussed as well.

2 Main Results

Let $\Omega \subset \mathbb{R}^d$, $d \in \{2, 3\}$, be a bounded polygonal (for $d = 2$) or polyhedral (for $d = 3$) Lipschitz domain with boundary $\Gamma := \partial\Omega$. We consider the bilinear form $a : H_0^1(\Omega) \times H_0^1(\Omega) \to \mathbb{R}$ associated with the Poisson problem, which is given by

$$a(u, v) := \langle \nabla u, \nabla v \rangle\,, \tag{1}$$

where $\langle \cdot, \cdot \rangle$ denotes the $L^2(\Omega)$-scalar product. For its discretization, we assume that Ω is triangulated by a quasiuniform mesh $\mathcal{T}_h = \{T_1, \ldots, T_{N_\mathcal{T}}\}$ of mesh width $h := \max_{T_j \in \mathcal{T}_h} \operatorname{diam}(T_j)$. The elements $T_j \in \mathcal{T}_h$ are triangles ($d = 2$) or tetrahedra ($d = 3$), and we assume that $\mathcal{T}_h$ is regular in the sense of Ciarlet. The nodes are denoted by $x_i \in \mathcal{N}_h$, for $i = 1, \ldots, N_\mathcal{N}$. Moreover, the mesh $\mathcal{T}_h$ is assumed to be γ-shape regular in the sense of $\operatorname{diam}(T_j) \le \gamma\, |T_j|^{1/d}$ for all $T_j \in \mathcal{T}_h$. In the following, the notation $\lesssim$ abbreviates $\le$ up to a constant $C > 0$ which depends only on Ω, the dimension d, and γ-shape regularity of $\mathcal{T}_h$. Moreover, we use $\simeq$ to abbreviate that both estimates $\lesssim$ and $\gtrsim$ hold.

For the sake of definiteness, we consider the lowest order Galerkin discretization of the bilinear form $a(\cdot, \cdot)$ by piecewise affine functions in $S_0^{1,1}(\mathcal{T}_h) := S^{1,1}(\mathcal{T}_h) \cap H_0^1(\Omega)$ with $S^{1,1}(\mathcal{T}_h) = \{u \in C(\Omega) \,:\, u|_{T_j} \in \mathcal{P}_1, \forall\, T_j \in \mathcal{T}_h\}$, taking as the basis of $S_0^{1,1}(\mathcal{T}_h)$ the classical hat-functions associated with the interior nodes of the triangulation. This basis is denoted by $\mathcal{B}_h := \{\psi_j \,:\, j = 1, \ldots, N\}$.

The Galerkin discretization of (1) results in a symmetric, positive definite matrix $\mathbf{A} \in \mathbb{R}^{N \times N}$ with

$$\mathbf{A}_{jk} = \left\langle \nabla\psi_j, \nabla\psi_k \right\rangle, \quad \psi_j, \psi_k \in \mathcal{B}_h.$$

Our goal is to derive an $\mathcal{H}$-matrix approximation $\mathbf{B}_{\mathcal{H}}$ of the inverse matrix $\mathbf{B} = \mathbf{A}^{-1}$. An $\mathcal{H}$-matrix $\mathbf{B}_{\mathcal{H}}$ is a blockwise low rank matrix based on the concept of "admissibility", which we now introduce:

Definition 1 (bounding boxes and η-admissibility). A *cluster* τ is a subset of the index set $\mathcal{I} = \{1, \ldots, N\}$. For a cluster $\tau \subset \mathcal{I}$, we say that $B_{R_\tau} \subset \mathbb{R}^d$ is a *bounding box* if:

(i) B_{R_τ} is a d-hypercube with side length R_τ,
(ii) $\psi_i \subset B_{R_\tau}$ for all $i \in \tau$.

Let $\eta > 0$. A pair of clusters (τ, σ) with $\tau, \sigma \subset \mathcal{I}$ is *η-admissible*, if there exist boxes B_{R_τ}, B_{R_σ} satisfying (i)–(ii) such that

$$\max\{\operatorname{diam} B_{R_\tau}, \operatorname{diam} B_{R_\sigma}\} \leq \eta \operatorname{dist}(B_{R_\tau}, B_{R_\sigma}). \tag{2}$$

Remark 1. By symmetry of $a(\cdot, \cdot)$, we can replace in (2) the "max" with "min" and the assertions of the present note still hold true. We refer to [5] for details.

Definition 2 (blockwise rank-r-matrices). Let P be a partition of $\mathcal{I} \times \mathcal{I}$ and $\eta > 0$. A matrix $\mathbf{B}_{\mathcal{H}} \in \mathbb{R}^{N \times N}$ is said to be a *blockwise rank-r matrix*, if for every η-admissible cluster pair $(\tau, \sigma) \in P$, the block $\mathbf{B}_{\mathcal{H}}|_{\tau \times \sigma}$ is a rank-r-matrix, i.e., it has the form $\mathbf{B}_{\mathcal{H}}|_{\tau \times \sigma} = \mathbf{X}_{\tau\sigma} \mathbf{Y}_{\tau\sigma}^T$ with $\mathbf{X}_{\tau\sigma} \in \mathbb{R}^{|\tau| \times r}$ and $\mathbf{Y}_{\tau\sigma} \in \mathbb{R}^{|\sigma| \times r}$. Here and below, $|\sigma|$ denotes the cardinality of a finite set σ.

The following theorems are the main results of this paper. Theorem 1 shows that admissible blocks can be approximated by rank-r-matrices:

Theorem 1. *Fix $\eta > 0$, $q \in (0, 1)$. Let the cluster pair (τ, σ) be η-admissible. Then, for $k \in \mathbb{N}$ there are matrices $\mathbf{X}_{\tau\sigma} \in \mathbb{R}^{|\tau| \times r}$, $\mathbf{Y}_{\tau\sigma} \in \mathbb{R}^{|\sigma| \times r}$ of rank $r \leq C_{\mathrm{dim}}(2 + \eta)^d q^{-d} k^{d+1}$ with*

$$\left\| \mathbf{A}^{-1}|_{\tau \times \sigma} - \mathbf{X}_{\tau\sigma} \mathbf{Y}_{\tau\sigma}^T \right\|_2 \leq C_{\mathrm{apx}}(1 + \eta) h^{-d} q^k. \tag{3}$$

Here, $C_{\mathrm{apx}}, C_{\mathrm{dim}} > 0$ depend only on Ω, d, and the γ-shape regularity of $\mathcal{T}_h$.

The approximations for the individual blocks can be combined to gauge the approximability of $\mathbf{A}^{-1}$ by blockwise rank-r matrices. Particularly satisfactory estimates are obtained if the blockwise rank-r-matrices have additional structure. To that end, we introduce the following definitions.

Definition 3 (cluster tree). A *cluster tree* with *leaf size* $n_{\mathrm{leaf}} \in \mathbb{N}$ is a binary tree $\mathbb{T}_{\mathcal{I}}$ with root $\mathcal{I}$ such that for each cluster $\tau \in \mathbb{T}_{\mathcal{I}}$ the following dichotomy holds: either τ is a leaf of the tree and $|\tau| \leq n_{\mathrm{leaf}}$, or there exist so called sons τ', $\tau'' \in \mathbb{T}_{\mathcal{I}}$, which are disjoint subsets of τ with $\tau = \tau' \cup \tau''$. The *level function* level : $\mathbb{T}_{\mathcal{I}} \to \mathbb{N}_0$ is inductively defined by $\operatorname{level}(\mathcal{I}) = 0$ and $\operatorname{level}(\tau') := \operatorname{level}(\tau) + 1$ for τ' a son of τ. The *depth* of a cluster tree is $\operatorname{depth}(\mathbb{T}_{\mathcal{I}}) := \max_{\tau \in \mathbb{T}_{\mathcal{I}}} \operatorname{level}(\tau)$.

Definition 4 (far field, near field, and sparsity constant). A partition P of $\mathcal{I} \times \mathcal{I}$ is said to be based on the cluster tree $\mathbb{T}_{\mathcal{I}}$, if $P \subset \mathbb{T}_{\mathcal{I}} \times \mathbb{T}_{\mathcal{I}}$. For such a partition P and fixed $\eta > 0$, we define the *far field* and the *near field* as

$$P_{\text{far}} := \{(\tau, \sigma) \in P \;:\; (\tau, \sigma) \text{ is } \eta\text{-admissible}\}, \qquad P_{\text{near}} := P \backslash P_{\text{far}}.$$

The *sparsity constant* C_{sp} of such a partition is defined by

$$C_{\text{sp}} := \max\left\{ \max_{\tau \in \mathbb{T}_{\mathcal{I}}} |\{\sigma \in \mathbb{T}_{\mathcal{I}} \;:\; \tau \times \sigma \in P_{\text{far}}\}| \,, \max_{\sigma \in \mathbb{T}_{\mathcal{I}}} |\{\tau \in \mathbb{T}_{\mathcal{I}} \;:\; \tau \times \sigma \in P_{\text{far}}\}| \right\}.$$

The following Theorem 2 shows that the matrix $\mathbf{A}^{-1}$ can be approximated by blockwise rank-r-matrices at an exponential rate in the block rank r:

Theorem 2. *Fix $\eta > 0$. Let a partition P of $\mathcal{I} \times \mathcal{I}$ be based on a cluster tree $\mathbb{T}_{\mathcal{I}}$. Then, there is a blockwise rank-r matrix $\mathbf{B}_{\mathcal{H}}$ such that*

$$\left\| \mathbf{A}^{-1} - \mathbf{B}_{\mathcal{H}} \right\|_2 \leq C_{\text{apx}} C_{\text{sp}} (1 + \eta) N \operatorname{depth}(\mathbb{T}_{\mathcal{I}}) e^{-br^{1/(d+1)}}. \tag{4}$$

The constant $C_{\text{apx}} > 0$ depends only on Ω, d, and the γ-shape regularity of $\mathcal{T}_h$, while $b > 0$ additionally depends on η.

Remark 2. Typical clustering strategies such as the "geometric clustering" described in [9] and applied to quasiuniform meshes with $\mathcal{O}(N)$ elements lead to fairly balanced cluster trees $\mathbb{T}_{\mathcal{I}}$ of depth $\mathcal{O}(\log N)$ and feature a sparsity constant C_{sp} that is bounded uniformly in N. We refer to [9] for the fact that the memory requirement to store $\mathbf{B}_{\mathcal{H}}$ is $\mathcal{O}\big((r + n_{\text{leaf}}) N \log N\big)$.

Remark 3. With the estimate $\frac{1}{\|\mathbf{A}^{-1}\|_2} \lesssim N^{-1}$, we get a bound for the relative error

$$\frac{\left\| \mathbf{A}^{-1} - \mathbf{B}_{\mathcal{H}} \right\|_2}{\|\mathbf{A}^{-1}\|_2} \lesssim C_{\text{apx}} C_{\text{sp}} (1 + \eta) \operatorname{depth}(\mathbb{T}_{\mathcal{I}}) e^{-br^{1/(d+1)}}. \tag{5}$$

3 Approximation of Galerkin Solution on Admissible Blocks

In terms of functions and function spaces, the question of approximating $\mathbf{A}^{-1}|_{\tau \times \sigma}$ by a low-rank factorization $\mathbf{X}_{\tau\sigma} \mathbf{Y}_{\tau\sigma}^T$ can be phrased as one of how well one can approximate, from low-dimensional spaces, the restriction to B_{R_τ} of the solution ϕ_h for data f that are supported by B_{R_σ}. In order to study this question, let $\operatorname{supp} f \subset B_{R_\sigma}$ and consider the question of finding $\phi_h \in S_0^{1,1}(\mathcal{T}_h)$ such that

$$a(\phi_h, \psi_h) = \langle \nabla \phi_h, \nabla \psi_h \rangle = \langle f, \psi_h \rangle \qquad \forall \psi_h \in S_0^{1,1}(\mathcal{T}_h). \tag{6}$$

By coercivity of $a(\cdot,\cdot)$, the solution ϕ_h is well-defined. In the following, we extend the Galerkin solution by zero outside of Ω and denote this extension by ϕ_h as well. Due to the boundary conditions, this extension belongs to $H^1(B_{R_\tau})$. For η-admissible cluster pairs (τ,σ), the restriction of the solution ϕ_h to B_{R_τ} can be approximated from a low-dimensional space. The heart of the matter is stated in the following:

Proposition 1. *Fix $\eta > 0$. Let the cluster pair (τ,σ) be η-admissible. Fix $q \in (0,1)$. Then, for each $k \in \mathbb{N}$ there exists a sequence V_k of spaces with $\dim V_k \leq C_{\rm dim}(2+\eta)^d q^{-d} k^{d+1}$ such that for arbitrary f with $\operatorname{supp} f \subset B_{R_\sigma} \cap \Omega$, the solution ϕ_h of* (6) *satisfies*

$$\min_{v \in V_k} \|\phi_h - v\|_{L^2(B_{R_\tau})} \leq C_{\rm box}(1+\eta) q^k \|f\|_{L^2(B_{R_\sigma}\cap\Omega)}. \tag{7}$$

The constant $C_{\rm dim} > 0$ depends only on Ω, d, and the γ-shape regularity of $\mathcal{T}_h$, while $C_{\rm box} > 0$ depends only on Ω.

The proof of Proposition 1 will be given at the end of this section. The basic steps are as follows: First, one observes that $\operatorname{supp} f \subset B_{R_\sigma} \cap \Omega$ as well as the admissibility condition $\operatorname{dist}(B_{R_\tau}, B_{R_\sigma}) \geq \eta^{-1}\max\{\operatorname{diam}(B_{R_\tau}), \operatorname{diam}(B_{R_\sigma})\} > 0$ imply the orthogonality condition

$$\langle \nabla\phi_h, \nabla\psi_h\rangle = \langle f, \psi_h\rangle_{L^2(B_{R_\sigma}\cap\Omega)} = 0 \quad \forall \psi_h \in S_0^{1,1}(\mathcal{T}_h), \operatorname{supp}\psi_h \subset B_{R_\tau} \tag{8}$$

i.e. ϕ_h is discrete harmonic on B_{R_τ}. Second, this observation will allow us to prove a Caccioppoli-type estimate (Lemma 1) in which stronger norms of ϕ_h are estimated by weaker norms of ϕ_h on slightly enlarged regions. Third, we proceed as in [2, 3] by iterating an approximation result (Lemma 2) derived from the Scott-Zhang interpolation of the Galerkin solution ϕ_h. This iteration argument accounts for the exponential convergence (Lemma 3).

3.1 The Space $\mathcal{H}_h(D)$ of Discrete Harmonic Functions

Let $D \subset \mathbb{R}^d$ be an open set. A function $u \in H^1(D)$ is called discrete harmonic on $D \cap \Omega$, if

$$\int_{D\cap\Omega} \nabla u \cdot \nabla\varphi_h \, dx = 0 \quad \forall \varphi_h \in S_0^{1,1}(\mathcal{T}_h), \quad \operatorname{supp}\varphi_h \subset D \cap \Omega. \tag{9}$$

For open sets D, we introduce a space of functions that are piecewise affine and discrete harmonic on $D \cap \Omega$:

$$\mathcal{H}_h(D) := \{u \in H^1(D) \colon \exists \tilde{u} \in S_0^{1,1}(\mathcal{T}_h) \text{ s.t. } u|_{D\cap\Omega} = \tilde{u}|_{D\cap\Omega},$$
$$\operatorname{supp} u \subset \overline{\Omega}, \quad u \text{ is discrete harmonic on } D\cap\Omega\}.$$

Clearly, the finite dimensional space $\mathcal{H}_h(D)$ is a closed subspace of $H^1(D)$, and we have $\phi_h \in \mathcal{H}_h(B_{R_\tau})$ for the solution ϕ_h of (6) with $\operatorname{supp} f \subset B_{R_\sigma}$ and bounding boxes B_{R_τ}, B_{R_σ} which satisfy the η-admissibility criterion (2).

A main tool in our proofs is the Scott-Zhang projection $J_h : H^1(\Omega) \to S^{1,1}(\mathcal{T}_h)$, introduced in [10], which preserves homogeneous Dirichlet boundary conditions, i.e., it maps $H_0^1(\Omega)$ to $S_0^{1,1}(\mathcal{T}_h)$. By $\omega_T := \bigcup\{T' \in \mathcal{T}_h \,:\, T \cap T' \neq \emptyset\}$, we denote the element patch of T, which contains T and all elements $T' \in \mathcal{T}_h$ that have a common node with T. Then, J_h has some local approximation property for $\mathcal{T}_h$-piecewise H^ℓ-functions $u \in H^\ell_{\text{pw}}(\Omega)$

$$\|u - J_h u\|^2_{H^m(T)} \leq C h^{2(\ell-m)} \sum_{T' \subset \omega_T} |u|^2_{H^\ell(T')}, \quad 0 \leq m \leq 1, \ m \leq \ell \leq 2; \tag{10}$$

here, $C > 0$ depends only on the γ-shape regularity of $\mathcal{T}_h$ and the dimension d.

For a box B_R with diameter R, we introduce the norm

$$|||u|||^2_{h,R} := \left(\frac{h}{R}\right)^2 \|\nabla u\|^2_{L^2(B_R)} + \frac{1}{R^2} \|u\|^2_{L^2(B_R)},$$

which is, for fixed h, equivalent to the H^1-norm. The following lemma states a Caccioppoli-type estimate for functions in $\mathcal{H}_h(B_{(1+\delta)R})$, where $B_{(1+\delta)R}$ is a box of side length $(1+\delta)R$ with the same barycenter as the box B_R obtained by a stretching of B_R by the factor $(1+\delta)$.

Lemma 1. *Let $\delta > 0$ and $\frac{h}{R} \leq \frac{\delta}{4}$. Let $u \in \mathcal{H}_h(B_{(1+\delta)R})$ for a box $B_{(1+\delta)R}$. Then, there exists a constant $C > 0$ which depends only on Ω, d, and the γ-shape regularity of $\mathcal{T}_h$, such that*

$$\|\nabla u\|_{L^2(B_R)} \leq C \frac{1+\delta}{\delta} |||u|||_{h,(1+\delta)R}. \tag{11}$$

Proof. Let η be a smooth cut-off function with $\operatorname{supp} \eta \subset B_{(1+\delta/2)R}$, $\eta \equiv 1$ on B_R, and $\|\nabla \eta\|_{L^\infty(B_{(1+\delta)R})} \lesssim \frac{1}{\delta R}$, $\left\|D^2 \eta\right\|_{L^\infty(B_{(1+\delta)R})} \lesssim \frac{1}{\delta^2 R^2}$. Recall that h is the maximal element width and $4h \leq \delta R$. Therefore, $T \subseteq B_{(1+\delta)R}$ for all $T \in \mathcal{T}_h$ with $T \cap \operatorname{supp} \eta \neq \emptyset$. With the abbreviate notation $B := B_{(1+\delta)R}$, we have

$$\|\nabla u\|^2_{L^2(B_R)} \leq \|\eta \nabla u\|^2_{L^2(B)} = \int_B \nabla u \cdot \nabla(\eta^2 u) - 2\eta u \nabla \eta \cdot \nabla u \, dx.$$

By locality of the Scott-Zhang projection $J_h : H^1(\Omega) \to S^{1,1}(\mathcal{T}_h)$, we observe $\operatorname{supp} J_h(\eta^2 u) \subset B$. The orthogonality relation (9) implies

$$\begin{aligned}\left|\int_B \nabla u \cdot \nabla(\eta^2 u) dx\right| &= \left|\int_B \nabla u \cdot \nabla(\eta^2 u - J_h(\eta^2 u)) dx\right| \\ &\leq \|\nabla u\|_{L^2(B)} \left\|\nabla(\eta^2 u - J_h(\eta^2 u))\right\|_{L^2(B)}.\end{aligned}$$

For the last term, we use the approximation property (10) and obtain

$$\left\| \nabla(\eta^2 u - J_h(\eta^2 u)) \right\|_{L^2(B)}^2 \lesssim h^2 \sum_{\substack{T \in \mathcal{T}_h \\ T \subseteq B}} \left\| D^2(\eta^2 u) \right\|_{L^2(T)}^2 \lesssim h^2 \left\| D^2(\eta^2 u) \right\|_{L^2(B)}^2$$

$$\lesssim h^2 \Big(\left\| D^2 \eta \right\|_{L^\infty(B)} \| \eta u \|_{L^2(B)} + \| \nabla \eta \|_{L^\infty(B)} \| \eta \nabla u \|_{L^2(B)}$$

$$+ \| D\eta \|_{L^\infty(B)}^2 \| u \|_{L^2(B)} \Big)^2$$

$$\lesssim \left(\frac{h}{\delta^2 R^2} \| u \|_{L^2(B)} + \frac{h}{\delta R} \| \eta \nabla u \|_{L^2(B)} \right)^2.$$

Finally, we combine these estimates and use the Young inequality to see

$$\| \eta \nabla u \|_{L^2(B)}^2 \lesssim \frac{h}{\delta R} \| \nabla u \|_{L^2(B)} \left(\frac{1}{\delta R} \| u \|_{L^2(B)} + \| \eta \nabla u \|_{L^2(B)} \right)$$

$$+ \frac{1}{\delta R} \| u \|_{L^2(B)} \| \eta \nabla u \|_{L^2(B)}$$

$$\le C \frac{h^2}{\delta^2 R^2} \| \nabla u \|_{L^2(B)}^2 + C \frac{1}{\delta^2 R^2} \| u \|_{L^2(B)}^2 + \frac{1}{2} \| \eta \nabla u \|_{L^2(B)}^2.$$

Moving the term $\frac{1}{2} \| \eta \nabla u \|_{L^2(B)}^2$ to the left-hand side, we conclude the proof. □

3.2 Low-Dimensional Approximation in $\mathcal{H}_h(D)$

Let $\Pi_{h,R} : (H^1(B_R), |||\cdot|||_{h,R}) \to (\mathcal{H}_h(B_R), |||\cdot|||_{h,R})$ be the orthogonal projection, which is well-defined since $\mathcal{H}_h(B_R) \subset H^1(B_R)$ is a closed subspace.

Lemma 2. *Let $\delta > 0$ and $u \in \mathcal{H}_h(B_{(1+2\delta)R})$. Assume $\frac{h}{R} \le \frac{\delta}{4}$. Let $\mathcal{K}_H$ be an (infinite) γ-shape regular triangulation of $\mathbb{R}^d$ of mesh width H and assume $\frac{H}{R} \le \frac{\delta}{4}$. Let $J_H : H^1(\mathbb{R}^d) \to S^{1,1}(\mathcal{K}_H)$ be the Scott-Zhang projection. Then, there exists a constant $C_{\text{app}} > 0$ which depends only on Ω, d, and γ, such that*

(i) $(u - \Pi_{h,R} J_H u)|_{B_R} \in \mathcal{H}_h(B_R)$ *and* $\Pi_{h,R}(u|_{B_R}) = u|_{B_R}$
(ii) $|||u - \Pi_{h,R} J_H u|||_{h,R} \le C_{\text{app}} \frac{1+2\delta}{\delta} \left(\frac{h}{R} + \frac{H}{R} \right) |||u|||_{h,(1+2\delta)R}$
(iii) $\dim W \le C_{\text{app}} \left(\frac{(1+2\delta)R}{H} \right)^d$, *where* $W := \Pi_{h,R} J_H \mathcal{H}_h(B_{(1+2\delta)R})$.

Proof. The statement (2) follows from the fact that $\dim J_H(\mathcal{H}_h(B_{(1+2\delta)R})) \simeq ((1 + 2\delta)R/H)^d$. For $u \in \mathcal{H}_h(B_{(1+2\delta)R})$, we have $u \in \mathcal{H}_h(B_R)$ as well and hence $\Pi_{h,R}(u|_{B_R}) = u|_{B_R}$, which gives (2). It remains to prove (2): The assumption

$\frac{H}{R} \leq \frac{\delta}{4}$ implies $\bigcup\{K \in \mathcal{K}_H \ : \ \omega_K \cap B_R \neq \emptyset\} \subseteq B_{(1+\delta)R}$. The locality and the approximation properties (10) of J_H yield

$$\frac{1}{H} \|u - J_H u\|_{L^2(B_R)} + \|\nabla(u - J_H u)\|_{L^2(B_R)} \lesssim \|\nabla u\|_{L^2(B_{(1+\delta)R})} .$$

We apply Lemma 1 with $\tilde{R} = (1 + \delta)R$ and $\tilde{\delta} = \frac{\delta}{1+\delta}$. Note that $(1 + \tilde{\delta})\tilde{R} = (1 + 2\delta)R$, and $\frac{h}{\tilde{R}} \leq \frac{\tilde{\delta}}{4}$ follows from $4h \leq \delta R = \tilde{\delta}\tilde{R}$. Hence, we obtain

$$\begin{aligned}
|||u - \Pi_{h,R} J_H u|||_{h,R}^2 &= |||\Pi_{h,R}(u - J_H u)|||_{h,R}^2 \leq |||u - J_H u|||_{h,R}^2 \\
&= \left(\frac{h}{R}\right)^2 \|\nabla(u - J_H u)\|_{L^2(B_R)}^2 + \frac{1}{R^2} \|u - J_H u\|_{L^2(B_R)}^2 \\
&\lesssim \frac{h^2}{R^2} \|\nabla u\|_{L^2(B_{(1+\delta)R})}^2 + \frac{H^2}{R^2} \|\nabla u\|_{L^2(B_{(1+\delta)R})}^2 \\
&\leq \left(C_{\text{app}} \frac{1 + 2\delta}{\delta} \left(\frac{h}{R} + \frac{H}{R}\right)\right)^2 |||u|||_{h,(1+2\delta)R}^2 .
\end{aligned}$$

□

By iterating this approximation result on suitable concentric boxes, we can construct a low-dimensional subspace of the space $\mathcal{H}_h$ and the best approximation from this space converges exponentially in the dimension, which is stated in the following lemma.

Lemma 3. *Let C_{app} be the constant of Lemma 2. Let $q \in (0, 1)$, κ, $R > 0$, $k \in \mathbb{N}$. Assume*

$$\frac{h}{R} \leq \frac{\kappa q}{8k \max\{1, C_{\text{app}}\}}. \tag{12}$$

Then, there exists a subspace V_k of $\mathcal{H}_h(B_R)$ with dimension

$$\dim V_k \leq C_{\text{dim}} \left(\frac{1 + \kappa^{-1}}{q}\right)^d k^{d+1},$$

such that for every $u \in \mathcal{H}_h(B_{(1+\kappa)R})$

$$\min_{v \in V_k} |||u - v|||_{h,R} \leq q^k \, |||u|||_{h,(1+\kappa)R} . \tag{13}$$

The constant $C_{\text{dim}} > 0$ depends only on Ω, d, and γ-shape regularity of $\mathcal{T}_h$.

Proof. We iterate the approximation result of Lemma 2 on boxes $B_{(1+\delta_j)R}$, with $\delta_j := \kappa\frac{k-j}{k}$ for $j = 0,\dots,k$. We note that $\kappa = \delta_0 > \delta_1 > \cdots > \delta_k = 0$. We choose $H = \frac{\kappa q R}{8k\max\{C_{\text{app}},1\}}$, where C_{app} is the constant in Lemma 2.

If $h \geq H$, then we select $V_k = \mathcal{H}_h(B_R)$. Due to the choice of H we have $\dim V_k \lesssim \left(\frac{R}{h}\right)^d \lesssim k\left(\frac{R}{H}\right)^d \simeq C_{\dim}\left(\frac{1+\kappa^{-1}}{q}\right)^d k^{d+1}$.

If $h < H$, we apply Lemma 2 with $\tilde{R} = (1+\delta_j)R$ and $\tilde{\delta}_j = \frac{1}{2k(1+\delta_j)} < \frac{1}{2}$. Note that $\delta_{j-1} = \delta_j + \frac{1}{k}$ gives $(1+\delta_{j-1})R = (1+2\tilde{\delta}_j)\tilde{R}$. The assumption $\frac{H}{\tilde{R}} \leq \frac{1}{4k(1+\delta_j)} = \frac{\tilde{\delta}_j}{4}$ is fulfilled due to our choice of H. For $j = 1$, Lemma 2 provides an approximation w_1 in a subspace W_1 of $\mathcal{H}_h(B_{(1+\delta_1)R})$ with $\dim W_1 \leq C\left(\frac{(1+\kappa)R}{H}\right)^d$ such that

$$\|u - w_1\|_{h,(1+\delta_1)R} \leq 2C\frac{H}{(1+\delta_1)R}\frac{1+2\tilde{\delta}_1}{\tilde{\delta}_1}\|u\|_{h,(1+\delta_0)R}$$
$$= 4C\frac{kH}{R}(1+2\tilde{\delta}_1)\|u\|_{h,(1+\kappa)R} \leq q\|u\|_{h,(1+\kappa)R}.$$

Since $u - w_1 \in \mathcal{H}_h(B_{(1+\delta_1)R})$, we can use Lemma 2 again and get an approximation w_2 of $u - w_1$ in a subspace W_2 of $\mathcal{H}_h(B_{(1+\delta_2)R})$ with $\dim W_2 \leq C\left(\frac{(1+\kappa)R}{H}\right)^d$. Arguing as for $j = 1$, we get

$$\|u - w_1 - w_2\|_{h,(1+\delta_2)R} \leq q\|u - w_1\|_{h,(1+\delta_1)R} \leq q^2\|u\|_{h,(1+\kappa)R}.$$

Continuing this process $k-2$ times leads to an approximation $v := \sum_{j=1}^k w_i$ in the space $V_k := \sum_{j=1}^k W_j$ of dimension $\dim V_k \leq C_{\dim}\left(\frac{1+\kappa^{-1}}{q}\right)^d k^{d+1}$. □

Now we are able to prove the main result of this section.

Proof (of Proposition 1). Choose $\kappa = \frac{1}{1+\eta}$, then the admissibility condition implies $\operatorname{dist}(B_{(1+\kappa)R_\tau}, B_{R_\sigma}) > 0$, and we have $\phi_h|_{B_{(1+\delta)R_\tau}} \in \mathcal{H}_h(B_{(1+\delta)R_\tau})$.

The Poincaré inequality implies

$$\|\phi_h\|_{H^1(\Omega)}^2 \lesssim \|\nabla\phi_h\|_{L^2(\Omega)}^2 = \langle f, \phi_h\rangle \lesssim \|f\|_{L^2(B_{R_\sigma}\cap\Omega)}\|\phi_h\|_{H^1(\Omega)}.$$

Furthermore, with $\frac{h}{R_\tau} < 1$, we get

$$\|\phi_h\|_{h,(1+\kappa)R_\tau} \lesssim \left(1 + \frac{1}{R_\tau}\right)\|\phi_h\|_{H^1(\Omega)} \lesssim \left(1 + \frac{1}{R_\tau}\right)\|f\|_{L^2(B_{R_\sigma}\cap\Omega)},$$

and we have a bound on the right-hand side of (13). If the condition (12) is not satisfied, we choose $V_k = S_0^{1,1}(\mathcal{T}_h)|_{B_{R_\tau}}$. If the condition (12) is satisfied, we get with the space V_k from Lemma 3 and the admissibility condition that

$$\min_{v \in V_k} \|\phi_h - v\|_{L^2(B_{R_\tau})} \leq R_\tau \min_{v \in V_k} \|\phi_h - v\|_{h,R_\tau} \lesssim (R_\tau + 1)q^k \|f\|_{L^2(B_{R_\sigma} \cap \Omega)}$$
$$\lesssim (\eta + 1)\mathrm{diam}(\Omega) q^k \|f\|_{L^2(B_{R_\sigma} \cap \Omega)} .$$

□

4 Proof of the Main Results

We use the approximation of ϕ_h from the low dimensional spaces, given in Proposition 1, to construct a blockwise low-rank approximation and consequently an $\mathcal{H}$-matrix approximation of the inverse FEM-matrix. The remaining steps of the proof of Theorem 1 follow the lines of [3]. Therefore, we only sketch the proof.

Proof (of Theorem 1). If $C_{\mathrm{dim}}(2+\eta)^d q^{-d} k^{d+1} \geq \min(|\tau|, |\sigma|)$, we use the exact matrix block $\mathbf{X}_{\tau\sigma} = \mathbf{A}^{-1}|_{\tau\times\sigma}$ and $\mathbf{Y}_{\tau\sigma} = I \in \mathbb{R}^{|\sigma|\times|\sigma|}$. If $C_{\mathrm{dim}}(2+\eta)^d q^{-d} k^{d+1} < \min(|\tau|, |\sigma|)$, let $\lambda_i : L^2(\Omega) \to \mathbb{R}$ be continuous linear functionals satisfying $\lambda_i(\psi_j) = \delta_{ij}$. We define the mappings

$$\Lambda_\tau : L^2(\Omega) \to \mathbb{R}^{|\tau|}, v \mapsto (\lambda_i(v))_{i\in\tau} \text{ and } \mathcal{J}_\tau : \mathbb{R}^{|\tau|} \to S_0^{1,1}(\mathcal{T}_h),\ \mathbf{x} \mapsto \sum_{j\in\tau} x_j \psi_j .$$

Let V_k be the finite dimensional space from Proposition 1. We define $\mathbf{X}_{\tau\sigma}$ as an orthogonal basis of the space $\mathcal{V} := \{\Lambda_\tau v \ :\ v \in V_k\}$ and $\mathbf{Y}_{\tau\sigma} := \mathbf{A}^{-1}|_{\tau\times\sigma}^T \mathbf{X}_{\tau\sigma}$. Then, the rank of $\mathbf{X}_{\tau\sigma}, \mathbf{Y}_{\tau\sigma}$ is bounded by $\dim V_k \leq C_{\mathrm{dim}}(2+\eta)^d q^{-d} k^{d+1}$. The error estimate follows from combining Proposition 1 with the stability estimate $h^{d/2}\|\mathbf{x}\|_2 \lesssim \|\mathcal{J}_\tau \mathbf{x}\|_{L^2(\Omega)} \lesssim h^{d/2}\|\mathbf{x}\|_2$, see [3, Theorem 2] for details. □

Now, the estimates on each block can be put together to prove our main result.

Proof (of Theorem 2). Theorem 1 provides matrices $\mathbf{X}_{\tau\sigma} \in \mathbb{R}^{|\tau|\times r}$, $\mathbf{Y}_{\tau\sigma} \in \mathbb{R}^{|\sigma|\times r}$, and we define the $\mathcal{H}$-matrix $\mathbf{B}_{\mathcal{H}}$ by

$$\mathbf{B}_{\mathcal{H}} = \begin{cases} \mathbf{X}_{\tau\sigma}\mathbf{Y}_{\tau\sigma}^T & \text{if } (\tau,\sigma) \in P_{\mathrm{far}}, \\ \mathbf{A}^{-1}|_{\tau\times\sigma} & \text{otherwise.} \end{cases}$$

On each admissible block $(\tau,\sigma) \in P_{\mathrm{far}}$, we use the blockwise estimate of Theorem 1. On the other blocks, the error is zero by definition. Now, an estimate for the global spectral norm by the local spectral norms from e.g. [7, 9] leads to

$$\left\|\mathbf{A}^{-1} - \mathbf{B}_{\mathcal{H}}\right\|_2 \leq C_{\mathrm{sp}}\Big(\sum_{\ell=0}^{\infty} \max\{\left\|(\mathbf{A}^{-1} - \mathbf{B}_{\mathcal{H}})|_{\tau\times\sigma}\right\|_2 : (\tau,\sigma) \in P, \mathrm{level}(\tau) = \ell\}\Big)$$
$$\leq C_{\mathrm{sp}} C_{\mathrm{apx}} (1+\eta) h^{-d} q^k \mathrm{depth}(\mathbb{T}_I).$$

Defining $b = -\frac{\ln(q)}{C_{\mathrm{dim}}^{1/(d+1)}} q^{d/(d+1)} > 0$, we obtain $q^k = e^{-br^{1/(d+1)}}$ and hence

$$\left\|\mathbf{A}^{-1} - \mathbf{B}_{\mathcal{H}}\right\|_2 \lesssim C_{\mathrm{apx}} C_{\mathrm{sp}} (1+\eta) N \operatorname{depth}(\mathbb{T}_{\mathcal{I}}) e^{-br^{1/(d+1)}}.$$

□

References

1. M. Bebendorf, *Hierarchical Matrices*, Lecture Notes in Computational Science and Engineering, vol. 63, Springer, Berlin, 2008.
2. M. Bebendorf and W. Hackbusch, *Existence of $\mathcal{H}$-matrix approximants to the inverse FE-matrix of elliptic operators with L^∞-coefficients*, Numer. Math. **95** (2003), no. 1, 1–28.
3. S. Börm, *Approximation of solution operators of elliptic partial differential equations by $\mathcal{H}$- and $\mathcal{H}^2$-matrices*, Numer. Math. **115** (2010), no. 2, 165–193.
4. S. Börm, *Efficient numerical methods for non-local operators*, EMS Tracts in Mathematics, vol. 14, European Mathematical Society (EMS), Zürich, 2010.
5. M. Faustmann, J.M. Melenk, and D. Praetorius, *$\mathcal{H}$-matrix approximability of the inverses of FEM matrices*, work in progress (2013).
6. L. Grasedyck and W. Hackbusch, *Construction and arithmetics of $\mathcal{H}$-matrices*, Computing **70** (2003), no. 4, 295–334.
7. L. Grasedyck, *Theorie und Anwendungen Hierarchischer Matrizen*, doctoral thesis (in German), Kiel, 2001.
8. W. Hackbusch, *A sparse matrix arithmetic based on $\mathcal{H}$-matrices. Introduction to $\mathcal{H}$-matrices*, Computing **62** (1999), no. 2, 89–108.
9. W. Hackbusch, *Hierarchische Matrizen: Algorithmen und Analysis*, Springer, Dordrecht, 2009.
10. L. R. Scott and S. Zhang, *Finite element interpolation of nonsmooth functions satisfying boundary conditions*, Math. Comp. **54** (1990), no. 190, 483–493.

Multidomain Extension of a Pseudospectral Algorithm for the Direct Simulation of Wall-Confined Rotating Flows

G. Fontaine, S. Poncet, and E. Serre

Abstract In this work, we improve an existing pseudospectral algorithm, in order to extend its properties to a multidomain patching of a rotating cavity. Viscous rotating flows have been widely studied over the last decades, either on industrial or academic approaches. Nevertheless, the range of Reynolds numbers reached in industrial devices implies very high resolutions of the spatial problem, which are clearly unreachable using a monodomain approach. Hence, we worked on the multidomain extension of the existing divergence-free Navier-Stokes solver with a Schur approach. The particularity of such an approach is that it does not require any subdomain superposition: the value of a variable on the boundary between two adjacent subdomains is treated as a boundary condition of a local Helmholtz solver. This value is computed on a direct way *via* a so-called *continuity influence matrix* and the derivative jump of an homogeneous solution computed independently on each subdomain. Such a method is known to have both good scalability and accuracy. It has been validated on two well documented three-dimensional rotating flows.

1 Numerical Modelling

Let's introduce the numerical fundamentals of the present method. A pseudospectral method is used to solve the Navier-Stokes PDE system in an annular cavity, where incompressibility is assured through a projection method.

G. Fontaine · S. Poncet (✉) · E. Serre
M2P2 Laboratory, Marseilles (France)
e-mail: guillaume.fontaine@l3m.univ-mrs.fr; poncet@l3m.univ-mrs.fr

M. Azaïez et al. (eds.), *Spectral and High Order Methods for Partial Differential Equations - ICOSAHOM 2012*, Lecture Notes in Computational Science and Engineering 95, DOI 10.1007/978-3-319-01601-6_21,

1.1 Pseudospectral Methods

Let Ω be the inner points of an annular cavity and Γ the domain boundary. The spatial approximation is of Chebyshev type in the axial z and radial r directions, of Fourier-Galerkin type in the azimuthal θ direction. Let Ψ be a variable (u, v, w, p, φ), which can be written for any point $(r, \theta, z) \in \Omega \cup \Gamma$ as:

$$\Psi_{N_r N_\theta N_z}(r, \theta, z) = \sum_{k=-N_\theta/2}^{N_\theta/2} \sum_{n=0}^{N_r-1} \sum_{m=0}^{N_z-1} \hat{\Psi}_{nkm} T_n(r)\, T_m(z)\, e^{ik\theta} \tag{1}$$

where T_n and T_m are the Chebyshev polynomials of degrees n and m respectively. N_r, N_z and N_θ are the approximation degrees in the radial, axial and azimuthal directions, respectively. $\hat{\Psi}_{nkm}$ is given by:

$$\hat{\Psi}_{nkm} = \frac{1}{K} \frac{1}{c_k c'_m} \sum_{q=0}^{K-1} \sum_{i=0}^{N} \sum_{j=0}^{M} \frac{1}{c_i c'_j} \Psi\left(r_i, \theta_q, z_j\right) \cos\left(\frac{in\pi}{N}\right) \cos\left(\frac{im\pi}{M}\right) e^{-ik\theta_q} \tag{2}$$

$c_0 = c'_0 = c_N = c'_M = 2$ and $c_k = c'_m = 1$ and for $n = 1, \ldots, N-1$ and $m = 1, \ldots, M-1$. This approximation is done using a Gauss-Lobatto point distribution in the radial and axial directions with Fourier-Galerkin points in the azimuthal direction. Thus derivatives can be estimated in the spectral space with a good precision, allowing the use of efficient FFT algorithms. These methods are called pseudospectral because of the non-linear diffusive terms, which must be computed in the physical space.

1.2 Geometry

The velocities are made dimensionless according to the Reynolds number $Re = \omega R_1^2/\nu$, where ω is the angular velocity, R_1 the outer radius of the domain, ν the kinematic viscosity. Hence the dimensionless geometry is defined by two parameters: the aspect ratio of the cavity $L = \frac{R_1 - R_0}{2h}$ and the curvature parameter $Rm = \frac{R_1 + R_0}{R_1 - R_0}$, where R_1 and R_0 are the outer and inner radii of the cavity, respectively, and $2h$ its height [3].

1.3 Projection Method

Let's consider the Navier-Stokes equations in primitive variables:

$$\left(\frac{\partial \mathbf{V}}{\partial t} + (\mathbf{V}.\nabla)\mathbf{V}\right) = -\frac{1}{\rho}\nabla p + \nu \Delta \mathbf{V} + \mathbf{F} \text{ in } \Omega \tag{3}$$

$$\nabla.\mathbf{V} = 0 \text{ in } \Omega \tag{4}$$

where $\mathbf{V}$ is the velocity vector with (u, v, w) its components in the cylindrical basis, p the pressure, ρ the density and $\mathbf{F}$ a force term.

1.3.1 Time Discretization

A semi-implicit second order scheme has been chosen for the time discretization with an implicit retarded Euler scheme of second order for the diffusive terms and an explicit Adams-Bashforth evaluation for the non-linar convective terms.

1.3.2 Projection Scheme

For incompressible viscous flows, an equation is missing to describe the pressure evolution. We deal with this particularity by using an improved projection method [2] based on the Goda's projection method. It requires three steps:

- First step: the computation of a preliminary pressure $\bar{p}^{n+1}$ through this Poisson equation:

$$\begin{cases} \nabla^2 \bar{p}^{n+1} = \nabla.[-2\mathbf{V}.\nabla\mathbf{V}^n + \mathbf{V}.\nabla\mathbf{V}^{n-1} + \mathbf{F}^{n+1}] \text{ in } \Omega \\ \dfrac{\partial \bar{p}^{n+1}}{\partial n} = \mathbf{n}.[\dfrac{-3\mathbf{W^{n+1}} + 4\mathbf{V^n} - \mathbf{V^{n-1}}}{2\delta t} - 2\mathbf{V}.\nabla\mathbf{V}^n + \mathbf{V}.\nabla\mathbf{V}^{n-1} + \\ \nu(2\Delta\mathbf{V}^n - \Delta\mathbf{V}^{n-1}) + \mathbf{F}^{n+1}] \text{ on } \Gamma \end{cases} \tag{5}$$

  where $\mathbf{W}$ represents the boundary conditions.
- Second step: prediction step. Using the gradient of $\bar{p}^{n+1}$, a preliminary velocity field $\mathbf{V}^*$ is computed through three Helmholtz solvers:

$$\begin{cases} \dfrac{3\mathbf{V}^* - 4\mathbf{V^n} + \mathbf{V^{n-1}}}{2\delta t} + 2\mathbf{V}.\nabla\mathbf{V}^n - \mathbf{V}.\nabla\mathbf{V}^{n-1} = -\nabla\bar{p}^{n+1} + \nu\Delta\mathbf{V}^* + \mathbf{F}^{n+1} \text{ in } \Omega \\ \mathbf{V}^* = \mathbf{W}^{n+1} \text{ on } \Gamma \end{cases} \tag{6}$$

  This velocity field does not a priori satisfy the incompressibility constraint in Ω. The principle of the projection method is namely to *project* this field on a divergence-free field.
- Third step: pseudo-pressure calculation and correction. An intermediate variable called *pseudo-pressure* φ is computed through a Poisson solver:

$$\varphi = \frac{2\delta t}{3}(p^{n+1} - \bar{p}^{n+1})$$

As $\nabla.\mathbf{V}^{n+1} = 0$, the Poisson problem to solve for φ is:

$$\begin{cases} \nabla^2\varphi = \nabla.\mathbf{V}^* \\ \dfrac{\partial\varphi}{\partial n} = 0 \end{cases} \tag{7}$$

Corrected pressure and velocity fields may be then evaluated at time step $n+1$:

$$\begin{cases} p^{n+1} = \overline{p}^{n+1} + \dfrac{3}{2\delta t}\varphi \\ \mathbf{V}^{\mathbf{n+1}} = \mathbf{V}^* - \nabla\varphi \end{cases} \tag{8}$$

1.4 Resolution Algorithm: Complete Matricial Diagonalisation Technique

For both Poisson and Helmholtz solvers, complete matricial diagonalisation technique is used. For the Poisson solver, this technique exhibits a null-eigenvalue problem, which is treated by a "source term reset" technique ([7]), for indexes in the operators corresponding to the null-eigenvalue index. According to the variable spatial evaluation, each Helmholtz/Poisson solving operation in the physical tridimensional domain is equivalent to N_θ bidimensional Helmholtz/Poisson solving operations in the (r,z) plane, each $k \in [\![1; N_\theta]\!]$ being the *azimuthal mode* in the flow spectrum [3]. For each mode k and for each variable $\Psi = (u, v, w, p, \varphi)$, the following bidimensional system has then to be solved:

$$\begin{cases} \Delta\hat{\Psi} - \sigma_k\hat{\Psi} = S_k \text{ in } [-1,1]\times[-1,1] \\ A\Psi = b \text{ on } \Gamma \end{cases} \tag{9}$$

$\hat{\Psi}$ being the Fourier transform of Ψ in the azimuthal direction. The azimuthal properties of the differentiation matrixes are treated in σ so as to allow the Δ operator to be independent of the azimuthal mode. The azimuthal dependance of what will follow is only linked to σ's one. Hence we will limit the multidomain approach to bidimensional problems, because the extension to tridimensional flows is quite immediate, thanks to spectral properties in the Fourier-Galerkin direction.

2 Multidomain Approach: Influence Matrix Technique

This technique is a direct Schur multidomain technique used by Raspo [1] for rotating flows using the vorticity-stream function formulation. It requires the patching of the cavity into subdomains without subdomain covering. The values of each variable on the frontier between two subdomains are treated as boundary conditions in the

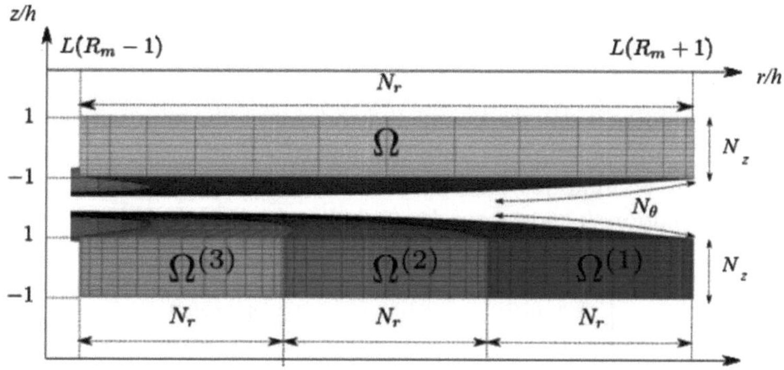

Fig. 1 Example of a multidomain decomposition in the radial direction with three subdomains

Helmholtz local solvers, the particularity of the influence matrix technique being the determination of this condition through a direct matrix computation.

2.1 *Multidomain Geometry*

We will limit the discussion to a radial multidomain decomposition (Fig. 1), because the curvature terms vary only along this direction. The generalization of this technique to an axial decomposition is quite immediate. We introduce the local aspect ratio of the subdomain m denoted $L^{(m)}$ and its local curvature parameter $Rm^{(m)}$ for $m \in [\![1; M]\!]$ (M the number of subdomains), which satisfy:

$$\sum_{m=1}^{M} L^{(m)} = L \tag{10}$$

$$Rm^{(m)} - 1 = Rm^{(m+1)} + 1 \tag{11}$$

Local derivation matrixes are deduced from these, in order to have a good approximation for the curvature terms from one subdomain to another and to adapt the local mapping to the one which would be used in a monodomain approach.

2.2 *Multidomain Decomposition of the Solutions*

Let Ψ be either (u, v, w, p, φ), $\Omega^{(m)}$ and $\Omega^{(n)}$ two adjacent subdomains, ξ the border between these two subdomains and λ the value of Ψ on ξ. For both subdomains, the local problems to be solved may be written as:

$$\begin{cases} \Delta^{(m)}\Psi^{(m)} - \sigma^{(m)}\Psi^{(m)} = S^{(m)} \text{ in } \Omega^{(m)} \\ A^{(m)}\Psi^{(m)} = b^{(m)} \text{ on } \Gamma^{(m)} \end{cases} \tag{12}$$

$$\begin{cases} \Delta^{(n)}\Psi^{(n)} - \sigma^{(n)}\Psi^{(n)} = S^{(n)} \text{ in } \Omega^{(n)} \\ A^{(n)}\Psi^{(n)} = b^{(n)} \text{ on } \Gamma^{(n)} \end{cases} \tag{13}$$

The resulting problem is that the boundary conditions to be imposed on the parts of $\Gamma^{(m)}$ and $\Gamma^{(n)}$ corresponding to ξ are unknown. To find λ, we choose to ensure both $\mathscr{C}^0$ and $\mathscr{C}^1$ continuities through ξ. Ψ is written as the combination of an *homogeneous* solution $\tilde{\Psi}$ and a *boundary* solution $\overline{\Psi}$: $\Psi = \tilde{\Psi} + \overline{\Psi}$.

On $\Omega^{(m)}$:

$$\begin{cases} \Delta^{(m)}\tilde{\Psi^{(m)}} - \sigma^{(m)}\tilde{\Psi^{(m)}} = S^{(m)} \text{ in } \Omega^{(m)} \\ A^{(m)}\tilde{\Psi^{(m)}} = b^{(m)} \text{ on } \Gamma^{(m)} \\ \tilde{\Psi^{(m)}} = 0 \text{ on } \xi^{(m)} \end{cases} \tag{14}$$

$$\begin{cases} \Delta^{(m)}\overline{\Psi^{(m)}} - \sigma^{(m)}\overline{\Psi^{(m)}} = 0 \text{ in } \Omega^{(m)} \\ A^{(m)}\overline{\Psi^{(m)}} = 0 \text{ on } \Gamma^{(m)} \\ \overline{\Psi^{(m)}} = \lambda^{(m)} \text{ on } \xi^{(m)} \end{cases} \tag{15}$$

On $\Omega^{(n)}$:

$$\begin{cases} \Delta^{(m+1)}\tilde{\Psi^{(n)}} - \sigma^{(n)}\tilde{\Psi^{(n)}} = S^{(n)} \text{ in } \Omega^{(n)} \\ A^{(m)}\tilde{\Psi^{(n)}} = b^{(n)} \text{ on } \Gamma^{(m)} \\ \tilde{\Psi^{(n)}} = 0 \text{ on } \xi \end{cases} \tag{16}$$

$$\begin{cases} \Delta^{(n)}\overline{\Psi^{(n)}} - \sigma^{(n)}\overline{\Psi^{(n)}} = 0 \text{ in } \Omega^{(m)} \\ A^{(n)}\overline{\Psi^{(n)}} = 0 \text{ on } \Gamma^{(n)} \\ \overline{\Psi^{(n)}} = \lambda \text{ on } \xi \end{cases} \tag{17}$$

If λ is assumed to be known, one can verify easily that (Eq. 12)=(Eq. 14)+(Eq. 15) and (Eq. 13) =(Eq. 16)+(Eq. 17).

2.3 The Influence Matrix Technique

Let's consider the boundary solution $\overline{\Psi}$. It can be written as the linear combination of *Green's elementary solutions* $\mathscr{G}_{k\xi}^{(m)}$, defined for each subdomain $\Omega^{(m)}$ by:

$$\begin{cases} \Delta^{(m)}\overline{\mathscr{G}_{k\xi}^{(m)}} - \sigma^{(m)}\overline{\mathscr{G}_{k\xi}^{(m)}} = 0 \text{in } \Omega^{(m)} \\ A^{(m)}\overline{\mathscr{G}_{k\xi}^{(m)}} = 0 \text{on } \Gamma^{(m)} \\ \overline{\mathscr{G}_{k\xi}^{(m)}}(\eta_l \in \xi^{(m)}) = \delta_{kl} \; \forall l \in [\![1; N_\xi]\!] \end{cases} \tag{18}$$

Assuming that ξ is the only boundary (i.e. there are only two subdomains $\Omega^{(m)}$ and $\Omega^{(n)}$), the boundary solution should be written as:

$$\begin{cases} \Psi^{(m)} = \tilde{\Psi^{(m)}} + \sum_{k=1}^{N_\xi} \lambda_k \mathcal{G}_{kE}^{(m)} \text{in } \Omega^{(m)} \\ \Psi^{(n)} = \tilde{\Psi^{(n)}} + \sum_{k=1}^{N_\xi} \lambda_k \mathcal{G}_{kS}^{(n)} \text{in } \Omega^{(n)} \end{cases} \tag{19}$$

$\tilde{\Psi^{(m)}} \cup \tilde{\Psi^{(n)}}$ is obviously continuous through ξ, but not $\frac{\partial \tilde{\Psi^{(m)}}}{\partial r} \cup \frac{\partial \tilde{\Psi^{(n)}}}{\partial r}$. As the boundary solution is continuous too, $\Psi^{(m)} \cup \Psi^{(n)}$ should be continuous for any value of λ. The influence matrix technique aims to find λ in order to make it $\mathcal{C}^0$ and $\mathcal{C}^1$ through ξ. We denote $\partial_r = \frac{\partial}{\partial r}$. This $\mathcal{C}^1$-continuity problem on ξ writes:

$$\widetilde{\partial_r \Psi^{(1)}}(\xi) - \widetilde{\partial_r \Psi^{(2)}}(\xi) = \sum_{l=1}^{N_\xi} \lambda_l [\partial_r \mathcal{G}_{lS}^{(2)}(\xi) - \partial_r \mathcal{G}_{lE}^{(1)}(\xi)] \tag{20}$$

This can be written in a matrix form:

$$\mathcal{D} = \mathcal{M}\lambda \tag{21}$$

where $\mathcal{D}_k$ is the derivative (time-dependent) jump vector and $\mathcal{M}$ the *continuity influence matrix* of the problem. Note that it depends only on the time-independent Green solutions, so it just has to be computed in pre-processing. This matrix is diagonal-dominant. If there are more than two frontiers in the domain, the influence matrix is built by blocks. The block dimension is then N_{front}, the number of frontiers. Each diagonal block is a Green derivative jump vector along each frontier. Some non-diagonal blocks appear, resulting locally of the *influence* of two frontiers on one another through a single subdomain, as shown on Fig. (2).
If $\mathcal{M}$ is inversible, we can find λ as:

$$\lambda = \mathcal{M}^{-1}\mathcal{D} \tag{22}$$

This is achieved using LAPACK subroutines. This vector is then used as a boundary condition on ξ in the local Helmholtz solvers to get a $\mathcal{C}^0$ and $\mathcal{C}^1$ $\Psi^{(m)} \cup \Psi^{(n)}$ solution.

2.4 *Singularity of the Poisson-Problem*

The Neuman-Poisson problem has an infinity of solutions defined up to an additive constant. As Dirichlet boundary conditions are implemented on the frontiers, this

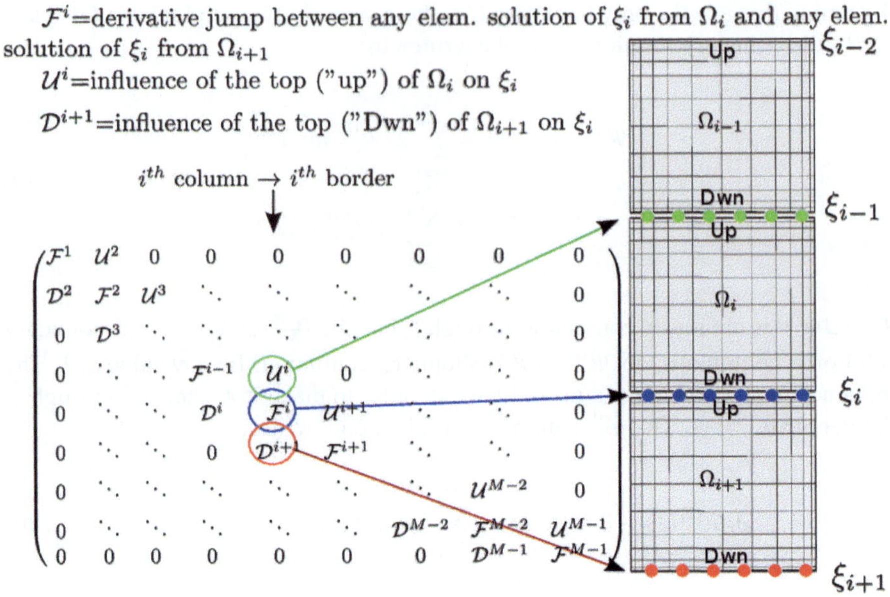

Fig. 2 Block definition of the influence matrix for an 1-D multidomain decomposition

problem no longer exists locally. Nevertheless, it is transposed to the influence matrix of the Poisson-problem of the $k = 0$ Fourier mode, which has a null-eigenvalue. It is treated by a diagonalisation technique of this modal matrix. The derivative jump is expressed in the diagonalisation basis and its i_0-th component is set to zero, if i_0 is the null eigenvalue index, as proposed by Abide and Viazzo [7].

3 Spatial Accuracy

Let's consider a domain Ω subdivided in M subdomains with N_r grid points in each subdomain. The total number of points may vary either with M or N_r. For $(r, z, \theta) \in [-1, 1] \times [-1, 1] \times [0, 2\pi]$, let's consider the divergence free analytical steady solution introduced by Raspo et al. [2]:

$$\begin{cases} u_{ana}(r,z,\theta) = \dfrac{1}{2\pi} sin(\pi r)^2 sin(2\pi z) cos(\theta) \\ v_{ana}(r,z,\theta) = -\dfrac{1}{2\pi} sin(\pi r)^2 sin(2\pi z) sin(\theta) \\ w_{ana}(r,z,\theta) = \dfrac{1}{2\pi L} sin(\pi z)^2 sin(2\pi r) cos(\theta) \\ p_{ana}(r,z,\theta) = [cos(\pi z) + cos(\pi r)] cos(\theta) \end{cases} \tag{23}$$

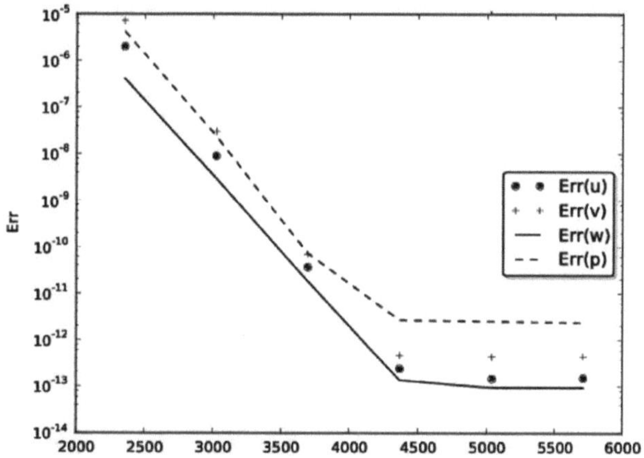

Fig. 3 Spatial accuracy evolution with the total number of points $(N_r N_z N_\theta M)$ when N_r varies: spectral convergence

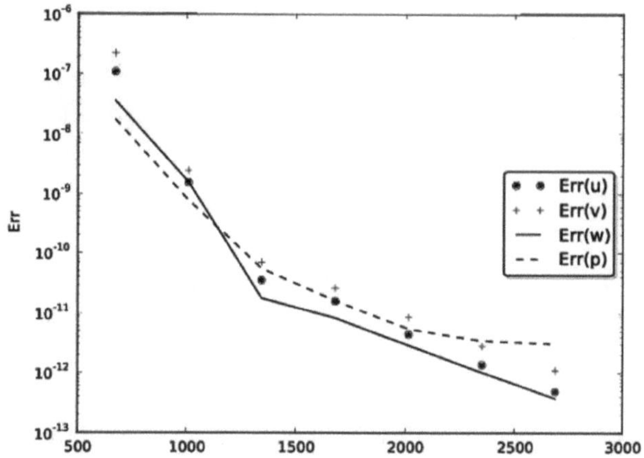

Fig. 4 Spatial accuracy evolution with the total number of points $(N_r N_z N_\theta M)$ when M varies: no spectral convergence

Figures 3 and 4 show the decrease of the quadratic truncature error L^2 when one increases N_r and M respectively. In the first case, the spectral convergence is obtained. As we impose only a $\mathscr{C}^1$ continuity through the frontier, the spatial accuracy increases indeed faster with N_r than with M, because in its last case, it multiplies the number of interfaces and so the number of $\mathscr{C}^d$ discontinuities $(d \geq 2)$.

4 Physical Results

We consider two multidomain rotating flow configurations: a high aspect ratio Taylor-Couette system and an interdisk rotor-stator flow.

Fig. 5 Iso-value $v = 0.5$ for a Wavy Vortex Flow at $\epsilon = 1.36$

Fig. 6 Iso-value $w = 0.005$ for a 3D rotor-stator flow at $Re = 25{,}000$: 17 spiral arm structures

4.1 Axial Decomposition: Taylor-Couette Flow

The numerical parameters are fixed to $M = 7$, $(N_r, N_\theta, N_z) = (35, 24, 85)$ (in each subdomain) with a time step set to $\delta t = 10^{-3}$. We consider a Taylor-Couette cavity with rotating terminal disks characterized by $L = 0.025$ and $Rm = 12.33$. We introduce the parameter $\epsilon = Ta/Ta_c$, where Ta_c is the critical Taylor number at which "Taylor rolls" appear. For $\epsilon = 1.36$, the flow becomes tridimensional (Wavy Vortex Flow, Fig. 6) with a dominant Fourier mode $k = 3$, which is in perfect agreement with the previous DNS results of Serre et al. [6].

4.2 Radial Decomposition: Interdisk Rotor-Stator Flow

We consider the rotor-stator interdisk cavity considered by Poncet et al. [5] for which $L = 6.26$ and $Rm = 1.8$. The domain is decomposed into four subdomains with $(N_r, N_\theta, N_z) = (45, 48, 45)$ in each subdomain and a time step equal to $\delta t = 10^{-4}$. The transition to tridimensional flow appears at $Re = 25{,}000$ with the appearance of 17 spiral arms, whose characteristics fully agree with the ones obtained by [5].

References

1. Raspo, I.: A direct spectral domain decomposition method for the computation of rotating flows in a T-shape geometry. Comput. Fluids **32**, 431–456 (2003)
2. Raspo, I., Hugues, S., Randriamampianina, A., Bontoux, P.: A spectral projection method for the simulation of complex three-dimensional rotating flows. Comput. Fluids **31**, 745–767 (2002)
3. Serre, E.: Instabilité de couche-limite dans les écoulements confinés en rotation. Simulation numérique directe par une méthode spectrale de comportements complexes. Phd thesis, Univ. Aix-Marseille II (2000)
4. Peyret, R.: Spectral methods for incompressible viscous flow. Springer, New-York (2002)
5. Poncet, S., Serre, E., Le Gal, P.: Revisiting the two first instabilities of the flow in an annular rotor-stator cavity. Phys. Fluids **21**, 064106 (2009)
6. Serre, E., Sprague, M., Lueptow, M.: Stability of Taylor-Couette flow with radial throughflow. Phys. Fluids **20**, 034106 (2008)
7. Abide, S., Viazzo S.: A 2D compact fourth-order projection decomposition method. J. Comp. Phys. **206**, 252–276 (2005)

A Comparison of High-Order Time Integrators for Highly Supercritical Thermal Convection in Rotating Spherical Shells

F. Garcia, M. Net, and J. Sánchez

Abstract The efficiency of implicit and semi-implicit time integration codes based on backward differentiation and extrapolation formulas for the solution of the three-dimensional Boussinesq thermal convection equations in rotating spherical shells was studied in Garcia et al. (J Comput Phys 229:7997–8010, 2010) at weakly supercritical Rayleigh numbers R, moderate (10^{-3}) and low (10^{-4}) Ekman numbers, E, and Prandtl number $\sigma = 1$. The results presented here extend the previous study and focus on the effect of σ and R by analyzing the efficiency of the methods for obtaining solutions at $E = 10^{-4}$, $\sigma = 0.1$ and low and high supercritical R. In the first case (quasiperiodic solutions) the decrease of one order of magnitude does not change the results significantly. In the second case (spatio-temporal chaotic solutions) the differences in the behavior of the semi-implicit codes due to the different treatment of the Coriolis term disappear because the integration is dominated by the nonlinear terms. As in Garcia et al. (J Comput Phys 229:7997–8010, 2010), high order methods, either with or without time step and order control, increase the efficiency of the time integrators and allow to obtain more accurate solutions.

1 Introduction

Thermal convection in rotating spherical geometries dominates the dynamics of several astrophysical and geophysical phenomena such as the generation of the magnetic field exhibited by celestial bodies or the cloud patterns and the differential rotation seen at the surface of the major planets.

F. Garcia (✉) · M. Net · J. Sánchez
Departament de Física Aplicada, Universitat Politènica de Catalunya, Jordi Girona Salgado s/n, Campus Nord. Mòdul B4, 08034 Barcelona, Spain
e-mail: ferran@fa.upc.edu; marta@fa.upc.edu; sanchez@fa.upc.edu

M. Azaïez et al. (eds.), *Spectral and High Order Methods for Partial Differential Equations - ICOSAHOM 2012*, Lecture Notes in Computational Science and Engineering 95, DOI 10.1007/978-3-319-01601-6_22,

There are several experimental and numerical difficulties in the study of thermal convection in spherical geometry. In the first case, the radial gravity can be reproduced by means of either an electrostatic radial field or by the centrifugal force. In the second case non-stationary tridimensional waves arise at the onset of convection due to the boundary curvature, and thus finding a solution requires very high resolutions. Frequently, as in [12], and [13], a two-dimensional annular geometry is used to approximate the real problem.

Due to the increase of the computing power, many numerical papers, [4, 8, 17, 18] among others, most of them based on pseudo-spectral methods and second order time integration, have been published. The most exhaustive tridimensional studies consist of numerical evolutions of periodic, quasi-periodic, and even turbulent flows, mainly with stress-free boundary conditions to avoid the formation of Ekman layers. These boundary conditions are inappropriate for comparison with laboratory studies and to model systems like the Earth's outer core, where thin Ekman boundary-layers exist near the rigid boundaries.

For a deeper understanding of the origin of the laminar flows and their dependence on parameters, pseudoarclength continuation methods [15, 16], and the linear stability analysis of the time dependent solutions [7, 11] have been successfully applied thanks to the use of high-order time integration methods which provide accurate enough solutions. On the other hand, high-order time integration can be also useful for evolving turbulent flows efficiently.

The performance of several high order implicit-explicit (IMEX) schemes, including those based on backward differentiation formulae (BDF), were exhaustively studied in [2] for the linear advection-diffusion one-dimensional problem. A stability analysis of the multistep methods up to fourth order were also performed. In that study the diffusive term is taken implicitly and the advection term explicitly. For the IMEX-BDF schemes, they showed that larger time-steps are allowed for the second order scheme when diffusion dominates the dynamics. In contrast, the third and fourth-order schemes can take larger time-steps when the explicit advection term becomes relevant. In addition, and in contrast to the widely used second order Crank-Nicolson and Adams-Bashforth scheme (CNAB2), the authors of [2] argued that IMEX-BDF methods are useful for reducing the aliasing effects when using pseudo-spectral methods [3], due to the strong damping of the high frequency modes, which appear when computing the nonlinear terms. Other similar class of IMEX methods with better stability regions are those based on Runge-Kutta (RK) schemes [1]. However, when compared with the multistep BDF, RK-based methods would require one additional nonlinear evaluation for each stage. This is not affordable in problems for which the evaluation of the nonlinear part is the most demanding task.

The efficiency of different time integration methods to solve the thermal convection equations in rotating spherical shells was studied in [6]. The same IMEX-BDF time integration pseudo-spectral codes, with the nonlinear terms of the equations taken explicitly in order to avoid solving nonlinear equations at each time step, are used in this study. The Coriolis term is treated either semi-implicitly or fully implicitly, giving rise to the different algorithms analyzed. The use of iterative methods facilitates the implementation of a suitable order and time stepsize control.

Two periodic solutions, of different E (the rest of parameters are the same) were integrated in [6] to highlight the influence of the Ekman number. Extending the previous study, a periodic and a quasiperiodic solution, computed with different σ are integrated to address the Prandtl number influence. In addition, by only varying the Rayleigh number, the efficiency of the time integration methods is studied when considering a spatio-temporal chaotic solution.

The rest of the article is organized as follows. In Sect. 2, the formulation of the problem and the spatial discretization of the equations are introduced. In Sect. 3, the time discretization schemes are described briefly. In Sect. 4 the differences between the constant stepsize methods are shown, and the study of the implicit and semi-implicit variable stepsize and variable order methods is reported. Finally, the paper closes in Sect. 5 with a brief summary of the main conclusions.

2 Mathematical Model and Spatial Discretization

The thermal convection of a spherical fluid shell differentially heated, rotating about an axis of symmetry with constant angular velocity $\Omega = \Omega\mathbf{k}$, and subject to radial gravity $\mathbf{g} = -\gamma\mathbf{r}$, where γ is a constant, and $\mathbf{r}$ the position vector, is considered. The mass, momentum and energy equations are written by using the same formulation and non-dimensional units as in [11]. The units are the gap width, $d = r_o - r_i$, for the distance, $\nu^2/\gamma\alpha d^4$ for the temperature, and d^2/ν for the time, ν being the kinematic viscosity, α the thermal expansion coefficient, and r_i and r_o the inner and outer radii, respectively. The velocity field $\mathbf{v}$ is expressed in terms of toroidal, Ψ, and poloidal, Φ, scalar potentials $\mathbf{v} = \nabla\times(\Psi\mathbf{r}) + \nabla\times\nabla\times(\Phi\mathbf{r})$, and $\Theta = T - T_c$ is the temperature perturbation from the conduction state $\mathbf{v} = \mathbf{0}$, $\quad T_c(r) = T_0 + R\eta/\sigma(1-\eta)^2 r$, with $r = ||\mathbf{r}||_2$.

With the functions $X = (\Psi, \Phi, \Theta)$ expanded in spherical harmonic series up to degree L, the equations written for their complex coefficients are

$$\partial_t \Psi_l^m = \mathcal{D}_l \Psi_l^m + \frac{1}{l(l+1)} \left[2E^{-1} \left(im\Psi_l^m - [Q\Phi]_l^m \right) - [\mathbf{r}\cdot\nabla\times(\mathbf{w}\times\mathbf{v})]_l^m \right], \quad (1)$$

$$\partial_t \mathcal{D}_l \Phi_l^m = \mathcal{D}_l^2 \Phi_l^m - \Theta_l^m + \frac{1}{l(l+1)} \left[2E^{-1} \left(im\mathcal{D}_l \Phi_l^m + [Q\Psi]_l^m \right) + [\mathbf{r}\cdot\nabla\times\nabla\times(\mathbf{w}\times\mathbf{v})]_l^m \right], \quad (2)$$

$$\partial_t \Theta_l^m = \sigma^{-1}\mathcal{D}_l \Theta_l^m + \sigma^{-1} l(l+1) R\eta(1-\eta)^{-2} r^{-3} \Phi_l^m - [(\mathbf{v}\cdot\nabla)\Theta]_l^m, \quad (3)$$

with boundary conditions

$$\Psi_l^m = \Phi_l^m = \partial_r \Phi_l^m = \Theta_l^m = 0, \quad (4)$$

corresponding to non-slip perfect thermally conducting boundaries, and where $\mathbf{w} = \nabla \times \mathbf{v}$ is the vorticity field.

The spherical harmonic coefficients of the operator $Q = Q^u + Q^l$ are

$$[Q^u f]_l^m = -\,l(l+2)c_{l+1}^m D_{l+2}^+ f_{l+1}^m, \quad [Q^l f]_l^m = -\,(l-1)(l+1)c_l^m D_{1-l}^+ f_{l-1}^m, \quad (5)$$

$$\text{with} \quad D_l^+ = \partial_r + \frac{l}{r}, \quad c_l^m = \left(\frac{l^2-m^2}{4l^2-1}\right)^{1/2}, \text{ and } \quad \mathscr{D}_l = \partial_{rr}^2 + \frac{2}{r}\partial_r - \frac{l(l+1)}{r^2}.$$

The governing parameters are the Rayleigh number R, the Prandtl number σ, the Ekman number E, and the radius ratio η. They are defined by

$$R = \frac{\gamma\alpha\Delta T d^4}{\kappa\nu}, \quad E = \frac{\nu}{\Omega d^2}, \quad \sigma = \frac{\nu}{\kappa}, \quad \eta = \frac{r_i}{r_o},$$

where κ is the thermal diffusivity, and ΔT the difference of temperature between the inner and outer boundaries.

The coefficients of the nonlinear terms of Eqs. (1)–(3) are obtained following [8]. In the radial direction, a collocation method on a Gauss-Lobatto mesh of $N_r + 1$ points is employed ($N_r - 1$ being the number of inner points). A large system of $N = (3L^2 + 6L + 1)(N_r - 1)$ ordinary differential equations must be advanced in time.

3 Time Integration Methods

The time integration methods used in this paper were described in detail in [6] so only the main ideas are exposed in the following. In order to simplify the notation, Eqs. (1)–(3) are written in the form

$$\mathscr{L}_0 \dot{u} = \mathscr{L} u + \mathscr{N}(u),$$

where $u = (\Psi_l^m(r_i), \Phi_l^m(r_i), \Theta_l^m(r_i))$, and $\mathscr{L}_0$ and $\mathscr{L}$ are linear operators including the boundary conditions. The former is invertible, and the latter, for any of the schemes used, includes the diffusive, the buoyancy, and part of the Coriolis terms to be specified below. The operator $\mathscr{N}$, which will be treated explicitly in the IMEX-BDF formulae, will always contain the nonlinear terms, and the rest of the Coriolis terms.

The IMEX-BDF formulae mentioned before are related to the BDF [5]. They obtain $u^{n+1} \approx u(t_{n+1})$ on a given time level t_{n+1}, $n = 0, 1, 2, \ldots,$ from the previous approximations u^{n-j}, $j = 0, 1, \ldots, k-1$, using the following k-steps formula

$$\left(\mathscr{I} - \frac{\Delta t_n}{\gamma_0(n)}\mathscr{L}_0^{-1}\mathscr{L}\right)u^{n+1} = \frac{\Delta t_n}{\gamma_0(n)}\mathscr{L}_0^{-1}p_{n,k-1}(t_{n+1}) - \frac{\dot{q}^0_{n,k}(t_{n+1})}{\dot{l}_{n,k}(t_{n+1})}, \quad (6)$$

where $q^0_{n,k}(t) = q_{n,k}(t) - u^{n+1}l_{n,k}(t)$, being $q_{n,k}$ the interpolating polynomial of degree at most k, such that $q_{n,k}(t_{n-j}) = u^{n-j}$, for $j = -1, 0, \dots, k-1$, and $l_{n,k}$ the polynomial of degree at most k taking the value 1 at t_{n+1}, and 0 at t_{n-j}, for $j = 0, 1, \dots, k-1$. Moreover $p_{n,k-1}$ is the interpolating polynomial of degree at most $k-1$, such that $p_{n,k-1}(t_{n-j}) = \mathscr{N}(u^{n-j})$, for $j = 0, 1, \dots, k-1$, $\mathscr{I}$ is the identity operator, and $\gamma_0(n) = \dot{l}_{n,k}(t_{n+1})\Delta t_n$, being $\Delta t_n = t_{n+1} - t_n$, $n = 0, 1, 2, \dots,$ the time step.

If the time step is constant, the IMEX-BDF formulae (6) reduces to

$$\left(\mathscr{I} - \frac{\Delta t}{\gamma_0}\mathscr{L}_0^{-1}\mathscr{L}\right)u^{n+1} = \sum_{i=0}^{k-1}\frac{\alpha_i}{\gamma_0}u^{n-i} + \sum_{i=0}^{k-1}\frac{\beta_i \Delta t}{\gamma_0}\mathscr{L}_0^{-1}\mathscr{N}(u^{n-i}), \quad (7)$$

where the coefficients α_i, β_j and γ_0 do not depend on n, and are listed, for instance, in [15]. In this case the matrix of the system to be solved does not change with n. On the other hand, changing the stepsize allows the use of formulas of different orders (step numbers) k, while maintaining accuracy. Then the integration can be started with $k = 1$ (and small Δt_0), when the lack of previously computed values precludes the use of higher order formulas, and then increase the order (and the step length) as the integration advances and previous approximations u^{n-j} are available. For the fixed-step-size codes, the starting values u^j, $j = 1, \dots, k-1$ are obtained by time integration from t_{j-1} to t_j with a VSVO (variable step-size variable order) code with sufficiently small tolerances ε^a and ε^r, which are, respectively, the tolerances below which the absolute and relative values of the local (time discretization) errors are required. The local error control of the IMEX VSVO methods is performed as usual, i.e. following [9]. If a time step is selected giving a point outside the stability region, the accumulation of local errors will enforce the method to select smaller time steps to ensure the stability. Details on the strategy carried out to control the stepsize and the order of the VSVO codes, such as the estimations of the local error of the k-order formula, are outlined in [6].

Once the nonlinear terms are evaluated, Eqs. (1)–(3) decouple for each azimuthal wave number m, thus, at every time step, $L + 1$ linear systems of the form $H^mU^m = V^m$, $m = 0, \dots, L$, have to be solved. The vectors U^m and V^m contain, respectively, the unknowns and the right hand side of the linear system derived from the IMEX-BDF formulae (6) or (7), with azimuthal wave number m. The dimension of the matrices H^m is $6(L-m+1)(N_r-1)$ and its structure depends on which terms of Eqs. (1)–(3) are treated implicitly (see the Appendix A of [6] for further details).

The inclusion of the diagonal parts of the Coriolis term containing $im\Psi_l^m$ and $im\Phi_l^m$ in $\mathscr{L}$, and of Q in $\mathscr{N}$, gives block-diagonal matrices H^m, with blocks of dimension $6(N_r - 1)$. The solution of these linear systems is performed by a direct LU method. From now on, the time discretization with this treatment of the operators will be called the Q-explicit method.

By adding Q^u or Q^d (see Eq. 5) to $\mathscr{L}$, the matrices H^m become upper or lower block-triangular matrices, respectively. They can be solved, with the same memory requirements and number of operations than the Q-explicit method, by using backward or forward block substitution. In order to implement this possibility in a symmetric way, the two options are used alternately, that is, one step is performed with Q^u implicit and Q^d explicit, and vice-versa in the following step. From now on, this time discretization will be called the Q-splitting method.

By setting Q totally implicit the operator $\mathscr{N}$ only includes the nonlinear terms, and then the matrices H^m become block-tridiagonal. From now on this method will be called the Q-implicit method. A direct block method for solving these linear systems involves about three times the memory storage required for solving the block-diagonal systems, and at least three times the computational cost of performing the LU decomposition. As the solution of the linear systems is not the most demanding task to advance one time step, we decided to solve it iteratively without storing the matrices to cope with higher resolutions, and to implement a variable size and order version, which requires updating the LU factorizations. Iterative methods based on Krylov techniques, can be used efficiently, if they are preconditioned with the block diagonal matrix. We have chosen the GMRES method because it is suitable for nonsymmetric linear systems and it has good convergence properties [14]. The initial approximation for solving the linear system is obtained by extrapolation from the previous steps. The increase in the cost of solving the linear systems may be offset by the increase of the time stepsize.

The integration with a constant time step can be unnecessarily expensive because the step must be short enough to cope with possible fast transients. To avoid this situation, a refined procedure is to use a VSVO method [9]. In the derivation of the VSVO IMEX-BDF formula, the matrices of the linear systems depend on the current time step. They can be solved efficiently, if a Krylov method with a preconditioning matrix depending on a fixed time step Δt^* is used. When the convergence of the iterative linear solver degrades doing more than 10 iterations, the preconditioning matrix can be updated with the current time step instead of restarting. In the case of the VSVO methods the tolerance for the GMRES residual is asked to be two orders of magnitude lower than that required for the time integration. The rate of convergence depends on the order k and the time step. For example, when integrating with low order $k = 2$ it converges in about 6–10 iterations, while with $k = 5$ only 2 iterations are required. All the semi-implicit methods described before have been implemented with constant time-step size, and with variable time-step size and order using our own codes (except the Q-splitting VSVO method). From now on, the VSVO implementations of the Q-explicit and Q-implicit methods will be called Q-explicit VSVO and Q-implicit VSVO, respectively.

The last option considered is a fully implicit treatment of the nonlinear terms with a VSVO formulation of the BDF. This leads to the solution of a nonlinear system of equations at each step. This solution is obtained by means of a Newton-Krylov method using GMRES to solve the Newton correction equations with zero initial seed, see [10] for further details. From now on this method will be called fully implicit method. We will use the DLSODPK code of the ODEPACK package [10]. The linear systems to be solved in this case depend not only on the current time step, but also on the current solution. As before, they can be preconditioned by the block-diagonal matrices computed with a fixed time step Δt^*. If during the integration the current time step is different from Δt^* by one order of magnitude the preconditioner is recomputed with the current time step. As in the semi-implicit VSVO methods, the tolerance imposed on the residuals depends on the tolerance imposed on the time integration. The cost of the nonlinear evaluations performed to approximate the Jacobian matrix products degrade the performance of the fully implicit methods.

4 Results

To study the effect of the Prandtl number σ and the Rayleigh number R on the efficiency of the time integration schemes presented in the previous section, we integrate three different cases with $\eta = 0.35$ and $E = 10^{-4}$. The first one (case S_1) was studied in [6] (there called C_2) and corresponds to a periodic traveling wave of wave number $m = 7$, computed at $\sigma = 1$ and $R = 800{,}000$. In the second case (S_2) we integrate a quasiperiodic three frequency wave with $m = 6$ computed at smaller Prandtl number $\sigma = 0.1$ and $R = 264{,}000$. On both cases R is slightly supercritical and the solutions are quasi-geostrophic and symmetric with respect to the equator. Finally, a highly supercritical spatio-temporal chaotic solution, computed with $\sigma = 0.1$ and $R = 2{,}000{,}000$, is considered for the third case (S_3). The numerical resolutions employed for the S_1, S_2 and S_3 cases are, $N_r = 48$ and $L = 63$ ($N = 577{,}442$ equations), $N_r = 32$ and $L = 54$ ($N = 281{,}263$ equations), and $N_r = 50$ and $L = 84$ ($N = 1{,}083{,}650$ equations), respectively.

To make our comparisons the initial transient has been discarded, and all the test runs are started with the same initial condition obtained after the solution has smoothed. To obtain it, the Q-implicit VSVO method with very low tolerances was used. Then the system (1–4) is evolved from the new initial condition to a final time t_f. In the S_1 and S_2 cases $t_f = 0.1$, while in the S_3 case $t_f = 0.01$.

To check the efficiency of the different schemes the relation between the relative error, and the run time is studied. The former is defined as

$$\varepsilon(u) = \frac{||u - u_r||_2}{||u_r||_2}, \tag{8}$$

where u is the solution we want to check, and u_r is an accurate reference solution obtained with the Q-implicit VSVO method. More precisely, u_r is obtained with tolerances $\varepsilon^a = \varepsilon^r$ equal to 10^{-13} in the S_1 and S_2 cases, and to 10^{-11} in the S_3 case. The decrease of the relative error (8) is achieved by decreasing the stepsize

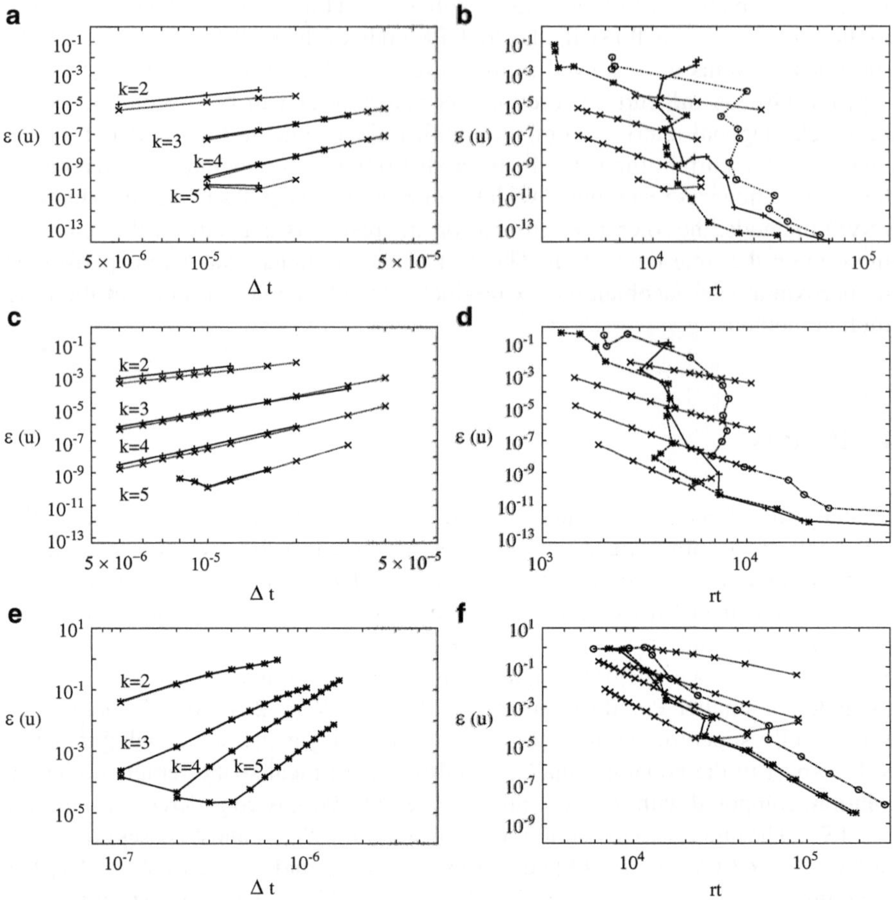

Fig. 1 (**a**) The relative error, $\varepsilon(u)$, plotted versus the time step Δt for constant time step integration, orders k from 2 to 5, and the S_1 case. (**c**), (**e**) Same as (**a**), for the S_2 and S_3 cases, respectively. (**b**) The relative error, $\varepsilon(u)$, plotted versus the run time for the Q-splitting and VSVO methods, and the S_1 case. (**d**), (**f**) Same as (**c**), for the S_2 and S_3 cases, respectively. The symbols mean: Q-explicit (+, *solid line*), Q-splitting (×, *dotted line*), Q-explicit VSVO (+, *solid line*), Q-implicit VSVO (∗, *dashed line*), and DLSODPK (∘, *dash-dotted line*)

in the case of fixed stepsize methods, or by decreasing the tolerances for the local errors in the case of the VSVO methods.

For the constant time stepsize methods of orders 2–5 (except the Q-implicit method for the sake of simplicity) the relative error $\varepsilon(u)$ is plotted against the time step in Fig. 1a,c,e. The efficiency curves are shown in Fig. 1b,d,f. In the latter, $\varepsilon(u)$ is plotted against the run time in seconds for the results of the VSVO codes together with the constant stepsize Q-splitting method for comparison purposes. Plots (a) and (b), (c) and (d), and (e) and (f), are for the S_1, S_2, and S_3 cases, respectively.

As mentioned previously, the study of the S_1 case (Fig. 1a–b) was performed in [6] so only a few words are commented here. For a given constant time stepsize, the Q-explicit and the Q-splitting methods of all the orders have almost the same computational cost, and therefore the higher order methods should be preferred. In addition, the Q-splitting method has shown itself to be more stable, allowing for larger time steps (see Fig. 1a), and hence, better efficiency. For the latter method, Fig. 1b shows that at approximately $\varepsilon(u) < 10^{-9}$, the order $k = 5$ is the most efficient but if $\varepsilon(u) > 10^{-9}$ the most efficient order is 4. The fully implicit method using DLSODPK is always more expensive than the Q-implicit VSVO methods because each iteration of the linear solver, and of the Newton's method requires an expensive evaluation of the non-linear terms. In all the results shown, it takes between 1 and 3 N iterations, and for each of them 1 or 2 GMRES iterations. The Q-explicit VSVO method is also less expensive than DLSODPK, except for the higher $\varepsilon(u)$, for which the cost of the former increases due to a decrease of the solver performance. The abrupt decrease of efficiency of the Q-implicit VSVO method close to $\varepsilon(u) = 10^{-3}$ was related in [6] with the shape of the stability regions of the BDF for constant stepsizes. A similar result was found in [2] in the framework of the one dimensional linear advection-diffusion problem. When the implicit term dominates (as occurs with the Q-implicit method at weak supercritical conditions) second order IMEX-BDF schemes allow larger time-steps than those of third or fourth order. The decrease of efficiency prevents the Q-implicit VSVO method from being as efficient as the Q-splitting method in the region of intermediate errors. However, for the latter, some previous experiments have to be performed to determine the optimal time step.

The behavior of the methods for integrating the S_2 case (Fig. 1c–d) is similar to that of the S_1 case, despite they are less accurate. Although their Prandtl numbers differ in one order of magnitude, the results seem reasonable because in both cases the solutions are smooth functions of time and, in all the methods, the terms of Eqs. (1–3) containing the Prandtl number are treated implicitly. However, for a given order, the error $\varepsilon(u)$ of the fixed step methods is almost two order of magnitude greater in the S_2 case. The same occurs for the VSVO methods. This differences could be due to S_2 being slightly more complicated, in the sense that it has two additional frequencies. Nevertheless, accurate solutions with $\varepsilon(u)$ down to 10^{-11} can be obtained in a reasonable time.

As expected, the accuracy for integrating the spatio-temporal chaotic solution of the S_3 case drastically decreases with respect to the other cases. This decrease is also due to the larger number of time steps needed by the methods to obtain the solution at the desired t_f. This is shown in Fig. 1e–f where $\varepsilon(u)$ down to 10^{-5}, and 10^{-9}, can be obtained with the fixed step, or the VSVO methods, respectively. Apart from the accuracy, the behavior of the methods is clearly different to that exhibited in the previous cases, at weakly supercritical regimes, where the Ekman number controls the dynamics. At highly supercritical R, the Ekman number plays minor role and the way the Coriolis term is treated becomes less important. According to this, the results obtained with the fixed time step Q-explicit, and the Q-splitting methods are nearly the same (Fig. 1e), and the same occurs for the results of the

VSVO Q-explicit and Q-implicit methods. Since the solution is strongly nonlinear, the efficiency of the fully implicit VSVO method becomes comparable to the semi-implicit VSVO methods, and better than the low order fixed step methods (Fig. 1f). As commented in [6] this is because the fully implicit method allows significantly larger time steps (nearly three times in this case), but which are computationally expensive. It is worth mentioning that in the S_3 case the VSVO methods obtain solutions up to four orders of magnitude more accurate than the fixed time methods, while in the previous cases with the VSVO methods the improvement is of two orders of magnitude. Despite these differences with respect to S_1 and S_2, in all cases the largest attainable values of the fixed Δt correspond to methods with order higher than two, obtaining therefore more accurate solutions in less time. Again, this behavior can be related with that observed in [2] for the one dimensional linear advection-diffusion problem. In the regime where the explicit term starts to dominate (and this occurs in the Q-explicit and Q-splitting methods for the cases S_1, S_2 and in the Q-implicit method for the case S_3), larger Δt were obtained for the IMEX-BDF schemes of orders 3 and 4 rather than for order 2.

5 Conclusions

In the time integration study [6] of the thermal convection in fast rotating fluid spherical shells, the possibility of handling implicitly the Coriolis term, and even the nonlinear term, thanks to the low memory requirements of the iterative Krylov methods used to solve the linear systems, was shown. That study focused on the influence of the Ekman number on the efficiency of the methods proposed. The present study extends the previous one by analyzing the influence of the Prandtl and the Rayleigh numbers.

The results presented here, computed at low $E = 10^{-4}$, show that the behavior of the methods for integrating a weakly supercritical oscillatory type of solution (periodic S_1 or quasiperiodic S_2) is very similar, despite their Prandtl number differ in one order of magnitude. At this regime, the Ekman number plays a major role, and a more implicit treatment of the Coriolis term becomes appropriate. In contrast, at strongly supercritical regime (S_3), an implicit treatment of the Coriolis term does not improve the integration, reflecting that the Ekman number plays a minor role. The solutions are obtained with less accuracy, reflecting their spatio-temporal chaotic character.

In all cases (S_1, S_2, and S_3) shown here (and also in the cases of [6]), the implementation of high order methods does not reduce the efficiency of the time integrators, and allows to obtain more accurate solutions. In addition, for the Q-splitting or Q-explicit fixed-step methods the largest time-steps are obtained with order higher than two, as occurs with the IMEX-BDF schemes applied in [2] to the one dimensional linear advection-diffusion problem when the dominant term of the equation is handled explicitly.

In practice the most efficient method depends strongly on R (also on E), but more weakly on σ, at least in the oscillatory regime. It depends also on the errors accepted for a solution, and on the type of solution. For instance, if one is just interested in obtaining solutions by direct numerical simulations (DNS), the best choice is to implement a fourth order Q-splitting method, and performing some previous experiments to determine the optimal time step. However, if the time integration is part of a continuation process, and/or one is interested in calculating the stability of the solutions, low errors must be requested to the time integration. Then the Q-implicit VSVO method will probably be the most efficient option. Moreover, since the lower run times correspond to the Q-implicit VSVO method with high tolerances, it might also be useful to pass long uninteresting transients, where having a control of the time stepsize might be important.

The results presented in this paper suggest that IMEX methods could also be efficiently used in other type of nonlinear problems with other spatial discretizations if the stiff part can be included in the implicit term of the scheme, and the cost of solving the corresponding linear systems, whatever their structure, becomes comparable to the evaluation of the explicit part.

Acknowledgements This research has been supported by Spain Ministerio de Ciencia e Innovación, and Generalitat de Catalunya under projects MTM2010-16930 and 2009-SGR-67, respectively.

References

1. Ascher, U.M., Ruuth, S.J., Spiteri, R.J.: Implicit-explicit Runge-Kutta methods for time-dependent partial differential equations. Applied Numerical Mathematics **25**, 151–167 (1997)
2. Ascher, U.M., Ruuth, S.J., Wetton, B.T.R.: Implicit-explicit methods for time-dependent partial differential equations. SIAM J. Numer. Anal. **32**(3), 797–823 (1995)
3. Canuto, C., Hussaini, M.Y., Quarteroni, A., Zang, T.A.: Spectral Methods in Fluid Dynamics. Springer (1988)
4. Christensen, U.: Zonal flow driven by strongly supercritical convection in rotating spherical shells. J. Fluid Mech. **470**, 115–133 (2002)
5. Curtiss, C.F., Hirschfelder, J.O.: Integration of stiff equations. PNASUSA **38**, 235–243 (1952)
6. Garcia, F., Net, M., García-Archilla, B., Sánchez, J.: A comparison of high-order time integrators for thermal convection in rotating spherical shells. J. Comput. Phys. **229**, 7997–8010 (2010)
7. Garcia, F., Sánchez, J., Net, M.: Antisymmetric polar modes of thermal convection in rotating spherical fluid shells at high Taylor numbers. Phys. Rev. Lett. **101**, 194,501–(1–4) (2008)
8. Glatzmaier, G.: Numerical simulations of stellar convective dynamos. I. The model and method. J. Comput. Phys. **55**, 461–484 (1984)
9. Hairer, E., Norsett, H.P., Wanner, G.: Solving Ordinary Differential Equations. I Nonstiff Problems (2nd. Revised Edition). Springer-Verlag (1993)
10. Hindmarsh, A.C.: ODEPACK, a systematized collection of ODE solvers. In: R.S.S. et al. (ed.) Scientific Computing, pp. 55–364. North-Holland, Amsterdam (1983)
11. Net, M., Garcia, F., Sánchez, J.: On the onset of low-Prandtl-number convection in rotating spherical shells: non-slip boundary conditions. J. Fluid Mech. **601**, 317–337 (2008)

12. Pino, D., Mercader, I., Net, M.: Thermal and inertial modes of convection in a rapidly rotating annulus. Phys. Rev. E **61**(2), 1507–1517 (2000)
13. Plaut, E., Busse, F.H.: Multicellular convection in rotating annuli. J. Fluid Mech. **528**, 119–133 (2005)
14. Saad, Y., Schultz, M.H.: GMRES: a generalized minimal residual algorithm for solving nonsymmetric linear systems. SIAM J. Sci. Stat. Comput. **7**, 865–869 (1986)
15. Sánchez, J., Net, M., García-Archilla, B., Simó, C.: Newton-Krylov continuation of periodic orbits for Navier-Stokes flows. J. Comput. Phys. **201**(1), 13–33 (2004)
16. Sánchez, J., Net, M., Simó, C.: Computation of invariant tori by Newton-Krylov methods in large-scale dissipative systems. Physica D **239**, 123–133 (2010)
17. Simitev, R., Busse, F.H.: Patterns of convection in rotating spherical shells. New Journal of Physics **5**, 97.1–97.20 (2003). DOI 10.1088/1367-2630/5/1/397
18. Tilgner, A.: Spectral methods for the simulation of incompressible flow in spherical shells. Int. J. Num. Meth. Fluids **30**, 713–724 (1999)

High Order Methods with Exact Conservation Properties

René Hiemstra and Marc Gerritsma

Abstract Conservation laws, in for example, electromagnetism, solid and fluid mechanics, allow an exact discrete representation in terms of line, surface and volume integrals. In this paper, we develop high order interpolants, from any basis that constitutes a partition of unity, which satisfy these integral relations exactly, at cell level. The resulting gradient, curl and divergence conforming spaces have the property that the conservation laws become completely independent of the basis functions. Hence, they are exactly satisfied at the coarsest level of discretization and on arbitrarily curved meshes. As an illustration we apply our approach to B-splines and compute a 2D Stokes flow inside a lid driven cavity, which displays, amongst others, a point-wise divergence-free velocity field.

1 Introduction

Conventional numerical methods, in particular finite difference and nodal finite element methods, expand their unknowns in terms of nodal interpolations only, and run into trouble when it comes to conservation. This can lead to instabilities, and perhaps more dangerously, to internal inconsistencies, such as the violation of fundamental conservation principles. Where instabilities lead to outright failure of

---

R. Hiemstra (✉)
The institute for computational Engineering and Sciences, The University of Texas at Austin, 201 East 24th St, Stop C0200 Austin, TX 78712-1229, USA
e-mail: rene@ices.utexas.edu

M. Gerritsma
Technical University Delft, Department of Aerospace engineering, Kluyverweg 2, 2629HT, Delft, The Netherlands
e-mail: m.i.gerritsma@tudelft.nl

M. Azaïez et al. (eds.), *Spectral and High Order Methods for Partial Differential Equations - ICOSAHOM 2012*, Lecture Notes in Computational Science and Engineering 95, DOI 10.1007/978-3-319-01601-6_23,

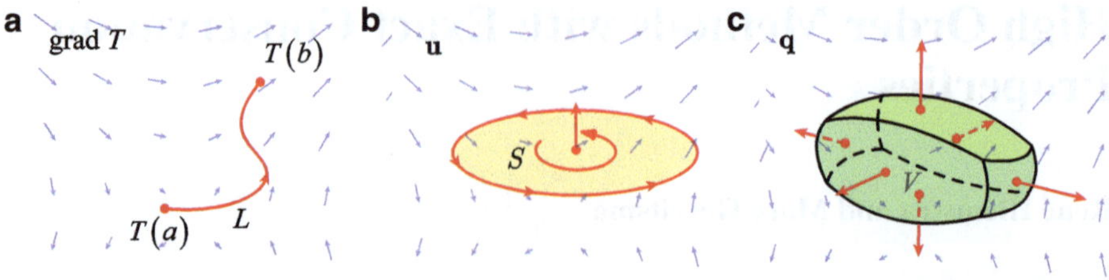

Fig. 1 Conservation, by definition, is a relation between a global 'measurable' quantity associated with a geometric object and another global 'measurable' quantity associated with its boundary. In $\mathbb{R}^3$ we distinguish between: (**a**) The fundamental theorem of calculus; (**b**) the Stokes circulation theorem; and (**c**) and the Gauss divergence theorem. Once we recognize that not all physical variables are associated with nodes, but to curves, surfaces and volumes as well, we can exactly satisfy the balance laws, that relate these different quantities, in the discrete setting [8]

a numerical method, inconsistencies often lead to nonphysical solutions that can go unnoticed by the human eye [12].

To capture the behavior of physical phenomena well, a discretization method should not only approximate the spaces of the infinite dimensional system, but should also follow the structure induced by the relations between them; in particular the structure induced by the fundamental conservation or balance laws, which in vector calculus are known as the fundamental theorem of calculus, the Stokes circulation theorem and the Gauss divergence theorem, depicted in Fig. 1.

We follow the recent advances in *Discrete Exterior Calculus* [4], *Finite Element Exterior Calculus* [1], *Compatible* [3, 5, 6] and *Mimetic Methods* [2, 7–11, 13, 14, 16]. These methods do not focus on one particular physical problem, but identify and discretize the underlying structure that constitutes a wide variety of physical field theories. They are said to be '*compatible*' with the geometric structure of the underlying physics; i.e. they '*mimic*' important properties of the physical system under consideration. This leads, amongst others, to naturally stable and consistent numerical schemes that have discrete conservation properties by construction and are applicable to a wide variety of physical theories. Furthermore, they offer insight into the properties of existing numerical schemes.

In this paper we develop arbitrary order interpolants – starting from any basis that constitutes a partition of unity – which satisfy the fundamental integral theorems exactly. The gradient, curl and divergence conforming spaces have the property that the conservation laws become completely independent of the basis functions. This means they are exactly satisfied at the coarsest level of discretization and on arbitrarily curved meshes. It is remarkable that inf-sup stability is automatically guaranteed when this physical structure is encoded in the discretization [5, 9, 10]. To illustrate our approach, we develop compatible spaces using B-splines, and use these to solve a Stokes flow inside a lid driven cavity.

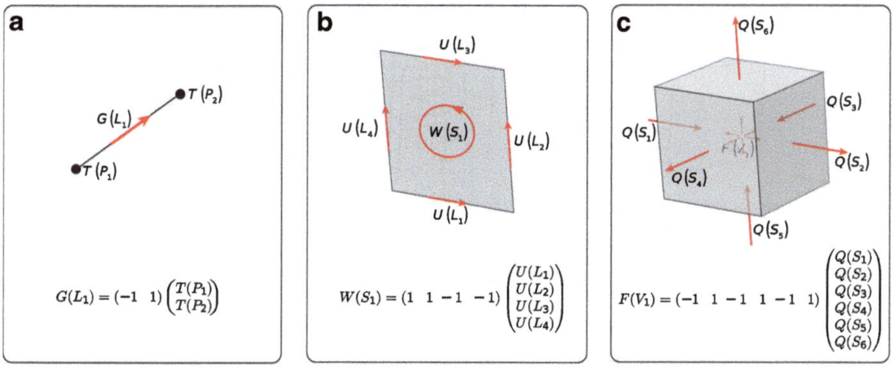

Fig. 2 Discrete differentiation. (**a**) Discrete gradient. (**b**) Discrete curl. (**c**) Discrete divergence

2 Discrete Conservation

Conservation, by definition, states a balance between a quantity associated with a geometric object and another quantity associated with its boundary. For instance, mass inside a control volume is only conserved if the in- and out going mass fluxes associated with its boundary surfaces cancel each other out. This is exactly the Gauss divergence theorem depicted in Fig. 1c. Important to note is, that the relation is discrete to begin with, since it is represented in terms of integral quantities. Furthermore, metric concepts such as size or shape of the control volume do not affect the discrete relation; hence the balance is purely topological. Perhaps this is more easily visualized in case of the fundamental theorem of calculus. Since the integral of the gradient of a temperature field along the curve in Fig. 1a, is equal to the difference between the discrete temperatures at the boundary points, the relation is independent of the specific path chosen to connect points a and b. Analogously, the rotation inside a surface is equal to the circulation around the bounding curve, according to the Stokes circulation theorem. Again this relation is purely topological since only the connectivity between the surface and the curve matters, not the size and shape of the surface. Figure 2 illustrates how the intrinsic discrete nature of the fundamental integral theorems can be used to our advantage in a numerical method. By associating discrete physical quantities to nodes, as well as edges, faces and volumes, we can obtain discrete matrix representations of the gradient, curl and divergence operators, which are completely free of approximation.

Remark 1. The global quantities associated with the discrete geometric elements in Fig. 2 are discrete real numbers. We actually do not require them to be of the integral kind. Hence, the discrete balance depicted in Fig. 2 is more general then the one depicted in Fig. 1.

By partitioning the computational domain into many of these sub-domains, i.e. in sub-volumes, faces, edges and nodes, as depicted in Fig. 3, we can obtain a matrix

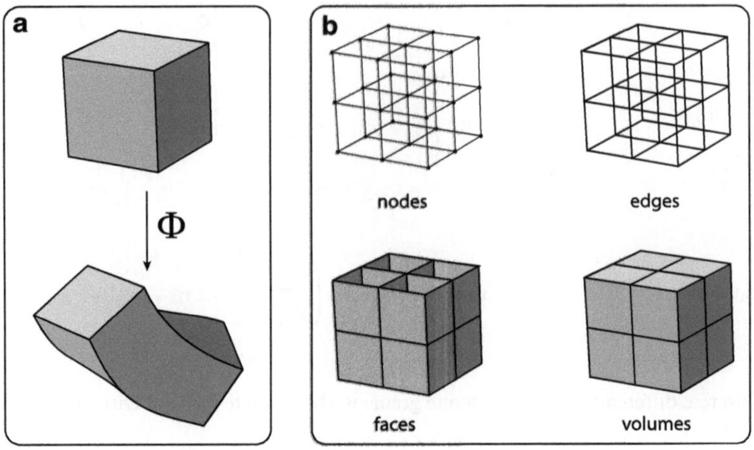

Fig. 3 The curved domain $\Phi(\Omega') = \Omega$ (**a**) has the same topology as the reference domain Ω'. Partitioning of Ω' into points, edges, faces and volumes (**b**)

representation of the gradient, curl and divergence for the whole computational domain. These sparse matrices contain solely the values zero, positive and negative one, denoting the connectivity between adjacent geometric elements, see [8]. Since these relations depend solely on mesh connectivity, they are the same on a square Cartesian mesh as on a highly curved domain, and remain unaltered under mesh deformation.

3 High Order Continuous Representations

In the previous section we have seen that we can represent the balance equations, represented by the fundamental integral theorems of vector calculus, without any approximation. Furthermore, we obtained a discrete matrix representation for the gradient, curl and divergence operators. Since the structure of these relations is inherently discrete and does not depend on metric considerations such as length, angle, area and volume, we do not expect that basis functions play any role here. Continuous representations are however required to be able to include metric and material dependent relations, i.e. the constitutive equations of a physical theory.

In this section we develop arbitrary order polynomial spaces which follow the discrete structure explained in the previous section. These spaces should thus conform to the rules of discrete differentiation, as delineated in Fig. 2.

Consider a one-dimensional partitioning in terms of a set of nodes $\{P_i\}_{i=0}^n$, connected by edges $L_i = [P_{i-1}, P_i]$, $i = 1$ to n, where $P_i > P_{i-1}$. Let T_h, for example a uni-variate temperature distribution, be expanded as,

$$T_h = \sum_{i=0}^{n} \alpha_i \, N_i(x) \qquad \text{with} \qquad \sum_{i=0}^{n} N_i(x) = 1, \tag{1}$$

where the degrees of freedom α_i and the basis functions $N_i(x)$ are associated with the nodes P_i, for $i = 0$ to n.

To take the gradient of T_h, as depicted in Fig. 2a, we seek a representation which looks like,

$$\frac{dT_h}{dx} = \sum_{i=1}^{n} \beta_i \, M_i(x) \qquad \text{where} \qquad \beta_i = \alpha_i - \alpha_{i-1}. \tag{2}$$

Here the degrees of freedom β_i and basis functions $M_i(x)$ can be associated with edges L_i, for $i = 1$ to n.

But what do these *edge* functions [7] actually look like? By working backwards from the desired representation we easily find the representation of the *edge* functions in terms of derivatives of the set $\{N_i(x)\}_{i=0}^n$:

$$\begin{aligned}
\frac{dT_h}{dx} = \sum_{i=0}^{n} \alpha_i \, \frac{dN_i}{dx} &= \sum_{i=1}^{n} (\alpha_i - \alpha_{i-1}) \, M_i(x) \\
&= \sum_{i=1}^{n} \alpha_i \, M_i(x) - \sum_{i=0}^{n-1} \alpha_i \, M_{i+1}(x) \\
&= \alpha_n \, M_n(x) + \sum_{i=1}^{n-1} \alpha_i \, (M_i(x) - M_{i+1}(x)) - \alpha_0 \, M_1(x)
\end{aligned}$$

Comparing the left and the right hand side we can conclude that the *edge* functions are defined according to,

$$M_1(x) = -\frac{dN_0}{dx}, \quad M_{i+1}(x) = M_i(x) - \frac{dN_i}{dx}, \quad M_n(x) = \frac{dN_n}{dx}. \tag{3}$$

Remark 2. Note that (3) is equivalent to $M_i(x) = -\sum_{j=0}^{i-1} \frac{dN_j}{dx} = \sum_{j=i}^{n} \frac{dN_j}{dx}$. Therefore, $\sum_{j=0}^{i-1} \frac{dN_j}{dx} + \sum_{j=i}^{n} \frac{dN_j}{dx} = \frac{d}{dx} \sum_{j=0}^{n} N_j(x) = 0$, which is satisfied by any basis in (1) with a partition of unity. Hence, (3) is valid for nodal interpolants such as

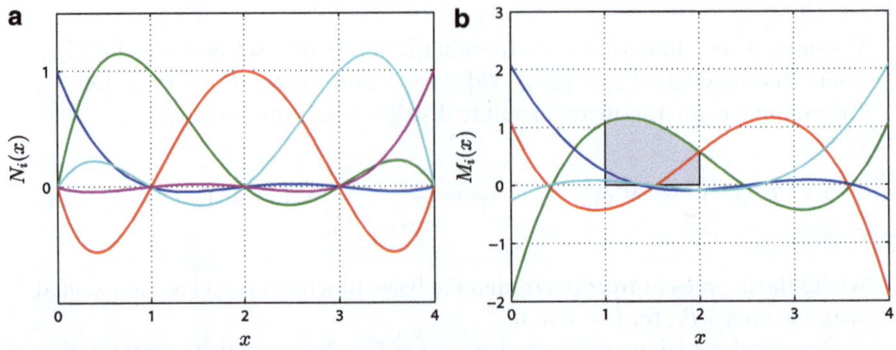

Fig. 4 Lagrange polynomials of degree 4 and derived *edge* functions [7] of degree 3 on the interval [0, 4]. While Lagrange polynomials satisfy $N_i(P_j) = \delta_{ij}$, the *edge* basis functions posses a unit integral property, $\int_{L_j} M_i(x)dx = \delta_{ij}$. (**a**) Lagrange nodal functions. (**b**) Lagrange *edge* functions

Lagrange polynomials, as well as any other basis with a partition of unity that does not have the interpolation property, examples of which are Bernstein polynomials and B-splines.

Remark 3. $\beta_i - (\alpha_i - \alpha_{i-1}) = 0$, from (2), represents a conservation law and is completely independent of the basis functions. This means these are exactly satisfied independent of the shape or coarseness of the mesh.

Remark 4. In the special case of a nodal interpolant, $N_i(P_j) = \delta_{ij}$, e.g. Lagrange polynomials, then α_i in (1) represents the value (e.g. temperature) in node P_i. Furthermore, it can be shown, see Fig. 4, that

$$\int_{P_{j-1}}^{P_j} M_i(x)dx = \delta_{ij}. \tag{4}$$

The coefficient β_i, associated with the *edge* function $M_i(x)$ in (2), represents the line integral over the edge L_i.

Remark 5. In the general case when the basis is not a nodal interpolant, $N_i(P_j) \neq \delta_{ij}$ (e.g. Bernstein polynomials or B-splines), then α_i is merely a discrete value associated with node P_i, and does not directly have a physical interpretation. Similarly, the coefficient β_i cannot be attributed any physical meaning either. As depicted in Fig. 5, a property similar to (4) exists however,

$$\int_{\text{span}} M_i(x)dx = 1. \tag{5}$$

These type of *edge* functions $\{M_i(x)\}_{i=1}^n$ were first derived by Gerritsma [7], from Lagrange polynomials $\{N_i(x)\}_{i=0}^n$, see Fig. 4. Buffa et. al in [3], on the other hand, made clever use of the fact that B-splines are naturally induced by such a discrete

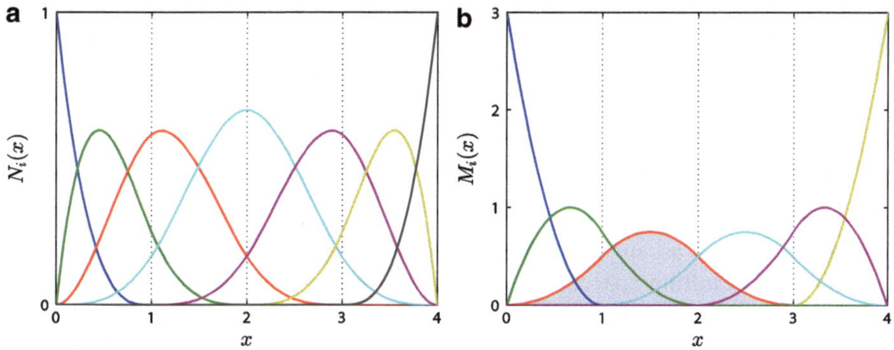

Fig. 5 Cubic B-splines and derived quadratic *edge* functions on the interval [0, 4]. The *edge* functions derived from B-splines are actually scaled B-splines of 1° less and have the unit integral property: $\int_{\text{supp}} M_i(x)dx = 1$. (**a**) B-spline *node* functions. (**b**) B-spline *edge* functions

structure; with B-splines $\{N_i(x)\}_{i=0}^{n}$ of order p, the *edge* functions $\{M_i(x)\}_{i=1}^{n}$ turn out to be scaled B-splines of order $p-1$, see Fig. 5. *Edge* functions based on Lagrange polynomials have been successfully employed in [10, 11, 13, 14, 16], while those based on B-splines, in [3, 5, 6].

The main contribution of this paper is the generalization of the property in (2) to any basis which constitutes a partition of unity. Hence, a common framework for high order compatible discretizations can be built, independent of the particular basis chosen.

Multivariate spaces can readily be constructed using tensor products of the univariate *node* $\{N_i(x)\}_{i=0}^{n}$, and *edge* functions $\{M_i(x)\}_{i=1}^{n}$. In $\Omega' \subset \mathbb{R}^2$ we can define the following finite dimensional spaces,

$$\mathscr{V}_h^{(P)}(\Omega') := \operatorname{span}\left\{N_i(x^1)N_j(x^2)\right\}_{i=0,j=0}^{n,m} \tag{6}$$

$$\mathscr{V}_h^{(L)}(\Omega') := \operatorname{span}\left\{N_i(x^1)M_j(x^2)\right\}_{i=0,j=1}^{n,m} \times \operatorname{span}\left\{M_i(x^1)N_j(x^2)\right\}_{i=1,j=0}^{n,m} \tag{7}$$

$$\mathscr{V}_h^{(S)}(\Omega') := \operatorname{span}\left\{M_i(x^1)M_j(x^2)\right\}_{i=1,j=1}^{n,m}. \tag{8}$$

Here the basis functions in $\mathscr{V}_h^{(P)}(\Omega') \subset H^1(\Omega')$ are associated with mesh nodes; those in $\mathscr{V}_h^{(L)}(\Omega') \subset \mathbf{H}(\div)(\Omega')$ with mesh edges and basis functions in $\mathscr{V}_h^{(S)}(\Omega') \subset L^2(\Omega')$ with mesh faces. The generalization to $\mathbb{R}^3$ is straightforward. In this case we also have tensor product basis functions associated with sub-volumes in the mesh.

Remark 6. The gradient, curl and divergence can be applied by differencing the degrees of freedom, as in Fig. 2, and subsequently taking the appropriate linear combination with *node*, *edge*, *face* or *volume* functions [6, 7].

4 Application to Stokes Flow in a Lid Driven Cavity

We take B-splines as an initial basis to derive *node*, *edge* and *face* functions, and apply the resulting spaces in (6)–(8) to the vorticity-velocity-pressure (VVP) Galerkin formulation of Stokes flow [6, 10, 11] in a lid driven cavity. Since no analytical solution to this problem exist, we compare with the benchmark results of [15].

Consider domain $\Omega = [0,1]^2$ filled with an in-compressible fluid of viscosity $\nu = 1$. On the top of the domain we apply a tangential velocity $u(x,1) = 1$, while on the other sides of the domain the tangential velocity is set to zero. The result is a clockwise rotating flow with small counter-rotating eddies in the two lower corners. Because of the discontinuity of the velocity in the two upper corners, both the vorticity and pressure are infinite at these places, which makes the lid driven cavity flow a challenging test case. Under the stated assumptions, the Stokes flow in Ω can be described by the following equations,

$$-\Delta \mathbf{u} + \text{grad}^\star\ p = \mathbf{f} \qquad \text{and} \qquad \div\ \mathbf{u} = 0, \tag{9}$$

where $\mathbf{u}$ is the velocity field, p the pressure, and $\mathbf{f}$ the right hand side forcing. Using the operator splitting $-\Delta\mathbf{u} = \text{rot curl}^\star\mathbf{u} - \text{grad}^\star \div\ \mathbf{u}$, the in-compressibility constraint, $\div\ \mathbf{u} = 0$, and introducing the vorticity ω, we obtain the first order system,

$$\omega = \text{curl}^\star\mathbf{u}, \qquad \text{rot}\ \omega + \text{grad}^\star p = \mathbf{f} \qquad \div\ \mathbf{u} = 0 \tag{10}$$

Remark 7. curl* and grad* are defined as the L^2 adjoint operators of rot and $\div$, respectively. While the former requires approximation, the latter allows an exact discrete representation in terms of a sparse difference matrix, see Fig. 2.

Multiplying the equations in (10) by test functions α, $\boldsymbol{\beta}$ and γ and using integration by parts to replace the curl* and grad* by their L^2 adjoint operators rot and $\div$ [6, 10, 11], we obtain the mixed problem: find $\{\omega \in H^1(\Omega), \mathbf{u} \in \mathbf{H}(\div)(\Omega), p \in L^2(\Omega)\}$ for all $\{\alpha \in H^1(\Omega), \boldsymbol{\beta} \in \mathbf{H}_0(\div)(\Omega), \gamma \in L^2(\Omega)\}$, such that

$$-(\alpha,\omega)_\Omega + (\text{rot}\alpha, \mathbf{u})_\Omega = \int_{\partial\Omega} \alpha\ \mathbf{u} \times \mathbf{n}\ d\Gamma \tag{11}$$

$$(\boldsymbol{\beta}, \text{rot}\ \omega)_\Omega + (\div\boldsymbol{\beta}, p)_\Omega = (\boldsymbol{\beta}, \mathbf{f})_\Omega \tag{12}$$

$$(\gamma, \div\ \mathbf{u})_\Omega = 0 \tag{13}$$

By substituting the infinite dimensional spaces by their finite dimensional counter parts, $\left\{\alpha_h, \omega_h \in \mathscr{V}_h^{(P)}(\Omega'), \boldsymbol{\beta}_h, \mathbf{u}_h \in \mathscr{V}_h^{(L)}(\Omega'), \gamma_h, p_h \in \mathscr{V}_h^{(S)}(\Omega')\right\}$, we obtain the following system of algebraic equations,

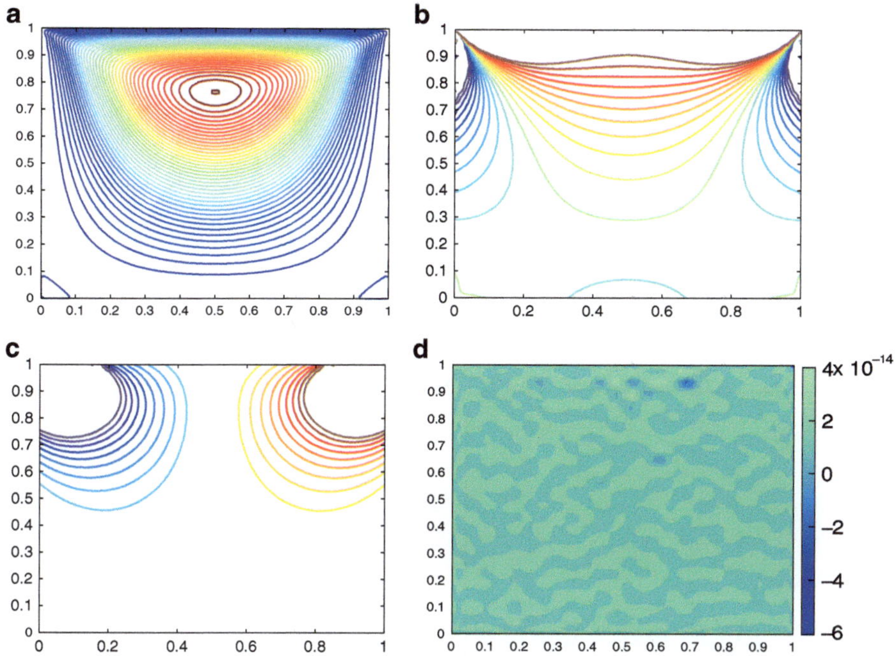

Fig. 6 Lid driven cavity flow using a 60 × 60 uniform bi-cubic B-spline mesh. Note that the divergence of the velocity field is at machine precision. (**a**) stream function. (**b**) vorticity. (**c**) pressure. (**d**) divergence of velocity

$$\begin{pmatrix} -\mathrm{M}_{(P)} & \mathrm{D}^T_{(\mathrm{rot})}\mathrm{M}_{(L)} & \oslash \\ \mathrm{M}_{(L)}\mathrm{D}_{(\mathrm{rot})} & \oslash & \mathrm{D}^T_{(\mathrm{div})}\mathrm{M}_{(S)} \\ \oslash & \mathrm{M}_{(S)}\mathrm{D}_{(\mathrm{div})} & \oslash \end{pmatrix} \begin{pmatrix} \boldsymbol{\omega}(P) \\ \mathbf{u}(L) \\ \mathbf{p}(S) \end{pmatrix} = \begin{pmatrix} \mathrm{B}_{(L)}(\mathbf{u}) \\ \mathrm{M}_{(L)}(\mathbf{f}) \\ \oslash \end{pmatrix} \tag{14}$$

Remark 8. The degrees of freedom of the vorticity $\boldsymbol{\omega}(P)$ are associated with mesh nodes P; the velocity degrees of freedom $\mathbf{u}(L)$ with mesh edges L; and the degrees of freedom of the pressure $\mathbf{p}(S)$ with mesh faces S.

Remark 9. $\mathrm{D}_{(\mathrm{rot})}$ and $\mathrm{D}_{(\div)}$ are sparse matrix representations of the rot and $\div$, see Fig. 2, which are exact and completely metric free.

Remark 10. $\mathrm{M}_{(P)}$, $\mathrm{M}_{(L)}$ and $\mathrm{M}_{(S)}$ are mass matrices resulting from an inner product on $\mathscr{V}_h^{(P)}$, $\mathscr{V}_h^{(L)}$ and $\mathscr{V}_h^{(S)}$, and are computed using Gauss numerical integration.

Remark 11. The problem is closed by imposing the normal velocity at the boundary strongly, while the tangential velocity is weakly enforced (boundary term $\mathrm{B}_{(L)}(\mathbf{u})$). Finally, given the small size of (14) we use a direct solver.

Figure 6 shows results for the stream function, vorticity, pressure and the divergence of the velocity, on a bi-cubic uniform B-spline mesh of maximum regularity and 60 × 60 degrees of freedom. Observe that the divergence of the

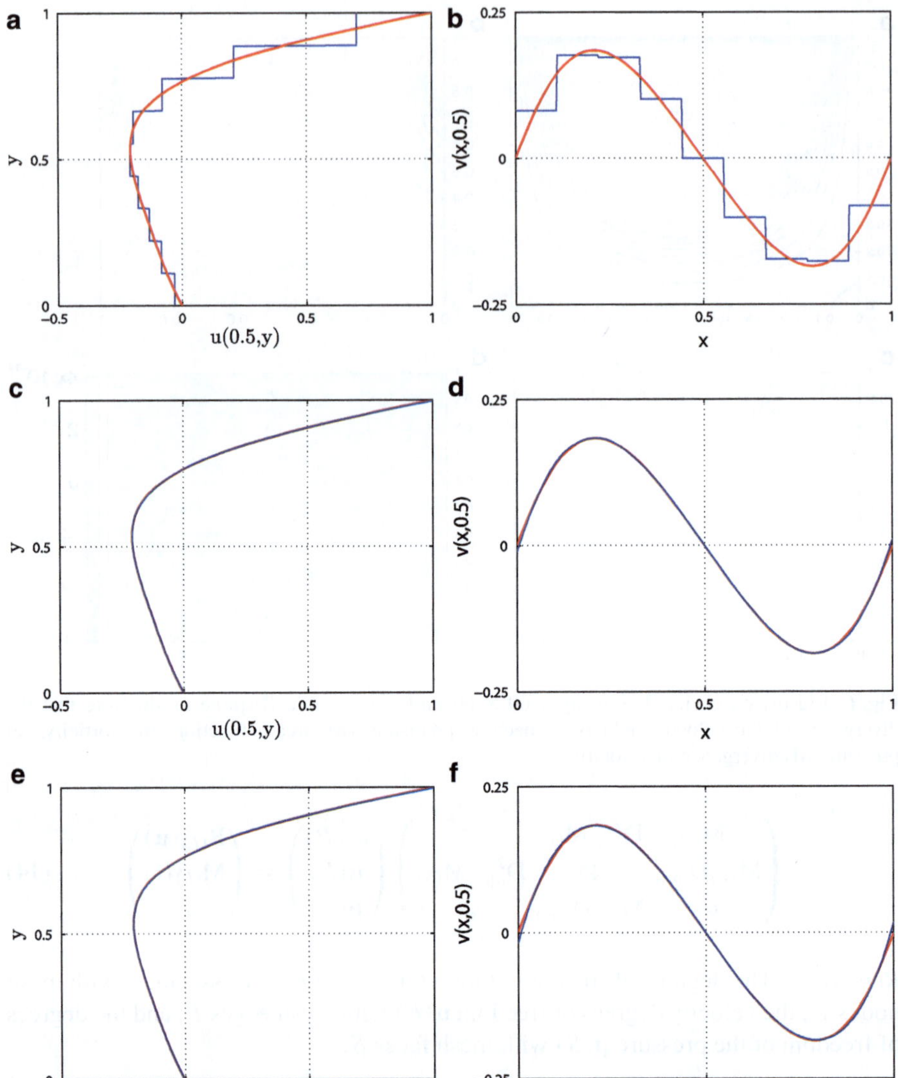

Fig. 7 Comparison of numerical approximation (*blue*) with benchmark results of [15] (*red*). Horizontal (**a,b,c**) and vertical (**d,e,f**) velocity profile at center-lines of cavity for a 9 × 9 uniform B-spline mesh of order 1, 3 and 5. (**a**) 9x9, $p = 1$. (**b**) 9x9, $p = 1$. (**c**) 9x9, $p = 3$. (**d**) 9x9, $p = 3$. (**e**) 9x9, $p = 5$. (**f**) 9x9, $p = 5$

velocity is point-wise zero in the whole domain. Furthermore no special treatment has been given to the corner singularities.

In Fig. 7 the horizontal component of the velocity has been plotted along the vertical center-line $(0.5, y)$ and the vertical component along the horizontal centerline $(x, 0.5)$. The results confirm very well with the benchmark results of [15] even though the mesh is very coarse (9×9 uniform B-spline grid of polynomial order 1, 3 and 5). Note that the integral values seem to match very well, already for the lowest order approximation. This is a direct consequence of the conservation properties that we have build into the basis. In fact, it does not depend on the specific type of basis functions used, but rather on the relations between the different function spaces. Similar results have been obtained in [11], where the spaces in (6,7,8) are derived from Lagrange polynomials.

References

1. D.N. Arnold, R.S. Falk, and R. Winther. Finite element exterior calculus, homological techniques, and applications. *Acta numerica*, 15:1–156, 2006.
2. P. Bochev and J. Hyman. Principles of mimetic discretizations of differential operators. *Compatible spatial discretizations*, pages 89–119, 2006.
3. A. Buffa, J. Rivas, G. Sangalli, and R. Vázquez. Isogeometric discrete differential forms in three dimensions. *SIAM Journal on Numerical Analysis*, 49:818, 2011.
4. M. Desbrun, A.N. Hirani, M. Leok, and J.E. Marsden. Discrete exterior calculus. *Arxiv preprint math/0508341*, 2005.
5. J.A. Evans and T.J. Hughes. Isogeometric divergence-conforming B-splines for the steady Navier-Stokes equations. *Technical report, DTIC Document*, 2012.
6. R.R. Hiemstra, R.H.M. Huijsmans and M.I. Gerritsma. High order gradient, curl and divergence conforming spaces, with an application to compatible IsoGeometric Analysis. *Arxiv preprint arXiv:1209.1793, submitted to Journal of Computational Physics*, 2012.
7. M. Gerritsma. Edge functions for spectral element methods. Spectral and High Order Methods for Partial Differential Equations *Lecture Notes in Computational Science and Engineering, Springer Berlin Heidelberg*, pages 199–207, 2011.
8. Marc Gerritsma, René Hiemstra, Jasper Kreeft, Artur Palha, Pedro Rebelo, Deepesh Toshniwal. The geometric basis of numerical methods. *Proceedings ICOSAHOM,* 2012.
9. J. Kreeft, A. Palha, and M. Gerritsma. Mimetic framework on curvilinear quadrilaterals of arbitrary order. *Arxiv preprint arXiv:1111.4304*, 2011.
10. Jasper Kreeft and Marc Gerritsma, Higher-order compatible discretization on hexahedrals. *Proceedings ICOSAHOM,* 2012.
11. J. Kreeft and M. Gerritsma. Mixed mimetic spectral element method for Stokes flow: A pointwise divergence-free solution. *Journal of Computational Physics*, 240,284–309, 2013.
12. J.B. Perot. Discrete conservation properties of unstructured mesh schemes. *Annual Review of Fluid Mechanics*, 43:299–318, 2011.
13. Artur Palha, Pedro Pinto Rebelo and Marc Gerritsma, Mimetic Spectral Element Advection. *Proceedings ICOSAHOM,* 2012.
14. Pedro Pinto Rebelo, Artur Palha and Marc Gerritsma, Mixed Mimetic Spectral Element method applied to Darcy's problem. *Proceedings ICOSAHOM,* 2012.
15. M. Sahin and R.G. Owens. A novel fully implicit finite volume method applied to the lid-driven cavity problem, part i: High Reynolds number flow calculations. *International journal for numerical methods in fluids*, 42(1):57–77, 2003.
16. Deepesh Toshniwal and Marc Gerritsma, A Geometric Approach Towards Momentum Conservation. *Proceedings ICOSAHOM,* 2012.

Spectral Element Discretization for the Vorticity, the Velocity and the Pressure Formulation of the Axisymmetric Navier-Stokes Problem

Chahira Jerbi and Nahla Abdellatif

Abstract We deal with the Navier-Stokes equations set in a three-dimensional axisymmetric bounded domain with non standard boundary conditions which involve the normal component of the velocity and tangential component of the vorticity. The axisymmetric property of the domain allows to reduce the three-dimensional problem into a two-dimensional one. We write a variational formulation with three independent unknowns: the vorticity, the velocity and the pressure. For the discretization, we use the spectral element methods, which are well-adapted here. We show the well-posedness of the obtained formulations and we establish error estimates for the three unknowns which proves the convergence of the method.

1 Introduction

We consider, in this paper, the Navier-Stokes problem set in a three-dimensional axisymmetric bounded domain and provided with non standard boundary conditions, which are given on the normal component of the velocity and tangential component of the vorticity. This problem reads:

C. Jerbi
Université El Manar, ENIT – LAMSIN, BP 137, Le Belvédère 1002, Tunis, Tunisia
e-mail: chahira20092009@hotmail.fr

N. Abdellatif (✉)
Université de Manouba, ENSI, Campus Universitaire, 2010 Manouba, Tunisia

Université El Manar, ENIT – LAMSIN, BP 137, Le Belvédère 1002, Tunis, Tunisia
e-mail: nahla.abdellatif@ensi.rnu.tn

M. Azaïez et al. (eds.), *Spectral and High Order Methods for Partial Differential Equations - ICOSAHOM 2012*, Lecture Notes in Computational Science and Engineering 95, DOI 10.1007/978-3-319-01601-6_24,

$$\begin{cases} -\nu\Delta\tilde{u} + (\tilde{u}.\nabla)\tilde{u} + \nabla\tilde{P} = \tilde{f} \text{ in } \tilde{\Omega}, \\ \text{div}\tilde{u} = 0 \text{ in } \tilde{\Omega}, \\ \tilde{u}.\tilde{n} = 0 \text{ on } \partial\tilde{\Omega}, \\ \text{curl}\tilde{u} \wedge \tilde{n} = 0 \text{ on } \partial\tilde{\Omega}. \end{cases} \tag{1}$$

where $\tilde{\Omega}$ is a bounded connected three-dimensional axisymmetric domain, the generic point in $\tilde{\Omega}$ is given by cylindrical components $(r, \theta, z) \in \mathbb{R}_+ \times]-\pi, \pi] \times \mathbb{R}$. ν is the viscosity of the fluid, $\tilde{u} = (u_r, u_\theta, u_z)$ the velocity, $\tilde{P}$ the pressure and $\tilde{f}$ is the data, which represent the density of body forces. When the data is axisymmetric, problem (1) is equivalent to two decoupled systems [9]. In the first one, the unknowns are the components u_r and u_z of the velocity and pressure P, we will focus on. The second is a Laplace problem where the unknown is the velocity component u_θ.

At first, this problem was studied in [1] but in an unspecified bounded domain, then it was taken again by Azaiez et al. [10] in a bounded domain included in $\mathbb{R}^2$ or $\mathbb{R}^3$ in formulation (u, p), though the formulation that we consider here deals with three unknowns: vorticity, velocity and pressure. The first numerical analysis relying on this formulation has been realized in [13] and [8] for finite element methods and it has been extended to the case of spectral methods in [3] and [10], using analogues of Nédélec's finite elements [6].

The discretization method which we use here is the spectral element methods, which are well adapted in domain decomposition. The main tool for the analysis of the nonlinear discrete problem is the theorem of Brezzi, Rappaz and Raviart [5]. We first prove the existence of a discrete solution. Then, by combining the results in [5, 11] and [7], we establish error estimates between the continuous solution and the discrete one, for the three unknowns.

The paper is organized as follows. In the next section, we introduce the variational formulation corresponding to the Navier-Stokes problem and we derive the existence of a solution. In Sect. 3, we study the discrete problem and we prove the well-posedness of this problem. We derive error estimates between the continuous solution and the discrete one in Sect. 4.

2 The Vorticity, Velocity and Pressure Formulation

The domain $\tilde{\Omega}$ is obtained by rotating a two-dimensional domain Ω around the axis $\{r = 0\}$. We note by Γ_0 the intersection of the boundary $\partial\Omega$ with the axis $r = 0$, $\Gamma = \partial\Omega \backslash \Gamma_0$ and by n the normal to Γ in the plane (r, z). We introduce the vorticity ω as a new unknown: $\omega = \text{curl}u$. The bidimensional problem resulting from (1) reads:

$$\begin{cases} \nu \text{curl}_r \omega + \omega \times u + \nabla P = f & \text{in } \Omega, \\ \text{div}_r u = 0 & \text{in } \Omega, \\ \omega = \text{curl} u & \text{in } \Omega, \\ u \cdot n = 0 & \text{on } \Gamma, \\ \omega = 0 & \text{on } \Gamma. \end{cases} \tag{2}$$

The operators div_r, curl and curl_r are given by: for $u = (u_r, u_z)$, $\text{div}_r u = \partial_r u_r + r^{-1} u_r + \partial_z u_z$ and $\text{curl} u = \partial_r u_z - \partial_z u_r$. And for any scalar function φ, we define $\text{curl}_r \varphi = \big(\partial_z \varphi, -r^{-1}\partial_r(r\varphi)\big)$. We refer to [11], for details.

In order to write the variational formulation of problem (2), we define the following weighted Sobolev spaces: For all s in $\mathbb{Z}$ and m in $\mathbb{N}$:

$$L_s^2(\Omega) = \left\{ v : \Omega \to \mathbb{R} \quad \text{measurable} \quad / \int_\Omega |v(r,z)|^2 r^s \, drdz < \infty \right\}$$

$$H_1^m(\Omega) = \left\{ v \in L_1^2(\Omega) \quad / \quad \partial_r^l \partial_z^{m-l} v \in L_1^2(\Omega) \quad \forall 0 \le l \le m \right\},$$

$$H_1(\text{curl}, \Omega) = \left\{ v \in L_1^2(\Omega)^2 \quad / \quad \text{curl} v \in L_1^2(\Omega) \right\},$$

$$H_1(\text{div}_r, \Omega) = \left\{ v \in L_1^2(\Omega)^2 \quad / \quad \text{div}_r v \in L_1^2(\Omega) \right\},$$

$$H_1^\diamond(\text{div}_r, \Omega) = \{ v \in H_1(\text{div}_r, \Omega) / \quad v \cdot n = 0 \quad \text{on} \quad \Gamma \},$$

$$H_1(\text{curl}_r, \Omega) = \left\{ \varphi \in L_1^2(\Omega) \quad / \quad \text{curl}_r \varphi \in L_1^2(\Omega)^2 \right\},$$

$$V_1^1(\Omega) = H_1^1(\Omega) \cap L_{-1}^2(\Omega) \quad \text{and} \quad V_{1\diamond}^1(\Omega) = \left\{ v \in V_1^1(\Omega) / \quad v = 0 \quad \text{on} \quad \Gamma \right\}.$$

The spaces $V_1^1(\Omega)$, $H_1(\text{div}_r, \Omega)$ and $H_1(\text{curl}_r, \Omega)$ are respectively provided with:

$$\|v\|_{V_1^1(\Omega)} = (\|\partial_r v\|^2_{L_1^2(\Omega)} + \|\partial_z v\|^2_{L_1^2(\Omega)} + \|v\|^2_{L_{-1}^2(\Omega)})^{\frac{1}{2}},$$

$$\|v\|_{H_1(\text{div}_r,\Omega)} = \left(\|v\|^2_{L_1^2(\Omega)} + \|\text{div}_r v\|^2_{L_1^2(\Omega)} \right)^{\frac{1}{2}},$$

$$\|\varphi\|_{H_1(\text{curl}_r,\Omega)} = \left(\|\varphi\|^2_{L_1^2(\Omega)} + \|\text{curl}_r \varphi\|^2_{L_1^2(\Omega)^2} \right)^{\frac{1}{2}}.$$

We note that the two norms $\|.\|_{H_1(\text{curl}_r,\Omega)}$ and $\|.\|_{V_1^1(\Omega)}$ are equivalent on $V_1^1(\Omega)$.
The variational problem reads:
Find $(\omega, u, p) \in V_{1\diamond}^1(\Omega) \times H_1^\diamond(\text{div}_r, \Omega) \times L_{1,0}^2(\Omega)$ such that:

$$\begin{cases} a(\omega, u; v) + K(\omega, u; v) + b(v, p) = \langle f, v \rangle, \ \forall v \in H_1^\diamond(\text{div}_r, \Omega), \\ b(u, q) = 0, \qquad \forall q \in L_{1,0}^2(\Omega), \\ c(\omega, u, \varphi) = 0, \qquad \forall \varphi \in V_{1\diamond}^1(\Omega). \end{cases} \tag{3}$$

where $\langle .,. \rangle$ is the duality pairing between $H_1^\diamond(\mathrm{div_r}, \Omega)$ and its dual space. The forms $a(.,.;.)$, $b(.,.)$ and $c(.,.;.)$ are defined by:

$$a(\omega, u, \theta) = \nu \int_\Omega (\theta.\mathrm{curl_r}\omega)(r,z) r dr dz, \quad b(v,q) = -\int_\Omega (\mathrm{div_r} v) q(r,z) r dr dz,$$

$$c(\omega, u, \varphi) = \int_\Omega \omega(r,z)\varphi(r,z) r dr dz - \int_\Omega (u.\mathrm{curl_r}\varphi)(r,z) r dr dz.$$

and K is the trilinear form given by: $K(\omega, u; v) = \int_\Omega (\omega \times u) .v(r,z) r dr dz$.
Using density results, we first prove that problems (2) and (3) are equivalent. To prove the existence and the uniqueness of the solution of problem (3), we define the two following kernels V and W:

$$V = \left\{ v \in H_1^\diamond(\mathrm{div_r}, \Omega), \quad \forall q \in L^2_{1,0}(\Omega) \quad / \quad b(v,q) = 0 \right\},$$

$$W = \left\{ (\vartheta, v) \in V^1_{1\diamond}(\Omega) \times V \quad / \quad \forall \varphi \in V^1_{1\diamond}(\Omega), \qquad c(\vartheta, v; \varphi) = 0 \right\},$$

and the reduced problem: Find (ω, u) in W such that:

$$\forall v \in V, \quad a(\omega, u; v) + K(\omega, u; v) = \langle f, v \rangle . \tag{4}$$

By using standard arguments and properties on the linear forms, proven in [3] and [11], we can prove the existence and uniqueness of a solution for problem (4). So for any function f in $H_1^\diamond(\mathrm{div_r}, \Omega)'$ such that

$$c_\diamond \nu^{-2} \|f\|_{H_1^\diamond(\mathrm{div_r}, \Omega)'} < 1, \tag{5}$$

Problem (3) admits a unique solution $(\omega, u; p)$ in $V^1_{1\diamond}(\Omega) \times H_1^\diamond(\mathrm{div_r}, \Omega) \times L^2_{1,0}(\Omega)$, such that

$$\|\omega\|_{V_1^1(\Omega)} + \|u\|_{H_1(\mathrm{div_r}, \Omega)} + \nu^{-1} \|p\|_{L_1^2(\Omega)} \leq c\nu^{-1} \|f\|_{H_1^\diamond(\mathrm{div_r}, \Omega)'} \left(1 + \nu^{-2} \|f\|_{H_1^\diamond(\mathrm{div_r}, \Omega)'}\right). \tag{6}$$

3 Discrete Navier-Stokes Problem

From now on, we assume that Ω is the rectangle $]0,1[\times]-1,1[$ and admits a partition without overlap into a finite number of subdomains:

$$\overline{\Omega} = \bigcup_{k=1}^{K} \Omega_k \quad \text{and} \quad \Omega_k \cap \Omega_{k'} = \emptyset \quad , \quad 1 \leq k < k' \leq K, \text{ such that:}$$

1. Each Ω_k , $1 \le k \le K$ is a rectangle.
2. The intersection between two subdomains $\overline{\Omega}_k$ and $\overline{\Omega}_{k'}$, $1 \le k < k' \le K$, if not empty, is either a vertex or a whole edge of both Ω_k and $\Omega_{k'}$.

The discrete spaces $\mathbb{D}_N$, $\mathbb{C}_N$ and $\mathbb{M}_N$ which approximate, respectively, $H_1^\diamond(\mathrm{div_r}, \Omega)$, $V_{1\diamond}^1(\Omega)$ and $L_{1,0}^2(\Omega)$ are defined from local discrete ones, for an integer $N \ge 2$ and $1 \le k \le K$, by:

$$\mathbb{D}_N = \{v_N \in H_1^\diamond(\mathrm{div_r}, \Omega);\ v_N|_{\Omega_k} \in \mathbb{P}_{N,N-1}(\Omega_k) \times \mathbb{P}_{N-1,N}(\Omega_k),\ 1 \le k \le K\},$$
$$\mathbb{C}_N = \{\varphi_N \in V_{1\diamond}^1(\Omega);\ \varphi_N|_{\Omega_k} \in \mathbb{P}_N(\Omega_k),\ 1 \le k \le K\} \text{ and}$$
$$\mathbb{M}_N = \{q_N \in L_{1,0}^2(\Omega);\ q_N|_{\Omega_k} \in \mathbb{P}_{N-1}(\Omega_k),\ 1 \le k \le K\}.$$

where $\mathbb{P}_{n,m}(\Omega_k)$ is the space of restrictions to Ω_k of polynomials with degree $\le n$ with respect to r and $\le m$ with respect to z, for any nonnegative integers n and m. To calculate the integrals involved in the discrete forms, we define (ξ_i, ρ_i), $0 \le i \le N$ the nodes and weights of the Gauss-Lobatto quadrature formula on $[-1, 1]$ for the measure $d\zeta$ and (ζ_j, ω_j), $1 \le j \le N+1$ their analogues for the measure $(1+\zeta)d\zeta$, see [9] for a more explicit definition, we need two different quadrature formulas. The quadrature formula on $[-1, 1]$ is given by:

$$\forall \phi \in \mathbb{P}_{2N-1}([-1,1]), \quad \int_{-1}^{1} \phi(\xi)d\xi = \sum_{i=0}^{N} \phi(\xi_i)\rho_i, \tag{7}$$

and by setting $r = \frac{1}{2}(1+\zeta)$, we define the quadrature formula with the measure rdr:

$$\forall \phi \in \mathbb{P}_{2N-1}([0,1]), \quad \int_{0}^{1} \phi(r)rdr = \frac{1}{4}\sum_{j=1}^{N+1} \phi(r_j)\omega_j. \tag{8}$$

We denote by $(\Omega_k)_{1 \le k \le K_0}$ the rectangles such that $\partial\overline{\Omega}_k \cap \Gamma_0 \neq \varnothing$ and by $(\Omega_k)_{K_0+1 \le k \le K}$ those such that $\partial\overline{\Omega}_k \cap \Gamma_0 = \varnothing$. Denoting by F_k the affine mapping that sends $]0,1[\times]-1,1[$ onto Ω_k, $1 \le k \le K_0$ and sends $]-1,1[^2$ onto Ω_k, $K_0+1 \le k \le K$. We define the discrete scalar product: For all functions u and v such that $u_k = u|_{\Omega_k}$ and $v_k = v|_{\Omega_k}$ are continuous on $\overline{\Omega}_k$, $1 \le k \le K$, by:

$$((u,v))_N = \sum_{k=1}^{K_0} \frac{mes(\Omega_k)}{4} \sum_{i=0}^{N}\sum_{j=1}^{N+1} u \circ F_k(r_j, \xi_i).v \circ F_k(r_j, \xi_i)\rho_i\omega_j.$$
$$+ \sum_{k=K_0+1}^{K} \frac{mes(\Omega_k)}{4} \sum_{i=0}^{N}\sum_{j=0}^{N} u \circ F_k(\xi_j, \xi_i).v \circ F_k(\xi_j, \xi_i)\rho_i\rho_j.$$

We denote by I_N^k, $1 \le k \le K$, the Lagrange interpolation operators associated with the nodes $F_k(r_j, \xi_i)_{1 \le j \le N+1, 0 \le i \le N}$ for $1 \le k \le K_0$ and with $F_k(\xi_j, \xi_i)_{0 \le j,i \le N}$ for $K_0 + 1 \le k \le K$, with values in $\mathbb{P}_N(\Omega_k)$, $1 \le k \le K$. For each function ϕ continuous on $\bar{\Omega}$, $I_N\phi$ denotes the function such that $I_N\phi|_{\Omega_k} = I_N^k\phi$,

$1 \leq k \leq K$. Using the Galerkin method with numerical integration, we build from the continuous problem (3) the following discrete problem:

Find $(\omega_N, u_N; p_N)$ in $\mathbb{C}_N \times \mathbb{D}_N \times \mathbb{M}_N$ such that

$$\begin{cases} a_N(\omega_N, u_N, v_N) + K_N(\omega_N, u_N, v_N), \\ \qquad\qquad +b_N(v_N, p_N) = ((f, v_N))_N, & \forall v_N \in \mathbb{D}_N, \\ \qquad\qquad b_N(u_N, q_N) = 0, & \forall q_N \in \mathbb{M}_N, \\ \qquad c_N(\omega_N, u_N, \varphi_N) = 0, & \forall \varphi_N \in \mathbb{C}_N. \end{cases} \tag{9}$$

where the bilinear forms $a_N(.,.;.)$, $b_N(.,.)$ and $c_N(.,.;.)$ are defined by: $a_N(\omega_N, u_N; v_N) = \nu((\text{curl}_r\omega_N, v_N))_N$, $\quad b_N(v_N, q_N) = -((\text{div}_r v_N, q_N))_N$, $c_N(\omega_N, u_N, \varphi_N) = ((\omega_N, \varphi_N))_N - ((u_N, \text{curl}_r\varphi_N))_N$, while the trilinear form $K_N(.,.;.)$ is given by: $K_N(\omega_N, u_N; v_N) = ((\omega_N \times u_N, v_N))_N$. In order to prove the well-posedness of the discrete problem, we need to introduce the kernels:

$$V_N = \{v_N \in \mathbb{D}_N / \forall q_N \in \mathbb{M}_N \quad , \quad b_N(vN, q_N) = 0\},$$
$$W_N = \{(\omega_N, u_N) \in \mathbb{C}_N \times V_N / \forall \theta_N \in \mathbb{C}_N, c_N(\omega_N, u_N, \theta_N) = 0\}.$$

We observe that, for any solution (ω_N, u_N, p_N) of problem (9), the pair (ω_N, u_N) is a solution of the reduced problem: Find $(\omega_N, u_N) \in W_N$ such that:

$$\forall v_N \in V_N \quad , \quad a_N(\omega_N, u_N; v_N) + K_N(\omega_N, u_N; v_N) = ((f, v_N))_N. \tag{10}$$

We recall from [4] and [7] that the bilinear form $a_N(.,.;.)$ satisfies, on the discrete spaces, a positivity property and an inf – sup condition with constants independent of N. We also refer to [4], for a discrete inf – sup condition on the form $b_N(.,.)$. Using the fixed point theorem of Brower, we can prove the wellposedness of problem (10) and then derive the:

Theorem 1. *For any data f continuous on $\overline{\Omega}$, the discrete problem (9) admits a solution $(\omega_N, u_N; p_N)$ in $\mathbb{C}_N \times \mathbb{D}_N \times \mathbb{M}_N$. Moreover, (ω_N, u_N) satisfies:*

$$\|\omega_N\|_{L^2_1(\Omega)} + \|u_N\|_{L^2_1(\Omega)^2} \leq c\nu^{-1} \|I_N f\|_{L^2_1(\Omega)^2}. \tag{11}$$

4 Error Estimates

We now intend to prove an error estimate between the solutions of problems (3) and (9). Since the error analysis of the discrete problem relies on the theory of Brezzi, Rappaz and Raviart [5], we express both problems (4) and (10) in a different form and we set $X = V^1_{1\diamond}(\Omega) \times (V \cap H_1(\text{curl}, \Omega))$. We denote by S the linear

operator of Stokes which for any f in the dual space of $H_1^\diamond(\mathrm{div}_r, \Omega)$, associates the solution (ω, u) of the following reduced problem:
Find $(\omega, u) \in W$ such that $\quad \forall v \in V, \quad a(\omega, u; v) = \langle f, v\rangle$.
We introduce the mapping G defined from X into the dual space of $H_1^\diamond(\mathrm{div}_r, \Omega)$ by: $\forall(\omega, u) \in X, \quad \forall v \in H_1^\diamond(\mathrm{div}_r, \Omega), \langle G(\omega, u), v\rangle = K(\omega, u; v) - \langle f, v\rangle$.
Then, problem (4) can be equivalently written as: Find $(\omega, u) \in X$ such that

$$(\omega, u) + SG(\omega, u) = 0. \tag{12}$$

Similarly, we define the discrete space $X_N = \mathbb{C}_N \times (V_N \cap H_1(\mathrm{curl}, \Omega))$. We thus define the discrete Stokes operator S_N: for any f in the dual space of $H_1^\diamond(\mathrm{div}_r, \Omega)$, $S_N f$ denotes the solution (ω_N, u_N) of problem: Find $(\omega_N, u_N) \in W_N$ such that

$$\forall v_N \in V_N, \quad a_N(\omega_N, u_N; v_N) = \langle f, v_N\rangle . \tag{13}$$

The well-posedness of problem (13) is proven in [4], for a slightly different right-hand side. Finally, we consider the mapping G_N defined from X_N in the dual space of $\mathbb{D}_N$ by $\quad \forall(\omega_N, u_N) \in X_N, \quad \forall v_N \in \mathbb{D}_N$

$$\langle G_N(\omega_N, u_N), v_N\rangle = K_N(\omega_N, u_N; v_N) - ((f, v_N))_N. \tag{14}$$

Problem (10) can equivalently be written as: $\quad$ Find $(\omega_N, u_N) \in X_N$ such that

$$(\omega_N, u_N) + S_N G_N (\omega_N, u_N) = 0. \tag{15}$$

Using analogous arguments to those in [4], we easily derive that the operator S_N satisfies a stability property, with a constant independent of N and that, the following error estimate holds for all f in $H_1^{s+1}(\Omega) \times H_1^s(\Omega)^2$, $s > 1$,

$$\|(S - S_N) f\|_X \leq cN^{-s} \|Sf\|_{H_1^{s+1}(\Omega)\times H_1^s(\Omega)^2} . \tag{16}$$

We are led to make the following assumptions. Here, D is the differential operator.

Assumption 1. The triplet (ω, u, p) is a solution of the problem (3) such that the operator $Id + SDG(\omega, u)$ is an isomorphism of X.

This assumption can equivalently be written as: For any data g in $H_1^\diamond(\mathrm{div}_r, \Omega)'$, the linearized problem
Find (ϑ, w, r) in $V_{1\diamond}^1(\Omega) \times \left(H_1^\diamond(\mathrm{div}_r, \Omega) \cap H_1(\mathrm{curl}, \Omega)\right) \times L_{1,0}^2(\Omega)$ such that:

$$\begin{cases} a(\vartheta, w, v) + K(\omega, w, v) + K(\vartheta, u; v) \\ \qquad\qquad + b(v, r) = \langle g, v\rangle , & \forall v \in H_1^\diamond(\mathrm{div}_r, \Omega) \cap H_1(\mathrm{curl}, \Omega), \\ \qquad\qquad b(w, q) = 0, & \forall q \in L_{1,0}^2(\Omega), \\ \qquad\qquad c(\vartheta, w, \varphi) = 0, & \forall \varphi \in V_{1\diamond}^1(\Omega). \end{cases} \tag{17}$$

has a unique solution with norm bounded by a constant times $\|g\|_{H_1^\diamond(\mathrm{div}_r,\Omega)}$. It yields the local uniqueness of the solution (ω, u, p) but is much less restrictive than the global uniqueness condition. We need to prove a few technical results in order to derive the error estimate. For this, we make the:

Assumption 2. The solution (ω, u, p) of problem (3) satisfying Assumption 1, belongs to $H_1^{s+1}(\Omega) \times H_1^s(\Omega)^2 \times H_1^s(\Omega)$, $s > 1$.

Then, we prove:

Lemma 1. *For any $(\omega_N, u_N; v_N)$ in $\mathbb{C}_N \times \mathbb{D}_N \times \mathbb{D}_N$,*

$$|K(\omega_N, u_N; v_N)| \leq c_1 N \, \|\omega_N\|_{V_1^1(\Omega)} \, \|u_N\|_{L_1^2(\Omega)^2} \, \|v_N\|_{L_1^2(\Omega)^2}, \tag{18}$$

$$|K_N(\omega_N, u_N; v_N)| \leq c_2 N \, \|\omega_N\|_{V_1^1(\Omega)} \, \|u_N\|_{L_1^2(\Omega)^2} \, \|v_N\|_{L_1^2(\Omega)^2}. \tag{19}$$

the constants c_1 and c_2 are independent of N.

Proof. According to the Cauchy-Schwarz inequality we have:

$$|K(\omega_N, u_N; v_N)| = \left|\int_\Omega (\omega_N \times u_N).v_N r dr dz\right| \leq \|\omega_N\|_{L_1^4(\Omega)} \, \|u_N\|_{L_1^4(\Omega)^2} \, \|v_N\|_{L_1^2(\Omega)^2}.$$

Using the inclusion of $V_1^1(\Omega)$ in $L_1^4(\Omega)$ and inequality:

$$\forall z_N \in \mathbb{P}_N(\Omega), \qquad \|z_N\|_{L_1^4(\Omega)} \leq cN \, \|z_N\|_{L_1^2(\Omega)}, \tag{20}$$

see [2], we have the first previous result. For the second one, we have with obvious notation,

$$\begin{aligned} K_N(\omega_N, u_N; v_N) &= ((\omega_N u_{Nr}, v_{Nz}))_N - ((\omega_N u_{Nz}, v_{Nr}))_N \\ &= ((I_N(\omega_N u_{Nr}), v_{Nz}))_N - ((I_N(\omega_N u_{Nz}), v_{Nr}))_N. \end{aligned}$$

By combining the Cauchy-Schwarz inequalities with inequality (3.7) in [7], we obtain

$$|K_N(\omega_N, u_N; v_N)| \leq \|I_N(\omega_N u_N)\|_{L_1^2(\Omega)^2} \, \|v_N\|_{L_1^2(\Omega)^2}.$$

Then, we use the following result which can be derived from its one-dimensional analogue [7],

$$\forall \varphi_M \in \mathbb{P}_M(\Omega_k), \quad \left\|I_N^k \varphi_M\right\|_{L_1^2(\Omega_k)} \leq c(1 + \frac{M}{m(N)})^2 \, \|\varphi_M\|_{L_1^2(\Omega_k)},$$

with $m(N) = E((1+\delta)N)$ and δ a real number between 0 and 1. We conclude, by using the inequalities (20) and $\|\omega_N u_N\|_{L^2_1(\Omega)^2} \leq \|\omega_N\|_{L^4_1(\Omega)} \|u_N\|_{L^4_1(\Omega)^2}$, together with the continuous inclusion of $V^1_1(\Omega)$ in $L^4_1(\Omega)$, that

$$\|\omega_N u_N\|_{L^2_1(\Omega)^2} \leq cN \|\omega_N\|_{V^1_1(\Omega)} \|u_N\|_{L^2_1(\Omega)^2}.$$

Remark 1. Similar arguments lead to estimate (18), if at most two of the three functions ω_N, u_N and v_N are replaced by their analogues ω in $V^1_{1\diamond}(\Omega)$, u and v in $\mathbb{D}(\Omega)$.

Remark 2. Under Assumption 2 and taking $\tilde{N} = E(2\delta N - 1)$, we can find $(\tilde{\omega}_N, \tilde{u}_N)$ in $\mathbb{C}_{\tilde{N}} \times V_{\tilde{N}}$ such that:

$$\|(\omega - \tilde{\omega}_N, u - \tilde{u}_N)\|_X \leq c\tilde{N}^{-s} \|(\omega, u)\|_{H^{s+1}_1(\Omega)\times H^s_1(\Omega)^2}, s > 1. \tag{21}$$

Note that estimate (21) makes sense only when $\tilde{N} \geq 2$.

Lemma 2. *If Assumptions 1 and 2 hold, there exists an integer N_0 such that, for all $N \geq N_0$, the operator $Id + S_N DG_N(\tilde{\omega}_N, \tilde{u}_N)$ is an isomorphism of X_N. Moreover, the norm of its inverse operator is bounded independently of N.*

Proof. We can write that:

$$\begin{aligned} Id + S_N DG_N(\tilde{\omega}_N, \tilde{u}_N) &= Id + SDG(\omega, u) - (S - S_N)DG(\omega, u) \\ &- S_N(DG(\omega, u) - DG(\tilde{\omega}_N, \tilde{u}_N)) - S_N(DG(\tilde{\omega}_N, \tilde{u}_N) - DG_N(\tilde{\omega}_N, \tilde{u}_N)). \end{aligned} \tag{22}$$

It follows from the definition of G and G_N that, for all (θ_N, ω_N) in X_N and v_N in V_N:
$\langle DG(\tilde{\omega}_N, \tilde{u}_N).(\theta_N, w_N), v_N\rangle = K(\tilde{\omega}_N, w_N; v_N) + K(\theta_N, \tilde{u}_N; v_N)$, and
$\langle DG_N(\tilde{\omega}_N, \tilde{u}_N).(\theta_N, w_N), v_N\rangle = K_N(\tilde{\omega}_N, w_N; v_N) + K_N(\theta_N, \tilde{u}_N; v_N)$.
Thanks to the choice of $(\tilde{\omega}_N, \tilde{u}_N)$, the term $S_N(DG(\tilde{\omega}_N, \tilde{u}_N) - DG_N(\tilde{\omega}_N, \tilde{u}_N))$ vanishes. Then, using the stability of S_N, we can derive that:

$$\begin{aligned} &\|S_N(DG(\omega, u) - DG(\tilde{\omega}_N.\tilde{u})).(\theta_N, w_N)\|_X \\ &\qquad \leq c \sup_{v_N \in V_N} \frac{K(\omega - \tilde{\omega}_N, w_N, v_N) + K(\theta_N, u - \tilde{u}_N, v_N)}{\|v_N\|_{L^2_1(\Omega)^2}}. \end{aligned}$$

By Lemma 1, we have:

$$\begin{aligned} &\|S_N(DG(\omega, u) - DG(\tilde{\omega}_N.\tilde{u}_N)).(\theta_N, w_N)\|_X \\ &\qquad \leq cN \left(\|\omega - \tilde{\omega}_N\|_{V^1_1(\Omega)} \|w_N\|_{L^2_1(\Omega)^2} + \|\theta_N\|_{V^1_1(\Omega)} \|u - \tilde{u}_N\|_{L^2_1(\Omega)^2}\right). \end{aligned} \tag{23}$$

Estimate (21) leads to

$$\lim_{N\to+\infty} \|S_N(DG(\omega,u) - DG(\tilde{\omega}_N.\tilde{u}_N))\|_{L(X_N)} = 0. \tag{24}$$

Finally, it follows from Assumption 2 that, when (θ, w) runs through the unit ball of X, $DG(\omega,u)(\theta,w)$ belongs to a compact subset of $L^2_1(\Omega)^2$. So, the next property is derived from the stability of S_N and from inequality (16) by standard arguments:

$$\lim_{N\to+\infty} \|(S - S_N)\, DG(\omega,u)\|_{L(X_N)} = 0. \tag{25}$$

Thanks to Assumption 1, for $\gamma = \left\|(Id + SDG(\omega,u))^{-1}\right\|_{L(X)}$, and by choosing N large enough so that the quantities in (24) and (25) are smaller than $\frac{1}{4\gamma}$, we obtain the desired property with $\left\|(Id + S_N DG_N(\tilde{\omega}_N,\tilde{u}_N))^{-1}\right\|_{L(X_N)} < 2\gamma$.

Lemma 3. *The following Lipschitz property holds:* $\forall(\omega_N^*, u_N^*) \in X_N$,

$$\left\|S_N\left(DG_N(\tilde{\omega}_N,\tilde{u}_N) - DG_N(\omega_N^*,u_N^*)\right)\right\|_{L(X_N)} \le cN\left\|\left(\tilde{\omega}_N - \omega_N^*, \tilde{u}_N - u_N^*\right)\right\|_X. \tag{26}$$

Proof. We just note that

$$\begin{aligned}&\left\langle\left(DG_N(\tilde{\omega}_N,\tilde{u}_N) - DG_N(\omega_N^*,u_N^*)\right).(\theta_N,w_N),v_N\right\rangle\\ &\qquad = K_N(\tilde{\omega}_N - \omega_N^*, w_N; v_N) + K_N(\theta_N, \tilde{u}_N - u_N^*; v_N).\end{aligned}$$

Lemma 1 leads to the desired property.

Lemma 4. *Assume that the data $f \in H_1^\sigma(\Omega)^2$, $\sigma > \frac{3}{2}$. Under Assumption 2,*

$$\begin{aligned}&\|(\tilde{\omega}_N,\tilde{u}_N) + S_N G_N(\tilde{\omega}_N,\tilde{u}_N)\|_X\\ &\qquad \le c(\omega,u)\left(N^{-s}\|(\omega,u)\|_{H_1^{s+1}(\Omega)\times H_1^s(\Omega)^2} + N^{-\sigma}\|f\|_{H_1^\sigma(\Omega)^2}\right),\end{aligned}$$

for a constant $c(\omega,u)$ only depending on the solution (ω,u).

Proof. From (12), we derive

$$\begin{aligned}\|(\tilde{\omega}_N,\tilde{u}_N) + S_N G_N(\tilde{\omega}_N,\tilde{u}_N)\|_X &\le \|(\omega - \tilde{\omega}_N, u - \tilde{u}_N)\|_X + \|(S - S_N)\, G(\omega,u)\|_X\\ &+ \|S_N(G(\omega,u) - G(\tilde{\omega}_N,\tilde{u}_N))\|_X + \|S_N(G(\tilde{\omega}_N,\tilde{u}_N) - G_N(\tilde{\omega}_N,\tilde{u}_N))\|_X\end{aligned}$$

The bound for the first term in the right-hand side obviously follows from (21). From estimate (16) with Assumption 2, we also derive

$$\|(S - S_N)\, G(\omega,u)\|_X \le cN^{-s}\|(\omega,u)\|_{H_1^{s+1}(\Omega)\times H_1^s(\Omega)^2}.$$

On the other hand,

$$K(\omega,u;v_N)-K(\tilde\omega_N,\tilde u_N;v_N)=K(\omega-\tilde\omega_N,u;v_N)+K(\omega,u-\tilde u_N;v_N)\\ -K(\omega-\tilde\omega_N,u-\tilde u_N;v_N).$$

So, we have from the stability property on S_N

$$\|S_N\left(G\left(\omega,u\right)-G\left(\tilde\omega_N,\tilde u_N\right)\right)\|_X\le c\sup_{v_N\in V_N}\frac{\langle K(\omega,u;v_N)-K(\tilde\omega_N,\tilde u_N;v_N),v_N\rangle}{\|v_N\|_{L^2_1(\Omega)^2}},$$

From (21), Remarks 1 and 2, we have

$$\|S_N\left(G\left(\omega,u\right)-G\left(\tilde\omega_N,\tilde u_N\right)\right)\|_X\le c(\omega,u)N^{-s}\,\|(\omega,u)\|_{H_1^{s+1}(\Omega)\times H_1^s(\Omega)^2}\,.$$

We note that $\forall v_N\in\mathbb{D}_N$, the quantities $K(\tilde\omega_N,\tilde u_N;v_N)$ and $K_N(\tilde\omega_N,\tilde u_N;v_N)$ coincide. Then, if Π_{N-1} denotes the orthogonal projection operator from $L^2_1(\Omega)$ onto the space of functions such that their restrictions to all Ω_k, $1\le k\le K$, belong to $\mathbb{P}_{N-1}(\Omega_k)$, and by adding and subtracting the quantity $\Pi_{N-1}f$ in $\|S_N\left(G\left(\tilde\omega_N,\tilde u_N\right)-G_N\left(\tilde\omega_N,\tilde u_N\right)\right)\|_X$, we can prove that $\|S_N\left(G\left(\tilde\omega_N,\tilde u_N\right)-G_N\left(\tilde\omega_N,\tilde u_N\right)\right)\|_X \le c\Big(\|f-\Pi_{N-1}f\|_{L^2_1(\Omega)^2}+\|f -I_Nf\|_{L^2_1(\Omega)^2}\Big)$.

Finally, the standard approximation properties of the operators Π_{N-1} and I_N, lead to

$$\|S_N\left(G\left(\tilde\omega_N,\tilde u_N\right)-G_N\left(\tilde\omega_N,\tilde u_N\right)\right)\|_X\le cN^{-\sigma}\,\|f\|_{H^\sigma(\Omega)^2}\,.$$

The desired bound is then derived by combining the previous estimates.

We are now in a position to prove the error estimate.

Theorem 2. *We assume that the data f is in $H_1^\sigma(\Omega)^2$, $\sigma>\frac{3}{2}$, and that the solution (ω,u,p), of problem (3) satisfies Assumptions 1 and 2.*

Then, there exists an integer $N_\diamond$ and a constant $c_\diamond$ such that for any $N\ge N_\diamond$, the problem (9) has a unique solution (ω_N,u_N,p_N) satisfying the following estimate:

$$\|\omega-\omega_N\|_{V_1^1(\Omega)}+\|u-u_N\|_{H_1(\mathrm{div}_r,\Omega)}+\|p-p_N\|_{L^2_1(\Omega)}\\ \le c(\omega,u)\Big[N^{1-s}\Big(\|\omega\|_{H_1^{s+1}(\Omega)}+\|u\|_{H_1^s(\Omega)^2}+\|p\|_{H_1^s(\Omega)}\Big)+N^{-\sigma}\,\|f\|_{H_1^\sigma(\Omega)^2}\Big]. \tag{27}$$

Proof. Combining Lemmas 2–4 with the Brezzi-Rappaz-Raviart theorem [5], yields that, for N sufficiently large, problem (10) has a unique solution (ω_N,u_N).

Moreover, thanks to the discrete inf-sup condition of $b_N(.,.)$, there exists a unique p_N in $\mathbb{M}_N$ such that

$$\forall v_N \in \mathbb{D}_N, \quad b_N(v_N, p_N) = ((f, v_N))_N - a_N(\omega_N, u_N; v_N) - K_N(\omega_N, u_N; v_N).$$

Hence, the existence and local uniqueness result follows. Moreover,

$$\begin{aligned} \forall q_N \text{ in } \mathbb{M}_N,\ b_N(v_N, p_N - q_N) &= b(v_N, p - q_N) - \langle f, v_N \rangle + ((f, v_N))_N \\ &+ a(\omega - \omega_N, u - u_N; v_N) + (a - a_N)(\omega_N, u_N; v_N) \quad (28) \\ &+ K(\omega, u; v_N) - K_N(\omega_N, u_N; v_N). \end{aligned}$$

so that the estimate for $\|p - p_N\|_{L^2_1(\Omega)}$ follows from the discrete inf-sup condition of $b_N(.,.)$, a triangle inequality and the same arguments as in the proof of Lemma 4.

To conclude, the vorticity-velocity and pressure formulation allows to decouple the calculus of the velocity and the pressure, to handle easily non standard boundary conditions and leads to a more accurate approximation of the pressure. The axisymmetric property of domain allows to move from a three-dimensional problem to a two-dimensional one, which reduces the cost of the resolution. In addition, the tensorization properties of the polynomial spaces, which characterize the spectral methods, enable to inverse the obtained system matrix with a raisonable cost.

References

1. C. Bègue, C. Conca, F. Murat, O. Pironneau : Les équations de Stokes et de Navier- Stokes avec conditions aux limites sur la pression. In Nonlinear Partial Differential Equations and their applications, Collège de France Seminar, Vol. Ix, (1988), pp. 179–264.
2. C. Bernardi, M. Dauge, Y. Maday: Polynomials in the Sobolev world, Rapport interne R03038, Laboratoire Jacques-Louis Lions,Université Pierre et Marie Curie (2003).
3. C. Bernardi, N. Chorfi: Spectral discretization of the vorticity, velocity and pressure formulation of the Stokes problem. SIAM J. Numer. Anal. Vol 44, No 2, 826–850.
4. C. Jerbi, N. Abdellatif: Spectral element discretization of the axisymmetric vorticity, velocity and pressure formulation of the Stokes problem, submitted.
5. F. Brezzi, J. Rappaz, P-A. Raviart: Finite dimensional approximation of nonlinear problems. Part I: Branches of nonsingular solutions, Numer. Math. 36 (1980), 1–25.
6. J.C. Nédélec: Mixed finite Elements in $\mathbb{R}^3$, Numer. Math. 35, 315–341 (1980).
7. K. Amoura, M. Azaiez, C. Bernardi, N. Chorfi, S. Saadi: Spectral element discretization of vorticity, velocity and pressure formulation of the Navier-Stokes problem, Cal.44(2007), 165–188.
8. M. Amara, D. Capatina-Papaghiuc, D. Trujillo: Stabilized finite element method for Navier-Stokes equations with physical boundary conditions, in Math. Comput 2007, n° 6, 1195–1217.
9. M. Azaïez, C. Bernardi, M. Dauge, Y. Maday: Spectral Methods for Axisymmetric Domains. Gauthier-Villars & North-Holland. *Ser. Appl. Math.* **3** (1999).
10. M. Azaiez, C. Bernardi, N. Chorfi: Spectral discretization of the vorticity, velocity and pressure formulation of the Navier-Stokes equations, Numer. Math. 104 (2006), 1–26.

11. N. Abdellatif, N. Chorfi, S. Trabelsi: Spectral discretization of the axisymmetric vorticity, velocity and pressure formulation of the NavierStokes problem, Journal of Computational and Applied Mathematics, Volume 241, 2013, Pages 1–18.
12. N. Abdellatif, N. Chorfi and S. Trabelsi: Spectral discretization of the vorticity, velocity and pressure formulation of the axisymmetric Stokes problem, J. of Sci. Comput., Vol 47, 3, (2011), 419–440.
13. S. Salmon: Développement numérique de la formulation tourbillon-vitesse-pression pour le problème de Stokes, Thèse, Université Paris VI, (1999).

11. N. Abdellatif, N. Chorfi, S. Trabelsi, Spectral discretization of the axisymmetric vorticity, velocity and pressure formulation of the Navier-Stokes problem, Journal of Computational and Applied Mathematics, Volume 2[illegible], 2014, Pages 1-18.
12. N. Abdellatif, N. Chorfi and S. Trabelsi: Spectral discretization of the vorticity, velocity and pressure formulation of the axisymmetric Stokes problem, J. of Sci. Comput. Vol 47, 3, (2011), 419-440.
13. S. Salmon, Développement numérique de la formulation tourbillon-vitesse-pression pour le problème de Stokes, Thèse, Université Paris VI (1999).

Higher-Order Compatible Discretization on Hexahedrals

Jasper Kreeft and Marc Gerritsma

Abstract We derive a compatible discretization method that relies heavily on the underlying geometric structure, and obeys the topological sequences and commuting properties that are constructed. As a sample problem we consider the vorticity-velocity-pressure formulation of the Stokes problem. We motivate the choice for a mixed variational formulation based on both geometric as well as physical arguments. Numerical tests confirm the theoretical results that we obtain a pointwise divergence-free solution for the Stokes problem and that the method obtains optimal convergence rates.

1 Introduction

As sample problem we consider the Stokes flow problem in its vorticity-velocity-pressure formulation,

$$\boldsymbol{\omega} - \operatorname{curl} \mathbf{u} = 0 \quad \text{in } \Omega, \tag{1a}$$

$$\operatorname{curl} \boldsymbol{\omega} + \operatorname{grad} p = \mathbf{f} \quad \text{in } \Omega, \tag{1b}$$

$$\operatorname{div} \mathbf{u} = 0 \quad \text{in } \Omega. \tag{1c}$$

In this article we consider prescribed velocity boundary conditions, $\mathbf{u} = 0$ on $\partial\Omega$, but the method holds for all admissible types of boundary conditions, see [8].

J. Kreeft (✉)
Shell Global Solutions, Grasweg 31, 1031 HW, Amsterdam, The Netherlands
e-mail: jasper.kreeft@shell.com; J.J.Kreeft@tudelft.nl

M. Gerritsma
Delft University of Technology, Kluyverweg 2, 2629 HT, Delft, The Netherlands
e-mail: m.i.gerritsma@tudelft.nl

M. Azaïez et al. (eds.), *Spectral and High Order Methods for Partial Differential Equations - ICOSAHOM 2012*, Lecture Notes in Computational Science and Engineering 95, DOI 10.1007/978-3-319-01601-6_25,

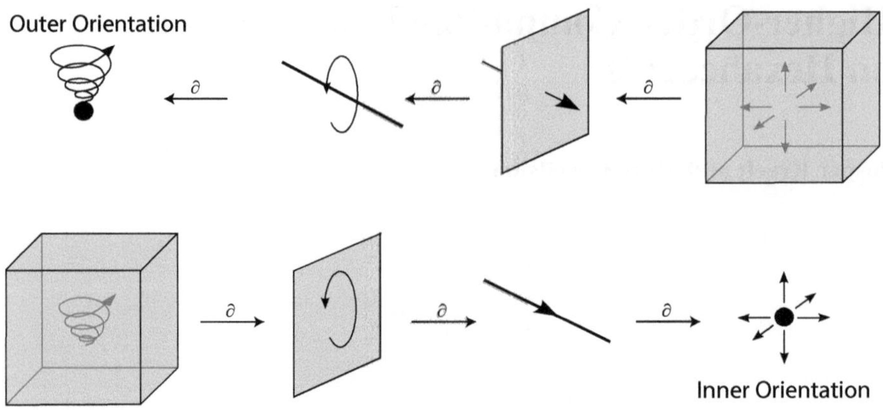

Fig. 1 The four geometric objects possible in $\mathbb{R}^3$, point, line, surface and volume, with outer- (*above*) and inner- (*below*) orientation. The boundary operator, ∂, maps k-dimensional objects to $(k-1)$-dimensional objects

Despite the simple appearance of Stokes flow model, there exists a large number of numerical methods to simulate Stokes flow. They all reduce to two classes, that is, either circumventing the LBB stability condition, like stabilized methods, e.g. [7], or satisfying this condition, as in compatible or mixed methods, e.g. [4]. The last requires the construction of dedicated discrete vector spaces. Best known are the curl conforming Nédélec and divergence conforming Raviart-Thomas spaces. Here, we consider a subclass of compatible methods, i.e. *mimetic methods*. Mimetic methods do not solely search for appropriate vector spaces, but aim to mimic structures and symmetries of the continuous problem, see [2, 3, 10, 11]. As a consequence of this mimicking, mimetic methods automatically preserve most of the physical and mathematical structures of the continuous formulation, among others the LBB condition and, most important, a pointwise divergence-free solution [8, 9].

At the heart of the mimetic method there are the well-known integral theorems of Newton-Leibniz, Stokes and Gauss, which couple the operators grad, curl and div, to the action of the boundary operator on a manifold. Therefore, obeying geometry and orientation will result in satisfying exactly the mentioned theorems, and consequently performing the vector operators exactly in a finite dimensional setting. In 3D we distinguish between four types of sub-manifolds, that is, points, lines, surfaces and volumes, and two types of orientation, namely, outer- and inner-orientation. Examples of sub-manifolds are shown in Fig. 1 together with the action of the boundary operator.

By creating a quadrilateral or hexahedral mesh, we divide the physical domain in a large number of these geometric objects, and to each geometric object we associate a discrete unknown. This implies that these discrete unknowns are *integral quantities*. Since the three earlier mentioned theorems are integral equations, it follows for example that taking a divergence in a volume is equivalent to taking the sum of the integral quantities associated to the surrounding surface elements, i.e. the fluxes. So using integral quantities as degrees of freedom to perform a grad, curl or div, is equivalent to taking the sum of the degrees of freedom located at its boundary.

These relations are of purely topological nature. They form a topological sequence or complex. This sequence is fundamental. It has a direct connection with the complexes that are related to the physical domain, the computational domain, the physical problem and the discretization.

Although the original work [9, 10], was presented in terms of differential geometry and algebraic topology, here we will use vector calculus because it is the more common mathematical language. Nevertheless, we will put emphasis on the distinction between topology and metric, on complexes and on commuting diagrams, which drives the former two languages.

We make use of spectral element interpolation functions as basis functions. In the past nodal spectral elements were mostly used in combination with Galerkin projection (GSEM). The GSEM satisfies the LBB condition by lowering the polynomial degree of the pressure by two with respect to the velocity. This results in a method that is only weakly divergence-free, meaning that the divergence of the velocity field only convergence to zero with mesh refinement. The present study uses mimetic spectral element interpolation or basis functions [10]. The mixed mimetic spectral element method (MMSEM) satisfies the LBB condition and gives a pointwise divergence-free solution for all mesh sizes.

2 Can We Really Discretize Exactly?

Since the Stokes flow model (1) should hold on a certain physical domain, we will include geometry by means of integration. In that case we can relate every physical quantity to a geometric object. Starting with the incompressibility constraint (1c) we have due to Gauss' divergence theorem,

$$\int_V \operatorname{div} \mathbf{u} \, \mathrm{d}V = \int_{\partial V} \mathbf{u} \cdot \mathbf{n} \, \mathrm{d}S = 0,$$

and using Stokes' circulation theorem the relation (1a) can be written as

$$\int_S \omega \times \mathbf{n} \, \mathrm{d}S = \int_S \operatorname{curl} \mathbf{u} \times \mathbf{n} \, \mathrm{d}S = \int_{\partial S} \mathbf{u} \cdot \mathbf{t} \, \mathrm{d}l.$$

From the first relation it follows that $\operatorname{div} \mathbf{u}$ is associated to volumes. The association to a geometric object for velocity $\mathbf{u}$ is less clear. In fact it can be associated to two different types of geometric objects. A representation of velocity compatible with the incompressibility constraint is given in terms of the velocity flux, $\mathbf{u} \cdot \mathbf{n}$, *through a surface* that bounds the volume, while in the circulation relation velocity, $\mathbf{u} \cdot \mathbf{t}$, is represented *along a line* that bounds the surface. We will call the velocity

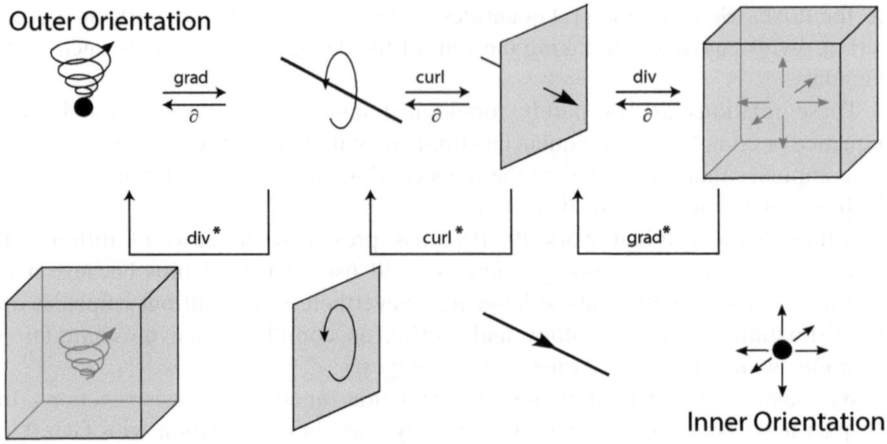

Fig. 2 Geometric interpretation of the action of the boundary operators, vector differential operators and their formal Hilbert adjoint operators

vector through a surface *outer-oriented* and the velocity along a line segment *inner-oriented*. A similar distinction can be made for vorticity, see [9].

The last equation to be considered is (1b). This equation shows that classical Newton-Leibniz, Stokes circulation and Gauss divergence theorems tell only half the story. From the perspective of the classical Newton-Leibniz theorem, the gradient acting on the pressure relates line values to their corresponding end point, while the Stokes circulation theorem shows that the curl acting on the vorticity vector relates surface values to the line segment enclosing it. So how does this fit into one equation? In fact, from a geometric perspective, there exists two gradients, two curls and two divergence operators. One of each is related to the mentioned integral theorems as explained above. The others are their formal adjoint operators. Let grad, curl and div be the original differential operators associated to the mentioned integral theorems, then the formal Hilbert adjoint operators grad*, curl* and div* are defined as,

$$\left(\mathbf{a},-\mathrm{grad}^*\, b\right)_\Omega := \left(\mathrm{div}\,\mathbf{a},b\right)_\Omega,\ \left(\mathbf{a},\mathrm{curl}^*\,\mathbf{b}\right)_\Omega := \left(\mathrm{curl}\,\mathbf{a},\mathbf{b}\right)_\Omega,\ \left(a,-\mathrm{div}^*\,\mathbf{b}\right)_\Omega := \left(\mathrm{grad}\,a,\mathbf{b}\right)_\Omega.$$

From a geometric interpretation, the adjoint operators detours via the opposite type of orientation. Where div relates a vector quantity associated to surfaces to a scalar quantity associated to a volume enclosed by these surfaces. Its adjoint operator, grad*, relates a scalar quantity associated with a volume to a vector quantity associated with its surrounding surfaces. This is illustrated in Fig. 2. Following Fig. 2, the adjoint operator grad* consists of three consecutive steps: First, switch from an outer oriented scalar associated to volumes to an inner oriented scalar associated to points, then take the derivative and finally switch from an inner oriented vector associated to lines to an outer oriented vector associated to surfaces. In a similar way we can describe the derivatives curl* and div*.

Since the horizontal relations are purely topological and the vertical relations purely metric, the operators grad, curl and div are purely topological operators, while grad*, curl* and div* are metric. This makes them much harder to discretize.

Now (1b) could then either be associated to an inner-oriented line segment by rewriting it as

$$\text{curl}^* \boldsymbol{\omega} + \text{grad}\, p = \mathbf{f},$$

or be associated to an outer-oriented surface by rewriting it as

$$\text{curl}\, \boldsymbol{\omega} + \text{grad}^*\, p = \mathbf{f}.$$

Without geometric considerations we could never make a distinction between grad, curl and div and their associated Hilbert adjoints div*, curl* and grad*.

Since our focus is on obtaining a pointwise divergence-free discretization, we decide to use the expression where the equations are associated to outer-oriented geometric objects,

$$\boldsymbol{\omega} - \text{curl}^* \mathbf{u} = 0 \quad \text{in } \Omega, \tag{2a}$$

$$\text{curl}\, \boldsymbol{\omega} + \text{grad}^*\, p = \mathbf{f} \quad \text{in } \Omega, \tag{2b}$$

$$\text{div}\, \mathbf{u} = 0 \quad \text{in } \Omega, \tag{2c}$$

where the first equation is associated to outer-oriented line segments, the second to outer-oriented surfaces and the third to outer-oriented volumes.

3 Complexes

Figure 2 reveals already a number of sequence or complex structures. Starting from geometry, we consider points, P, lines, L, surfaces, S, and volumes, V. They possess a sequence in combination with the boundary operator, ∂. The boundary of a volume is a surface, the boundary of a surface is a line and the boundary of a line are its two end points. This results in the following complex,

$$0 \stackrel{\partial}{\longleftarrow} P \stackrel{\partial}{\longleftarrow} L \stackrel{\partial}{\longleftarrow} S \stackrel{\partial}{\longleftarrow} V. \tag{3}$$

An important property of the complex is that if we apply the boundary operator twice, we always find an empty set, e.g. if $S = \partial V$, then $\partial S = \emptyset$. As follows directly from the previously mentioned integral theorems, it follows, as a consequence of $\partial\partial = \emptyset$, that $\text{curl}\,\text{grad} = 0$ and $\text{div}\,\text{curl} = 0$. The derivatives themselves also form a complex. In a Hilbert setting this becomes,

$$H^1(\Omega) \stackrel{\text{grad}}{\longrightarrow} H(\Omega, \text{curl}) \stackrel{\text{curl}}{\longrightarrow} H(\Omega, \text{div}) \stackrel{\text{div}}{\longrightarrow} L^2(\Omega), \tag{4}$$

and using the Hilbert adjoint relations we also obtain the adjoint complex with properties $\text{curl}^*\text{grad}^* = 0$ and $\text{div}^*\text{curl}^* = 0$,

$$L_0^2(\Omega) \overset{-\text{div}^*}{\longleftarrow} H_0(\Omega, \text{div}^*) \overset{\text{curl}^*}{\longleftarrow} H_0(\Omega, \text{curl}^*) \overset{-\text{grad}^*}{\longleftarrow} H_0^1(\Omega) \tag{5}$$

In the Hilbert setting, the variables of the Stokes problem are in the following spaces, $\boldsymbol{\omega} \in H(\Omega, \text{curl}) \cap H_0(\Omega, \text{div}^*)$, $\mathbf{u} \in H(\Omega, \text{div}) \cap H_0(\Omega, \text{curl}^*)$ and $p \in L^2(\Omega) \cap H_0(\Omega, \text{grad}^*)$. It is hard, if even possible at all, to find discrete vector spaces that are subsets of these function spaces and simultaneously satisfy the complex properties. Instead, the Stokes problem can be cast into an equivalent variational or mixed formulation where we make use of the Hilbert adjoint properties. This simplifies the function spaces of the flow variables. The mixed formulation reads;

Find $(\boldsymbol{\omega}, \mathbf{u}, p) \in \{H(\Omega, \text{curl}) \times H(\Omega, \text{div}) \times L^2(\Omega)\}$ with $\mathbf{f} \in H(\Omega, \text{div})$ given, for all $(\boldsymbol{\sigma}, \mathbf{v}, q) \in \{H(\Omega, \text{curl}) \times H(\Omega, \text{div}) \times L^2(\Omega)\}$, such that,

$$(\boldsymbol{\sigma}, \boldsymbol{\omega})_\Omega - (\text{curl}\, \boldsymbol{\sigma}, \mathbf{u})_\Omega = 0, \tag{6a}$$

$$(\mathbf{v}, \text{curl}\, \boldsymbol{\omega})_\Omega - (\text{div}\, \mathbf{v}, p)_\Omega = (\mathbf{v}, \mathbf{f})_\Omega, \tag{6b}$$

$$(q, \text{div}\, \mathbf{u})_\Omega = 0. \tag{6c}$$

With the formulation and corresponding function spaces, we are able to construct compatible discrete vector spaces. Note that we now completely avoid the metric dependent derivatives grad^* and curl^*, and their corresponding complex.

4 Discretization of Stokes Problem

Degrees of freedom. In many numerical methods, especially in finite difference and finite element methods, the discrete coefficients are point values. In the proposed mimetic structure, the discrete unknowns represent integral values on k-dimensional submanifolds, ranging from points to volumes, so $0 \leq k \leq 3$. These k-dimensional submanifolds are oriented, constitute the computational domain and span the physical domain. The concept of orientation shown in Figs. 1 and 2 gave rise to the boundary operator, ∂, which can be represented by connectivities consisting only of -1, 0 and 1, see also [9].

The space of degrees of freedom are given by $\mathscr{P}$, $\mathscr{L}$, $\mathscr{S}$ and $\mathscr{V}$. These spaces form a duality pairing with the geometric spaces P, L, S and V. The degrees of freedom are integral values, i.e.

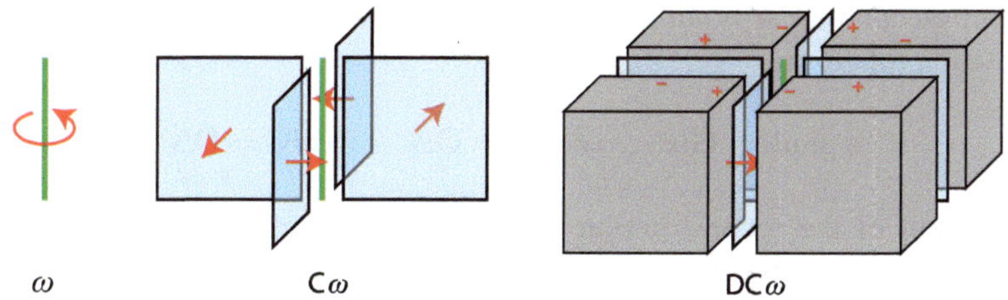

Fig. 3 The action of twice the coboundary operator δ on a vorticity d.o.f. has a zero net result on its surrounding volumes, because they all have both a positive and a negative contribution from its neighboring velocity faces

$$\int_l \mathbf{w}\cdot\mathbf{t}\,\mathrm{d}l \in \mathscr{L}, \quad \int_S \mathbf{u}\cdot\mathbf{n}\,\mathrm{d}S \in \mathscr{S}, \quad \int_V p\,\mathrm{d}V \in \mathscr{V}. \tag{7}$$

By the definition of the degrees of freedom spaces and the previously mentioned integral theorems, we can define the formal adjoint of the boundary operator, i.e. the coboundary operator, δ. The coboundary operator is the discrete representation of the topological derivatives grad, curl and div. Since $\partial\partial = \emptyset$, it follows from a discrete Newton-Leibniz, Stokes and Gauss theorem that applying the coboundary operator twice is always zero, $\delta\delta = \emptyset$ (see [2, 9]). The coboundary operator also has matrix representations, G, C and D, that are the transpose of the connectivity matrices. We obtain the following topological sequence,

$$\mathscr{P} \xrightarrow{\mathsf{G}} \mathscr{L} \xrightarrow{\mathsf{C}} \mathscr{S} \xrightarrow{\mathsf{D}} \mathscr{V}, \tag{8}$$

where $\mathsf{CG} = \mathbf{0}$ and $\mathsf{DC} = \mathbf{0}$. These matrices will explicitly appear in the final matrix system. An illustration of $\mathsf{DC} = \mathbf{0}$ is given in Fig. 3. More details on the structure of geometry, orientation and degrees of freedom can be found in Gerritsma et al. [6].

Mimetic Operators. Let $W = H(\Omega, \text{curl})$, $V = H(\Omega, \text{div})$ and $Q = L^2(\Omega)$. The discretization of the flow variables involves a projection operator, π_h, from the complete vector spaces W, V and Q, to the discrete vector spaces W_h, V_h and Q_h. Here the flow variables are expressed in terms of d.o.f. defined on k-cells, and corresponding interpolation functions (also called basis-functions). The projection operator actually consists of two steps, a reduction operator, $\mathscr{R}$, that integrates the flow variables on k-cells, and a reconstruction operator, $\mathscr{I}$, that interpolates the d.o.f. using the appropriate basis-functions. These mimetic operators were defined in [2, 10]. A composition of the two operators gives the projection operator $\pi_h = \mathscr{I} \circ \mathscr{R}$.[1]

[1]For completeness, in a Hilbert setting the projection needs an additional smoothing argument. This step is ignored here to increase readibility. See [8] for more details.

Reduction operator $\mathscr{R}$ is simply defined by integration. It possesses the following commutation relations,

$$\mathscr{R}\mathrm{grad} = \mathsf{G}\mathscr{R}, \quad \mathscr{R}\mathrm{curl} = \mathsf{C}\mathscr{R}, \quad \mathscr{R}\mathrm{div} = \mathsf{D}\mathscr{R}. \tag{9}$$

The treatment of the reconstruction operator leaves some freedom, as long as it satisfies the following properties: be the right inverse of the reduction, $\mathscr{R}\mathscr{I} = Id$, be the approximate left inverse of the reduction, $\mathscr{I}\mathscr{R} = Id + \mathscr{O}(h^p)$, and it should possess the following commutation relations,

$$\mathrm{grad}\,\mathscr{I} = \mathscr{I}\mathsf{G}, \quad \mathrm{curl}\,\mathscr{I} = \mathscr{I}\mathsf{C}, \quad \mathrm{div}\,\mathscr{I} = \mathscr{I}\mathsf{D}. \tag{10}$$

When both the reduction and reconstruction operators commute with continuous and discrete differentiation, than also the projection operator π_h possesses a commutation relation with differentiation. In case of the divergence operator, which is relevant to obtain a pointwise divergence-free solution, the commutation relation is given by,

$$\mathrm{div}\,\pi_h = \mathrm{div}\,\mathscr{I}\mathscr{R} = \mathscr{I}\mathsf{D}\mathscr{R} = \mathscr{I}\mathscr{R}\mathrm{div} = \pi_h\mathrm{div}\,. \tag{11}$$

The commutation relations in case of divergence are illustrated below,

$$\begin{array}{ccc} V & \xrightarrow{\mathrm{div}} & Q \\ \downarrow \mathscr{R} & & \downarrow \mathscr{R} \\ \mathscr{S} & \xrightarrow{\mathsf{D}} & \mathscr{V}. \end{array} \quad + \quad \begin{array}{ccc} \mathscr{S} & \xrightarrow{\mathsf{D}} & \mathscr{V} \\ \downarrow \mathscr{I} & & \downarrow \mathscr{I} \\ V_h & \xrightarrow{\mathrm{div}} & Q_h. \end{array} \quad = \quad \begin{array}{ccc} V & \xrightarrow{\mathrm{div}} & Q \\ \downarrow \pi_h & & \downarrow \pi_h \\ V_h & \xrightarrow{\mathrm{div}} & Q_h. \end{array}$$

Since property (11) also holds for the grad and curl, we obtain the following complex for discrete vector spaces,

$$\Phi_h \xrightarrow{\mathrm{grad}} W_h \xrightarrow{\mathrm{curl}} V_h \xrightarrow{\mathrm{div}} Q_h. \tag{12}$$

In practice we use $\mathscr{I}\mathsf{D}\mathscr{R}$ from (11) in computations. Relation (11) implies among others that it satisfies the discrete LBB condition,

$$\beta_h := \inf_{q_h \in Q_h} \sup_{v_h \in V_h} \frac{\left(q_h, \mathrm{div}\,\mathbf{v_h}\right)_\Omega}{\|q_h\|_Q \|v_h\|_V} > \beta > 0, \tag{13}$$

where β is the inf-sup constant of the continuous problem (1). Whereas the LBB condition is a measurement for numerical stability, the commutation relation indicates physical correctness of the numerical method. This last is a much stronger statement, which includes also the former.

The conditions on the reconstruction operator have led to the construction of mimetic spectral element basis-functions [5, 10]. Since we use a tensor-based construction of point, line, surface and volume corresponding basis-functions, we only need nodal and edge interpolation functions. The nodal interpolation functions are the well-known Lagrange polynomials. The edge polynomials were derived from the Lagrange polynomials, based on the given conditions. For a set of Lagrange polynomials, $l_i(x)$, $i = 0, \ldots, N$, the edge polynomials, $e_i(x)$, $i = 1, \ldots, N$, are given by,

$$e_i(x) = -\sum_{k=0}^{i-1} \frac{\mathrm{d}l_k(x)}{\mathrm{d}x}. \tag{14}$$

The Lagrange and edge polynomials possess the condition $\mathscr{R}\mathscr{I} = Id$, i.e.,

$$l_i(x_j) = \delta_{i,j}, \qquad \int_{x_{j-1}}^{x_j} e_i(x)\,\mathrm{d}x = \delta_{i,j}, \tag{15}$$

where $\delta_{i,j}$ is the Kronecker delta. The interpolation function for a variable associated to a surface, for example, is given by, $s_{i,j,k}(x, y, z) = \{l_i(x)e_j(y)e_k(z), e_i(x)l_j(y)e_k(z),\ e_i(x)e_j(y)l_k(z)\}$.

Example 1 (Divergence operator in 2D). One of the most interesting properties of the mimetic method presented in this paper, is that within our weak formulation, the divergence-free constraint is satisfied pointwise. Let $u_h \in V_h$ be the velocity flux defined as

$$\mathbf{u}_h = \begin{pmatrix} \sum_{i=0}^{N} \sum_{j=1}^{N} u_{i,j} l_i(x) e_j(y) \\ \sum_{i=1}^{N} \sum_{j=0}^{N} v_{i,j} e_i(x) l_j(y) \end{pmatrix}. \tag{16}$$

Then the change of mass, $m_h \in Q_h$, is equal to the divergence of $\mathbf{u}_h$,

$$\begin{aligned} m_h = \operatorname{div} \mathbf{u}_h &= \sum_{i=1}^{N} \sum_{j=1}^{N} (u_{i,j} - u_{i-1,j} + v_{i,j} - v_{i,j-1}) e_i(x) e_j(y). \\ &= \sum_{i=1}^{N} \sum_{j=1}^{N} m_{i,j} e_i(x) e_j(y), \end{aligned} \tag{17}$$

(continued)

Example 1 (continued)

where $m_{i,j} = u_{i,j} - u_{i-1,j} + v_{i,j} - v_{i,j-1}$ can be compactly written as $\mathsf{m} = \mathsf{Du}$. Note that if the mass production is zero, as in our model problem (1c), the incompressibility constraint can already be satisfied at the discrete level. Interpolation using $e_i(x)e_j(y)$ then results in a solution of velocity $\mathbf{u}_h$ that is pointwise divergence-free.

5 A Priori Error Estimates

By standard interpolation theory it follows that we obtain the following h-convergence rates for the interpolation errors of the flow variables,

$$\|\boldsymbol{\omega}-\pi_h\boldsymbol{\omega}\|_{H(\mathrm{curl})} = \mathcal{O}(h^N), \quad \|\mathbf{u}-\pi_h\mathbf{u}\|_{H(\mathrm{div})} = \mathcal{O}(h^N), \quad \|p-\pi_h p\|_{L^2} = \mathcal{O}(h^N), \tag{18}$$

and that $\|\mathrm{div}\,\mathbf{u} - \mathrm{div}\,\pi_h\mathbf{u}\|_{L^2} = 0$ due to the commuting property.

In cases with empty harmonic vector spaces, we have that the discrete vector spaces are conforming, i.e., $W_h \subset W$, $V_h \subset V$ and $Q_h \subset Q$. Moreover, due to the commuting property, it follows that these spaces are compatible, i.e., curl $W_h \subset V_h$ and div $V_h = Q_h$. Finally they possess a Helmholtz-Hodge decomposition, $\boldsymbol{\sigma} = \mathrm{grad}\,\phi + \mathrm{curl}^*\mathbf{v}$ and $\mathbf{v} = \mathrm{curl}\,\boldsymbol{\sigma} + \mathrm{grad}^* q$. In terms of vector spaces, this is, $W_h = Z_{W_h} \oplus Z^{\perp}_{W_h}$ and $V_h = Z_{V_h} \oplus Z^{\perp}_{V_h}$, where Z refers to the kernel or nullspace and $Z^{\perp}$ to its orthogonal complement. Having all these properties, a priori error estimates are derived in [8] that show optimal convergence rates for all admissible boundary conditions, including the no-slip boundary condition, which is non-trivial in mixed finite element methods. The a priori error estimates are given by

$$\|\boldsymbol{\omega} - \boldsymbol{\omega}_h\|_W \leq C \inf_{\boldsymbol{\sigma}_h \in W_h} \|\boldsymbol{\omega} - \boldsymbol{\sigma}_h\|_W, \tag{19}$$

$$\|\mathbf{u} - \mathbf{u}_h\|_V \leq C \inf_{\mathbf{v}_h \in V_h} \|\mathbf{u} - \mathbf{v}_h\|_V + C \inf_{\boldsymbol{\sigma}_h \in W_h} \|\boldsymbol{\omega} - \boldsymbol{\sigma}_h\|_W, \tag{20}$$

$$\|p - p_h\|_Q \leq C \inf_{q_h \in Q_h} \|p - q_h\|_Q + C \inf_{\mathbf{v}_h \in V_h} \|\mathbf{u} - \mathbf{v}_h\|_V + C \inf_{\boldsymbol{\sigma}_h \in W_h} \|\boldsymbol{\omega} - \boldsymbol{\sigma}_h\|_W, \tag{21}$$

where the constants C will differ in each case and are independent of h. It shows that the rate of convergence of the approximation errors are the same as those of the interpolation errors.

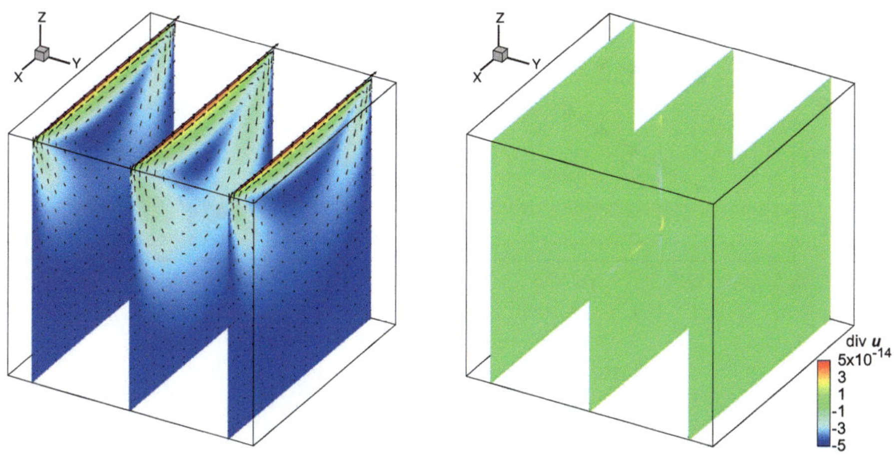

Fig. 4 *Left*: slices of magnitude of the velocity field of a three dimensional lid-driven cavity Stokes problem obtained on a $2 \times 2 \times 2$ element mesh with $N = 8$. *Right*: slices of the divergence of velocity. Is confirms a divergence-free velocity field

6 Numerical Results

For many years, the lid-driven cavity flow was considered one of the classical benchmark cases for the assessment of numerical methods and the verification of incompressible (Navier)-Stokes codes. The 3D lid-driven cavity test case deals with a flow in a unit box with five solid boundaries and moving lid as the top boundary, moving with constant velocity equal to minus one in x-direction. Especially the two line singularities make the lid-driven cavity problem a challenging test case.

The left plot in Fig. 4 shows slices of the magnitude of the velocity field in a three dimensional lid-driven cavity Stokes problem, obtained on a $2 \times 2 \times 2$ element mesh, where each element contains a Gauss-Lobatto mesh of $N = 8$. The slices are taken at 10, 50 and 90 % of the y-axis. The right plot in Fig. 4 shows slices of divergence of the velocity field. Figure 4 confirms that the mixed mimetic spectral element method leads to an accurate result with a divergence-free solution.

The second testcase shows the optimal convergence behavior for a 2D Stokes problem with no-slip boundary conditions. The testcase originates from a recent paper by Arnold et al. [1], where sub-optimal convergence is shown and proven for no-slip boundary conditions when using Raviart-Thomas elements. Since Raviart-Thomas elements are the most popular $H(\text{div}, \Omega)$ conforming elements, we compare our method to these results.

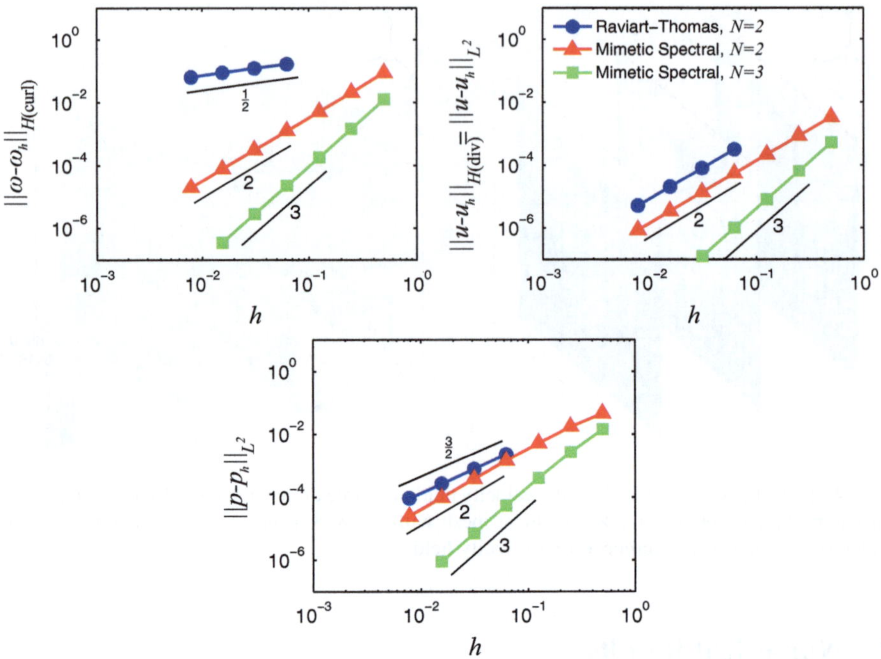

Fig. 5 Comparison of the h-convergence between Raviart-Thomas and Mimetic spectral element projections for the 2D Stokes problem with no-slip boundary conditions

Figure 5 shows the results of the Stokes problem on a unit square with velocity and pressure fields given by $\mathbf{u} = \left[-2x^2(x-1)^2y(2y-1)(y-1), 2y^2(y-1)^2x(2x-1)(x-1)\right]^T$, $p = (x-\frac{1}{2})^5+(y-\frac{1}{2})^5$. While for velocity both methods show optimal convergence, for pressure a difference of $\frac{1}{2}$ is noticed in the rate of convergence and for vorticity a difference in rate of convergence of $\frac{3}{2}$ is revealed.

References

1. D. Arnold, R. Falk, and J. Gopalakrishnan. Mixed finite element approximation of the vector Laplacian with Dirichlet boundary conditions. *Mathematical Models & Methods in Applied Sciences*, 22(9):1250024, 2012.
2. P. Bochev and J. Hyman. Principles of mimetic discretizations of differential operators. In D. Arnold, P. Bochev, R. Lehoucq, R. Nicolaides, and M. Shashkov, editors, *Compatible Discretizations*, volume 142 of *IMA Volumes in Mathematics and its Applications*, pages 89–119. Springer, 2006.
3. F. Brezzi and A. Buffa. Innovative mimetic discretizations for electromagnetic problems. *Journal of computational and applied mathematics*, 234:1980–1987, 2010.
4. F. Brezzi and M. Fortin. *Mixed and hybrid finite element methods, volume 15 of Springer Series in Computational Mathematics*. Springer-Verlag, New York, 1991.

5. M. Gerritsma. Edge functions for spectral element methods. In J. Hesthaven and E. Rønquist, editors, *Spectral and High Order Methods for Partial Differential Equations*, pages 199–208. Springer, 2011.
6. M. Gerritsma, R. Hiemstra, J. Kreeft, A. Palha, P. Rebelo, and D. Toshniwal. The geometric basis of mimetic spectral approximations. In *(this issue)*, 2013.
7. T. Hughes, P. Franca, and M. Balestra. A new finite element formulation for computational fluid dynamics: V. Circumventing the Babuška-Brezzi condition: A stable Petrov-Galerkin formulation of the Stokes problem accomodating equal-order interpolations. *Computer Methods in Applied Mechanics and Engineering*, 59:85–99, 1986.
8. J. Kreeft and M. Gerritsma. A priori error estimates for compatible spectral discretization of the Stokes problem for all admissible boundary conditions. *submitted*, arXiv:1206.2812, 2013.
9. J. Kreeft and M. Gerritsma. Mixed mimetic spectral element method for Stokes flow: a pointwise divergence-free solution. *Journal Computational Physics*, 240:284–309, 2013.
10. J. Kreeft, A. Palha, and M. Gerritsma. Mimetic framework on curvilinear quadrilaterals of arbitrary order. *submitted*, arXiv:1111.4304, 2013.
11. J. B. Perot. Discrete conservation properties of unstructured mesh schemes. *Annual review of fluid mechanics*, 43:299–318, 2011.

Mimetic Spectral Element Advection

Artur Palha, Pedro Pinto Rebelo, and Marc Gerritsma

Abstract We present a discretization of the linear advection of differential forms on bounded domains. The framework established in [4] is extended to incorporate the Lie derivative, $\mathcal{L}$, by means of Cartan's homotopy formula. The method is based on a physics-compatible discretization with spectral accuracy. It will be shown that the derived scheme has spectral convergence with local mass conservation. Artificial dispersion depends on the order of time integration.

1 Introduction

Consider the classical advection problem for a scalar function in conservation form,

$$\frac{\partial \rho}{\partial t} + \nabla \cdot (\mathbf{v}\rho) = 0, \tag{1}$$

where $\mathbf{v}$ is a prescribed uniformly Lipschitz continuous vector field and ρ the advected scalar function. The method presented in this work is based on the approximation of the differential operators with the focus on the spatial discretization and a time-stepping scheme that distinguishes between quantities evaluated at time instants and quantities evaluated over time intervals.

The mimetic framework, presented in [4], showed that using the differential geometric approach for the representation of physical laws clarifies the underlying structures. One clearly identifies to what kind of geometrical object a certain physical quantity is associated and this determines how its discretization must be done

A. Palha (✉) · P.P. Rebelo · M. Gerritsma
Faculty of Aerospace Engineering, Aerodynamics Group, Delft University of Technology, 2600GB – Delft, The Netherlands
e-mail: A.Palha@tudelft.nl; P.J.PintoRebelo@tudelft.nl; M.I.Gerritsma@tudelft.nl

M. Azaïez et al. (eds.), *Spectral and High Order Methods for Partial Differential Equations - ICOSAHOM 2012*, Lecture Notes in Computational Science and Engineering 95, DOI 10.1007/978-3-319-01601-6_26,

(e.g.: evaluation at points, integration over lines, surfaces or volumes). Additionally, a well defined, metric free, representation of differential operators is obtained, together with their metric dependent Hilbert adjoints. For these reasons, the authors followed this approach for the advection equation. It is known, see [1, p. 317], that (1) is a particular case of the generalized advection equation which can be written in terms of differential geometry as,

$$\frac{\partial \alpha^{(k)}}{\partial t} + \mathcal{L}_{\mathbf{v}}\, \alpha^{(k)} = 0. \tag{2}$$

The advection operator, $\mathcal{L}_{\mathbf{v}}$, is the Lie derivative for the prescribed velocity field $\mathbf{v}$ and the advected quantity is given by the k-differential form $\alpha^{(k)}$. Depending on the index k the quantity $\alpha^{(k)}$ can represent scalar, vector and higher dimensional quantities.

2 Differential Geometry

In this section a brief introduction to differential geometry is given. For a more detailed introduction the reader is directed to [1]. Given an n-dimensional smooth orientable manifold Ω it is possible to define in each point a tangent vector space E of dimension n. The space of smooth vector fields on a manifold is the space, Γ, of smooth assignments of elements of E to each point of the manifold. We denote by Λ^k, k an integer $0 \leq k \leq n$, the space of differentiable k-forms, i.e. all smooth k-linear, antisymmetric maps $\omega^{(k)} : E \times \cdots \times E \to \mathbb{R}$, at every point of the manifold. We recall the wedge product $\wedge : \Lambda^k \times \Lambda^l \to \Lambda^{k+l}$ for $k + l \leq n$ with the property that $\alpha^{(k)} \wedge \beta^{(l)} = (-1)^{kl} \beta^{(l)} \wedge \alpha^{(k)}$. The inner product $(\cdot,\cdot)$ on E induces at each point of the manifold an inner product $(\cdot,\cdot)$ on Λ^1. In turn, this can be extended to a local inner product on Λ^k [8, p. 149]. The local inner product gives rise to a unique metric operator, Hodge-$\star$, $\star : \Lambda^k \to \Lambda^{n-k}$, defined by $\alpha^{(k)} \wedge \star\beta^{(k)} = \left(\alpha^{(k)}, \beta^{(k)}\right) \omega^{(n)}$, where $\omega^{(n)} = \star 1$ is the standard volume form. By integration, one can define an inner product on Ω as $(\cdot,\cdot)_{L^2} := \int_\Omega (\cdot,\cdot)\, \omega^{(n)}$. The exterior derivative $\mathrm{d} : \Lambda^k \to \Lambda^{k+1}$ satisfies the following rule, $\mathrm{d}\left(\alpha^{(k)} \wedge \beta^{(l)}\right) = \mathrm{d}\alpha^{(k)} \wedge \beta^{(l)} + (-1)^k \alpha^{(k)} \wedge \mathrm{d}\beta^{(l)}$ and by definition $\mathrm{d}\alpha^{(n)} = 0$. The flat operator, $\flat$, is a mapping $\flat : \Gamma \mapsto \Lambda^1$.

The Lie derivative along a tangent vector field, $\mathbf{v}$, is denoted by $\mathcal{L}_{\mathbf{v}}$ and represents the advection operator in differential geometry. It is a mapping $\mathcal{L}_{\mathbf{v}} : \Lambda^k \mapsto \Lambda^k$. From Cartan's homotopy formula the Lie derivative can be written as

$$\mathcal{L}_{\mathbf{v}}\, \alpha^{(k)}() := \mathrm{d}\iota_{\mathbf{v}}\alpha^{(k)} + \iota_{\mathbf{v}}\mathrm{d}\alpha^{(k)}\ ,$$

where the interior product of a tangent vector field, $\mathbf{v}$, with a k-form, $\alpha^{(k)}$, is a mapping $\iota_{\mathbf{v}}\alpha^{(k)} : \Lambda^k \to \Lambda^{k-1}$ given by:

$$\iota_{\mathbf{v}}\alpha^{(k)}(\mathbf{X}_2, \cdots, \mathbf{X}_k) := \alpha^{(k)}(\mathbf{v}, \mathbf{X}_2, \cdots, \mathbf{X}_k)\,, \quad \forall \mathbf{X}_i \in \Gamma \quad \text{and} \quad \iota_{\mathbf{v}}\alpha^{(0)} = 0, \quad \forall \mathbf{v} \in \Gamma\ .$$

The interior product is the adjoint of the wedge product, made explicit by:

$$\left(\iota_{\mathbf{v}}\alpha^{(k)},\beta^{(k-1)}\right)_{L^2\Lambda^{k-1}} = \left(\alpha^{(k)},\mathbf{v}^\flat\wedge\beta^{(k-1)}\right)_{L^2\Lambda^k}, \quad \forall\beta^{(k-1)}\in\Lambda^{k-1} \tag{3}$$

where $\mathbf{v}^\flat = \nu^{(1)} \in \Lambda^1$ and $\alpha_h^{(k)} \in \Lambda^k$.

The relevance of this adjoint relation between the interior product and the wedge product lies in the fact that it shows how a physical quantity represented by an interior product with a vector field can be represented by its dual differential 1-form.

For a volume form $\rho^{(n)}$ the Lie derivative is simply $\mathcal{L}_{\mathbf{v}}\rho^{(n)} = \mathrm{d}\iota_{\mathbf{v}}\rho^{(n)}$ and for a 0-form, $\mathcal{L}_{\mathbf{v}}\alpha^{(0)} = \iota_{\mathbf{v}}\mathrm{d}\alpha^{(0)}$.

3 Mimetic Discretization

In this section a brief introduction to the discretization of physical quantities and to the discretization of the exterior derivative is presented. For a more detailed presentation the reader is directed to [2,4,6].

Consider a three dimensional domain Ω and an associated grid consisting of a collection of points, $\tau_{(0),i}$, line segments connecting the points, $\tau_{(1),i}$, surfaces bounded by these line segments, $\tau_{(2),i}$, and volumes bounded by these surfaces, $\tau_{(3),i}$.

Let Λ^k be the space of smooth differentiable k-forms. Additionally, let the finite dimensional space of differentiable forms be defined as $\Lambda_h^k = \mathrm{span}(\{\epsilon_i^{(k)}\})$, $i = 1,\cdots,\dim(\Lambda_h^k)$, where $\epsilon_i^{(k)} \in \Lambda^k$ are basis k-forms. Under these conditions it is possible, see [4, 6], to define a projection operator π_h which projects elements of Λ^k onto elements of Λ_h^k which satisfies:

$$\pi_h\mathrm{d} = \mathrm{d}\pi_h\ . \tag{4}$$

It is possible to write:

$$\pi_h\alpha^{(k)} = \alpha_h^{(k)} = \sum_i \alpha_i\epsilon_i^{(k)}\ ,$$

where

$$\alpha_i = \int_{\tau_{(k),i}}\alpha^{(k)} \quad \text{and} \quad \int_{\tau_{(k),i}}\epsilon_j^{(k)} = \delta_{ij}, \quad k = 0,1,\cdots,n\ .$$

A set of basis functions yielding a projection operator π_h that satisfies (4) can be constructed using piecewise polynomial expansions on the quadrilateral elements using tensor products. Thus, it suffices to derive the basis forms in one dimension on a reference interval and generalize them in n dimensions.

In one dimension take a 0-form, $\alpha^{(0)} \in \Lambda^0\left(Q_{ref}\right)$, where $Q_{ref} := [-1, 1]$. Define on Q_{ref} a cell complex D of order p consisting of $(p + 1)$ nodes $\tau_{(0),i} = \xi_i$ with $i = 0, \cdots, p$, where $-1 \leq \xi_0 < \cdots < \xi_i < \cdots \xi_p \leq 1$ are the Gauss-Lobatto quadrature nodes, and p edges, $\tau_{(1),i} = [\xi_{i-1}, \xi_i]$ with $i = 1, \cdots, p$. The projection operator π_h reads:

$$\pi_h \alpha^{(0)}(\xi) = \sum_{i=0}^{p} \alpha_i \epsilon_i^{(0)}(\xi) \, , \tag{5}$$

where $\epsilon_i^{(0)}(\xi) = l_i(\xi)$ are the pth order *Lagrange polynomials* and $\alpha_i = \alpha^0(\xi_i)$. Similarly in one dimension for the projection of 1-forms Gerritsma [3] and Robidoux [7] derived 1-form polynomials called *edge polynomials*, $\epsilon_i^{(1)} \in \Lambda_h^1\left(Q_{ref}\right)$,

$$\epsilon_i^{(1)}(\xi) = e_i(\xi)\,\mathrm{d}\xi, \quad \text{with} \quad e_i(\xi) = -\sum_{k=0}^{i-1} \frac{dl_k}{d\xi} \, . \tag{6}$$

Note that in this way we have:

$$\int_{\xi_{j-1}}^{\xi_j} \epsilon_i^{(1)} = \int_{\xi_{j-1}}^{\xi_j} e_i(\xi)\,\mathrm{d}\xi = \delta_{ij} \, . \tag{7}$$

Moreover, the exterior derivative of the basis 0-forms is given by:

$$\mathrm{d}\epsilon_i^{(0)} = \frac{\mathrm{d}l_i}{\mathrm{d}\xi}\,\mathrm{d}\xi = -\sum_{k=0}^{i-1} \frac{\mathrm{d}l_k}{\mathrm{d}\xi}\,\mathrm{d}\xi - \left(-\sum_{k=0}^{i} \frac{\mathrm{d}l_k}{\mathrm{d}\xi}\,\mathrm{d}\xi\right) = \epsilon_i^{(1)} - \epsilon_{i+1}^{(1)}, \quad i = 1, \cdots, p-1 \, . \tag{8}$$

In this way, the exterior derivative of a discrete 0-form can be written as:

$$\mathrm{d}\alpha_h^{(0)} = \mathrm{d}\sum_{i=0}^{p} \alpha_i \epsilon_i^{(0)} = \sum_{i=0,j=1}^{p} \mathsf{E}_{ij}^{(1,0)} \alpha_j \epsilon_i^{(1)} \, , \tag{9}$$

where, $\mathsf{E}_{ij}^{(1,0)}$ is the incidence matrix containing only the values, 0, 1 and -1, see [2, 4, 6] for more details. This idea can be extended to higher dimensions, giving rise to k-incidence matrices, $\mathsf{E}_{ij}^{(k+1,k)}$, which represent the discrete exterior derivative on discrete k-forms, see [2, 6].

4 Mimetic Spectral Advection: An Application to 1D Advection

In this section we want to illustrate how to discretize the advection equation. Take the Lie advection of a 1-form,

$$\frac{\partial \rho^{(1)}}{\partial t} + \mathrm{d}\iota_{\mathbf{v}}\rho^{(1)} = 0 \Leftrightarrow \begin{cases} \frac{\partial \rho^{(1)}}{\partial t} = -\mathrm{d}\varsigma^{(0)} \\ \iota_{\mathbf{v}}\,\rho^{(1)} = \varsigma^{(0)} \end{cases} . \tag{10}$$

Here $\rho^{(1)}$ is the advected quantity, say mass density, and $\varsigma^{(0)}$ represents the instantaneous fluxes of the advected quantity under the vector field $\mathbf{v}$, which are discretized in space as,

$$\varsigma_h^{(0)} = \sum_{i=0}^{p} \varsigma_i(t)\epsilon_i^{(0)} \quad \text{and} \quad \rho_h^{(1)} = \sum_{i=1}^{p} \rho_i(t)\epsilon_i^{(1)} . \tag{11}$$

For the sake of clarity in the method presentation we first introduce the time treatment, then the interior product discretization and finally their combination for a numerical solution of the advection problem.

4.1 Time Integration

The time integrator used for solving the time evolution part of the advection equation is the canonical mimetic one, an arbitrary order symplectic operator derived in [5], which is connected to canonical Gauss collocation integrators. Take an ordinary differential equation of the unknown function $y(t)$:

$$\frac{\mathrm{d}y}{\mathrm{d}t} = h(y,t), \quad t \in I \subset \mathbb{R} . \tag{12}$$

Discretizing $y(t)$ as $y_h = \sum_{k=0}^{p} y^k l_k(t)$, one gets:

$$\frac{\mathrm{d}y_h}{\mathrm{d}t} = \sum_{k=0}^{p} y^k \frac{\mathrm{d}l_k(t)}{\mathrm{d}t} = \sum_{k=1}^{p} (y^k - y^{k-1})\, e_k(t) ,$$

where the superscript k denotes the time level.

The approximated solution, $y_h(t)$, is a polynomial of order p determined by means of $(p+1)$ degrees of freedom such as its values at the Gauss-Lobatto nodes, red dots in Fig. 1. On the other hand, $\frac{\mathrm{d}y_h}{\mathrm{d}t}$ is a polynomial of order $(p-1)$ defined by only p degrees of freedom. One can set these degrees of freedom to be the values

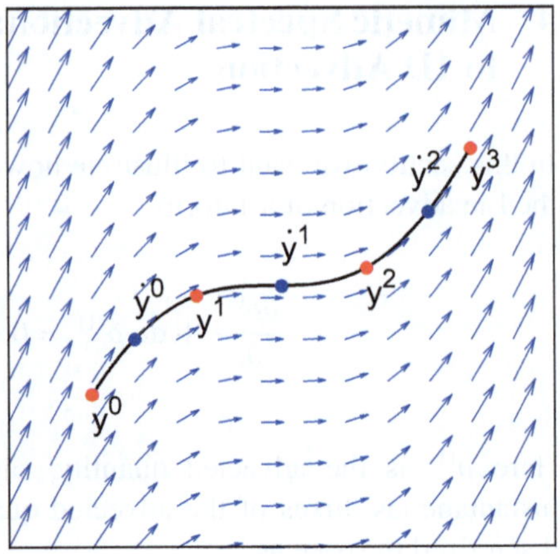

Fig. 1 Geometric interpretation of the solution of (12) as given by (13): $(t, y^{(0)}(t))$. In *red* the Gauss-Lobatto nodes where the trajectory is discretized. In *blue*, the Gauss nodes where its derivative is discretized. The flow field, represented by *arrows*, is tangent to the curve at the Gauss nodes. That is, the derivative of the approximate trajectory is exactly equal to the flow field at the Gauss nodes

of the derivative in one point inside each of the p intervals $[t^k, t^{k+1}]$. A choice that results in a symplectic integrator of order $2p$ is to select these points as the Gauss nodes of order $(p-1)$, the blue nodes of Fig. 1. Notice that along the trajectory these nodes will not show the usual Gauss-Lobatto and Gauss distribution patterns, since in general the velocity field is not constant. In this way the discrete integrator becomes:

$$\sum_{k=1}^{p}(y^k - y^{k-1})\, e_k(\tilde{t}^q) = h\left(\sum_{k=0}^{p} y^k l^k(\tilde{t}^q), \tilde{t}^q\right), \qquad q = 1, 2, \cdots, p\,, \quad (13)$$

with $\tilde{t}^j$ the p nodes of a Gauss quadrature formula. The fact that the instants in time, t^k, where the y_i^k are defined alternate with the instants in time, $\tilde{t}^j$, where the h_i are evaluated (see Fig. 1), corresponds to a staggering in time. This staggering also appears in leap-frog methods and in the implicit midpoint rule, for instance.

The first equation in (10) using the discretization (11) and (9) can be written as:

$$\frac{\sum_i \mathrm{d}\rho_i(t)\epsilon_i^{(1)}}{\mathrm{d}t} = -\sum_{il} \mathsf{E}_{il}^{(1,0)} \varsigma_l(t)\epsilon_i^{(1)} \quad \Rightarrow \quad \frac{\mathrm{d}\rho_i(t)}{\mathrm{d}t} = -\sum_{l} \mathsf{E}_{il}^{(1,0)} \varsigma_l(t)\,. \quad (14)$$

This equation has a similar form as (12), but now as a system of equations, therefore one can apply the mimetic integrator, yielding:

$$\sum_{k}(\rho_i^{k+1} - \rho_i^k)\, e_k(\tilde{t}^q) = -\sum_{l} \mathsf{E}_{il}^{(1,0)} \varsigma_l^q\,. \quad (15)$$

Recall that ρ_i^k is the discrete degree of freedom of the advected quantity at the t^k instants of time associated to Gauss-Lobatto nodes and ς_l^q is the discrete degree of freedom of the fluxes of the advected quantity at the $\tilde{t}^q$ instants of time associated to the Gauss nodes, just as stated for the systems of ordinary differential equations.

4.2 Interior Product

The discretization of the interior product is done using (3), in the following way:

Definition 1 (Discrete interior product). In one dimension, the discrete interior product $\iota_{\mathbf{v},h} : \Lambda_h^1 \rightarrow \Lambda_h^0$ is such that:

$$\left(\iota_{\mathbf{v},h}\, \alpha_h^{(1)}, \epsilon_i^{(0)}\right)_{L^2} = \left(\alpha_h^{(1)}, \mathbf{v}^\flat \wedge \epsilon_i^{(0)}\right)_{L^2}, \quad \forall \epsilon_i^{(0)} \in \Lambda_h^0 \tag{16}$$

where $\mathbf{v}^\flat = \nu^{(1)} \in \Lambda^1$ and $\alpha_h^{(1)} \in \Lambda_h^1$.

In this way one satisfies the duality pairing between the interior product and the wedge product in the discrete setting.

Partitioning the domain Ω in a spectral element cell complex one can apply the discretization of the interior product in each spectral element, obtaining:

$$\sum_i \rho_i(t) \left(\epsilon_i^{(1)}, \nu^{(1)} \wedge \epsilon_j^{(0)}\right)_{L^2} = \sum_i \varsigma_i(t) \left(\epsilon_i^{(0)}, \epsilon_j^{(0)}\right)_{L^2}, \quad \forall \epsilon_j^{(0)} \in \Lambda_h^0 . \tag{17}$$

4.3 Putting Things Together: Advection

The complete discrete systems becomes:

$$\begin{cases} \sum_k (\rho_i^{k+1} - \rho_i^k)\tilde{e}_k(\tilde{t}^q) = -\sum_l \mathsf{E}_{il}^{(1,0)} \varsigma_l^q \\ \sum_{i,k} \rho_i^k \epsilon_k^{(0)}(\tilde{t}^q) \left(\epsilon_i^{(1)}, \star\nu^{(0)} \wedge \epsilon_j^{(1)}\right)_{L^2\Lambda^1(\Omega_m)} = \sum_i \varsigma_i^q \left(\epsilon_i^{(0)}, \epsilon_j^{(0)}\right)_{L^2\Lambda^0(\Omega_m)}, \ \forall \epsilon_j^{(0)} \in \Lambda^0(\Omega_m) \end{cases} \tag{18}$$

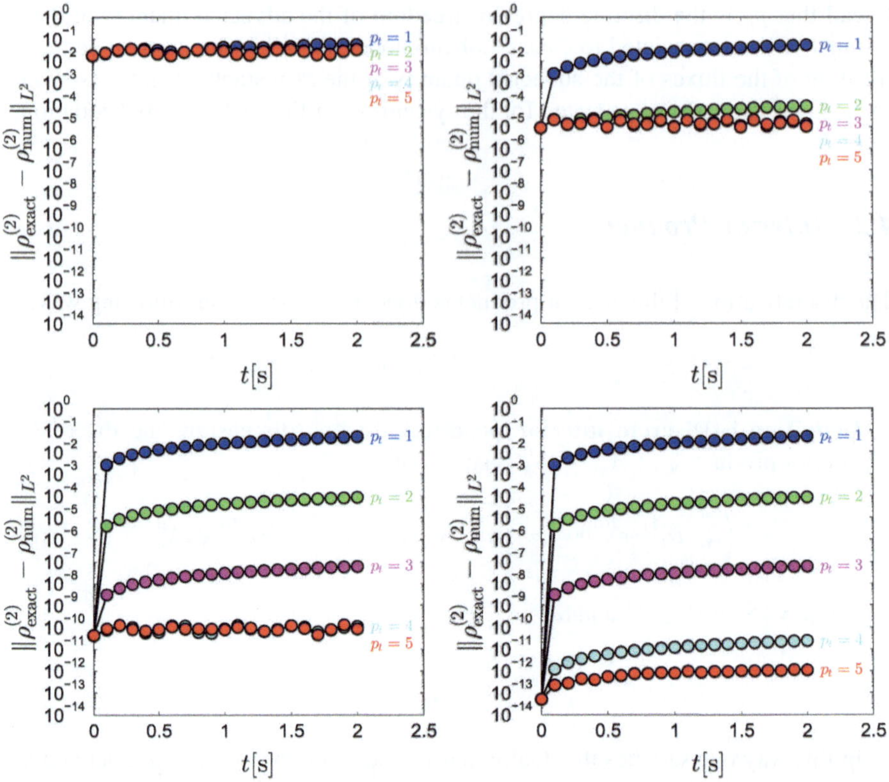

Fig. 2 Error in time of the numerical solution of (10) with $\mathbf{v} = \mathbf{e}_x$ and 4×4 elements of order $p = 3$, $p = 6$, $p = 10$ and $p = 12$ (from *left* to *right* and *top* to *bottom*) and $\Delta t = 0.1$ s, for the sine wave $\rho^{(2)}(x, y) = \sin(\pi x)\sin(\pi y)\mathrm{d}x\mathrm{d}y$. As shown, the error in the solution increases with time due to the inaccuracy of time integration. When time integration is accurate enough the error in the initial state is preserved. Here p_t denotes the polynomial degree in time

5 Numerical Results

This approach was applied to the two dimensional solution of an advected sine wave and a sine bell in a constant velocity field $\mathbf{v} = \mathbf{e}_x$: $\rho^{(2)}(x, y) = \sin(\pi x)\sin(\pi y)\mathrm{d}x\mathrm{d}y$ (sine wave) and $\rho^{(2)}(x, y) = \sin(2\pi x)\sin(2\pi y)\mathrm{d}x\mathrm{d}y$ if $(x, y) \in [0, 0.5] \times [0, 0.5]$ and $\rho^{(2)}(x, y) = 0$ in $(x, y) \in \mathbb{R}^2\backslash[0, 0.5] \times [0, 0.5]$ (sine bell), on a domain with periodic boundary conditions.

In Fig. 2 the error in time of the numerical solution of (10) for a mesh of 4×4 elements with a $\Delta t = 0.1$ s and various polynomial orders in space, p, and time, p_t, is presented. The initial error, due to the discretization, is conserved, as long as the time integration is sufficiently accurate.

In Fig. 3, the h- and p-convergence plots are shown for different values of the order of the time integration scheme, p_t, and $\Delta t = 0.1$ s. It is possible to see that the

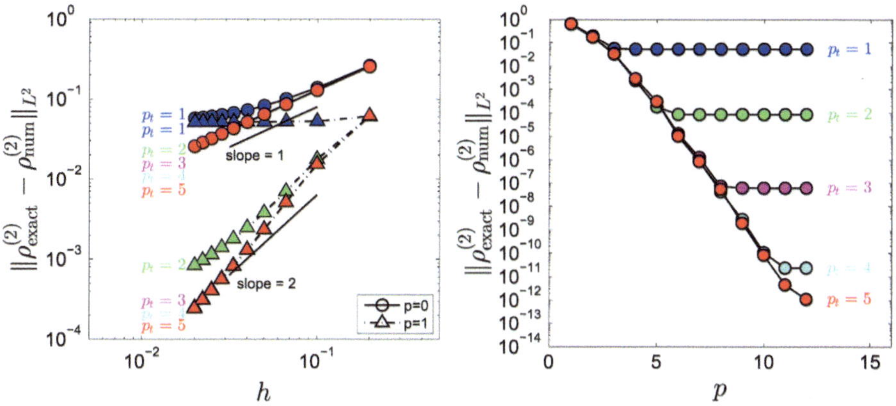

Fig. 3 *Left*: h convergence in space for the advection of a sine wave with $\Delta t = 0.1$ s. *Right*: p convergence in space for the advection of a sine wave, 4×4 elements and $\Delta t = 0.1$ s

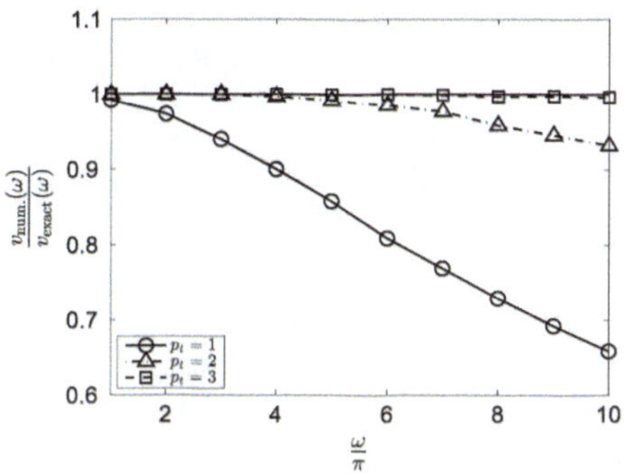

Fig. 4 Error in velocity as a function of the frequency of the advected sine wave: numerical dispersion. $p = 10$, $\Delta t = 0.1$ s and $n = 4 \times 4$ elements

method presents algebraic h-convergence rates of order $(p + 1)$ as long as the time integration error does not dominate the spatial one. The method shows a spectral p-convergence as soon as the time integration is accurate enough.

In Fig. 4 the error on the velocity is presented as a function of the advected sine wave frequency. This figure shows that the numerical method introduces an artificial dispersion if the time scheme is not accurate enough.

Another fundamental aspect is the conservation of the advected quantity. Figure 5 shows the mass error in time, that is: $\int_{\Omega} \rho_t^{(2)} - \int_{\Omega} \rho_{t_0}^{(2)}$. The error goes from the zero machine in the first 10^3 time steps while thereafter it steadily increases. Notice that even after 2×10^4 time steps the error is still below 10^{-12}.

In Fig. 6 one can see the advection of a sine wave of frequency $\omega = \pi$ in a Rudman vortex for 100 time steps after which the direction of the flow is reversed and the calculation is continued for another 100 time steps, with 4×4 curved

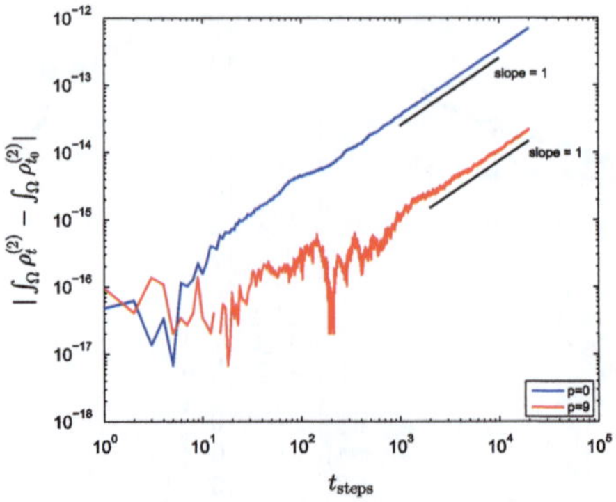

Fig. 5 Sum of the local errors of the advected 2-form for a sine bell in a velocity field $\mathbf{v} = \mathbf{e}_x$, with 50×50 elements of order $p = 0$ (*blue*) and 4×4 elements of order $p = 9$ (*red*), $\Delta t = 0.01$ and $p_t = 2$

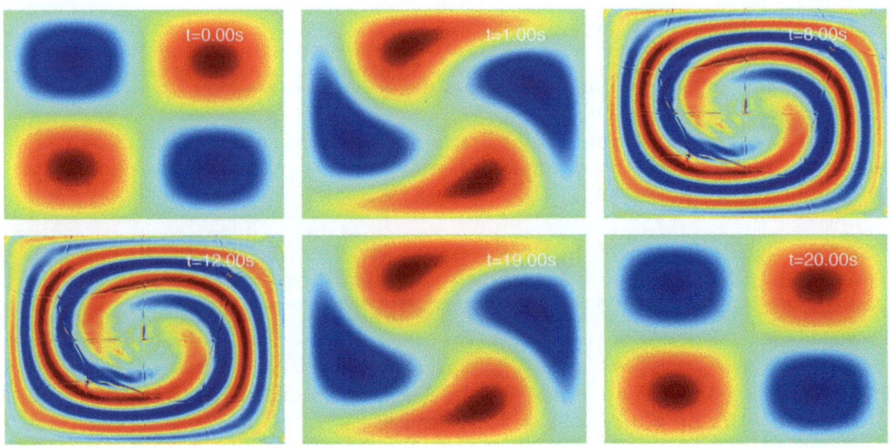

Fig. 6 From *left* to *right* and from *top* to *bottom*: advection of a sine wave of frequency $\omega = \pi$ on a Rudman vortex with 4×4 curved elements of order $p = 9$, $\Delta t = 0.1$ s and time integration of order $p_t = 2$, on a distorted mesh. At time $t = 10$, the flow field is reversed. At times $t = 8.0$ s and $t = 12.0$ s the mesh is visible in the solution. See http://www.youtube.com/watch?v=QmoJyqtk9YA for animation

elements of order $p = 9$, $\Delta t = 0.1$ s and time integration of order $p_t = 2$, on a distorted mesh. The mimetic advection enables one to recover the initial solution, thus demonstrating that the integration method is reversible.

Acknowledgements The authors would like to thank the valuable comments of both reviewers and the funding received by FCT – Foundation for science and technology Portugal through SRF/BD/36093/2007 and SFRH/BD/79866/2011.

Reference

1. R. Abraham, J. E. Marsden, and T. Ratiu. *Manifolds, Tensor Analysis, and Applications*, volume 75 of *Applied Mathematical Sciences*. Springer, 2001.
2. M. I. Gerritsma, R. Hiemstra, J. J. Kreeft, A. Palha, P. Rebelo, and D. Toshniwal. The geometric basis of numerical methods. *Proceedings of ICOSAHOM 2012 (this issue)*.
3. Marc Gerritsma. Edge functions for spectral element methods. In Jan S. Hesthaven and Einar M. Rønquist, editors, *Spectral and High Order Methods for Partial Differential Equations*, volume 76 of *Lecture Notes in Computational Science and Engineering*, pages 199–207. Springer Berlin Heidelberg, 2011.
4. J. Kreeft, A. Palha, and M. I. Gerritsma. Mimetic framework on curvilinear quadrilaterals of arbitrary order. *Arxiv preprint arXiv:1111.4304*, 2011.
5. A. Palha. *High order mimetic discretization: Development and application to Laplace and convection-diffusion problems in arbitrary quadrilaterals*. PhD thesis, TUDelft, 2013.
6. A. Palha, P. Pinto Rebelo, R. Hiemstra, J. Kreeft, and M. I. Gerritsma. Physics-compatible discretization techniques on single and dual grids, with application to the poisson equation of volume forms. *Submitted to J. Comp Phys.*, 2012.
7. N. Robidoux. Polynomial Histopolation, Superconvergent Degrees Of Freedom, And Pseudospectral Discrete Hodge Operators. *Unpublished: http://www.cs.laurentian.ca/nrobidoux/prints/super/histogram.pdf*, 2008.
8. Morita Shigeyuki. *Geometry of differential forms*, volume 201 of *Translations of mathematical monographs*. American mathematical society, 2001.

Large Eddy Simulation of a Muffler with the High-Order Spectral Difference Method

Matteo Parsani, Michael Bilka, and Chris Lacor

Abstract The combination of the high-order accurate spectral difference discretization on unstructured grids with subgrid-scale modelling is investigated for large eddy simulation of a muffler at $Re = 4.64 \cdot 10^4$ and $M = 0.05$. The subgrid-scale stress tensor is modelled by the wall-adapting local eddy-viscosity model with a cut-off length which is a decreasing function of the order of accuracy of the scheme. Numerical results indicate that even when a fourth-order accurate scheme is used, the coupling with a subgrid-scale model improves the quality of the results.

1 Introduction

Throughout the past two decades, the development of high-order accurate spatial discretization has been one of the major fields of research in computational fluid dynamics (CFD), computational aeroacoustics (CAA), computational electromagnetism (CEM) and in general computational physics characterized

M. Parsani (✉)
Computational Aerosciences Branch, NASA Langley Research Center, Hampton, VA 23681, USA

Division of Computer, Electrical and Mathematical Sciences & Engineering, King Abdullah University of Science and Technology, Thuwal, 23955-6900, Saudi Arabia
e-mail: matteo.parsani@nasa.gov; parsani.matteo@gmail.com

M. Bilka
University of Notre Dame, Department of Aerospace and Mechanical Engineering, 365 Fitzpatrick Hall, Notre Dame, IN 46556-5637, USA
e-mail: michael.bilka.1@nd.edu

C. Lacor
Department of Mechanical Engineering, Vrije Universiteit Brussel, Pleinlaan 2, 1050 Brussels, Belgium
e-mail: chris.lacor@vub.ac.be

M. Azaïez et al. (eds.), *Spectral and High Order Methods for Partial Differential Equations - ICOSAHOM 2012*, Lecture Notes in Computational Science and Engineering 95, DOI 10.1007/978-3-319-01601-6_27,

by linear and nonlinear wave propagation phenomena. High-order accurate discretizations have the potential to improve the computational efficiency required to achieve a desired error level. In fact, compared with low order schemes, high order methods offer better wave propagation properties and increased accuracy for a comparable number of degrees of freedom (DOFs). Therefore, it may be advantageous to use such schemes for problems that require very low numerical dissipation and small error levels [1]. Moreover, since computational science is increasingly used as an industrial design and analysis tool, high accuracy must be achieved on unstructured grids which are required for efficient meshing. These needs have been the driving force for the development of a variety of higher order schemes for unstructured meshes such as the Discontinuous Galerkin (DG) method [2, 3], the Spectral Volume (SV) method [4], the Spectral Difference (SD) method [5, 6], the Energy Stable Flux Reconstruction [7] and many others.

In this study we focus on a SD solver for unstructured hexahedral grids (tensorial cells). The SD method has been proposed as an alternative high order collocation-based method using local interpolation of the strong form of the equations. Therefore, the SD scheme has an important advantage over classical DG and SV methods, that no integrals have to be evaluated to compute the residuals, thus avoiding the need for costly high-order accurate quadrature formulas.

Although the formulation of high-order accurate spatial discretization is now fairly mature, their application for the simulation of general turbulent flows implies that particular attention has still to be paid to subgrid-scale (SGS) models. So far, the combination of the SD method with SGS models for LES has not been widely investigated. In 2010, Parsani et al. [8] reported the first implementation in study of a two-dimensional (2D) third-order accurate SD solver coupled with the Wall-Adapting Local Eddy-viscosity (WALE) model [14] and a cut-off length which is a decreasing function of the order of accuracy. A successful extension of that approach to a three-dimensional (3D) second-order accurate SD solver has been reported in [12]. Very recently, Lodato and Jameson [13] have presented an alternative technique to model the unresolved scales in the flow field: A structural SGS approach with the WALE Similarity Mixed model (WSM), where constrained explicit filtering represents a key element to approximate subgrid-scale interactions. The performance of such an algorithm has been also satisfactory.

In this study, we couple for the first time the approach proposed in [8] with a 3D fourth-order accurate SD solver, for the simulation of the turbulent flow in an industrial-type muffler at $Re = 4.64 \cdot 10^4$. The goal is to investigate if the coupling of a high-order SD scheme with a sub-grid closure model improves the quality of the results when the grid resolution is relatively low. The latter requirement is often desirable when a high-order accurate spatial discretization is used.

2 Physical Model and Numerical Algorithm

In this study the system of the Navier-Stokes equations for a compressible flow are discretized in space using the SD method and the subgrid-scale stress tensor is modelled by the WALE approach.

2.1 Filtered Navier-Stokes Equations

The three physical conservation laws for a general Newtonian fluid, i.e., the continuity, the momentum and energy equations, are introduced using the following notation: ρ for the mass density, $\mathbf{u} \in \mathbb{R}^{dim}$ for the velocity vector in a physical space with dim dimensions, P for the static pressure and E for the specific total energy which is related to the pressure and the velocity vector field by $E = \frac{1}{\gamma-1}\frac{P}{\rho} + \frac{|\mathbf{u}|^2}{2}$, where γ is the constant ratio of specific heats and it is 1.4 for air in standard conditions.

The system, written in divergence form and equipped with suitable initial-boundary conditions, is

$$\frac{\partial \mathbf{w}}{\partial t} + \nabla \cdot (\mathbf{f}_C(\mathbf{w}) - \mathbf{f}_D(\mathbf{w}, \nabla \mathbf{w})) = \frac{\partial \mathbf{w}}{\partial t} + \nabla \cdot \mathbf{f} = 0, \tag{1}$$

where $\mathbf{w} = \left(\overline{\rho}, \overline{\rho}\tilde{\mathbf{u}}, \overline{\rho}\tilde{E}\right)^T$ is the vector of the filtered conservative variables and $\mathbf{f}_C = \mathbf{f}_C(\mathbf{w})$ and $\mathbf{f}_D = \mathbf{f}_D(\mathbf{w}, \nabla \mathbf{w})$ represent the convective and the diffusive fluxes, respectively. Here the symbols $\overline{(\cdot)}$ and $\widetilde{(\cdot)}$ represent the spatially filtered field and the Favre filtered field defined as $\tilde{\mathbf{u}} = \overline{\rho \mathbf{u}}/\overline{\rho}$.

In a general 3D ($dim = 3$) Cartesian space, $\mathbf{x} = [x_1, x_2, x_3]^T$, the components of the flux vector $\mathbf{f}(\mathbf{w}, \nabla \mathbf{w}) = [\mathbf{f}_1, \mathbf{f}_2, \mathbf{f}_3]^T$ are given by

$$\mathbf{f}_1 = \begin{pmatrix} \overline{\rho}\tilde{u}_1 \\ \overline{\rho}\tilde{u}_1^2 + \overline{P} - \tilde{\sigma}_{11} + \tau_{11}^{sgs} \\ \overline{\rho}\tilde{u}_1\tilde{u}_2 - \tilde{\sigma}_{21} + \tau_{21}^{sgs} \\ \overline{\rho}\tilde{u}_1\tilde{u}_3 - \tilde{\sigma}_{31} + \tau_{31}^{sgs} \\ \tilde{u}_1\left(\overline{\rho}\tilde{E} + \overline{P}\right) - \tilde{u}_1\left(\tilde{\sigma}_{11} - \tau_{11}^{sgs}\right) - \tilde{u}_2\left(\tilde{\sigma}_{21} - \tau_{21}^{sgs}\right) - \tilde{u}_3\left(\tilde{\sigma}_{31} - \tau_{31}^{sgs}\right) - c_P \frac{\mu}{Pr}\frac{\partial \tilde{T}}{\partial x_1} + q_1^{sgs} \end{pmatrix},$$

$$\mathbf{f}_2 = \begin{pmatrix} \overline{\rho}\tilde{u}_2 \\ \overline{\rho}\tilde{u}_1\tilde{u}_2 - \tilde{\sigma}_{12} + \tau_{12}^{sgs} \\ \overline{\rho}\tilde{u}_2^2 + \overline{P} - \tilde{\sigma}_{22} + \tau_{22}^{sgs} \\ \overline{\rho}\tilde{u}_2\tilde{u}_3 - \tilde{\sigma}_{32} + \tau_{32}^{sgs} \\ \tilde{u}_2\left(\overline{\rho}\tilde{E} + \overline{P}\right) - \tilde{u}_1\left(\tilde{\sigma}_{12} - \tau_{12}^{sgs}\right) - \tilde{u}_2\left(\tilde{\sigma}_{22} - \tau_{22}^{sgs}\right) - \tilde{u}_3\left(\tilde{\sigma}_{32} - \tau_{32}^{sgs}\right) - c_P \frac{\mu}{Pr}\frac{\partial \tilde{T}}{\partial x_2} + q_2^{sgs} \end{pmatrix},$$

$$\mathbf{f}_3 = \begin{pmatrix} \overline{\rho}\tilde{u}_3 \\ \overline{\rho}\tilde{u}_1\tilde{u}_3 - \tilde{\sigma}_{13} + \tau_{13}^{sgs} \\ \overline{\rho}\tilde{u}_2\tilde{u}_3 - \tilde{\sigma}_{23} + \tau_{23}^{sgs} \\ \overline{\rho}\tilde{u}_3^2 + \overline{P} - \tilde{\sigma}_{33} + \tau_{33}^{sgs} \\ \tilde{u}_3\left(\overline{\rho}\tilde{E} + \overline{P}\right) - \tilde{u}_1\left(\tilde{\sigma}_{13} - \tau_{13}^{sgs}\right) - \tilde{u}_2\left(\tilde{\sigma}_{23} - \tau_{23}^{sgs}\right) - \tilde{u}_3\left(\tilde{\sigma}_{33} - \tau_{33}^{sgs}\right) - c_P \frac{\mu}{Pr}\frac{\partial \tilde{T}}{\partial x_3} + q_3^{sgs} \end{pmatrix},$$

where c_P, μ, Pr and T represent respectively the specific heat capacity at constant pressure, the dynamic viscosity, the Prandtl number and the temperature of the fluid. Moreover, σ_{ij} represents the ij−component of the resolved viscous stress tensor [15].

Both momentum and energy equations differ from the classical fluid dynamic equations only for two terms which take into account the contributions from the unresolved scales. These contributions, represented by the specific subgrid-scale stress tensor τ_{ij}^{sgs} and by the subgrid heat flux vector defined q_i^{sgs}, appear when the spatial filter is applied to the convective terms [15]. The interactions of τ_{ij}^{sgs} and q_i^{sgs} with the resolved scales have to be modeled through a subgrid-scale closure model because they cannot be determined using only the resolved flow field **w**.

2.1.1 The Wall-Adapted Local Eddy-Viscosity Closure Model

The smallest scales present in the flow field and their interaction with the resolved scales have to be modeled through the subgrid-scale term τ_{ij}^{sgs}. The most common approach to model such a tensor is based on the eddy-viscosity concept in which one assumes that the residual stress is proportional to a measure of the filtered local strain rate [15], which is defined as follows:

$$\tau_{ij}^{sgs} - \tau_{kk}^{sgs}\delta_{ij} = -2\,\overline{\rho}\,\nu_t \left(\tilde{S}_{ij} - \frac{\delta_{ij}}{3}\tilde{S}_{kk}\right). \tag{2}$$

In the WALE model, it is assumed that the eddy-viscosity ν_t is proportional to the square of the length scale of the cut-off length (or width of the grid filter) and the filtered local rate of strain. Although the model was originally developed for incompressible flows, it can also be used for variable density flows by giving the formulation as follows

$$\nu_t = (C\Delta)^2 \left|\tilde{S}\right|. \tag{3}$$

Here $\left|\tilde{S}\right|$ is defined as

$$\left|\tilde{S}\right| = \frac{\left[\tilde{S}_{ij}^d\,\tilde{S}_{ij}^d\right]^{3/2}}{\left[\tilde{S}_{ij}\,\tilde{S}_{ij}\right]^{5/2} + \left[\tilde{S}_{ij}^d\,\tilde{S}_{ij}^d\right]^{5/4}}, \tag{4}$$

where $\tilde{S}_{ij}^d$ is the traceless symmetric part of the square of the resolved velocity gradient tensor $\tilde{g}_{ij} = \frac{\partial \tilde{u}_i}{\partial x_j}$. Note that in Eq. (3) Δ, i.e., the cut-off length, is an unknown function. Often the cut-off length is taken proportional to the smallest resolvable length scale of the discretization. In the present work, the definition of the grid filter function is given in Sect. 2.2, where the SD method is discussed.

2.2 Spectral Difference Method

Consider a problem governed by a general system of conservation laws given by Eq. (1) and valid on a domain $\Omega \subset \mathbb{R}^{dim}$ with boundary $\partial\Omega$ and completed with consistent initial and boundary conditions. The domain is divided into N non-overlapping cells, with cell index i.

In order to achieve an efficient implementation of the SD method, all hexahedral cells in the physical domain are mapped into cubic elements using local coordinates $\boldsymbol{\xi} = [\xi_1, \xi_2, \xi_3]^T$. Such a transformation is characterized by the Jacobian matrix $\mathbf{J}_i$ with determinant $det(\mathbf{J}_i)$. Therefore, system (1) can be written in the mapped coordinate system as

$$\frac{\partial \mathbf{w}_i^{\xi}}{\partial t} = -\frac{\partial \mathbf{f}_{1,i}^{\xi}}{\partial \xi_1} - \frac{\partial \mathbf{f}_{2,i}^{\xi}}{\partial \xi_2} - \frac{\partial \mathbf{f}_{3,i}^{\xi}}{\partial \xi_3} = -\nabla^{\xi} \cdot \mathbf{f}_i^{\xi}, \tag{5}$$

where $\mathbf{w}_i^{\xi} \equiv det(\mathbf{J}_i)\,\mathbf{w}$ and ∇^{ξ} are the conserved variables and the generalized divergence differential operator in the mapped coordinate system, respectively.

For a $(p+1)$-th-order accurate dim-dimensional scheme, N^s *solution collocation points* with index j are introduced at positions $\boldsymbol{\xi}_j^s$ in each cell i, with N^s given by $N^s = (p+1)^{dim}$. Given the values at these points, a polynomial approximation of degree p of the solution in cell i can be constructed. This polynomial is called the *solution polynomial* and is usually composed of a set of Lagrangian basis polynomial $L_j^s(\boldsymbol{\xi})$ of degree p:

$$\mathbf{W}_i(\boldsymbol{\xi}) = \sum_{j=1}^{N^s} \mathbf{W}_{i,j}\, L_j^s(\boldsymbol{\xi}). \tag{6}$$

Therefore, the unknowns of the SD method are the interpolation coefficients $\mathbf{W}_{i,j} = \mathbf{W}_i\left(\boldsymbol{\xi}_j^s\right)$ which are the approximated values of the conserved variables $\mathbf{w}_i$ at the solution points.

The divergence of the mapped fluxes $\nabla^{\xi} \cdot \mathbf{f}^{\xi}$ at the solution points is computed by introducing a set of N^f flux collocation points with index l and at positions $\boldsymbol{\xi}_l^f$, supporting a polynomial of degree $p+1$. The evolution of the mapped flux vector $\mathbf{f}^{\xi}$ in cell i is then approximated by a flux polynomial $\mathbf{F}_i^{\xi}$, which is obtained by reconstructing the solution variables at the flux points and evaluating the fluxes $\mathbf{F}_{i,l}^{\xi}$ at these points. The flux is also represented by a Lagrange polynomial:

$$\mathbf{F}_i^{\xi}(\boldsymbol{\xi}) = \sum_{l=1}^{N^f} \mathbf{F}_{i,l}^{\xi}\, L_l^f(\boldsymbol{\xi}), \tag{7}$$

where the coefficients of the flux interpolation are defined as

$$\mathbf{F}^{\xi}_{i,l} = \begin{cases} \mathbf{F}^{\xi}_{i}\left(\xi^{f}_{l}\right), & \xi^{f}_{l} \in \Omega_i, \\ \mathbf{F}^{\xi}_{\text{num}}\left(\xi^{f}_{l}\right), & \xi^{f}_{l} \in \partial\Omega_i. \end{cases} \tag{8}$$

Here $\mathbf{F}^{\xi}_{\text{num}}$ is the numerical flux vector at the cell interface. In fact, the solution at a face is in general not continuous and requires the solution of a Riemann problem to maintain conservation at a cell level (i.e., the flux component normal to a face $\mathbf{F}^{\xi}_{\text{num}}\cdot\mathbf{n}^{\xi}$ must be continuous between two neighboring cells). Approximate Riemann solvers are typically used (e.g. Rusanov Riemann solver). The tangential component of $\mathbf{F}^{\xi}_{\text{num}}$ is usually taken from the interior cell.

Taking the divergence of the flux polynomial $\nabla^{\xi}\cdot\mathbf{F}^{\xi}_{i}$ in the solution points results in the following modified form of (5), describing the evolution of the conservative variables at the solution points:

$$\frac{d\mathbf{W}_{i,j}}{dt} = -\left.\nabla\cdot\mathbf{F}_{i}\right|_{j} = -\frac{1}{J_{i,j}}\left.\nabla^{\xi}\cdot\mathbf{F}^{\xi}_{i}\right|_{j} = \mathbf{R}_{i,j}, \tag{9}$$

where $\mathbf{F}_i$ is the flux polynomial vector in the physical space whereas $\mathbf{R}_{i,j}$ is the SD residual associated with $\mathbf{W}_{i,j}$. This is a system of ODEs, in time, for the unknowns $\mathbf{W}_{i,j}$. In this work, the optimized explicit eighteen-stages fourth-order Runge-Kutta schemes presented in [16] is used to solve such a system at each time step.

2.2.1 Solution and Flux Points Distributions

In 2007, Huynh [9] showed that for quadrilateral and hexahedral cells, tensor product flux point distributions based on a one-dimensional flux point distribution consisting of the end points and the Legendre-Gauss quadrature points lead to stable schemes for arbitrary order of accuracy. In 2008, Van den Abeele et al. [10] showed an interesting property of the SD method, namely that it is independent of the positions of its solution points in most general circumstances. This property implies an important improvement in efficiency, since the solution points can be placed at flux point positions and thus a significant number of solution reconstructions can be avoided. Recently, this property has been proved by Jameson [11].

2.2.2 Cut-Off Length Δ

In Sect. 2.1.1 we have seen that in the WALE model the cut-off length Δ is used to compute the turbulent eddy-viscosity ν_t, i.e., Eq. (3). Following the approach presented in [8], for each cell with index i and each flux points with index l and positions ξ^{f}_{l}, we use the following definition of filter width

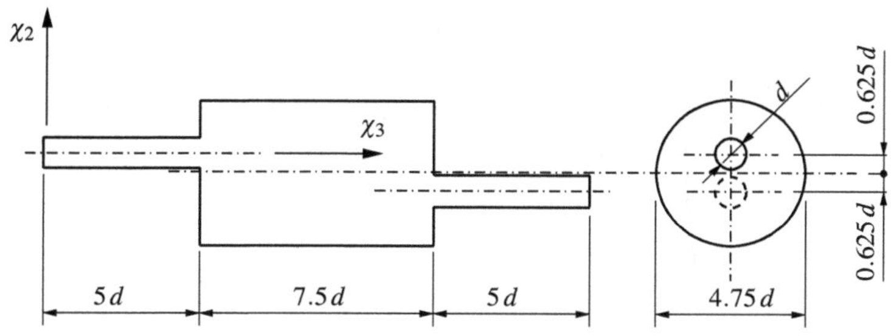

Fig. 1 Configuration of the 3D muffler test case

$$\Delta_{i,l} = \left[\frac{1}{N^s} det\left(\mathbf{J}_i|_{\xi_l^f}\right)\right]^{1/dim} = \left(\frac{det(J_{i,l})}{N^s}\right)^{1/dim}. \tag{10}$$

Notice that the cell filter width is not constant in one cell, but it varies because the Jacobian matrix is a function of the positions of the flux points. Moreover, for a given mesh, the number of solution points depends on the order of the SD scheme, so that the grid filter width decreases by increasing the polynomial order of the approximation.

3 Numerical Results

The main purpose of this section is to evaluate the accuracy and the reliability of the fourth-order SD-LES solver for simulating a 3D turbulent flow in an industrial-type muffler. The results are compared with the particle image velocimetry (PIV) measurement performed at the Department of Environmental and Applied Fluid Dynamics of the von Karman Institute for Fluid Dynamics [17]. In Fig. 1, the geometry of the muffler and its characteristic dimensions are illustrated, where the flow is from left to right.

At the inlet, mass density and velocity profiles are imposed. The inlet velocity profile in the x_3 direction is given by

$$u_3 = u_{max}\left\{\frac{1}{2} - \frac{1}{2}\tanh\left[2.2\left(\frac{r}{d/2} - \frac{d/2}{r}\right)\right]\right\}.$$

At the outlet only the pressure is prescribed. In accordance to the experiments, the inlet Mach number and the Reynolds number, based on maximum velocity at the inlet u_{max} and the diameter of the inlet/outlet d ($d = 4\,cm$), are set respectively to $M_{inlet} = 0.05$ and $Re = 4.64 \cdot 10^4$.

The flow is computed using fourth-order ($p = 3$) SD scheme on a grid with 36,612 hexahedral elements which was generated with the open source software Gmsh [18]. Second-order boundary elements are used to approximate the curved

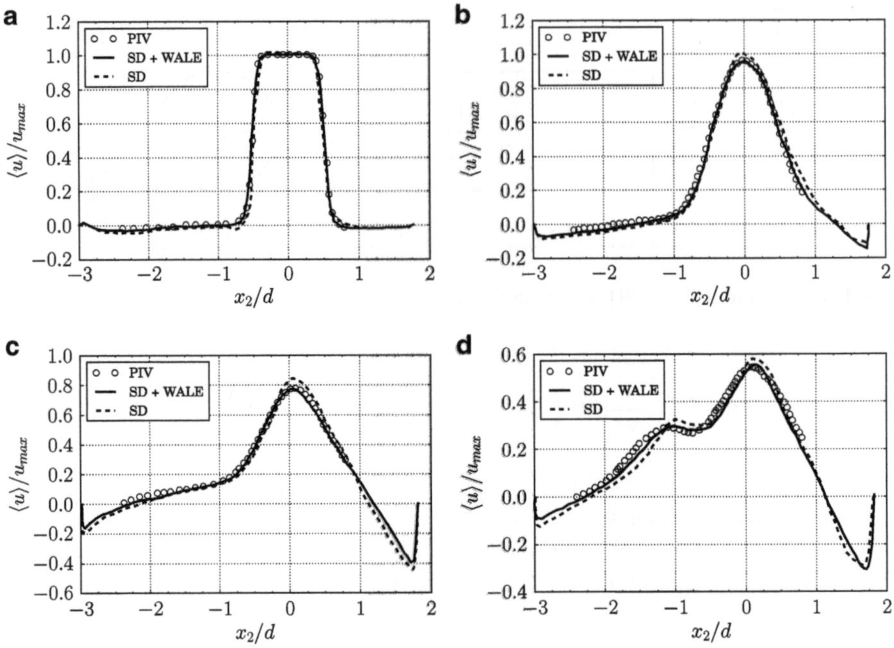

Fig. 2 Time-averaged velocity profile in the axial direction $\langle \tilde{u}_3 \rangle / u_{max}$ at four cross sections in the expansion chamber, obtained with fourth-order ($p = 3$) SD-LES method. Comparison with experimental measurements (PIV) [17]. (**a**) 1d downstream. (**b**) 4d downstream. (**c**) 6d downstream. (**d**) 7d downstream

geometry. The total number of DOFs is approximately $2.3 \cdot 10^6$ (i.e., $36,612 \cdot (p + 1)^3$). The maximum CFL number used for the computations started from 0.1 and increased up to 0.65. After the flow field was fully developed, time averaging is performed for a period corresponding to about 25 flow-through times.

The computation is validated on the center plane of the expansion coinciding with the center planes of the inlet and outlet pipes using the PIV results from [17]. All of the measurements are taken on the symmetrical center plane of the muffler. The reference cross section corresponds to the entrance of the expansion chamber. It should be noted that the circular nature of the geometry acts as a lens causing a change in magnification in the radial direction (x_2) which prevents from capturing images close to the wall. It is found that outside 1 cm from the wall the magnification effect is negligible and as the mean stream-wise direction is in the direction of constant magnification and has only little effect on the particle correlations no corrections are deemed necessary.

In Fig. 2, the non-dimensional mean velocity profile in the axial direction $\langle \tilde{u}_3 \rangle / u_{max}$ is shown for four different cross sections in the expansion chamber, where the PIV measurements were done. In this figure, the PIV data are also plotted for comparison. Figure 3 shows the non-dimensional Reynolds stress $\langle u'_2 u'_3 \rangle / u^2_{max}$ at

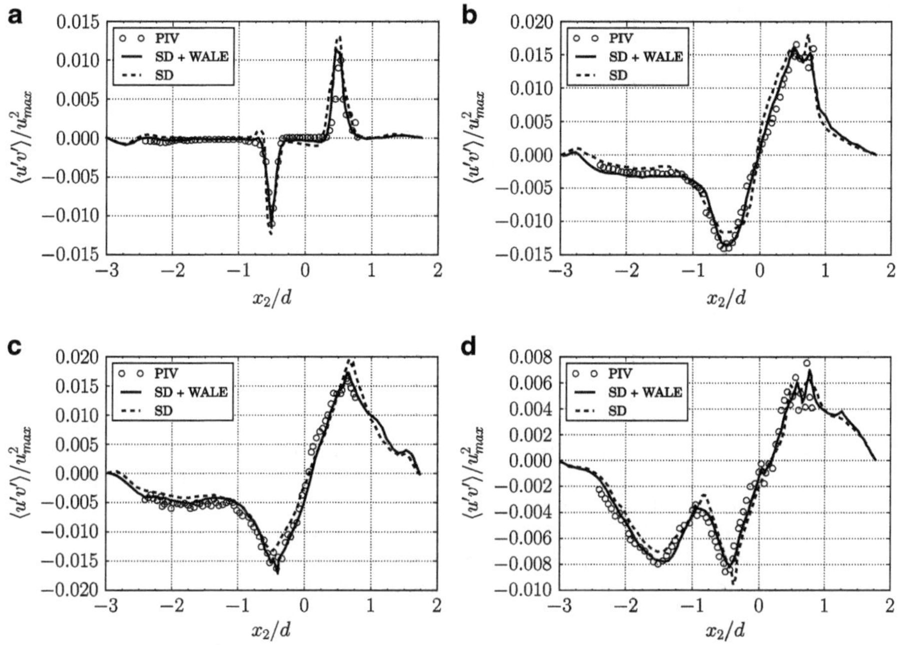

Fig. 3 Reynolds stress $\langle u_2' u_3' \rangle / u_{max}^2$ in the axial direction at four cross sections in the expansion chamber, obtained with fourth-order ($p = 3$) SD-LES method. Comparison with experimental measurements (PIV) [17]. (**a**) $1d$ downstream. (**b**) $4d$ downstream. (**c**) $6d$ downstream. (**d**) $7d$ downstream

the same cross sections. Although the high-order implicit LES is already able to capture well the features of the flow field, the use of the WALE model improves the results. In particular, when the SGS model is active, the local extrema of the time-averaged velocity profiles and the second-order statistical moment (which get fairly oscillatory by moving far away from the inlet pipe) are better captured.

4 Conclusions

The fourth-order SD method in combination with the WALE model and the variable filter width performs well. The numerical results confirm that the model is correctly accounting for the unresolved shear stress computed from the resolved field, for the present internal flow. However, it should be noted that the solver without subgrid-scale modelling also works rather well, for the grid resolution used in this study.

Work is currently under way to test the two approaches for other realistic turbulent flows, varying the order of accuracy and the grid resolution. We believe that the flexibility of the high-order SD scheme on unstructured grids together with probably the development of robust sub-grid closure models and efficient grid generators for high-order accurate schemes will allow to perform LES of industrial-type flows in the near future.

Acknowledgements The authors would like to thank Professor David I. Ketcheson and Professor Mark H. Carpenter for their support. This research used partially the resources of the KAUST Supercomputing Laboratory and was supported in part by an appointment to the NASA Postdoctoral Program at Langley Research Center, administered by Oak Ridge Associates Universities. These supports are gratefully acknowledged.

References

1. Wang, Z. J.: Adaptive High-order Methods in Computational Fluid Dynamics (Advances in Computational Fluid Dynamics), World Scientific Publishing Company (2011).
2. Busch, K., König, M., and Niegemann, J.: Discontinuous Galerkin methods in nanophotonics. Laser Photonics Rev. **5**(6), 773–809 (2011).
3. Hesthaven, Jan S., and Warburton, Tim: Nodal Discontinuous Galerkin Methods: Algorithms, Analysis, and Applications. Texts in Applied Mathematics, **54**. Computational Science & Engineering, Springer Publishing Company, Incorporated (2007).
4. Sun, Y., Wang, Z. J., and Liu Y.: Spectral (finite) volume method for conservation laws on unstructured grids VI: extension to viscous flow. J. of Comput. Phys., **215**(1), 41–58 (2006).
5. May, G., and Jameson, A.: A spectral difference method for the Euler and Navier-Stokes equations on unstructured meshes. *AIAA paper* 2006–304. 44th AIAA Aerospace Sciences Meeting, Reno, Nevada, USA, January 9–12, 2006.
6. Sun, Y., Wang, Z. J., and Liu, Y.: High-order multidomain spectral difference method for the Navier-Stokes equations on unstructured hexahedral grids. Commun. Comput. Phys, **2**(2), 310–333 (2007).
7. Castonguay, P., Vincent P., and Jameson, A.: Application of High-Order Energy Stable Flux Reconstruction Schemes to the Euler Equations. *AIAA paper* 2011-686. 49th AIAA Aerospace Sciences Meeting including the New Horizons Forum and Aerospace Exposition, Orlando, Florida, USA, January 4–7, 2011.
8. Parsani, M., Ghorbaniasl, G., Lacor C., and Turkel, E.: An implicit high-order spectral difference approach for large eddy simulation. J. Comput. Phys., **229**(14), 5373–5393 (2010).
9. Huynh, H. T.: A flux reconstruction approach to high-order schemes including discontinuous Galerkin methods. *AIAA paper* 2007–4079. 18th AIAA Computational Fluid Dynamics Conference, Miami, Florida, USA, June 25–28, 2007.
10. Van den Abeele, K., Lacor, C., and Wang, Z. J.: On the stability and accuracy of the spectral difference method, J. Sci. Comput., **37**(2), 162–188 (2008).
11. Jameson, A.: A proof of the stability of the spectral difference method for all orders of accuracy, J. Sci. Comput., **45**(1–3), 348–358 (2010).
12. Parsani, M., Ghorbaniasl, G., and Lacor C.: Validation and application of an high-order spectral difference method for flow induced noise simulation. J. Comput. Acoust., **19**(3), 241–268 (2011).
13. Lodato, G., and Jameson, A.: LES modeling with high-order flux reconstruction and spectral difference schemes. *ICCFD paper* 2201. 7th ICCFD Conference, Big Island, Hawaii, July 9–13, 2012.
14. Nicoud, F., and Ducros, F.: Subgrid-scale stress modelling based on the square of the velocity gradient tensor. Flow Turbul. Combust., **62**(3), 183–200 (1999).
15. Pope, Stephen B.: Turbulent flows. Cambridge University Press (2003).

16. Parsani, M., Ketcheson, David I., and Deconinck, W.: Optimized explicit Runge-Kutta schemes for the spectral difference method applied to wave propagation problems, SIAM J. Sci. Comput, 35(2):A957–A986 (2013).
17. Bilka, M., and Anthoine, J.: Experimental investigation of flow noise in a circular expansion using PIV and acoustic measurements. *AIAA paper* 2008–2952. 14th AIAA/CEAS Aeroacoustics Conference, Vancouver, British Columbia, Canada, May 5–6, 2008.
18. Geuzaine, C., and Remacle J.-F.: Gmsh: a three-dimensional finite element mesh generator with built-in pre- and post-processing facilities. Int. J. Numer. Meth. Eng., **79**(11), 1309–1331 (2009).

Stability Tools for the Spectral-Element Code Nek5000: Application to Jet-in-Crossflow

A. Peplinski, P. Schlatter, P.F. Fischer, and D.S. Henningson

Abstract We demonstrate the use of advanced linear stability tools developed for the spectral-element code `Nek5000` to investigate the dynamics of nonlinear flows in moderately complex geometries. The aim of stability calculations is to identify the driving mechanism as well as the region most sensitive to the instability: the *wavemaker*. We concentrate on global linear stability analysis, which considers the linearised Navier–Stokes equations and searches for growing small disturbances, i.e. so-called linear global modes. In the structural sensitivity analysis these modes are associated to the eigenmodes of the direct and adjoint linearised Navier–Stokes operators, and the wavemaker is defined as the overlap of the strongest direct and adjoint eigenmodes. The large eigenvalue problems are solved using matrix-free methods adopting the time-stepping Arnoldi approach. We present here our implementation in `Nek5000` with the `ARPACK` library on a number of test cases.

1 Introduction

The flow of fluids can be either laminar, characterised by smooth patterns, or turbulent, appearing chaotic and unpredictable. Understanding the physics of laminar-turbulent flow transition has been originally motivated by aerodynamic applications, but has become more widespread since. Initially, hydrodynamic

A. Peplinski (✉) · P. Schlatter · D.S. Henningson
Linné FLOW Centre and Swedish e-Science Research Centre (SeRC), KTH Mechanics, Royal Institute of Technology, SE-100 44 Stockholm, Sweden
e-mail: adam@mech.kth.se; pschlatt@mech.kth.se; henning@mech.kth.se

P.F. Fischer
MCS, Argonne National Laboratory, 9700 S. Cass Avenue, Argonne, IL 60439, USA
e-mail: fischer@mcs.anl.gov

M. Azaïez et al. (eds.), *Spectral and High Order Methods for Partial Differential Equations - ICOSAHOM 2012*, Lecture Notes in Computational Science and Engineering 95, DOI 10.1007/978-3-319-01601-6_28,

stability was studied by means of the classical linear stability theory investigating the behaviour of small disturbances in space and time around some basic flow state. The exponential growth of linear perturbations is studied at each streamwise position and the distinction between local convective and absolute stability is made [12]. This local treatment is legitimate for parallel and weakly non-parallel flows, but many of the flow configurations developing strong instabilities and eventually exhibiting transition to turbulence are strongly non-parallel and may belong to the open flow category, where fluid particles continuously enter and leave the considered domain. Such unstable open flows require global analysis where the evolution of perturbations is considered in the whole physical domain [6]. The global behaviour of the flow depends on the competition between local instability and basic advection. The extensive work on global stability in the past years has been reviewed e.g. in [17].

However, such linear modal analysis often fails in predicting the transitional Reynolds number determined experimentally, and the more accurate transition scenario based on receptivity has to be considered. In this case the non-normality of the linearised Navier-Stokes (LNS) operator has to be taken into account [6, 18] and the global modes of the adjoint operator have to be calculated. This kind of analysis has been performed for the 2D cases e.g. the flow past a circular cylinder [9]. The limitations of structural sensitivity analysis are discussed by Chomaz [6], where it was pointed out that this method is better suited for strongly non-parallel flows than for almost parallel flows, as the very high degree of the operator non-normality can lead to wrong predictions of the dynamics.

Although global analysis allows to avoid the limitation of local theory, it is computationally much more expensive, as linear global modes have been associated to the eigenmodes of the LNS operator [10] involving large eigenvalue problems. For sizes of order $\dim(\mathbf{A}) \sim 10^7$ special matrix-free methods using time-steppers are required [2, 3]. Recent advances in numerical methods, in particular tools for solving very large eigenvalue problems [15], make it possible to use linear stability theory for global analysis of 2D and 3D flows with nearly arbitrary complexity, based on only minimal modifications of existing numerical simulation codes [4]. A number of authors have determined the spectrum of the LNS operator for different 2D flows, however, the first calculations for the fully 3D base flow were done by Bagheri et al. [3, 16] for a jet in crossflow (JCF). This work has been later extended in [13] by calculating 3D adjoint global eigenmodes.

The objective of the present paper is to demonstrate the use of global linear stability tools developed for the spectral-element code `Nek5000` [7,8] to investigate the dynamics of flows in moderately complex geometry. As the final case we consider the so-called jet in crossflow which refers to a jet of fluid exiting through a nozzle and interacts with the surrounding cross-flow fluid. It is a canonical flow with complex, fully 3D dynamics which allows for a test of the simulation capabilities and the methods for studying the flow stability. The previous results for this flow [3, 13, 16] were obtained for simplified setups, in which the inflow jet was represented by a Dirichlet boundary condition due to the limitations of the applied pseudo-spectral simulation method. We avoid this limitation using the more flexible spectral-element method (SEM), which provides spectral accuracy while allowing for complex geometries.

2 Direct and Adjoint Global Modes

Structural sensitivity analysis determines the instability mechanism that initiates the transition to an unsteady flow. It combines global linear stability with receptivity looking into the eigenmodes of the LNS (direct) operator $\mathbf{A}$ and its adjoint $\mathbf{A}^\dagger$, where the adjoint operator is defined by the property $\langle \mathbf{u}^\dagger, \mathbf{A}\mathbf{u}\rangle = \langle \mathbf{A}^\dagger\mathbf{u}^\dagger, \mathbf{u}\rangle$, with $\mathbf{u}^\dagger, \mathbf{u}$ and $\langle \cdot, \cdot \rangle$ being vector functions and inner product, respectively. The linear stability analysis of the direct problem let us determine several characteristics: the parameters (e.g. Reynolds number) at which the flow first becomes unstable, and the frequencies ω_r, growth rate ω_i and spatial structure of the linear perturbations. On the other hand, the adjoint system provides information on the optimal way to excite the instability, as the perturbation in receptive region amplify more due to forcing. In combination the two types of modes can be used to locate the most sensitive region in the flow known as *wavemaker*, which is defined as the overlap η of the direct $\hat{\mathbf{u}}$ and adjoint $\hat{\mathbf{u}}^\dagger$ strongest global modes [6, 9, 11] (see Figs. 2 and 6),

$$\eta(\mathbf{x}_0) = \frac{|\hat{\mathbf{u}}^\dagger(\mathbf{x}_0)| \cdot |\hat{\mathbf{u}}(\mathbf{x}_0)|}{\langle \hat{\mathbf{u}}^\dagger, \hat{\mathbf{u}} \rangle} . \tag{1}$$

The wavemaker is the region in the flow where a variation in the flow structure provides the largest drift of the eigenvalues and therefore pinpoints the most likely region in the flow for the inception of the global instability.

We consider the incompressible Navier–Stokes equations linearised about a base flow $\mathbf{U}_b$ in non-dimensional form with $\mathbf{u}$, p and Re being velocity and pressure perturbation and the Reynolds number, respectively,

$$\frac{\partial \mathbf{u}}{\partial t} + \mathbf{u} \cdot \nabla \mathbf{U}_b + \mathbf{U}_b \cdot \nabla \mathbf{u} - \frac{1}{Re}\nabla^2 \mathbf{u} + \nabla p = \mathbf{f} , \tag{2}$$

$$\nabla \cdot \mathbf{u} = 0 \quad \text{in } \Omega , \tag{3}$$

$$\mathbf{u} = 0 \quad \text{on } \partial\Omega_v , \tag{4}$$

$$p\mathbf{n} - \frac{1}{Re}\nabla \mathbf{u} \cdot \mathbf{n} = 0 \quad \text{on } \partial\Omega_o . \tag{5}$$

Two last equations are the boundary conditions (BC) on the surface of the computational domain Ω. Subscripts v and o stand for regions where either *velocity* (Dirichlet) or *outflow* BC are specified, and $\mathbf{n}$ denotes the outward normal. The forcing $\mathbf{f}$ usually vanishes inside Ω, but may be used as sponge layers at inflow/outflow boundary. The corresponding set of adjoint equations reads

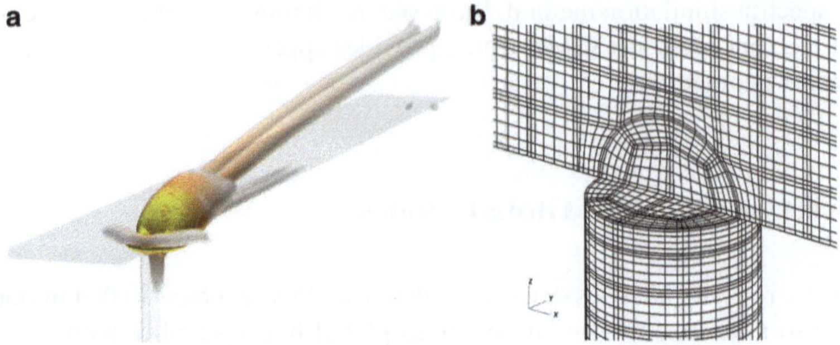

Fig. 1 (**a**) Vortical structures (λ_2 isolevels) of the base flow for JCF setup including the pipe. (**b**) Mesh structure at the connection of the circular pipe with the rectangular box. The element boundary and the position of the GLL points are shown

$$\frac{\partial \mathbf{u}^\dagger}{\partial t} + (\nabla \mathbf{U}_b)^T \mathbf{u}^\dagger - \mathbf{U}_b \cdot \nabla \mathbf{u}^\dagger + \frac{1}{Re} \nabla^2 \mathbf{u}^\dagger + \nabla p^\dagger = \mathbf{f}\,, \tag{6}$$

$$\nabla \cdot \mathbf{u}^\dagger = 0 \quad \text{in } \Omega\,, \tag{7}$$

$$\mathbf{u}^\dagger = 0 \quad \text{on } \partial\Omega_v\,, \tag{8}$$

$$p^\dagger \mathbf{n} + \frac{1}{Re} \nabla \mathbf{u}^\dagger \cdot \mathbf{n} = (\mathbf{U}_b \cdot \mathbf{n})\mathbf{u}^\dagger \quad \text{on } \partial\Omega_o\,, \tag{9}$$

where $\mathbf{u}^\dagger$ and $p^\dagger$ are adjoint perturbations. Notice the change of sign in the equations and the fact that *outflow* BC are inhomogeneous.

The solution to the direct and adjoint problem is computed using a Legendre polynomial based SEM implemented in `Nek5000` [8]. In this method the governing equations are cast into weak form and discretised in space by the Galerkin approximation, following the $P_N - P_{N-2}$ approach. The velocity space is spanned by Nth-order Lagrange polynomial interpolants, based on tensor-product arrays of Gauss–Lobatto–Legendre (GLL) quadrature points in a local element. The individual elements take the shape of hexahedra which can then be transformed using general coordinate mapping as shown in Fig. 1b.

`Nek5000` does not support the general inhomogeneous BC as given above. Therefore, to keep direct and adjoint problems consistent we set homogeneous Dirichlet BC on all $\partial\Omega$. To avoid reflections we use a sponge forcing $\mathbf{f} = \lambda(\mathbf{x})$ $(\mathbf{U}_b - \mathbf{v})$ at the inflow/outflow boundaries, where $\mathbf{v}$ stands for $\mathbf{u}$ or $\mathbf{u}^\dagger$ and $\lambda(\mathbf{x})$ is a smooth step function [5]. The dependency of the operator spectra on the applied BC for the flow past circular cylinder case is discussed in Sect. 4.

To obtain the base flow one has to find the steady state solution of the non-linear Navier–Stokes equations, which in many of the considered cases is unstable, in particular for strongly convectively unstable flows (e.g. JCF). We compute the base flow using selective frequency damping (SFD) [1], which damps the oscillations of the unsteady part of the solution using a temporal low-pass filter by setting

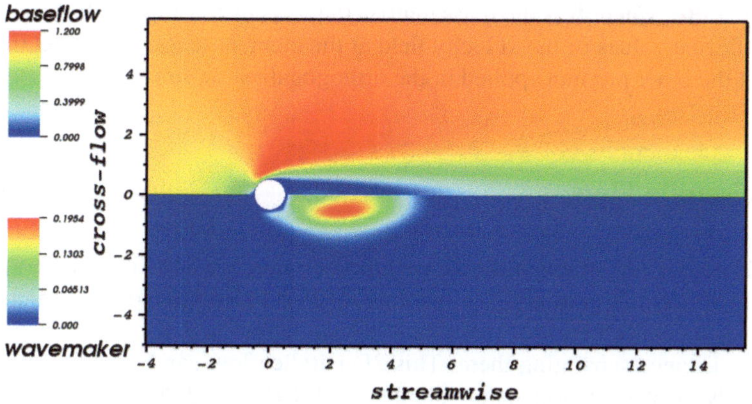

Fig. 2 Two-dimensional flow past a circular cylinder at $Re = 50$. The upper part shows the velocity magnitude of the base flow, and the lower part presents the overlap function η for the strongest direct and adjoint modes. This plot can be compared with Figs. 8 and 17 in Ref. [9]

$\mathbf{f} = -\chi(\mathbf{u} - \mathbf{w})$, where $\mathbf{u}$ is the flow solution and $\mathbf{w}$ its temporally low-pass-filtered counterpart obtained by a differential exponential filter $\mathbf{w}_t = (\mathbf{u} - \mathbf{w})/\Delta$. The amplitude of the forcing $\epsilon = \|(\mathbf{u} - \mathbf{w})\|$ in Ω is a good indicator of convergence; 2D test cases reached levels $\epsilon \approx 10^{-13}$, whereas the computationally more expensive 3D runs were stopped at 10^{-10} or 10^{-7} depending on the resolution, which is lower than the tolerance used for eigenvalue calculation (10^{-6}). An example of such a SFD base flows for JCF and cylinder flow are presented in Figs. 1a and 2.

The eigenvalue problem is then constructed rewriting the LNS equations in operator form $\mathbf{u}_t = \mathbf{A}\mathbf{u}$ and assuming $\mathbf{u}(\mathbf{x}, t) = \hat{\mathbf{u}}(\mathbf{x}) \exp(-i\omega t)$, where $\hat{\mathbf{u}}(\mathbf{x})$ is the global mode and ω its complex eigenvalue. For general 3D flows the size of the matrix $\mathbf{A}$ prohibits its explicit construction, so only the action of $\mathbf{A}$ on the vector $\mathbf{u}$ can be calculated. Solving the eigenvalue problem can then be achieved using the so-called time-stepper method, i.e. an iterative technique based on orthogonal projection of $\mathbf{A}$ onto a lower-dimensional subspace, in which the Arnoldi algorithm is applied and the Krylov subspace is constructed using snapshots taken from the evolution of the flow field $\mathbf{u}$ separated by a constant time interval Δt. To avoid frequency aliasing Δt must be small enough such that at least two sampling points in one period of the highest frequency mode are included (see e.g. Ref. [2]). In presented studies we do not consider dependency of calculated spectra on the size of Krylov space and Δt. The other important parameter is the actual time step δt of the simulation, which is related to the CFL condition. In our simulations we have tested Courant numbers in the range 0.05–0.2 and found significant dependency of the operator $\mathbf{A}$ spectra if the cases were marginally resolved in space. On the other hand, the fully resolved 2D simulations show little dependence of the spectra on the Courant number.

In our implementation we use the implicitly restarted Arnoldi method (IRAM) from the ARPACK library [15]. We solve for the generalised eigenvalue problem

$\mathbf{A}\hat{\mathbf{u}} = -i\omega\mathbf{B}\hat{\mathbf{u}}$, where $\mathbf{B}$ is the mass matrix. It allows us to simplify the treatment of the duplicated values of the velocity field at the element faces, and to get the exact value of the inner product applied in the orthogonalisation step.

3 Poiseuille Flow

To validate our implementation we performed a number of tests corresponding to different flow configurations. The first one is plane Poiseuille flow, in which a fluid is moving laterally between two plates whose length and width is much greater than the distance separating them. This 2D parallel flow can be treated exactly by local analysis. We performed our calculations for $Re = 2{,}000$ (based on centreline velocity and channel half height) on a rectangular grid of streamwise length 2π with periodic BC, built of 6×6 spectral elements with polynomial order N ranging from 11 to 17. We compared our result with local analysis (O. Tammisola, private communication) and found very good match for the first 100 eigenvalues calculated for $N = 17$ (Fig. 3). Although only three modes with the highest frequency $\omega_r > \omega_{max} \approx 21$ are visibly displaced, we can see slow decrease of accuracy with growing ω_r. This becomes more pronounced with decreasing N as the maximum frequency of the well resolved waves ($\omega_{max} \approx 16$ for $N = 11$) is getting lower and thus a small number of spurious modes appears. Similar conclusions can be drawn comparing direct and adjoint modes, however the threshold frequency for fast relative error growth appears to be lower ($\omega_{max} \approx 12$ for $N = 17$). There is also a number of modes clustered around $\omega_r = 0$, with relative error of order 10^{-6} which all correspond to highly damped modes.

4 Flow Past Circular Cylinder

The next case is the plane wake behind a circular cylinder, which is a canonical 2D, non-parallel flow extensively studied in the literature. Its structural sensitivity was investigated in Ref. [9] and we compare our results against this work adopting the grid from [14], where the cylinder of unit diameter was placed at (0,0) in the grid extending from -15 to 35 and from -15 to 15 in streamwise and cross-flow directions, respectively. We performed a number of runs for $Re = 40$, 45 and 50 (based on diameter and incoming velocity) calculating the most unstable global modes and their overlap function η. Very good agreement with [9] is obtained. An example of a base flow and a wavemaker for $Re = 50$ is presented in Fig. 2 which can readily be compared with Figs. 8 and 17 in Ref. [9]. The left plot in Fig. 4 (negative ω_r) presents the comparison of the calculated spectra of the direct operator for $Re = 50$ with the results of CPL code, which is a modified version of the code employed in [9] (I. Lashgari, private communication). There is good agreement for the least damped, low frequency modes, but we observe a relative

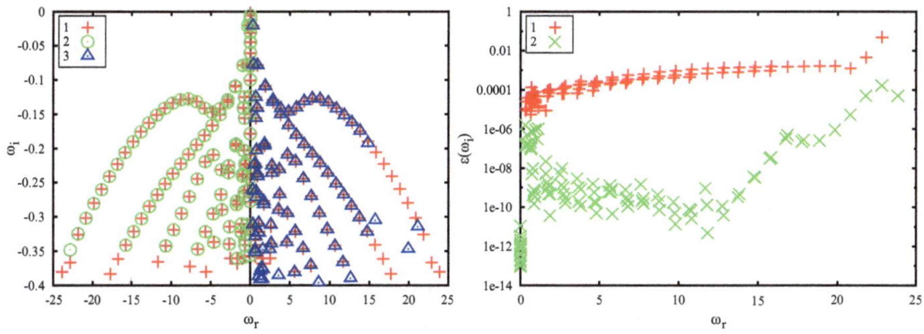

Fig. 3 Spectra (growth rate ω_i versus frequency ω_r) for the different simulations of plane Poiseuille flow at $Re = 2{,}000$. We utilise the negative and positive ω_r parts to compare different cases. *Left*: Comparison of the spectra of `Nek5000` run with $N = 17$ (*1*) with results of the local analysis (*2*) and the low resolution $N = 11$ (*3*). *Right*: Relative error of the growth rate $\epsilon(\omega_i)$ for $N = 17$ as a function of ω_r with respect to local analysis (*1*) and the spectra of the adjoint operator (*2*)

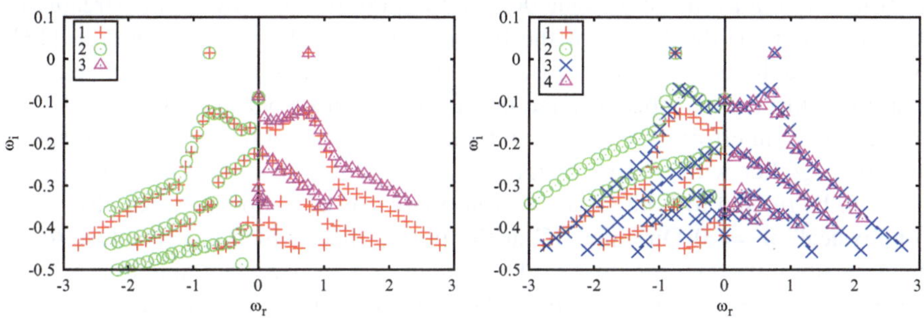

Fig. 4 Spectra of the direct operator for the flow past circular cylinder at $Re = 50$. *Left*: Case with outflow BC in the non-symmetric box (*1*) compared with results of the CPL code (*2*) and run in the symmetric box (*3*). *Right*: Cases in non-symmetric box with different BC: (*1*) outflow (the same as (*1*) on *left plot*), (*2*) Dirichlet without sponge, (*3*) Dirichlet with sponge applied to inflow/outflow regions. (*4*) symmetric box with Dirichlet BC and sponge

shift of the modes growths with increasing ω_r and decreasing ω_i. These weaker modes are very sensitive to the exact simulations details such as grid size and boundary conditions. Furthermore, we present results of the runs with outflow BC (Eq. 5) for non-symmetric ($x \in [-15, 35]$; crosses) and symmetric ($x \in [-35, 35]$; triangles) mesh; the two meshes differ in the extent of the upstream (inflow) part which is important for the adjoint simulations. There is clearly sensitivity of the spectra to these details as the position of the branches is moved. The dependency of the spectra on the applied BC and grid size is illustrated in Fig. 4. We found the Dirichlet BC combined with the sponge forcing $\mathbf{u}$ and $\mathbf{u}^{\dagger}$ at inflow/outflow to be the least sensitive to the grid and we thus use these settings in the remainder of our study. The relative error of the growth rate pertaining to the direct and adjoint modes for this setup is lower than 10^{-6} (not shown). The resolution studies for this case showed strong dependency of the spectra on the polynomial order N with the frequency of the poorly resolved modes shifted towards $\omega_r = 0$.

5 Jet in Crossflow

The most complex flow case considered in this study is the jet in crossflow (JCF), which is a non-parallel and fully 3D flow referring to a jet of fluid exiting a pipe and interacting with the boundary layer perpendicular to the pipe orifice (see Fig. 1). For the detailed description of the case we refer to Refs. [3, 13]. We consider two different setups of this flow: one corresponding the simplified setup of Ilak et al. [13], in which the inflow jet was represented by a Dirichlet boundary condition, and the more realistic one with the pipe included in the domain as shown in Fig. 1. For the simplified setup, we made two major changes compared to Ref. [13]: (i) no fringe region as the SEM code does not require periodic BC in the streamwise direction, and (ii) the length of the box is longer (150 versus 75 units in [13]). We increased the box length because we found the result to be very sensitive to any kind of disturbances; especially the proper treatment of outflow BC proved to be crucial, so in an effort to reduce its influence we increased the downstream part of the grid together with a sponge region. This extreme sensitivity of the simplified JCF can be related to both strong non-normality of LNS operator, but also to the "unphysical" $\mathbf{u} = 0$ Dirichlet BC at the pipe orifice, which is very close to the dynamically important region. For the same reason we were not able to reproduce the results of [13]. In our runs we set the jet to free-stream velocity ratio R to 1.5, and the Reynolds number at the jet position $Re = 178.2$ (based on free-stream velocity and cross-flow displacement thickness). The jet diameter and pipe length are equal 3 and 20 units, respectively.

Most of our runs we performed studying the simplified setup and we found it very sensitive to the grid resolution. On the left plot in Fig. 5 the results of the higher resolution (polynomial order $N = 9$, crosses) are compared with the lower resolution ($N = 6$, circles) and the spectra of the adjoint operator ($N = 9$, triangles). The increased resolution causes the initially unstable flow (positive growth rate of the strongest mode) to stabilise, as the whole spectra shifts down. Comparison of the direct and adjoint spectra also shows that the even $N = 9$ resolution is only marginal, as the lowest value of the growth rate error $\epsilon(\omega_i)$ is on the order of 10^{-4}.

Similar conclusions can be drawn comparing the direct (crosses) and adjoint (circles) spectra of the high resolution ($N = 9$) run including the pipe (right plot in Fig. 5). One can see a clustering of the poorly resolved adjoint modes around $\omega_r = 0$, which is similar to what we already observed in the cylinder case (Sect. 4). However, the least stable modes agree well between direct and adjoint simulation indicating that the main dynamics is captured well. Figure 6 presents a 2D cut

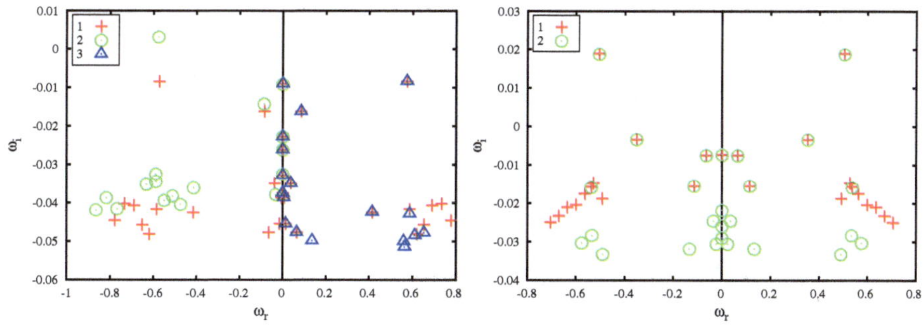

Fig. 5 The spectra of the direct and adjoint operator for JCF for velocity ratio $R = 1.5$. Left plot corresponds to the simplified setup discarding the pipe and presents the direct operator spectra for high (*1*) and low (*2*) resolution runs together with the adjoint operator spectra of the high resolution run (*3*). The right plot shows the spectra of the direct (*1*) and adjoint (*2*) operator for the setup including pipe in the consideration

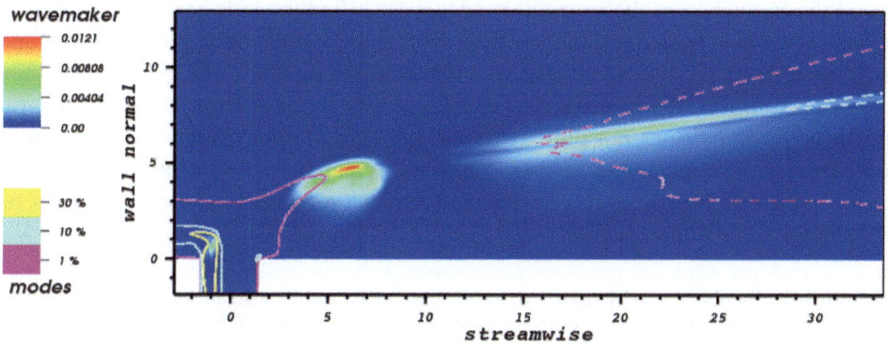

Fig. 6 Two-dimensional cut through the symmetry plane of the grid for the JCF setup including the pipe for $R = 1$. The colours shows the value of the strongest overlap η. Isolevels of the direct (*dashed line*) and adjoint (*continuous line*) strongest modes at 1, 10 and 30 % are also shown

through the symmetry plane of the grid showing isosurfaces of the direct (dashed) and adjoint (continuous line) strongest modes as well as their overlap η (colour). The isocontours are placed at 1, 10 and 30 % of the maximum value of the modes showing their spatial extent and illustrating considerable separation of their maxima, which is related to the strong non-normality of the operator. As the plot covers the region close to the pipe orifice, only the adjoint mode maximum is visible. The overlap function η features a total of three maxima of which one is clearly related to the adjoint mode, located close to the steady horseshoe vortex upstream the jet. The other two maxima appear in the shear layer downstream of the jet forming the wavemaker.

The analysis of 3D flows is computationally very expensive, and the computation of the single JCF spectrum with $N = 9$ takes about 2–3 weeks on 1,024 cores.

6 Conclusions

In this work we investigated the use of linear stability tools implemented in the SEM code `Nek5000` for studying the stability and sensitivity of 3D flows with moderately complex geometry. We validated our implementation on a number of 2D parallel and non-parallel flow cases against the local stability analysis as well as literature data. Resolution studies show that the calculated spectra are very sensitive to the grid resolution and proper treatment of boundary conditions. In our simulations we adopted Dirichlet boundary conditions together with sponge layers to keep direct and adjoint problems consistent, however another possible solution would be to adopt correct direct and adjoint outflow boundary according to Eqs. 5 and 9. The grid spacing defines the shortest wave length that can be properly resolved, which corresponds to setting the maximum possible frequency of the calculated modes. Higher modes then appear as spurious modes in the spectrum. In the case of the flow past cylinder and a jet in crossflow (including the inflow pipe) the frequency of the spurious modes was shifted towards zero giving clusters of spurious low frequency modes with low growth rates. Comparing direct and adjoint spectra is shown to help in identifying spurious modes, but even then careful resolution studies are necessary. The dependency of the spectra on the grid resolution usually does not play a crucial role for 2D simulations, but it becomes an important issue when moving to 3D, in particular in regions of active dynamics and complex geometry. In this case adaptive mesh refinement algorithms might be instrumental for future studies.

Acknowledgements We would like to thank O. Tammisola and I. Lashgari for help in developing and testing our implementation. Computer time was provided by Swedish National Infrastructure for Computing (SNIC). The third author was supported by the U.S. Department of Energy under Contract DE-AC02-06CH11357.

References

1. Åkervik, E., Brandt, L., Henningson, D.S., Hœpffner, J., Marxen, O., Schlatter, P.: Steady solutions of the Navier-Stokes equations by selective frequency damping. Phys. Fluids **18**(068102), 1–4 (2006)
2. Bagheri, S., Åkervik, E., Brandt, L., Henningson, D.S.: Matrix-free methods for the stability and control of boundary layers. AIAA J. **47**, 1057–1068 (2009)
3. Bagheri, S., Schlatter, P., Schmid, P.J., Henningson, D.S.: Global stability of a jet in crossflow. J. Fluid Mech. **624**, 33–43 (2009)
4. Barkley, D., Gomes, M.G.M., Henderson, R.D.: Three-dimensional instability in flow over a backward-facing step. J. Fluid Mech. **473**, 167–190 (2002)
5. Chevalier, M., Schlatter, P., Lundbladh, A., Henningson, D.S.: SIMSON: a pseudo-spectral solver for incompressible boundary layer flows. Tech. Rep. 2007:07, KTH Mechanics (2007)
6. Chomaz, J.M.: Global Instabilities in Spatially Developing Flows: Non-Normality and Nonlinearity. Annu. Rev. Fluid Mech. **37**, 357–392 (2005)
7. Fischer, P., Lottes, J., Kerkemeier, S.: nek5000 Web page (2008). http://nek5000.mcs.anl.gov

8. Fischer, P.F.: An Overlapping Schwarz Method for Spectral Element Solution of the Incompressible Navier Stokes Equations. J. Comput. Phys. **133**, 84–101 (1997)
9. Giannetti, F., Luchini, P.: Structural sensitivity of the first instability of the cylinder wake. J. Fluid Mech. **581**, 167–197 (2007)
10. Henningson, D.S., Åkervik, E.: The use of global modes to understand transition and perform flow control. Phys. Fluids **20**(3), 031,302 (2008)
11. Hill, D.C.: Adjoint systems and their role in the receptivity problem for boundary layers. J. Fluid Mech. **292**, 183–204 (1995)
12. Huerre, P., Monkewitz, P.A.: Local and global instabilities in spatially developing flows. Annu. Rev. Fluid Mech. **22**, 473–537 (1990)
13. Ilak, M., Schlatter, P., Bagheri, S., Henningson, D.S.: Bifurcation and stability analysis of a jet in cross-flow: onset of global instability at a low velocity ratio. J. Fluid Mech. **696**, 94–121 (2012)
14. Lashgari, I., Pralits, J.O., Giannetti, F., Brandt, L.: First instability of the flow of shear-thinning and shear-thickening fluids past a circular cylinder. J. Fluid Mech. **701**, 201–227 (2012)
15. Lehoucq, R.B., Sorensen, D.C., Yang, C.: ARPACK users guide: Solution of large scale eigenvalue problems by implicitly restarted Arnoldi methods. (1997)
16. Schlatter, P., Bagheri, S., Henningson, D.S.: Self-sustained global oscillations in a jet in crossflow. Theor. Comput. Fluid Dyn. **25**, 129–146 (2011)
17. Theofilis, V.: Global Linear Instability. Annu. Rev. Fluid Mech. **43**, 319–352 (2011)
18. Trefethen, L.N., Trefethen, A.E., Reddy, S.C., Driscoll, T.A.: Hydrodynamic stability without eigenvalues. Science **261**, 578–584 (1993)

A High-Order Discontinuous Galerkin Method for Viscoelastic Wave Propagation

Fabien Peyrusse, Nathalie Glinsky, Céline Gélis, and Stéphane Lanteri

Abstract We present a high-order discontinuous Galerkin method for the simulation of P-SV seismic wave propagation in heterogeneous media and two dimensions of space. The first-order velocity-stress system is obtained by assuming that the medium is linear, isotropic and viscoelastic, thus considering intrinsic attenuation. The associated stress-strain relation in the time domain being a convolution, which is numerically intractable, we consider the rheology of a generalized Maxwell body replacing the convolution by differential equations. This results in a velocity-stress system which contains additional equations for the anelastic functions including the strain history of the material. Our numerical method, suitable for complex triangular unstructured meshes, is based on a centered numerical flux and a leap-frog time-discretization. The extension to high order in space is realized by Lagrange polynomial functions, defined locally on each element. The inversion of a global mass matrix is avoided since an explicit scheme in time is used. The method is validated through numerical simulations including comparisons with a finite difference scheme.

F. Peyrusse (✉) · S. Lanteri
Inria Sophia Antipolis – Méditerranée, Nachos project-team 2004 Route des Lucioles – BP 93, 06902 Sophia Antipolis Cedex, France
e-mail: fabien.peyrusse@inria.fr; stephane.lanteri@inria.fr

N. Glinsky
Ifsttar/CETE, Laboratoire régional de Nice, 56 bd Stalingrad, 06359 Nice Cedex 4, France
Inria Sophia Antipolis – Méditerranée, Nachos project-team 2004 Route des Lucioles – BP 93, 06902 Sophia Antipolis Cedex, France
e-mail: nathalie.glinsky@inria.fr

C. Gélis
IRSN, BP 17, 92262 Fontenay-aux-Roses Cedex, France
e-mail: celine.gelis@irsn.fr

M. Azaïez et al. (eds.), *Spectral and High Order Methods for Partial Differential Equations - ICOSAHOM 2012*, Lecture Notes in Computational Science and Engineering 95, DOI 10.1007/978-3-319-01601-6_29,

1 Formulation of the Viscoelastic System

Computational seismology has become a very important discipline for estimates of ground motion thanks to accurate numerical solvers and a better understanding of physical phenomena. In realistic media such as sedimentary basins where incident waves are trapped, site effects due to local geological and geotechnical conditions can be observed, as first described by Singh et al. [13] for the 19 September 1985 Michoacan earthquake in Mexico city; these effects result in a strong increase in amplification and duration of the ground motion at some particular locations. For such estimates, the basic assumption of linear elasticity is no more valid because it results in a severe overestimation of amplitude and duration of the ground motion since attenuation is not taken into account.

We study here the P-SV wave propagation by solving the two-dimensional velocity-stress formulation of the elastodynamic equations. In order to include realistic attenuation, we suppose that the medium is linear, isotropic and viscoelastic, combining the behavior of both elastic solids and viscous fluids. Then, the associated stress-strain relation, given by the causality principle, establishes that the stress, at a given time t, is a function of the entire strain history until t; in the time domain, this relation is the convolution of a relaxation function and the strain rate, which is numerically intractable. Therefore, since we study this problem in the time domain, we follow the method, presented by Day and Minster [2], replacing the convolution by additional differential equation and the improvement proposed by Emmerich and Korn [5] which consider the rheology of a generalized Maxwell body (GMB) with L *relaxation frequencies* ω_l, chosen logarithmically equidistant in the frequency band of interest and allows the fitting of any Q-law.

Following Kristek and Moczo [9], the *anelastic functions* depending on the strain ϵ are defined by

$$\zeta^l(t) = \omega_l \int_{-\infty}^{t} e^{-\omega_l (t-\tau)} \, \epsilon(\tau) \, d\tau, \qquad l = 1, \ldots, L \tag{1}$$

and allow to turn the convolution into L differential equations

$$\frac{\partial}{\partial t}\zeta^l(t) + \omega_l \, \zeta^l(t) = \omega_l \, \epsilon(t), \qquad l = 1, \ldots, L \tag{2}$$

which complete the anelastic wave system. Thus, the velocity-stress system we consider is

$$\begin{cases} \rho \dfrac{\partial \mathbf{v}}{\partial t} = \nabla \cdot \sigma \,, \\ \dfrac{\partial \sigma}{\partial t} = \lambda \, (\nabla \cdot \mathbf{v}) \, I + \mu \, (\nabla \mathbf{v} + \nabla \mathbf{v}^T) - \displaystyle\sum_{l=1}^{L} \left(\lambda \, \Upsilon^{\lambda,l} \, tr(\xi^l) \, I + 2\mu \, \Upsilon^{\mu,l} \, \xi^l \right) , \\ \dfrac{\partial \xi^l}{\partial t} = \dfrac{\omega_l}{2} (\nabla \mathbf{v} + \nabla \mathbf{v}^T) - \omega_l \, \xi^l , \qquad l = 1, \ldots, L \end{cases} \tag{3}$$

where ρ is the density, λ and μ are the Lamé parameters related to the P- and S-wave velocities which write $v_p = \sqrt{(\lambda + 2\mu)/\rho}$ and $v_s = \sqrt{\mu/\rho}$ for an unrelaxed purely elastic medium; $\mathbf{v}$ is the velocity vector and σ is the stress tensor. The L tensors ξ^l are the derivatives of the memory variables ζ defined in (2) and $\Upsilon^{M,l}$ are their associated anelastic coefficients, with M equal to λ or μ. These coefficients are determined by solving an overdetermined system of equations using desired values of the quality factors Q_M at frequencies $\tilde{\omega}_k$ [9]:

$$\sum_{l=1}^{L} \frac{\omega_l \tilde{\omega}_k + \omega_l^2 Q_M^{-1}(\tilde{\omega}_k)}{\omega_l^2 + \tilde{\omega}_k^2} \Upsilon^{M,l} = Q_M^{-1}(\tilde{\omega}_k), \; k = 1, \ldots, 2L-1. \quad (4)$$

Since observations show that in the Earth, the internal friction is nearly constant over the seismic frequency range, we solve (4) assuming Q is frequency-independent.

If we define a single vector $\mathbf{W} = (\mathbf{v}, \mathbf{W}_{\sigma,\xi})^T$, where $\mathbf{W}_{\sigma,\xi}$ is the $(3 + 3\,L)$-vector of stress and anelastic unknowns, the system (3) can be written in the following compact form

$$\frac{\partial \mathbf{W}}{\partial t} + \sum_{\alpha \in \{x,z\}} B_\alpha(\rho, \lambda, \mu)\, \partial_\alpha \mathbf{W} = E\, \mathbf{W}. \quad (5)$$

The detail of the matrices B_α and E can be found in [8].

2 Discretization

Many different numerical methods have been developed within the last few decades to solve these equations. Among all of them, we consider here a non-dissipative high-order discontinuous Galerkin method (DG) applicable to unstructured triangular meshes. Initially introduced by Reed and Hill [11] for the solution of neutron transport problems, the DG method became quite recently very popular to solve hyperbolic systems. For seismic wave propagation problems, especially in viscoelastic media, Käser et al. [8] have proposed a DG finite element method based on an upwind scheme and the ADER approach. In the following, we present the spatial and time discretization of the system (5), which is an extension of the method proposed in [3] for elastic media.

We consider a polygonal domain Ω divided into N_T triangles. For a given element T_i, we denote by $S_{ik} = T_i \cap T_k$ its internal faces – indexed by $\mathscr{V}(i)$ – and by $S_k^{b_i} = T_i \cap \partial\Omega$ the faces which are common to the boundary of Ω, indexed by $\mathscr{E}(i)$. We first multiply the system (5) by a scalar test function $\phi_l^{T_i}$ and integrate on T_i

$$\int_{T_i} \left[\frac{\partial \mathbf{W}}{\partial t} + \sum_{\alpha \in \{x,z\}} B_\alpha(\rho, \lambda, \mu)\, \partial_\alpha \mathbf{W} - E\, \mathbf{W} \right] \phi_l^{T_i}\, dV = 0. \quad (6)$$

Assuming that the parameters ρ, λ and μ are constant on T_i and using Green's formula for the second term of Eq. (6), we get

$$\int_{T_i} \frac{\partial \mathbf{W}}{\partial t} \phi_l^{T_i} \, dV - \sum_{\alpha \in \{x,z\}} B_\alpha^{T_i} \int_{T_i} \mathbf{W} \, \partial_\alpha \phi_l^{T_i} \, dV + B_{\mathbf{n}}^{T_i} \int_{\partial T_i} \mathbf{W} \, \phi_l^{T_i} \, ds = E^{T_i} \int_{T_i} \mathbf{W} \, \phi_l^{T_i} \, dV, \tag{7}$$

where $\mathbf{n}$ is the outward normal vector to T_i and $B_{\mathbf{n}}(\rho, \lambda, \mu) = \sum_{\alpha \in \{x,z\}} B_\alpha(\rho, \lambda, \mu) \, n_\alpha$. In (7), $B_{\mathbf{n}}^{T_i}$ (respectively E^{T_i}) is the restriction of $B_{\mathbf{n}}$ (respectively E) to T_i.

We now write the approximation of $\mathbf{W}$ using Lagrange interpolation polynomials of degree m on T_i

$$\mathbf{W}_{|T_i}(x, t) = \sum_{j=1}^{d_i} \mathbf{W}_j^{T_i}(t) \, \phi_j^{T_i}(x), \tag{8}$$

where d_i is the number of degrees of freedom on T_i and $\phi_j^{T_i}$ ($j = 1, \ldots, d_i$) are the basis functions. For the first term of (7), we obtain

$$\forall \, l = 1, \ldots, d_i, \quad \int_{T_i} \frac{\partial}{\partial t} \mathbf{W} \, \phi_l^{T_i} \, dV = \sum_{j=1}^{d_i} M_{lj}^{T_i} \frac{d}{dt} \mathbf{W}_j^{T_i}, \tag{9}$$

where $M^{T_i} = \left(\int_{T_i} \phi_j^{T_i} \, \phi_l^{T_i} \, dV \right)_{1 \le j,l \le d_i}$ is the mass matrix on T_i, and for the second term, we get

$$\forall \, l = 1, \ldots, d_i, \quad E^{T_i} \int_{T_i} \mathbf{W} \, \phi_l^{T_i} \, dV = E^{T_i} \sum_{j=1}^{d_i} M_{lj}^{T_i} \, \mathbf{W}_j^{T_i}. \tag{10}$$

The second integral of (7) is approximated by

$$\forall \, l = 1, \ldots, d_i, \quad \sum_{\alpha \in \{x,z\}} B_\alpha^{T_i} \int_{T_i} \mathbf{W} \, \partial_\alpha \phi_l^{T_i} \, dV = \sum_{\alpha \in \{x,z\}} B_\alpha^{T_i} \sum_{j=1}^{d_i} G_{\alpha,lj}^{T_i} \, \mathbf{W}_j^{T_i}, \tag{11}$$

where $G_\alpha^{T_i} = \left(\int_{T_i} \phi_j^{T_i} \, \partial_\alpha \phi_l^{T_i} \, dV \right)_{1 \le j,l \le d_i}$.

In order to calculate the boundary integral in (7), we split ∂T_i into internal and boundary parts

$$\forall \, l = 1, \ldots, d_i, \quad B_{\mathbf{n}}^{T_i} \sum_{k \in \mathscr{V}(i)} \int_{S_{ik}} \mathbf{W} \, \phi_l^{T_i} \, ds + B_{\mathbf{n}}^{T_i} \sum_{k \in \mathscr{E}(i)} \int_{S_k^{bi}} \mathbf{W} \, \phi_l^{T_i} \, ds. \tag{12}$$

For the internal faces S_{ik}, we introduce a centered flux

$$\mathbf{W}_{|S_{ik}} = \frac{1}{2}\left(\mathbf{W}^{T_i} + \mathbf{W}^{T_k}\right). \tag{13}$$

Hence, the corresponding integral writes: $\forall\, l = 1, \ldots, d_i$

$$B_{\mathbf{n}}^{T_i} \sum_{k \in \mathscr{V}(i)} \int_{S_{ik}} \mathbf{W}\, \phi_l^{T_i}\, ds = \frac{1}{2}\, B_{\mathbf{n}}^{T_i} \sum_{k \in \mathscr{V}(i)} \sum_{j=1}^{d_i} \left[\left(R_{|S_{ik}}^{T_i}\right)_{lj} \mathbf{W}_j^{T_i} + \left(R_{|S_{ik}}^{T_k}\right)_{lj} \mathbf{W}_j^{T_k} \right], \tag{14}$$

where $R_{|S_{ik}}^{T_i} = \left(\int_{S_{ik}} \phi_j^{T_i} \phi_l^{T_i}\, ds\right)_{1 \le j,l \le d_i}$ and $R_{|S_{ik}}^{T_k} = \left(\int_{S_{ik}} \phi_j^{T_k} \phi_l^{T_i}\, ds\right)_{1 \le j,l \le d_i}$.
In the following numerical simulations, we will consider two types of boundary conditions. Firstly, the free surface condition $\sigma \cdot \mathbf{n} = 0$ which we introduce weakly in the integral and, secondly, periodicity conditions for the lateral boundaries of the domain.

For the time discretization, we use a second-order leap-frog scheme, so that $\mathbf{v}$ on one hand and $\mathbf{W}_{\sigma,\xi}$ on the other hand are computed at staggered times. We denote by $\mathscr{F}^{T_i}$ and $\mathscr{G}^{T_i}$ the discrete operators taking into account the integrals on T_i and ∂T_i. Then, we obtain the system of discrete equations for each element T_i

$$M^{T_i}\, \frac{\mathbf{v}^{n+1} - \mathbf{v}^n}{\Delta t} = \mathscr{F}^{T_i}\left(\mathbf{W}_{\sigma,\xi}^{n+\frac{1}{2}}\right), \tag{15}$$

$$M^{T_i}\, \frac{\mathbf{W}_{\sigma,\xi}^{n+\frac{3}{2}} - \mathbf{W}_{\sigma,\xi}^{n+\frac{1}{2}}}{\Delta t} = \mathscr{G}^{T_i}\left(v^{n+1}, \mathbf{W}_{\sigma,\xi}^{n+\frac{1}{2}}\right), \tag{16}$$

where $\Delta t = t^{n+1} - t^n$ is the time step and $\mathbf{v}^n = \mathbf{v}(t^n)$.

3 Numerical Results

3.1 Propagation of a Plane Wave in a Layered Medium

First, in order to accurately validate the proposed DG method, we consider the propagation of a vertical plane wave in a layered medium containing seven different sedimentary deposits over a bedrock, as depicted in Fig. 1. The material properties of the different layers, given in Table 1, exhibit high constrasts between the media. The P-wave velocity v_p and the S-wave velocity v_s are given at a reference frequency f_r since the media are dispersive. The source excitation is a vertically incident plane SV wave, introduced in the bedrock (as a source term for the horizontal velocity equation) and which propagates vertically across the various layers up to the surface.

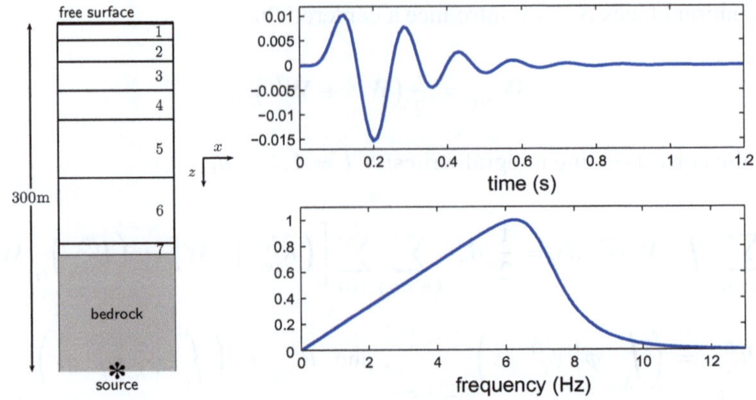

Fig. 1 Description of the layered medium (*left picture*); the material properties of the media are given in Table 1. Time (*upper*) and frequency (*lower*) dependence of the vertical incident SV plane wave

Table 1 Material properties for the different layers of the heterogeneous column

	$v_p(f_r)$ [m/s]	$v_s(f_r)$ [m/s]	ρ [kg/m^3]	Q_p	Q_s
1	1,500	130	2,050	75	15
2	1,500	200	2,150	75	20
3	1,650	300	2,075	83	30
4	2,050	450	2,100	103	40
5	2,450	600	2,155	123	60
6	2,550	700	2,200	140	70
7	3,500	1,250	2,500	200	100
Bedrock	4,500	2,600	2,600	50,000	50,000

Its frequency content is in the band [0,10 Hz] (as shown in Fig. 1). We apply a free surface condition on the top of the numerical domain and periodic conditions at lateral boundaries, as this problem is 1D (Fig. 1).

We compare the results of our DG solver to those of two different reference methods: first, a finite difference (FD) method, detailed in Gélis et al. [6], based on the rotated staggered grid of Saenger [12] and the technique developed by Liu and Archuleta [10] to take into account anelastic losses (viscoelastic model) and, secondly, the Haskell-Thomson method (HT) [7]. For a precise comparison of the solutions, we plot, in Fig. 2, the spectral ratio of the horizontal velocity at the surface (or transfer function), i.e. the ratios in the frequency domain between the solution computed at the surface of the heterogeneous column and the corresponding solution for a homogeneous elastic medium (rocky medium only). The DG solutions have been computed using a second-order Lagrange interpolation and, in the viscoelastic case, eight mechanisms ($L = 8$) have been used. The time step Δt is constrained by a CFL condition depending on the highest velocity.

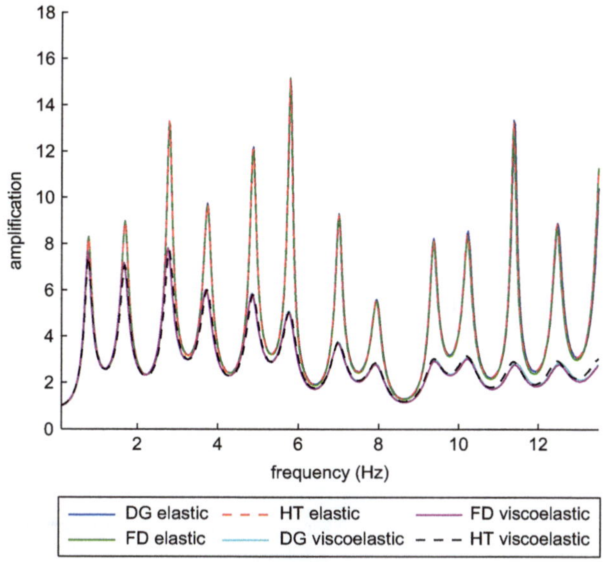

Fig. 2 Spectral ratios of the horizontal velocity v_x for the heterogeneous column test case. Comparison between discontinuous Galerkin (DG), finite difference (FD) and Haskell-Thomson (HT) methods for elastic and viscoelastic cases

When observing the results in Fig. 2, we first remark a perfect accordance of the DG solution with the reference solutions, in both elastic and viscoelastic cases. Note the complex profiles of the spectral ratio and the high values of the amplification (between 6 and 15) at the fundamental frequency and higher modes, resulting from the successive constructive downgoing and upgoing waves trapped in the column. Lastly, the comparison between the elastic and viscoelastic cases highlights the effect of the attenuation which results in lower values of the amplification, especially at high frequencies.

3.2 Propagation in a Realistic Basin

We now consider a more realistic problem and study the propagation of a vertical SV plane wave in a 2D basin extracted from the 3D model of the Nice area (south of France) [1]. The location of the 2D profile and its description are given in Figs. 3 and 4 respectively. As it is seen on the figure describing the basin, the 1,000 m wide model of the basin is extended laterally by two homogeneous flat parts; this model is complex and highly heterogeneous as proven by the media properties of Table 2. As previously, the solutions of the DG method are compared to the results of the FD solver. The numerical domain is 2,100 m wide and up to 75 m deep. An unstructured triangular mesh respecting the interfaces between the different media, has been generated with Simail [4]; the mesh size is approximatively 1 m in the basin and 4 m in the bedrock and the mesh contains 10,707 triangles. The number of mechanisms is here $L = 3$. The FD calculations are done with a uniform cartesian

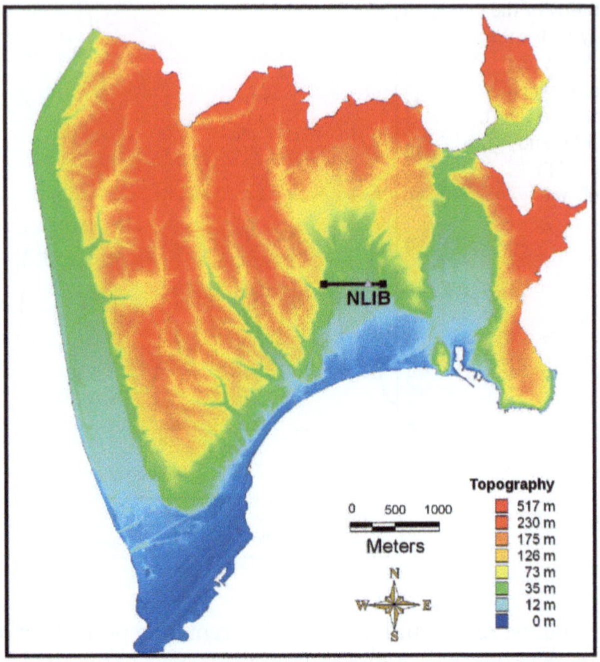

Fig. 3 Topography of the area of Nice and localization of the 2D profile (*black line*)

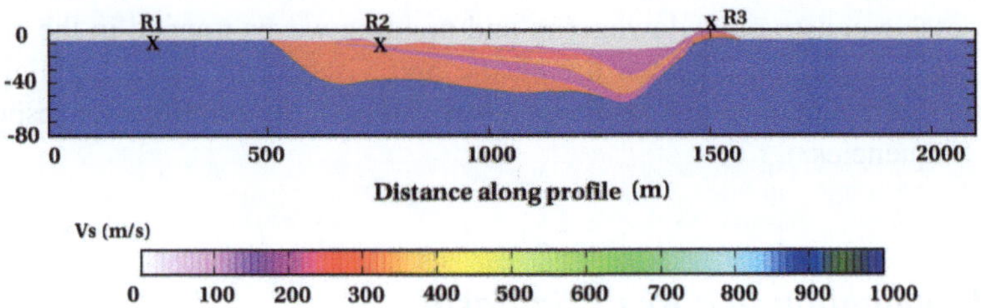

Fig. 4 Model of the 2D basin used for the simulations and location of the receivers R1, R2 and R3; color map corresponds to the value of velocity v_s

grid ($\Delta x = \Delta y = 0.125$ m, which corresponds to 60 points by wavelength, as required by the stencil of Saenger in the presence of free surface condition) and vacuum conditions are applied at the topography [14]. The solution is recorded at receivers located every 5 m at the surface and three of them are selected (see R1, R2 and R3 on Fig. 4) outside, inside and at the edge of the basin. The source excitation is the same as for the previous test case. We plot on Fig. 5 the evolution of the horizontal velocity v_x as a function of time calculated at the receivers R1, R2 and R3 with both DG and FD methods. Note that, also for this complex case, the results of the two numerical methods are in perfect accordance which proves that, for this frequency range, computations can be done using three mechanisms only. The trapped waves inside the basin are clearly visible on the solutions at receiver R2 for

Table 2 Material properties of the different media of the basin. $Q_P = v_p/10$ and $Q_S = v_s/10$

	$v_p(f_r)$ [m/s]	$v_s(f_r)$ [m/s]	ρ [kg/m^3]
1	440	180	1,900
2	710	290	1,900
3	489	200	1,700
4	808	330	2,100
5	612	250	1,800
6	734	300	2,100
7	538	220	1,800
8	710	290	2,000
9	734	300	2,000
Bedrock	2,449	1,000	2,100

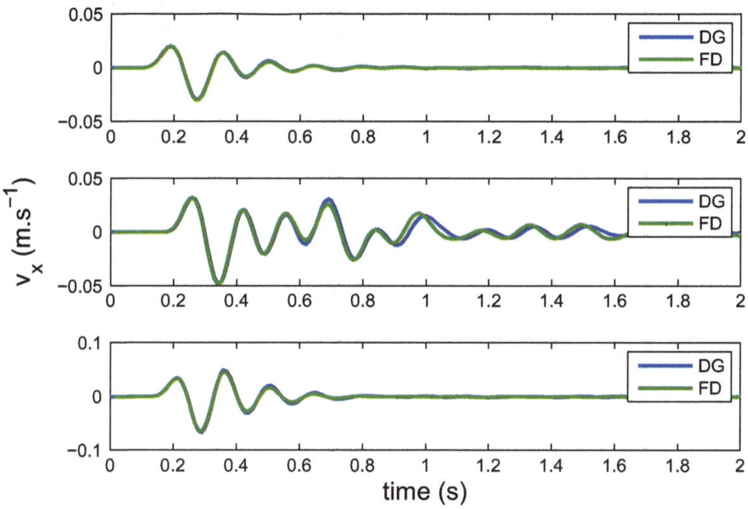

Fig. 5 Horizontal velocity v_x as a function of time at the three surface receivers: R1 (*top*) located outside the basin, R2 (*medium*), in the basin and R3 (*bottom*), at the edge. Comparison between discontinuous Galerkin (DG) and finite difference (FD) methods

which the velocity amplitude is higher and the signal duration significantly longer. This is a typical feature of site effects. Finally, we present in Fig. 6 the transfer function at the surface i.e. the spectral ratio of the horizontal velocity as a function of the frequency, for all the receivers on the topography. On this picture, we note the high amplification essentially in the basin at a frequency approximatively equal to 2 Hz which is in accordance with the real measurements at station NLIB [1] (see Fig. 3). Moreover, amplification for higher frequencies can be observed at the right boundary of the basin (for coordinates about 1,500), for receivers located above the thinnest sediment layers in the basin and at the highest altitude.

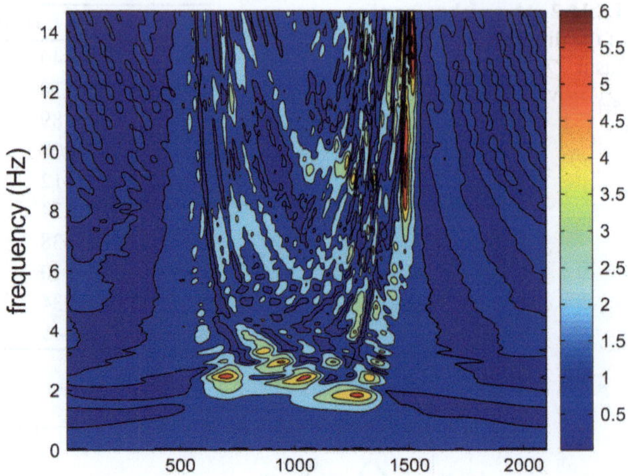

Fig. 6 Transfer function of the horizontal velocity v_x at the surface as a function of the frequency. Ratios are calculated with respect to solutions of a flat homogeneous medium (rocky medium only). Distance between the surface receivers is 5 m

In conclusion of this study, we have proposed a high-order discontinuous Galerkin method for the wave propagation in viscoelastic media. It has been validated through simulations in heterogeneous realistic configurations. A mathematical analysis of the method is in progress and the extension to 3D is underway.

Acknowledgements The authors acknowledge the CETE Méditerranée – LRPC de Nice for its collaboration and support, and thank, in particular, E. Bertrand for providing the geotechnical model of Nice and the results of the related site effect assessment. The PhD fellowship of Fabien Peyrusse is funded by Région Provence-Alpes-Côte d'Azur and Ifsttar.

References

1. E. Bertrand, A.-M. Duval, M. Castan, and S. Vidal. 3D geotechnical soil model of Nice, France, inferred from seismic noise measurements for seismic hazard assessment. *AGU Fall Meeting, San Francisco*, 2007.
2. S. M. Day and J. B. Minster. Numerical simulation of attenuated wavefields using a Padé approximant method. *Geophys. J. R. astr. Soc.*, 78:105–118, 1984.
3. S. Delcourte, L. Fezoui, and N. Glinsky-Olivier. A high-order discontinuous Galerkin method for the seismic wave propagation. In *ESAIM: Proceedings*, pages 70–89, 2009.
4. Distene. Simail software. http://www.distene.com/fr/create/simail.html.
5. H. Emmerich and M. Korn. Incorporation of attenuation into time-domain computations of seismic wave fields. *Geophysics*, 52:1252–1264, 1987.
6. C. Gélis, D. Leparoux, J. Virieux, A. Bitri, S. Operto, and G. Grandjean. Numerical modeling of surface waves over shallow cavities. *J. Environ. Eng. Geophys.*, 10:49–59, 2005.
7. N. A. Haskell. The dispersion of surface waves on multilayered media. *Bull. Seism. Soc. Am.*, 43(1):17–34, 1951.
8. M. Käser, M. Dumbser, J. de la Puente, and H. Igel. An arbitrary high-order discontinuous Galerkin method for elastic waves on unstructured meshes – III. Viscoelastic attenuation. *Geophys. J. Int.*, 168:224–242, 2007.

9. J. Kristek and P. Moczo. Seismic wave propagation in viscoelastic media with material discontinuities: a 3D fourth-order staggered-grid finite-difference modeling. *Bull. Seism. Soc. Am.*, 93:2273–2280, 2003.
10. P. C. Liu and R. J. Archuleta. Efficient modeling of Q for 3D numerical simulation of wave propagation. *Bull. Seism. Soc. Am.*, 96(4A):1352–1358, 2006.
11. W. H. Reed and T. R. Hill. Triangular mesh methods for the neutron transport equation. Technical report, Los Alamos Scientific Laboratory, 1973.
12. E. H. Saenger, N. Gold, and S. A. Shapiro. Modeling the propagation of elastic waves using a modified finite-difference grid. *Wave Motion*, 31:77–92, 2000.
13. S. K. Singh, E. Mena, and R. Castro. Some aspects of source characteristics of the 19 September 1985 Michoacan earthquake and ground motion amplification in and near Mexico city from strong motion data. *Bull. Seism. Soc. Am.*, 78:451–477, 1988.
14. J. Zahradnik, P. Moczo, and F. Hron. Testing for elastic finite-difference schemes for behaviour at discontinuities. *Bull. Seism. Soc. Am.*, 83(1):107–129, 1993.

9. J. Kristek and P. Moczo. Seismic-wave propagation in viscoelastic media with material discontinuities: a 3D fourth-order staggered-grid finite-difference modeling. Bull. Seism. Soc. Am., 93(5):2273–2280, 2003.
10. P. C. Liu and R. J. Archuleta. Efficient modeling of Q for 3D numerical simulation of wave propagation. Bull. Seism. Soc. Am., 96(4A):1352–1358, 2006.
11. W. H. Reed and T. R. Hill. Triangular mesh methods for the neutron transport equation. Technical report, Los Alamos Scientific Laboratory, 1973.
12. E. H. Saenger, N. Gold, and S. A. Shapiro. Modeling the propagation of elastic waves using a modified finite-difference grid. Wave Motion, 31:77–92, 2000.
13. S. K. Singh, E. Mena, and R. Castro. Some aspects of source characteristics of the 19 September 1985 Michoacan earthquake and ground motion amplification in and near Mexico City from strong motion data. Bull. Seism. Soc. Am., 78:451–477, 1988.
14. J. Zahradník, P. Moczo and F. Hron. Testing four elastic finite-difference schemes for behavior at discontinuities. Bull. Seism. Soc. Am., 83(1):107–129, 1993.

Mixed Mimetic Spectral Element Method Applied to Darcy's Problem

Pedro Pinto Rebelo, Artur Palha, and Marc Gerritsma

Abstract We present a discretization for Darcy's problem using the recently developed *Mimetic Spectral Element Method* (Kreeft et al. (2011) Mimetic framework on curvilinear quadrilaterals of arbitrary order. Submitted to FoCM, Arxiv preprint arXiv:1111.4304). The gist lies in the exact discrete representation of integral relations. In this paper, an anisotropic flow through a porous medium is considered and a discretization of a full permeability tensor is presented. The performance of the method is evaluated on standard test problems, converging at the same rate as the best possible approximation.

1 Darcy Flow

Anisotropic heterogeneous diffusion problems are ubiquitous across different scientific fields, such as, hydrogeology, oil reservoir simulation, plasma physics, biology, etc [10]. Darcy's equation describes a steady pressure-driven flow through a porous medium where fluxes and pressure are linearly related,

$$\operatorname{div} \frac{\mathbb{K}}{\mu} \operatorname{grad} p = \phi \stackrel{\mu=1}{\longrightarrow} \begin{cases} \mathbf{u} - \operatorname{grad} p = 0 \text{ in } \Omega & \text{(1a)} \\ \operatorname{div} \mathbf{q} = \phi \qquad \text{in } \Omega & \text{(1b)} \\ \mathbf{q} = \mathbb{K}\mathbf{u} \qquad\; \text{in } \Omega & \text{(1c)} \\ \mathbf{q} = \mathbf{q}_0 \qquad\; \text{in } \partial\Omega & \text{(1d)} \end{cases}$$

P.P. Rebelo (✉) · A. Palha · M. Gerritsma
Faculty of Aerospace Engineering, Aerodynamics Group, Delft University of Technology, Delft, The Netherlands
e-mail: P.J.PintoRebelo@tudelft.nl; A.Palha@tudelft.nl; M.I.Gerritsma@tudelft.nl

M. Azaïez et al. (eds.), *Spectral and High Order Methods for Partial Differential Equations - ICOSAHOM 2012*, Lecture Notes in Computational Science and Engineering 95, DOI 10.1007/978-3-319-01601-6_30,

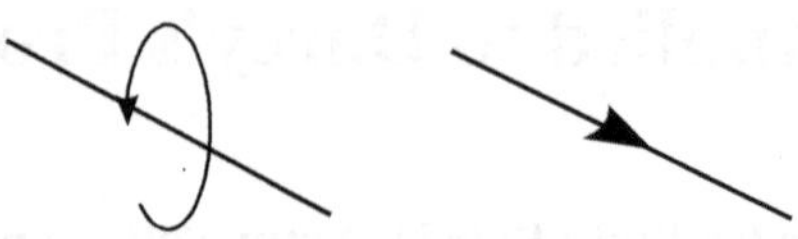

Fig. 1 Consider a line where we can have two types of orientation: *Outer* – around the line; *Inner* – along the line

where $\mathbf{u}$ is the fluid velocity, p the pressure, $\mathbf{q}$ the mass flux and ϕ the prescribed source term. Without loss of generality let the viscosity, $\mu = 1$, and consider a permeability symmetric, positive definite tensor denoted by $\mathbb{K}$.

In a three-dimensional setting are: four types of submanifolds (*points, lines, surfaces and volumes*); and two orientations (*outer and inner*, as an example see Fig. 1). Tessellation divides the physical domain in a set of these geometric objects to which we associate discrete variables, i.e. *integral quantities*. Thus, associated with every physical variable is a correspondent geometric object, this symbiotic relation between physics and geometry is the core of *mimetic methods*. Many scholars are aware of this relationship [3, 5, 7, 23].

Starting from the mass balance equation, (1b),

$$\int_V \operatorname{div} \mathbf{q} \, \mathrm{d}V = \int_{\partial V} \mathbf{q} \cdot \mathbf{n} \, \mathrm{d}S = \int_V \phi \, \mathrm{d}V, \tag{2}$$

it is clear that the *divergence in a volume* is equal to the sum of the *surface integral quantities*, i.e. *oriented fluxes*. Thus, we will associate mass fluxes, $\mathbf{q}$, with quantities that go *through* surfaces. This equation therefore tells us that the right hand side term ϕ is associated to *outer-oriented volumes*.

Similarly, using Newton-Leibniz relation for Eq. (1a),

$$\int_C \operatorname{grad} p \, \mathrm{d}C = \int_{\partial C} p = p(B) - p(A) = \int_C \mathbf{u} \, \mathrm{d}C, \tag{3}$$

the fluid velocity, $\mathbf{u}$, is represented *along* lines and p is represented by the values in points. From (3) we deduces that u and p are *inner-oriented variables*.

The *constitutive/material relation* relation (1c) is given by,

$$\mathbf{q} = \mathbb{K}\mathbf{u}, \tag{4}$$

which defines how quantities associated to inner-oriented lines relate to quantities associated to outer-oriented surfaces. Whereas Eqs. (2) and (3) can be exactly satisfied on a finite grid the constitutive equation (4) needs to be approximated.

The importance of respecting the geometric nature in physics is discussed in [9]. Figure 2 summarizes the geometric character of the Darcy's problem.

We will denote the space of variables associated to outer-oriented k-dimensional objects by $\Lambda^k(\mathcal{M})$ and the space of variables associated to inner-oriented k-dimensional objects by $\tilde{\Lambda}^k(\mathcal{M})$ as indicated in Fig. 2.

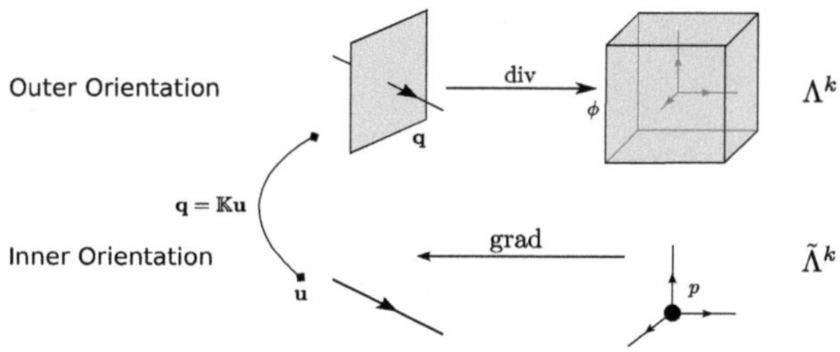

Fig. 2 Darcy's flow problem geometric characterization. Fluxes, $\mathbf{q}$, are associated with outer oriented surfaces; ϕ, is associated with outer oriented volumes; velocity, $\mathbf{u}$, is associated with inner oriented lines; pressure, p, is associated with inner oriented points

In this paper we will make use of the spectral element method described in [9,19], application of these ideas to Stokes' flow see [16–18]; Poisson equation for volume forms [22]; advection equation [21]; derivation of a momentum conservation scheme [24]. Extension to compatible isogeometric methods see [11, 12]. For applications of these ideas in a finite difference setting see Brezzi et al. [4]. In the context of finite element methods Arnold et al. [1] proposed a *Finite Element Exterior Calculus*. In a more geometric spirit Desbrun et al. [6] and Hirani [13] developed the *discrete exterior calculus* (DEC). An application of the latter to Darcy flow can be found in [14].

2 Discretization of Equations

In this section we will describe the discretization by defining the weak formulation. The approached followed here is similar to [2, 15].

For vectors associated with outer-oriented surfaces, $\Lambda^2(\mathcal{M})$, we define the weighted inner product

$$(\mathbf{a},\mathbf{b})_{\mathcal{M},\mathbb{K}} = \int_{\mathcal{M}} \mathbf{a}\mathbb{K}^{-1}\mathbf{b}\,\mathrm{d}V. \tag{5}$$

Furthermore, we define bilinear maps $((\cdot,\cdot))_{\mathcal{M}} : \Lambda^1(\mathcal{M}) \times \tilde{\Lambda}^{n-1}(\mathcal{M}) \to \mathbb{R}$

$$((\mathbf{u},\tilde{\mathbf{v}}))_{\mathcal{M}} := \int_{\mathcal{M}} \mathbf{u}\cdot\tilde{\mathbf{v}}\,\mathrm{d}V. \tag{6}$$

and $((\cdot,\cdot))_{\mathcal{M},\mathbb{K}} : \Lambda^k(\mathcal{M}) \times \tilde{\Lambda}^{n-k}(\mathcal{M}) \to \mathbb{R}$ given by

(continued)

(continued)

$$((\mathbf{u}, \tilde{\mathbf{v}}))_{\mathcal{M},\mathbb{K}} := \int_{\mathcal{M}} \mathbf{u} \cdot \mathbb{K}^{-1} \tilde{\mathbf{v}} \, \mathrm{d}V. \tag{7}$$

For $\mathbf{q} \in \Lambda^2(\mathcal{M})$ and $p \in \tilde{\Lambda}^0(\mathcal{M})$ and homogeneous boundary values we have

$$\begin{aligned} ((\mathrm{div}\mathbf{q},\ p))_{\mathcal{M}} &= -((\mathbf{q}, \mathrm{grad}\, p))_{\mathcal{M}} \\ &= -\left((\mathbf{q}, \mathbb{K}^{-1}[\mathbb{K}\ \mathrm{grad}\, p])\right)_{\mathcal{M}} \\ &= \left((\mathbf{q}, \mathrm{grad}^*_{\mathbb{K}}\ p)\right)_{\mathcal{M},\mathbb{K}} . \end{aligned} \tag{8}$$

It is possible to define a new gradient operator,

$$\mathrm{grad}^*_{\mathbb{K}} = -\mathbb{K}\ \mathrm{grad}. \tag{9}$$

2.1 Mixed Formulation

Starting from (1) and making use of the bilinear maps defined above we have for all vectors $\boldsymbol{\tau} \in \Lambda^{n-1}(\mathcal{M})$ associated to outer-oriented surfaces

$$\begin{gathered} ((\boldsymbol{\tau}, \mathbf{u} - \mathrm{grad}\, p))_{\mathcal{M}} = 0 \\ \Longleftrightarrow \\ ((\boldsymbol{\tau}, \mathbb{K}\mathbf{u} - \mathbb{K}\mathrm{grad}\, p))_{\mathcal{M},\mathbb{K}} = 0 \\ \overset{(9)}{\Longleftrightarrow} \\ ((\boldsymbol{\tau}, \mathbb{K}\mathbf{u}))_{\mathcal{M},\mathbb{K}} + \left(\boldsymbol{\tau}, \mathrm{grad}^*_{\mathbb{K}} p\right)_{\mathcal{M},\mathbb{K}} = 0 \\ \overset{\text{(1c) and (8)}}{\Longleftrightarrow} \\ (\boldsymbol{\tau}, \mathbf{q})_{\mathcal{M},\mathbb{K}} + ((\mathrm{div}\ \boldsymbol{\tau},\ p))_{\mathcal{M}} = 0 \end{gathered} \tag{10}$$

The constitutive equation is included in the last step by converting the bilinear form to a weighted inner product on $\Lambda^2(\mathcal{M})$ as defined in (5). For (1b) we take the bilinear map for the divergence of a vector $\mathbf{q}$ associated with outer-oriented surfaces and an arbitrary scalar function defined in inner-oriented points, $\gamma \in \tilde{\Lambda}^0(\mathcal{M})$,

$$((\mathrm{div}\mathbf{q}, \gamma))_{\mathcal{M}} = ((\phi, \gamma))_{\mathcal{M}} . \tag{11}$$

The mixed formulation becomes: Find $(\mathbf{q}, p) \in \left\{\Lambda^2(\mathcal{M}) \times \tilde{\Lambda}^0(\mathcal{M})\right\}$, given $\phi \in \Lambda^3(\mathcal{M})$, for all $(\boldsymbol{\tau}, \gamma) \in \left\{\Lambda^2(\mathcal{M}) \times \tilde{\Lambda}^0(\mathcal{M})\right\}$ such that,

$$(\tau, \mathbf{q})_{\mathcal{M},\mathbb{K}} + ((\text{div}\, \boldsymbol{\tau},\ p))_{\mathcal{M}} = 0 \tag{12}$$

$$((\text{div}\mathbf{q}, \gamma))_{\mathcal{M}} = ((\phi, \gamma))_{\mathcal{M}}\ . \tag{13}$$

2.2 Basis Functions

For the high order representation we use Lagrange, $l_i(\xi)$, and edge functions, $e_i(\xi)$. Lagrange polynomials interpolate nodal values. The edge functions, derived by Gerritsma [8] are constructed such that when integrating over a line segment it gives one for the corresponding element and zero for any other line segment,

$$l_i(\xi_j) = \delta_{i,j} \qquad \int_{\xi_{j-1}}^{\xi_j} e_i(\xi) = \delta_{i,j}\,. \tag{14}$$

The relation between the Lagrange and the edge functions is given by,

$$e_i(\xi) = \epsilon_i(\xi)\, d\xi, \quad \text{with} \quad \epsilon_i(\xi) = -\sum_{k=0}^{i-1} \frac{d l_k}{d\xi}. \tag{15}$$

Note that this definition implies

$$\frac{d l_i}{d\xi} = e_i(\xi) - e_{i+1}(\xi)\ . \tag{16}$$

Extension to the multidimensional is obtained by means of tensor products. For more details see [19].

2.3 Mimetic Discretization in 2D

2.3.1 Expansion of Unknowns in $\mathbb{R}^2$

Let $\mathbf{q} \in \Lambda^1(\mathcal{M})$ be expanded as,

$$\mathbf{q}_h = \begin{bmatrix} \sum_{i=0}^{N} \sum_{j=1}^{N} q_{i,j}^x l_i(\xi)\, e_j(\eta) \\ \sum_{i=1}^{N} \sum_{j=0}^{N} q_{i,j}^y e_i(\xi)\, l_j(\eta) \end{bmatrix}, \tag{17}$$

and the pressure, $p_h \in \tilde{\Lambda}^0(\mathcal{M})$ as,

$$p_h = \sum_{i=1}^{N}\sum_{j=1}^{N} p_{i,j}\epsilon_i(\xi)\,\epsilon_j(\eta)\,. \tag{18}$$

2.3.2 Discrete Divergence in $\mathbb{R}^2$

The divergence of $\mathbf{q}_h$ is then given by

$$\operatorname{div}\mathbf{q}_h = \sum_{i=1}^{N}\sum_{j=1}^{N}\left(q^x_{i,j} - q^x_{i-1,j} + q^y_{i,j} - q^y_{i,j-1}\right) e_i(\xi)\, e_j(\eta)\;, \tag{19}$$

where we repeatedly used (16). The scalar $\phi \in \Lambda^2(\mathcal{M})$ associated with outer-oriented volumes is expanded as

$$\phi_h = \sum_{i=1}^{N}\sum_{j=1}^{N} \phi_{i,j} e_i(\xi)\, e_j(\eta)\;. \tag{20}$$

Equating (19) and (20) yields

$$\sum_{i=1}^{N}\sum_{j=1}^{N} \phi_{i,j}\cancel{e_i(\xi)\, e_j(\eta)} = \sum_{i=1}^{N}\sum_{j=1}^{N}\left(q^x_{i,j} - q^x_{i-1,j} + q^y_{i,j} - q^y_{i,j-1}\right)\cancel{e_i(\xi)\, e_j(\eta)} \tag{21}$$

$$[\phi] = \mathsf{E}^{(2,1)}\,[\mathbf{q}]\,. \tag{22}$$

We see that the basis functions cancel from this relation. The matrix $\mathsf{E}^{(2,1)}$ relates the fluxes $q^x_{i,j}$ and $q^y_{i,j}$ to the volume integral $\phi_{i,j}$, as depicted in Fig. 3. This fully discrete equation is a restatement of the integral relation (2). The matrix $\mathsf{E}^{(2,1)}$ only contains the values -1, 0 and 1 and is fully determined by the grid, see [19]. This is an incidence matrix showing the topological nature of the discrete divergence.

If we insert the expansions of our unknowns in (12) and (13) we obtain in $\mathbb{R}^n$ the saddle point problem given by,

$$\begin{bmatrix} M^{(n-1)}_{\mathbb{K}} & \left(\mathsf{E}^{(n,n-1)}\right)^T M^{(n)} \\ M^{(n)}\mathsf{E}^{(n,n-1)} & 0 \end{bmatrix}\begin{bmatrix}\mathbf{q}\\ p\end{bmatrix} = \begin{bmatrix} 0 \\ M^{(n)}\phi\end{bmatrix}, \tag{23}$$

(continued)

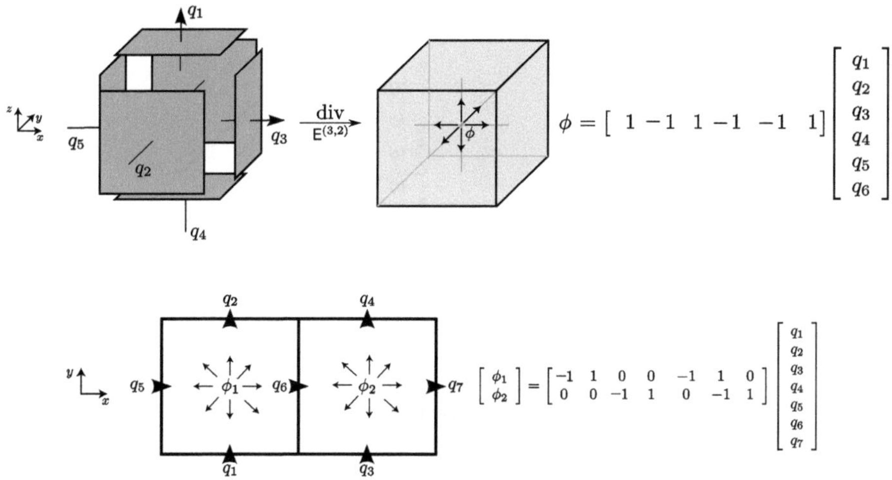

Fig. 3 Discrete representation of the action of the divergence in $\mathbb{R}^3$ and $\mathbb{R}^2$

(continued)

where $M^{(k)}$ is the symmetric mass matrix obtain from the bilinear pairing between variables associated with outer-orientation and inner-orientation, (6), $M_{\mathbb{K}}^{(n-1)}$ is the mass matrix obtained from the weighted inner product (5) and $\mathsf{E}^{(n,n-1)}$ the incidence matrix which relates fluxes over surfaces to volumes. The resulting system (23) is symmetric.

The pressure which is represented on an inner-oriented grid (which is not explicitly constructed in this single grid approach) is pre-multiplied by $M^{(n)}$ to represent it on the outer-oriented grid.

3 Numerical Results

The method derived in this paper respects the geometric nature of the problem. However, it is crucial to verify the numerical benefits of this approach. This section presents hp-convergence studies for anisotropic permeability.

3.1 Manufactured Solution: Anisotropic Permeability

The first test case assesses the convergence for h- and p-refinement of the mixed mimetic spectral element method applied to the Darcy model. This is a benchmark

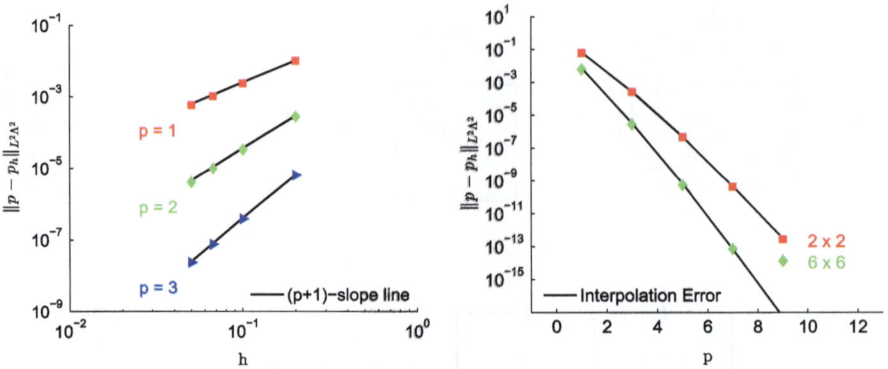

Fig. 4 Plots of the h- and p-convergence for anisotropic permeability given in (24)

problem presented in [15]. The problem is defined on a unit square, $\Omega = [-1, 1]^2$, with Cartesian coordinates with permeability given by,

$$\mathbb{K} = \begin{bmatrix} 2 & 1 \\ 1 & 2 \end{bmatrix} \tag{24}$$

and the right hand side, $\phi \in L^2(\mathcal{M})$ given by,

$$\phi^{(2)} = 2\left(1 + x^2 + xy + y^2\right) e^{xy} \, \mathrm{d}x\mathrm{d}y. \tag{25}$$

This results in an exact solution for pressure $p \in \tilde{\Lambda}^0(\mathcal{M})$ given by,

$$p^{(0)} = e^{xy} \tag{26}$$

Figure 4 shows the h- and p-convergence for the pressure in straight mesh. For the h-convergence the expected rate of convergence is of $(p + 1)$, where p is the polynomial degree. The solid line for the interpolation error in the p-convergence plot is the L^2-error from interpolating the exact solution, the solution converges exponentially. Both the numerical solution and the interpolated exact solution converge exponentially.

3.2 Layered Medium

A classical benchmark for Darcy flow codes is the piecewise constant permeability in a square [20]. Such a medium is called *layered* medium.

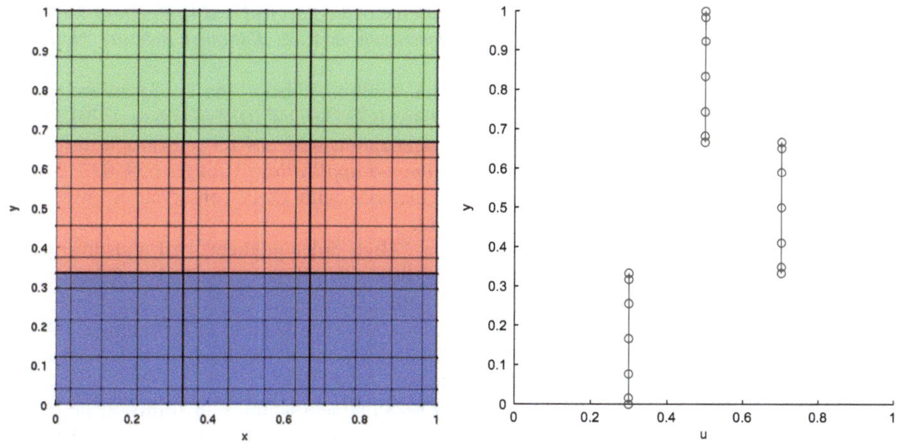

Fig. 5 Layered medium with 3 different permeabilities, α ranges top to bottom from 0.5, 0.7 and 0.3. The *left figure* shows a velocity profile at an arbitrary x. Note the numerical solution in the Gauss-Lobatto nodes

$$\mathbb{K} = \begin{bmatrix} \alpha & 0 \\ 0 & \alpha \end{bmatrix} \quad \alpha = \begin{cases} 0.3 & \text{if } y \leq \frac{1}{3} \\ 0.7 & \text{if } \frac{1}{3} < y \leq \frac{2}{3} \\ 0.5 & \text{if } y > \frac{2}{3} \end{cases} \tag{27}$$

The fluid comes into the domain from the left to the right. Since the pressure depends linearly on x, horizontal constant velocity is expected in each layer, Fig. 5.

Acknowledgements The authors gratefully acknowledge the funding received by FCT – Foundation for science and technology Portugal through SRF/BD/36093/2007 and SFRH/BD/79866/2011 and the anonymous reviewers for their helpful comments.

References

1. D. Arnold, R. Falk, and R. Winther. Finite element exterior calculus: from Hodge theory to numerical stability. *American Mathematical Society*, 47(2):281–354, 2010.
2. J. Bonelle and A. Ern. Analysis of compatible discrete operator schemes for elliptic problems on polyhedral meshes. *arXiv preprint arXiv:1211.3354*, 2012.
3. A Bossavit. Discretization of electromagnetic problems. *Handbook of numerical analysis*, 13:105–197, 2005.
4. F Brezzi, A Buffa, and K Lipnikov. Mimetic finite differences for elliptic problems. *Mathematical Modelling and Numerical Analysis*, 43(2):277–296, 2009.
5. W. L. Burke. *Applied differential geometry*. Cambridge Univ Pr, 1985.
6. M. Desbrun, A. Hirani, M. Leok, and J. Marsden. Discrete exterior calculus. *Arxiv preprint math/0508341*, 2005.
7. T. Frankel. *The Geometry of Physics*. Cambridge University Press, 2nd edition, 2004.

8. M Gerritsma. Edge functions for spectral element methods. *Spectral and High Order Methods for Partial differential equations, Eds J.S. Hesthaven & E.M. Rønquist, Lecture Notes in Computational Science and Engineering*, 76.
9. M. Gerritsma, R. Hiemstra, J. Kreeft, A. Palha, P. Pinto Rebelo, and D. Toshniwal. The geometric basis of numerical methods. *Proceedings ICOSAHOM 2012 (this issue)*, 2012.
10. R. Herbin and F. Hubert. Benchmark on discretization schemes for anisotropic diffusion problems on general grids. *Finite volumes for complex applications V*, pages 659–692, 2008.
11. R. Hiemstra and M. Gerritsma. High order methods with exact conservation properties. *Proceedings ICOSAHOM 2012 (this issue)*, 2012.
12. R. Hiemstra, R. Huijsmans, and M. Gerritsma. High order gradient, curl and divergence conforming spaces, with an application to compatible isogeometric analysis. *Submitted to J. Comp Phys., arXiv preprint arXiv:1209.1793*, 2012.
13. A. Hirani. *Discrete Exterior Calculus*. PhD thesis, California Institute of Technology, 2003.
14. A. Hirani, K. Nakshatrala, and J. Chaudhry. Numerical method for Darcy flow derived using discrete exterior calculus. *arXiv preprint arXiv:0810.3434*, 2008.
15. J. Hyman, M. Shashkov, and S. Steinberg. The numerical solution of diffusion problems in strongly heterogeneous non-isotropic materials. *Journal of Computational Physics*, 132(1):130–148, 1997.
16. J. Kreeft and M. Gerritsma. Higher-order compatible discretization on hexahedrals. *Proceedings ICOSAHOM 2012 (this issue)*, 2012.
17. J. Kreeft and M. Gerritsma. Mixed mimetic spectral element method for stokes flow: a pointwise divergence-free solution. *Journal of Computational Physics*, 2012.
18. J. Kreeft and M. Gerritsma. A priori error estimates for compatible spectral discretization of the stokes problem for all admissible boundary conditions. *arXiv preprint arXiv:1206.2812*, 2012.
19. J. Kreeft, A. Palha, and M. Gerritsma. Mimetic framework on curvilinear quadrilaterals of arbitrary order. *Submitted to FoCM, Arxiv preprint arXiv:1111.4304*, 2011.
20. A. Masud and T.J.R. Hughes. A stabilized mixed finite element method for darcy flow. *Computer Methods in Applied Mechanics and Engineering*, 191(39):4341–4370, 2002.
21. A. Palha, P. Pinto Rebelo, and M. Gerritsma. Mimetic spectral element solution for conservative advection. *Proceedings ICOSAHOM 2012 (this issue)*, 2012.
22. A. Palha, P. Pinto Rebelo, R. Hiemstra, J. Kreeft, and M. Gerritsma. Physics-compatible discretization techniques on single and dual grids, with application to the Poisson equation of volume forms. *Submitted to J. Comp Phys.*, 2012.
23. E Tonti. On the formal structure of physical theories. *preprint of the Italian National Research Council*, 1975.
24. D. Toshniwal, R.H.M. Huijsmans, and M. Gerritsma. A geometric approach towards momentum conservation. *Proceedings ICOSAHOM 2012 (this issue)*, 2012.

Novel Outflow Boundary Conditions for Spectral Direct Numerical Simulation of Rotating Flows

Stéphanie Rodriguez, Bertrand Viaud, and Eric Serre

Abstract In this paper we introduce new outlet boundary conditions to simulate 3D rotating flows within an interdisk cavity. The boundary conditions have been implemented in a multidomain pseudo-spectral algorithm and a quantitative method has been developped to qualify them. Numerical results show a reduction of about 50 % of the refracted wave amplitude compared to the convective boundary conditions that are commonly used.

1 Introduction

The study of the flow between two rotating disks presents multiple interests whether it be for technological devices such as turbomachinery, or for the study of geofluids for instance as discussed by Hide [1]. It is also of academical interest as it provides an example of a three-dimensional boundary layer. Indeed, the existence of a known laminar solution, makes it a good candidate for stability analysis. The open cavity introduces in addition to the rotation rate, a second control parameter: the mass flow rate. It has thus been used by Viaud et al. [2, 3] to study transition to turbulence scenarii. The small numerical dissipation and dispersion to accurately investigate the dynamics of infinitesimal perturbations and the statistics in turbulent regimes, respectively, require the use of high-order methods like the spectral ones. Due to the extreme sensitivity of the transition process to the numerical noise, and the use of a

S. Rodriguez (✉) · B. Viaud
Centre de Recherche de l'arme de l'Air, CReA BA 701, 13661 Salon air, France
e-mail: stephanie.rodriguez@inet.air.defense.gouv.fr; bertrand.viaud@inet.air.defense.gouv.fr

E. Serre
M2P2, CNRS Universits Aix Marseille, IMT Chteau Gombert, 13452 Marseille, France
e-mail: eric.serre@l3m.univ-mrs.fr

M. Azaïez et al. (eds.), *Spectral and High Order Methods for Partial Differential Equations - ICOSAHOM 2012*, Lecture Notes in Computational Science and Engineering 95, DOI 10.1007/978-3-319-01601-6_31,

spectral method that propagates any given perturbation, these investigations depend on the quality of the outflow boundary conditions that must not induce refractions in the computational domain.

2 Geometrical and Mathematical Modelling

The configuration is that of a rotating cavity formed by two parallel disks with a radial throughflow at the hub (Fig. 1).

The annular cavity formed by two parallel disks is defined by its inlet and outlet radii (respectively R_2^* and R_1^*) and the distance $2h^*$ between the disks. Two global geometry parameters are thus defined: a curvature parameter Rm and an aspect ratio L

$$Rm = \frac{R_1^* + R_2^*}{R_1^* - R_2^*},\ L = \frac{R_1^* - R_2^*}{2h^*} \tag{1}$$

For a rotating disk and in the limit of high rotation rates, the Navier-Stokes equations admit an asymptotic solution, the so-called Ekman boundary layer where inertial nonlinear terms are dominated by Coriolis forces. The flow has a characteristic length $\delta = \sqrt{\Omega/\nu}$, called Ekman scale. Two global parameters are thus defined, a Reynolds number for the rotation Re_Ω, and a radial mass flow, Cw

$$Re_\Omega = \frac{\Omega R_1^{*2}}{\nu},\ Cw = \frac{Q^*}{\nu R_1^*} \tag{2}$$

The azimuthal velocity in the core is called geostrophic velocity, and we introduce two local parameters: a local Reynolds number Re_δ, and a Rossby number Ro

$$Re_\delta = \frac{V_g^* \delta^*}{\nu},\ Ro = \frac{V_g^*}{\Omega r^*} \tag{3}$$

where ν is the kinematic viscosity and Ω the rotation velocity, V_g^* is the geostrophic velocity and r^* the radius. The flow is governed by the three-dimensional Navier Stokes equations considered in the primitive variables formulation in the cylindrical coordinates (r, θ, z):

$$\begin{cases} \partial_t V^* + V^*.\nabla V^* = -\nabla p^* + \nu \Delta V^* + F^* & in\ \Omega \\ V^* = W^* & on\ \Gamma \\ \nabla.V^* = 0 & in\ \Omega \end{cases} \tag{4}$$

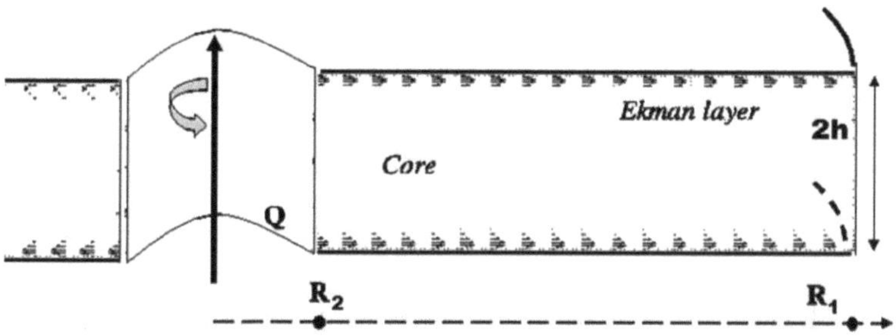

Fig. 1 Configuration of the cavity showing the baseflow composed by two Ekman layers at the disks separated by a geostrophic non viscous core. R_2^* and R_1^* respectively designate the radius of the inlet at the hub and the outlet at the rim of the $2h^*$high cavity

where V is the velocity of components (u, v, w) respectively in the radial, azimuthal and axial directions, p is the pressure, F represents body forces and Δ is the Laplacian operator written in the cylindrical coordinates.

3 Numerical Modelling

3.1 Discretisation

The discretisation of the Navier Stokes equations is based on the projection algorithm presented in of Raspo et al. [4]. The three-dimensional time-dependent incompressible Navier-Stokes equations are discretised through a semi-implicit 2nd-order time scheme that is a combination of an Adams Basforth treatment of the non linear terms, and a 2nd-order backward differentiation formula for the linear terms:

$$\begin{cases} \frac{3V^{n+1}-4V^n+V^{n-1}}{2\delta t} + 2(V.\nabla V)^n - (V.\nabla V)^{n-1} = -\nabla p^{n+1} + \nu \Delta V^{n+1} + F^{n+1} & in\ \Omega \\ V^{n+1} = W^{n+1} & on\ \Gamma \\ \nabla . V^{n+1} = 0 & in\ \bar{\Omega} \end{cases} \quad (5)$$

Due to the cylindrical configuration, the solution presents a 2π-periodicity in the azimuthal direction. The space discretisation in this direction is based on the Fourier Galerkin method. The discretisation in the non periodic directions is made through a Chebyshev-collocation method.

The velocity-pressure coupling is solved through a prediction-correction algorithm [4]. It involves the successive resolution of bidimensional Helmoltz problems for each Fourier mode at each time step.

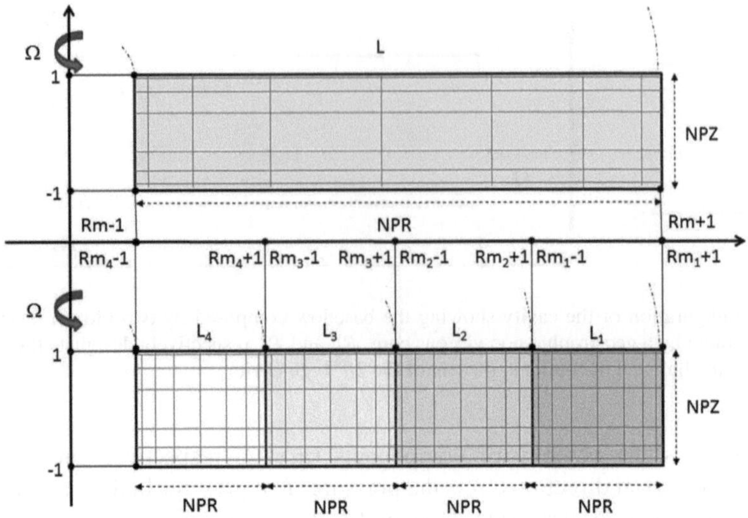

Fig. 2 Multidomain decomposition

3.2 One Dimensional Multidomain Decomposition

The multidomain decomposition in the radial direction is based on the influence matrix technique. The cylindrical configuration imposes the definition of local geometry parameters that are compatible with the global ones defined previously (Fig. 2).

As in the monodomain configuration, each subdomain i presents the following geometry parameters:

$$Rm_i = \frac{R_i^1 + R_i^2}{R_i^1 - R_i^2}, \; L_i = \frac{R_i^1 - R_i^2}{2h} \tag{6}$$

So as to simplify the notations, let us consider a decomposition with two domains and let Φ_i be one of the variables in subdomain i. The spatio-temporal discretisation of the Navier Stokes equations leads to the following Helmoltz problems in each subdomain and for each Fourier mode:

$$\begin{cases} \Delta\Phi_i - \sigma\Phi_i = S_i \; in & \Omega_i \\ A_i\Phi_i = b_i & on \quad \Gamma_i \cap \Gamma \\ \Phi_i = \Phi & on \; \Gamma_i \cap \Omega = \xi \end{cases} \tag{7}$$

with $A_i = I$ for the velocity components and $A_i = \partial_n$ for the pressure, and σ a function of r, the Fourier mode k and the time step δt.

The value of Φ on the interface ξ is unknown and is computed through the influence matrix technique. The linearity of the Helmoltz problems allows us to search the solution as the sum of two functions $\tilde{\Phi}_i$ and $\bar{\Phi}_i$ solutions of the following problems:

- A homogeneous problem: $\begin{cases} \Delta\tilde{\Phi}_i - \sigma\tilde{\Phi}_i = S_i & in \quad \Omega_i \\ A_i\tilde{\Phi}_i = b_i & on\ \Gamma_i \cap \Gamma \\ \tilde{\Phi}_i = 0 & on \quad \xi \end{cases}$
- A stationary problem: $\begin{cases} \Delta\bar{\Phi}_i - \sigma\bar{\Phi}_i = 0 & in \quad \Omega_i \\ A_i\bar{\Phi}_i = 0 & on\ \Gamma_i \cap \Gamma \\ \bar{\Phi}_i = \Phi & on \quad \xi \end{cases}$

Due to the Dirichlet boundary conditions at the border, the homogeneous solution is continuous but presents a derivative jump H. There is only one vector λ that nullifies H. This derivative jump is corrected by the influence matrix M which is computed from the derivatives of the stationary problem solutions. By inverting the influence matrix M, the value of Φ that has to be imposed at each collocation point on the border is obtained through matricial product: $\lambda = MH^{-1}$.

4 Outflow Boundary Conditions

In the case of open flows, non-reflective boundary conditions are particularly important, all the more where spectral accuracy is involved seeing as it propagates any perturbation. Halpern and Schatzman [5] established that completely transparent boundary conditions are not applicable locally. Approaching conditions have thus been investigated and Orlanski [6] introduced convective boundary conditions, satisfied by waves propagating perpendicularly to the boundary, but that caused instabilities when ingoing and outgoing waves are combined. Ruith et al. [7] proposed an improvement to these convective boundary conditions that are now commonly used:

$$\partial_t u + C\partial_x u = 0 \tag{8}$$

where C designates the advection velocity that is a constant fixed arbitrarily so as to ensure the flow rate conservation for example. Equation (8) is solved locally at the outflow at each time-step to obtain the unsteady Dirichlet conditions for the velocity field.

Fournier et al. [8] proposed an alternative to these convective conditions in the case of a two-dimensional incompressible laminar wall-bounded flow. They proposed to substitute the Navier Stokes equations at the outlet by the boundary layer equations. Indeed, ideally we would want to solve the Navier Stokes equations at the outlet, but these equations are elliptic in space. On the opposite, the boundary layer equations, which only retains one direction in the second order derivatives of the laplacian, are parabolic and can thus be solved at the outlet. Moreover the boundary-layer solution supports the same instability waves as the full NS solution, ensuring a low refraction. In the case of a Blasius boundary layer, the boundary conditions introduced by Fournier et al. are written:

$$\frac{\partial u}{\partial t} + u_x \frac{\partial u}{\partial x} + u_y \frac{\partial u}{\partial y} - \nu \frac{\partial^2 u}{\partial y^2} = 0 \tag{9}$$

In our configuration, outflow boundary conditions are obtained from the axisymmetric linearised Navier Stokes equations and with a boundary layer type hypothesis then of geostrophic equilibrium:

$$\begin{cases} \partial_t u^* - 2\Omega(v^* - V_g^*) - \nu\left(\frac{1}{r^*}\partial_r u^* + \partial_{zz}u^* - \frac{u^*}{r^{*2}}\right) = 0 \\ \partial_t v^* - 2\Omega u^* - \nu\left(\frac{1}{r^*}\partial_r v^* + \partial_{zz}v^* - \frac{v^*}{r^{*2}}\right) = 0 \\ \partial_t w^* - \nu\left(\frac{1}{r^*}\partial_r w^* + \partial_{zz}w^*\right) = 0 \end{cases} \tag{10}$$

These equations are made dimensionless by using $h^*, \Omega^{-1}, \Omega R_1^*$, as length, time and velocity scales, which gives:

$$\begin{cases} \partial_t u - 2(v - V_g) - \frac{4}{Re_h}\left(\frac{1}{r}\partial_r u + \partial_{zz}u - \frac{u}{r^2}\right) = 0 \\ \partial_t v - 2u - \frac{4}{Re_h}\left(\frac{1}{r}\partial_r v + \partial_{zz}v - \frac{v}{r^2}\right) = 0 \\ \partial_t w - \frac{4}{Re_h}\left(\frac{1}{r}\partial_r w + \partial_{zz}w\right) = 0 \end{cases} \tag{11}$$

where $Re_h = 4\left(\frac{h}{\delta}\right)^2$ designates the Reynolds number cothe quality of the dierent types of boundary conditions i.e. the refracted wave induced at the cavity outlet is measured (Fig. 3).

When the radial mass flow is increased, nonlinear inertial forces cannot be neglected. Indeed, while for $Cw = 500$, there are 3–4 orders of magnitude separating diffusive and convective effects, there are only 2 orders of magnitude when $Cw = 2{,}000$. The local boundary layer equations used have to be modified in some way to retain these terms, giving:

$$\begin{cases} \partial_t u + R_1\left(u\partial_r u + w\partial_z u - \frac{v^2}{r}\right) - 2(v - V_g) - \frac{4}{Re_h}\left(\frac{1}{r}\partial_r u + \partial_{zz}u - \frac{u}{r^2}\right) = 0 \\ \partial_t v + R_1\left(u\partial_r v + w\partial_z v - \frac{uv}{r}\right) - 2u - \frac{4}{Re_h}\left(\frac{1}{r}\partial_r v + \partial_{zz}v - \frac{v}{r^2}\right) = 0 \\ \partial_t w + R_1\left(u\partial_r w + w\partial_z w\right) - \frac{4}{Re_h}\left(\frac{1}{r}\partial_r w + \partial_{zz}w\right) = 0 \end{cases} \tag{12}$$

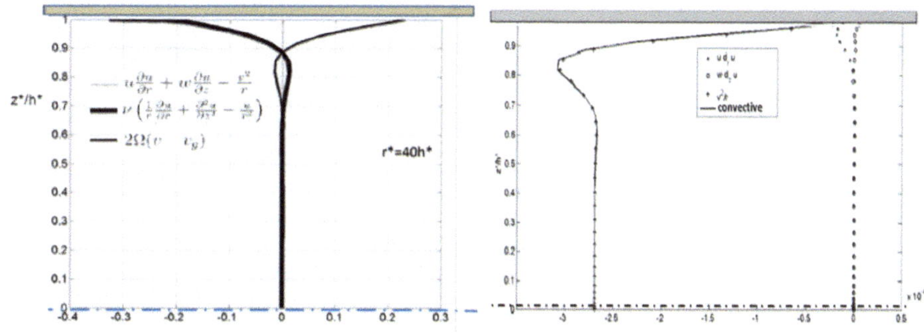

Fig. 3 Order of magnitude of the different terms of the momentum equation (in the radial direction) for the laminar solution in a cavity between $R_2^* = 30h^*$ and $R_1^* = 50h^*$with $Re_\Omega = 487{,}500$ and $Cw = 500$. (**a**) The *bold line* represents the diffusive terms, the *solid line* is the Coriolis force and the *dashed line* the convective terms. Computation is made at $r^* = 40h^*$ which corresponds to $Ro = 0, 114$ (**b**) Contribution of the different terms part of the convective terms. v^2/r is dominant

The discretisation of these equations is made through a 1st order Euler scheme for stability concerns. Derivatives in the axial direction are obtained through a spectral differentiation matrix in the spectral space, while in the radial direction finite differences are used since [7] has shown that a local approximation gave better results.

5 Evaluation of the Quality of the Boundary Conditions

5.1 Method

At the outlet, the unstationary field is the sum of two distinct waves : an incident wave that corresponds to the passing of a perturbation, and a refracted wave of smaller amplitude and quickly deadened, induced by the boundary conditions. Let us suppose that over a small enough distance, both the wavelength λ and the amplitude A of the wave packet crossing the outlet can be considered constant. The idea is to reconstruct a wave that fits the wave packet as it approaches the boundary, using the general form $Acos(r/\lambda + \phi)$. The wavelength is obtained through a Fast Fourier Transformation of the crossing wave. The phase is then obtained through a correlation function and finally the amplitude A of the wave is fitted using the least-square method. It is assumed that the wave constructed is the incident one, by difference between the total wave and incident wave we thus obtain the reflected wave induced by the different boundary conditions. Figure 4a shows a wave packet crossing the outlet. Incident and reflected waves are reconstructed on an interval of length λ on Fig. 4b.

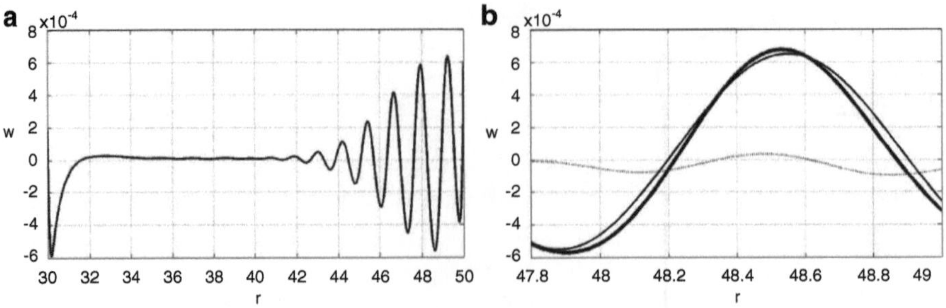

Fig. 4 (**a**) Wave packet crossing the border, visualised through the axial velocity profiles for θ and z fixed arbitrarily. (**b**) Crossing wave (in *bold*) separated in an incident wave (*solid line*) and reflected wave (*dashed line*). Computation made with convective boundary conditions

5.2 Comparison of the Boundary Conditions

This quantitative method has been used to evaluate the performances of the convective and boundary layer conditions with and without the non-linear terms. Calculations have been made starting from the same initial solution presenting a developed perturbation in the middle of the cavity, corresponding to a type II convective instability, obtained as the impulsional response of the flow to a perturbation localised in time and space. With the initial stationary flow control parameters being $Re_\Omega = 487{,}500$ and $Cw = 500$, the flow sustains only type II convective instability. The computation has then been pursued with each type of boundary conditions until the wave packet reached and crossed the outlet. In particular, the influence of the non-linear terms in system (11) has been investigated, even for moderate values of the imposed mass flow rate. Figure 3b shows that the terms in v^2/r represent the main contribution to the convective terms. Furthermore, it has appeared that the crossed terms including derivatives such as $u\partial_r u$ presented numerical stability problems. It has thus been decided to keep only the terms v^2/r and uv/r in the convective terms.

The reflected waves induced by the three sets of boundary conditions have been represented in Fig. 6. The amplitude $\tilde{A}$ of the reflected wave induced by the new boundary conditions is smaller than the amplitude $\widetilde{A_c}$ induced by the convective boundary conditions, with $\tilde{A}/\widetilde{A_C} = 0.78$. When the convective terms are taken into account, the amplitude of the reflected waved is again lowered with $\tilde{A}/\widetilde{A_C} = 0.47$. The wavelength of the reflected wave has been found to be about half the wavelength of the incident wave in both cases, with less than 1 % margin (Fig. 5).

Regarding the convective boundary conditions, the influence of the choice of the advection velocity has been investigated. Several values of the advection velocity have been used ranging from 0.05 to $10V_\Phi$ where V_ϕ designates the phase velocity of the wave packet. The ratio between the amplitude of the incident wave and the amplitude of the reflected wave has been plotted in Fig. 6. The amplitude of the reflected wave is sensitively constant and about 5.7 % of the amplitude of

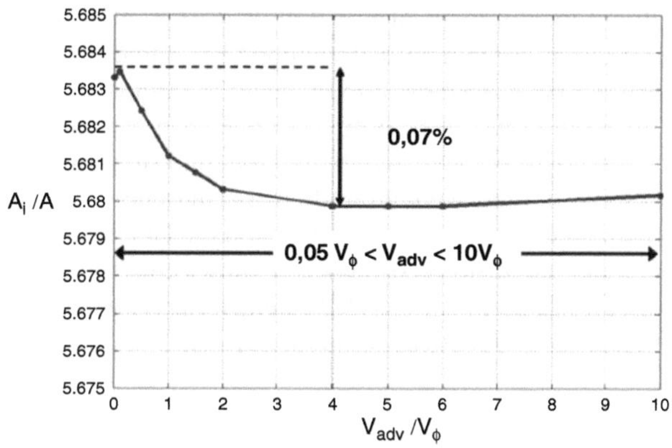

Fig. 5 Influence of the choice of the advection velocity on the amplitude of the reflected wave

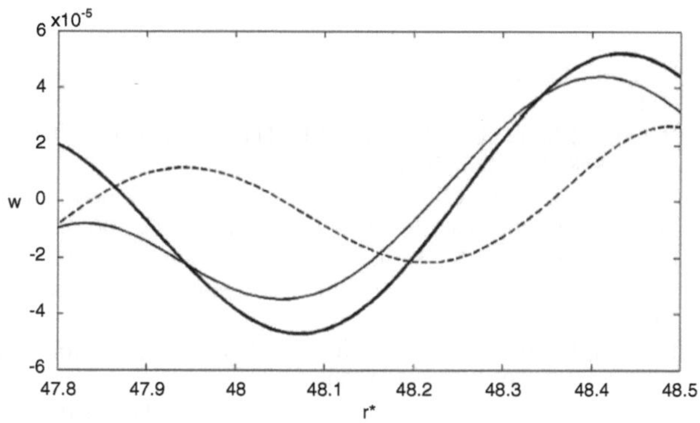

Fig. 6 Comparison of the reflected waves induced by: the convective boundary conditions in *bold*, the linearised boundary layer equations in *solid* and the boundary layer equations with convective terms in *dashes*

the incident wave, which shows that the choice of the advection velocity has no impact on the quality of the boundary conditions, as qualititavely observed by Ruith et al. [7].

In the case of a higher mass flow rate, where inertial forces cannot be neglected the linearised boundary layer equations deteriorate the solution in the long term, compared to the convective boundary conditions. Introducing the terms in v^2/r ensure the same behaviour as the convective boundary conditions. The quantitative measure to the reflected wave, in this case, requires the use of a very thin mesh so as to avoid aliasing due to the heavily intense instabilities that develop within the cavity. These heavy simulations are still ongoing.

6 Conclusion

Novel outflow boundary conditions have been developped on a spectral code dedicated to the study of rotating flows, providing an alternative solution to the convective boundary conditions commonly used. They have been obtained by solving locally a set of parabolic equations that derive from those governing the flow in the domain so as to support the same instabilities. A quantitative method has then been developed to evaluate the transparency of the boundary conditions. This method has been used to compare in the same situations the reflected wave induced by the convective boundary conditions, and the new boundary conditions, showing an improvement in their transparency. The new conditions introducing no other velocity, they have no impact on the CFL stability as opposed to the convective boundary conditions. It has been shown that the new boundary conditions have a better transparency (78 or 47 % depending on whether the convective terms are kept or not), for an identical computation cost.

References

1. Hide, R.: On source-sink flows stratified in a rotating annulus. J. Fluid Mech. **32**, 737–764 (1968)
2. Viaud, B., Serre, E., Chomaz, J.M.: The elephant mode between two rotating disks. J. Fluid Mech. **598**, 451–464 (2008)
3. Viaud, B., Serre, E., Chomaz, J.M.: Transition to turbulence through steep global-modes cascade in an open rotating cavity. J. Fluid Mech. **688**, 493–506 (2011)
4. Raspo, I., Hughes, S., Serre, E., Randriamampianina, A., Bontoux, P.: A spectral projection method for the simulation of complex three-dimensional rotating flows. Comp. and Fluids. **31**, 745–767 (2002)
5. Halpern, L., Schatzman, M.: Artificial boundary conditions for incompressible viscous flows. J. Math Anal. **20**, 308–353 (1989)
6. Orlanski, I.: A simple boundary condition for unbounded hyperbolic flows. J. of Comp. Phys. **21**, 251–270 (1976)
7. Ruith, M.T., Chen, P., Meiburg, E.: Development of boundary conditions for DNS of three-dimensional vortex breakdown phenomena in semi-infinite domains. Comp. and Fluids. **33**, 1225–1250 (2004)
8. Fournier, G., Golanski, F., Pollard, A.: Novel outflow boundary conditions for incompressible laminar wall-bounded flows. J. of Comp Phys. **227**, 7077–7082 (2008)

A Geometric Approach Towards Momentum Conservation

Deepesh Toshniwal, R.H.M. Huijsmans, and M.I. Gerritsma

Abstract In this work, a geometric discretization of the Navier-Stokes equations is sought by treating momentum as a covector-valued volume-form. The novelty of this approach is that we treat conservation of momentum as a tensor equation and describe a higher order approximation to this tensor equation. The resulting scheme satisfies mass and momentum conservation laws exactly, and resembles a staggered-mesh finite-volume method. Numerical test-cases to which the discretization scheme is applied are the Kovasznay flow, and lid-driven cavity flow.

1 Navier-Stokes Equations

Mimetic discretizations aim to represent physics in a discrete sense, in contrast to differential formulations, which are concerned with the limit $h \rightarrow 0$. For the case in which $h \neq 0$ geometrical considerations play an important role in the correct discrete formulation, [1, 2, 4, 7, 10]. Application of these ideas to continuum models are described in [5, Appendix A] and [8, 13]. The novel aspect in this paper is that continuum ideas are applied to incompressible, viscous flows using spectral basis functions.

D. Toshniwal (✉)
EPFL, Lausanne, Switzerland
e-mail: toshniwald.iitkgp@gmail.com; deepesh.toshniwal@epfl.ch

R.H.M. Huijsmans
Maritime Engineering, TU Delft, The Netherlands
e-mail: R.H.M.Huijsmans@tudelft.nl

M.I. Gerritsma
Aerospaee Engineering, TU Delft, The Netherlands
e-mail: M.I.Gerritsma@tudelft.nl

M. Azaïez et al. (eds.), *Spectral and High Order Methods for Partial Differential Equations - ICOSAHOM 2012*, Lecture Notes in Computational Science and Engineering 95, DOI 10.1007/978-3-319-01601-6_32,

We start with the incompressible Navier-Stokes equations ($\rho = 1$), written in the integral formulation, as given in many textbooks and we try to make precise what these statements mean. It is important to give an accurate meaning to all variables, because when we want to represent these physical quantities on finite grids, we want to preserve the main structure of the equations. Conservation of mass ($\rho = 1$) is usually given by

$$\int_{\partial\Omega} \mathbf{v} \cdot \mathbf{n} \, \mathrm{d}S = 0 \,, \tag{1}$$

and conservation of momentum,

$$\int_{\Omega} \frac{\partial \mathbf{v}}{\partial t} \, dV + \int_{\partial\Omega} \mathbf{v} \otimes \mathbf{v} \cdot \mathbf{n} \, dS = \int_{\partial\Omega} \boldsymbol{\sigma} \cdot \mathbf{n} \, dS \tag{2}$$

and Newtonian stress relation

$$\boldsymbol{\sigma} = -p\mathbb{I} + \mu \left(\nabla\mathbf{v} + (\nabla\mathbf{v})^T \right) \,. \tag{3}$$

Here $\mathbf{v}, p, \sigma$ and μ denote velocity, pressure, total stress tensor and dynamic viscosity, respectively; $\mathbb{I}$ and $\mathbf{n}$ are the identity matrix and the outward unit normal to the boundary, respectively. The above are balance equations for volumetric quantities that depend on their fluxes through surfaces and are more physical than their differential counterparts.

1.1 Momentum and Velocity

The first term in (2) indicates that velocity (and its time derivative) can be integrated over a volume. But velocity is generally *not* associated to volumes, but is defined as the tangent vector at a given point along the trajectory of a particle. Velocity is therefore a vector-valued 0-form. This statement means that to every point in space-time (a zero-dimensional object) we associate a vector. Let V be the linear vector-space of all possible vectors at a given point in space, then we can define the space V^* of all linear functionals on V. Elements of V^* are called *covectors*. The spaces V and V^* are isomorphic, but there is no canonical isomorphism which relates an element $v \in V$ to an element $\alpha \in V^*$. Once a metric is defined, one can associate with every vector at a point a corresponding covector. This map is called the flat operator: $\flat : V \rightarrow V^*$. The covector associated with a vector v is then denoted by $v^\flat$.

The linear vector space V associated to a point p is called the *tangent space* at p, denoted by $T_p\Omega$. The corresponding dual space is called the *cotangent space* at p denoted by $T_p^*\Omega$. The collection of all tangent spaces in the domain Ω is called the *tangent bundle*, $T\Omega$ and the collection of cotangent spaces is called the *cotangent* *bundle*, $T^*\Omega$. Let $\alpha \in T^*\Omega$ and $\mathbf{v} \in T\Omega$, then $\langle \alpha, \mathbf{v} \rangle$ associates to each point p in Ω the value $\alpha|_p\,(\mathbf{v}|_p)$.

With every k-form we can associate a $(n-k)$-form with a different type of orientation, see [2]. The collection of all k-forms on Ω is denoted by $\Lambda^k(\Omega)$. The metric dependent operator which establishes this connection is the Hodge-$\star$ operator. For continuum models we need to combine the $\flat$ and Hodge-$\star$ into the operator $\star^\flat$, (see also [13] for such operations)

$$\star^\flat : T\Omega \otimes \Lambda^k(\Omega) \to T^*\Omega \otimes \Lambda^{n-k}(\Omega) .$$

If we apply this operator to velocity $\mathbf{v} \in T\Omega \otimes \Lambda^0(\Omega)$ we obtain

$$m := \star^\flat(\mathbf{v}) \in T^*\Omega \otimes \Lambda^n\Omega .$$

Similarly, we can define $\star^\sharp : T^*\Omega \otimes \Lambda^k(\Omega) \to T\Omega \otimes \Lambda^{n-k}(\Omega)$. The physical quantity m is called *momentum density* or the *momentum per unit volume*. This is a *covector-valued volume form*. So instead of integrating 'velocity' over the domain we are tempted to write

$$\int_\Omega m = \int_\Omega \star^\flat(\mathbf{v}) .$$

This integral is not defined, because it assumes that we can integrate over the tangent spaces in Ω. The basis in each tangent space, however, may differ from point to point. In order to define the momentum integral we introduce the operator $\dot{\wedge}$

$$\dot{\wedge}\colon \left(T^*\Omega \otimes \Lambda^k(\Omega)\right) \otimes \left(T\Omega \otimes \Lambda^l(\Omega)\right) \to \Lambda^{k+l}(\Omega) ,$$

given by $\alpha \in T^*\Omega \otimes \Lambda^k(\Omega)$ and $\mathbf{w} \in T\Omega \otimes \Lambda^l(\Omega)$

$$\alpha \,\dot{\wedge}\, \mathbf{w} = \langle \alpha, \mathbf{w} \rangle \mathrm{d}x^{(k)} \wedge \mathrm{d}x^{(l)} .$$

This operation yields a $(k+l)$-form which can be integrated over $(k+l)$-dimensional submanifolds.

If we apply momentum density m to any vector field $\mathbf{w}$ (not necessarily a velocity field) using this operator we get $m \,\dot{\wedge}\, \mathbf{w} \in \Lambda^n(\Omega)$ and this can be integrated over a volume. So the proper way to interpret the time rate of change of momentum should be

$$\int_\Omega \frac{\partial}{\partial t} \star^\flat(\mathbf{v}) \,\dot{\wedge}\, \mathbf{w} , \quad \forall \mathbf{w} \in T\Omega \otimes \Lambda^0(\Omega) . \tag{4}$$

In many textbooks on fluid dynamics the distinction between momentum density (usually called 'momentum') and velocity is ignored; one is just a scalar multiple of the other, $m = \rho\mathbf{v}$, but the use of the vector $\mathbf{w}$ in (4) is generally incorporated. The textbooks then say: 'We consider this equation for each component separately ...'.

This is a strange sentence, because components have no physical relevance, only vectors, i.e. components plus associated basis vectors are physically relevant. But what is meant by this statement is that for the vector field $\mathbf{w}$ in (4) a uniform vector field in the x^i-direction is taken. The generality 'all vector fields' is in these textbooks compensated by the fact that momentum conservation should hold for 'all volumes'.

1.2 Convection

Now that we understand how momentum density should be integrated over a volume, we can also define convection of momentum density. After pairing with an arbitrary vector field, $\mathbf{w}$, we obtain a volume form and we apply the Lie derivative to this volume form, see [11]. The Lie derivative for a volume form, $\beta^{(n)}$, is given by

$$\mathcal{L}_{\mathbf{v}}\beta^{(n)} = \mathrm{d}\mathrm{i}_{\mathbf{v}}\beta^{(n)} \ ,$$

and then the generalized Stokes theorem converts this exact form to a boundary integral

$$\int_{\Omega} \mathcal{L}_{\mathbf{v}} m \dot{\wedge} \mathbf{w} = \int_{\partial\Omega} \mathrm{i}_{\mathbf{v}}(m \dot{\wedge} \mathbf{w}) \ . \tag{5}$$

Compare this expression with the convective term in (2) and note that it does not require an inner product nor the definition of an outward unit normal. The inner product is avoided since we work with differential forms and duality pairing is metric-free and the orientation of the elements in the mesh, [7], avoids the use of explicitly defined normals.

1.3 Stress Tensor and Surface Force Density

The last term in (2) denotes the action of the viscous forces on the flow represented by the stress tensor $\boldsymbol{\sigma}$. The stress tensor is an infinitesimal quantity in the limit for $h \to 0$. On a finite mesh we can identify volumes over which we integrate the momentum density and the boundary of these volumes where surface forces act. In continuum mechanics forces are 'smeared out', so we introduce the *surface force density* given by $\mathrm{t} \in T^*\Omega \otimes \Lambda^{n-1}(\Omega)$. This is a *covector-valued* $(n-1)$*-form*. Forces are generally associated with covectors, [2, 12], and in the current setting need to be covectors in order to equate them to the time rate of change of momentum which was also covector-valued. It is furthermore a $(n-1)$-form since it acts on the boundary of n-dimensional volumes, see also [5, Appendix A] and [8, 13]. Again, covector-valued forms cannot be integrated, so the proper way is to pair it with an arbitrary vector field $\mathbf{w}$ before integration over surfaces is possible. The momentum equation then becomes

$$\frac{d}{dt}\int_{\Omega} \star^{\flat}(v) \dot{\wedge} \mathbf{w} + \int_{\partial\Omega} \mathrm{i}_{\mathbf{v}}(\star^{\flat}(v) \dot{\wedge} \mathbf{w}) = \int_{\partial\Omega} \mathrm{t} \dot{\wedge} \mathbf{w} \,, \quad \forall \mathbf{w} \in T\Omega \otimes \Lambda^0(\Omega) \,. \tag{6}$$

1.4 Newtonian Stress Relation

The pressure scalar is an outer-oriented volume form, $p^{(n)}$. Pressure force density is represented as a covector-valued $(n-1)$-form

$$\mathbf{p} = (\star p)\ \mathrm{d}x^i \otimes \mathrm{d}x^1 \wedge \ldots \widehat{\mathrm{d}x^i} \ldots \wedge \mathrm{d}x^n \ ,$$

where the notation $\widehat{\cdot}$ indicates that this term is omitted and $\mathrm{d}x^i \otimes \mathrm{d}x^1 \wedge \ldots \widehat{\mathrm{d}x^i} \ldots \wedge \mathrm{d}x^n$ is the identity tensor, see also example [5, §9.3a]. This description agrees with [9] for Stokes flow. Note that $\mathbf{p} \dot{\wedge} \mathbf{w} = \mathrm{i}_{\mathbf{w}} p^{(n)}$.

The velocity gradient is represented as the covariant differential of the velocity vector field, $\nabla\mathbf{v}$ which is a vector-valued 1-form, see [5, §9.3b]. In this paper we restrict ourselves to Euclidean space for which the connection 1-forms vanish. Applying $\star^{\flat}_{\mu}(\nabla\mathbf{v})$ transforms the vector-valued 1-form into a covector-valued $(n-1)$-form, where the diffusion coefficient is contained in the Hodge-$\star$ operator. In this paper we assume μ to be constant.

1.5 Conservation of Mass

Let $\omega^{(n)}$ be the standard volume form, then the divergence of a vector field is defined as $(\operatorname{div}\mathbf{v})\omega^{(n)} = \mathscr{L}_{\mathbf{v}}\omega^{(n)} = \mathrm{d}\mathrm{i}_{\mathbf{v}}\omega^{(n)}$. Integration over a volume and applying Stokes theorem gives

$$\int_{\Omega} \mathrm{d}\mathrm{i}_{\mathbf{v}}\omega^{(n)} = \int_{\partial\Omega} \mathrm{i}_{\mathbf{v}}\omega^{(n)} \ .$$

This is the proper translation of (1) as found in textbooks on incompressible flow. The velocity flux field, $\mathrm{i}_{\mathbf{v}}\omega^{(n)}$, is isomorphic to the velocity vector field. The velocity flux field will be used in the discrete representation of velocity in the Navier-Stokes equations. The relation between the velocity fluxes $\mathrm{i}_{\mathbf{v}}\omega^{(n)}$ and $\star^{\flat}(\mathbf{v})$ is given by

$$\star^{\flat}(\mathbf{v}) \dot{\wedge} \mathbf{w} = \mathbf{w}^{\flat} \wedge \mathrm{i}_{\mathbf{v}}\omega^{(n)} \ , \quad \forall \mathbf{w} \in T\Omega \otimes \Lambda^0(\Omega) \ . \tag{7}$$

The volume forms and $(n-1)$-forms appearing in all integrals are all *outer-oriented.*

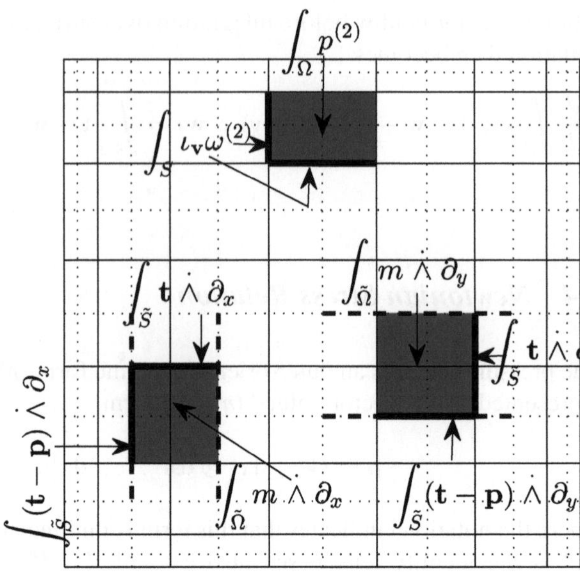

Fig. 1 The velocities are discretized as outer-oriented mass-fluxes, and live on surfaces (S) of the Gauss-Lobatto grid shown above, while pressure is discretized on volumes (Ω). Momenta are discretized on staggered volumes ($\tilde{\Omega}$) and their fluxes on the surfaces ($\tilde{S}$) surrounding these staggered volumes

2 Discrete Representation

In the full differential geometric setting as described above, the integration only makes sense when paired with all vector fields $\mathbf{w}$. Here we choose the uniform vector field in the x- and y-direction only and impose that conservation should hold for all volumes in our spectral elements. These volumes are generated by the Gauss-Lobatto grid in the spectral element and will be denoted by Ω_{ij}. So in this section $\mathbf{w}$ is either $\boldsymbol{\partial}_x$ or $\boldsymbol{\partial}_y$.

Figure 1 displays one spectral element and its Gauss-Lobatto grid (solid lines) in 2D. The dotted gray lines represent the dual grid, see [7, 10].

Momentum is reduced onto a volume consisting of a primal $(n-1)$-chain and a dual 1-chain. In 2D these volumes consist of tensor products of primal and dual edge as shown in Fig. 1 by volumes enclosed by solid (primal) and dashed (dual) lines. The location of the unknowns coincides with those in staggered finite volume methods. The difference is that in this formulation the unknowns represent integral values, whereas in finite volume methods the unknowns either represent average or nodal values. Let us denote the primal surfaces by S_i, then discrete velocity is given by

$$\bar{v}_i = \int_{S_i} \mathrm{i}_{\mathbf{v}} \omega^{(n)} .$$

This yields a metric-free description of conservation as mass as shown in [6,9]. The reduction of the pressure field is on outer-oriented volumes, see also [9].

Integrals of momentum flux, $\mathscr{F}^{(n-1)}$, pressure force, $\mathrm{i}_{\mathbf{w}} p^{(n)}$, and velocity gradients are represented on the boundary of the momentum volumes indicated in Fig. 1.

Once we have the discrete variables for mass flux, momentum and pressure, we use the spectral element functions described in [6, 10], to interpolate these values in such a way that the integral values are preserved.

Using (7) we can write the relation between momentum and velocity flux as

$$\int_{\tilde{\Omega}_{ij}} m \dot{\wedge} \mathbf{w} - \int_{\tilde{\Omega}_{ij}} \mathbf{w}^{\flat} \wedge \mathrm{i}_{\mathbf{v}} \omega^{(n)} = 0 \longrightarrow \bar{m}_{\mathbf{w}} - P^{m}_{\mathbf{w}} \bar{u} = 0 \,, \tag{8}$$

where $\tilde{\Omega}_{ij}$ are the volume where momentum is reduced, see Fig. 1, and $P^{m}_{\mathbf{w}}$ is the matrix which maps discrete velocity (which is discretized as mass-fluxes) to discrete momentum (on the staggered-grid). The discrete representation of momentum-flux, pressure force $\mathrm{i}_{\mathbf{w}} p^{(n)}$ and traction forces, $\star^{\flat}_{\mu}(\nabla_{\mathbf{w}} \mathbf{v})$ (which can be equivalently written as $\star^{\flat}_{\mu}(\nabla \star^{\sharp} \star^{\flat} \mathbf{v}) \dot{\wedge} \mathbf{w} = \star^{\flat}_{\mu}(\nabla \star^{\sharp} m) \dot{\wedge} \mathbf{w}$, which, in Cartesian coordinates and with constant $\mathbf{w}$ becomes $d^{\star}_{\mu}(m \dot{\wedge} \mathbf{w})$), are given by

- Convective-flux, see [11], $\mathscr{F}^{(1)}_{\mathbf{w}} = \mathrm{i}_{\mathbf{v}}(m \dot{\wedge} \mathbf{w})$:

$$\begin{aligned} \left(\mathscr{F}^{(1)}_{\mathbf{w}}, \beta^{(1)}\right)_{\Omega} - \left(m \dot{\wedge} \mathbf{w}, \mathbf{v}^{\flat} \wedge \beta^{(1)}\right)_{\Omega}, &= 0 \,, \\ \longrightarrow \tilde{M}_{11} \bar{\mathscr{F}}_{\mathbf{w}} - \tilde{C}_{\mathbf{v}} \bar{m}_{\mathbf{w}} &= 0 \,. \end{aligned} \tag{9}$$

- Pressure-force, $\mathscr{H}^{(1)}_{\mathbf{w}} = \mathrm{i}_{\mathbf{w}} p^{(2)}$:

$$\mathscr{H}^{(1)}_{\mathbf{w}} - p^{(2)}(\mathbf{w}) = 0 \longrightarrow \bar{\mathscr{H}}_{\mathbf{w}} - P^{p}_{\mathbf{w}} \bar{p} = \tilde{B}_{P} \,. \tag{10}$$

- Diffusive-fluxes, $\mathscr{T}^{(1)}_{\mathbf{w}} = d^{\star}_{\mu}(m \dot{\wedge} \mathbf{w})$:

$$\begin{aligned} \left(\mathscr{T}^{(1)}_{\mathbf{w}}, \beta^{(1)}\right)_{\Omega} - \left(m \dot{\wedge} \mathbf{w}, \mathrm{d}\beta^{(1)}\right)_{\Omega} &= -\int_{\partial\Omega} \beta^{(1)} \wedge \star(m \dot{\wedge} \mathbf{w}) \,, \\ \longrightarrow \tilde{M}_{11} \bar{\mathscr{T}}_{\mathbf{w}} - \tilde{D}^{T}_{21} \tilde{M}_{22} \bar{m} &= \tilde{B}_{T} \,. \end{aligned} \tag{11}$$

The discrete continuity equation is given by

$$D_{21} \bar{u} = 0 \,. \tag{12}$$

In the above, $\beta^{(1)}$ is an arbitrary 1-form; $\tilde{M}_{11}$ and $\tilde{M}_{22}$ are mass-matrices for 1- and 2-forms on the staggered mesh; $C_{\mathbf{v}}$ is the convection matrix and depends on $\mathbf{v}$ (which can be retrieved from reconstruction of $\bar{v}$ using the edge basis, [7]); $P^{p}_{\mathbf{w}}$ is the matrix which converts the scalar $\bar{p}$ to pressure-force 1-forms; $\tilde{B}_{P}$ and $\tilde{B}_{T}$ are the boundary integrals for pressure and stress, respectively, obtained from integration

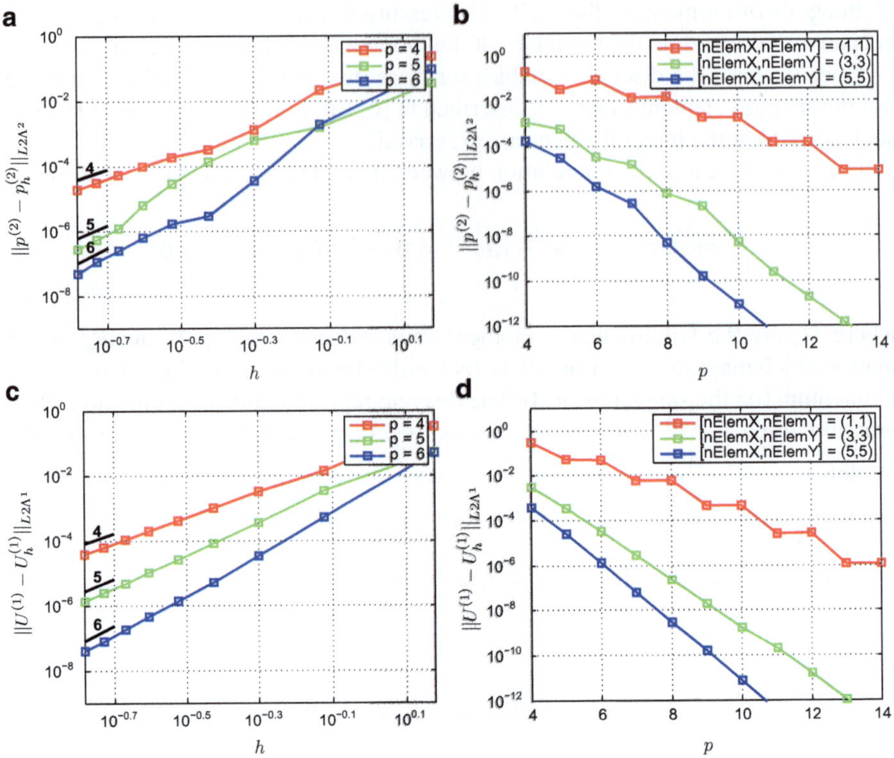

Fig. 2 Convergence plots for Kovasznay flow with mesh size, h, and order, p. Optimal rates are shown (▬) for the h-refinement cases. $nElemX$ and $nElemY$ refer to the number of elements in X and Y directions, and p refers to the order of elements used. **(a)** Pressure, h-refinement. **(b)** Pressure, p-refinement. **(c)** Velocity, h-refinement. **(d)** Velocity, p-refinement

by parts; and D_{21} and $\tilde{D}_{21}$ are incidence matrices which discretely represent the exterior-derivative with entries containing only $\{-1, 0, 1\}$. The algebraic system thus obtained is solved for $\bar{v}$ and $\bar{p}$ for $\mathbf{w} = \{\boldsymbol{\partial}_x, \boldsymbol{\partial}_y\}$.

3 Results

3.1 Kovasznay Flow

Kovasznay flow is an analytical solution to Navier-Stokes' equations. The solution is $u = 1 - e^{\lambda x}cos(2\pi y)$, $v = \frac{\lambda}{2\pi}e^{\lambda x}sin(2\pi y)$ and $p = \frac{1}{2}(1 - e^{2\lambda x})$, where $\lambda = \frac{1}{2\nu} - \sqrt{\frac{1}{4\nu^2} + 4\pi^2}$. The kinematic-viscosity chosen for this flow was $\nu = \frac{\mu}{\rho} = \frac{1}{40}$ and the computational domain considered was $\Omega = [-0.5\ 1] \times [-0.5\ \ -0.5]$. The h, p-adaptivity plots for this problem are given in Fig. 2 for pressures (Fig. 2a, b)

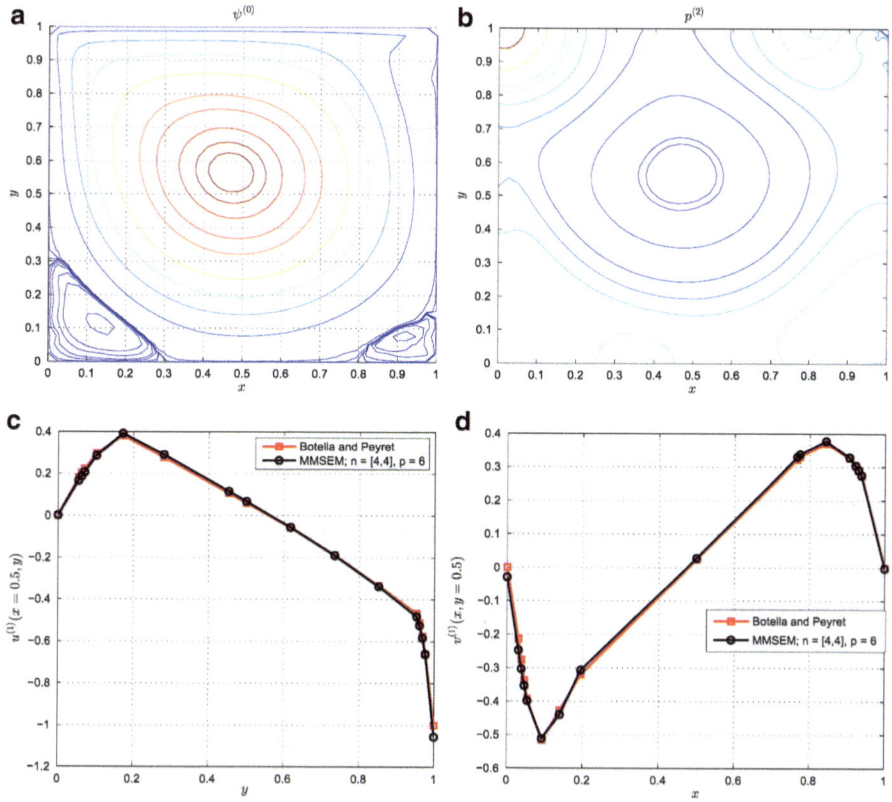

Fig. 3 (*Top*) Streamfunction and pressure contours with a single spectral element of order 16. (*Bottom*) Centerline velocities are plotted (▬) and compared with the solutions of [3] (▬), and the solutions are found to be reasonably close. Mesh size is 4×4 and made up of elements of order 6. **(a)** Stream-function contours. **(b)** Pressure contours. **(c)** X-velocity at $x = 0.5$. **(d)** Y-velocity at $y = 0.5$

and velocities (Fig. 2c, d). It can be seen that the solutions converge exponentially and optimally. Some oscillatory behaviour is observed in convergence for a mesh with a single element, and this is attributed to the fact that our basis may not be capturing certain modes (even/odd).

3.2 *Lid-Driven Cavity Flow*

The second numerical test-case chosen was the classic lid-driven cavity flow on a unit square domain with the top-lid velocity, $u_L = -1$ and a Reynolds number of 1000. The solutions for the pressure and streamfunction contours calculated for a single spectral element of order $p = 16$ are shown in the top-half of Fig. 3.

Centerline-velocity solutions with a lower order of $p = 6$ but with multiple elements (4×4 mesh) and comparisons with the results of [3] are also shown in the bottom-half of Fig. 3. Good agreement is seen between the benchmark results and our results.

References

1. P.B. Bochev and J.M. Hyman. Principles of mimetic discretizations of differential operators. *IMA Volumes In Mathematics and its Applications*, 142:89, 2006.
2. A. Bossavit. *Handbook of Numerical Analysis*, volume 13, chapter Discretization of electromagnetic problems, pages 105–197. Elsevier, 2005.
3. O. Botella and R. Peyret. Benchmark spectral results on the lid-driven cavity flow. *Computers & Fluids*, 27(4):421–433, 1998.
4. M. Desbrun, A.N. Hirani, M. Leok, and J.E. Marsden. Discrete exterior calculus. *Arxiv preprint arXiv:math/0508341*, 2005. URL http://arxiv.org/abs/math/0508341.
5. Th. Frankel. *The geometry of physics: An introduction*. Cambridge University Press, 2011.
6. M. Gerritsma. Edge functions for spectral element methods. In *Spectral and High Order Methods for Partial Differential Equations*, pages 199–207. Springer, 2011.
7. M. Gerritsma, R. Hiemstra, J. Kreeft, A. Palha, P. Rebelo, and D. Toshniwal. The geometric basis of mimetic spectral approximations. *Proceedings of ICOSAHOM 2012–2013 (this issue)*, 2012.
8. E. Kanso, M. Arroyo, Y. Tong, A. Yavari, Marsden J.E., and M. Desbrun. On the geometric character of stress in continuum mechanics. *Mathematik und Physik*, 58:843–856, 2007.
9. J. Kreeft and M. Gerritsma. Mixed mimetic spectral element method for stokes flow: A pointwise divergence-free solution. *Journal of Computational Physics*, 240:284–309, 2013.
10. Jasper Kreeft, Artur Palha, and Marc Gerritsma. Mimetic framework on curvilinear quadrilaterals of arbitrary order. *Arxiv preprint arXiv:1111.4304*, page 69, November 2011. URL http://arxiv.org/abs/1111.4304.
11. A. Palha, P. Rebelo, and M. Gerritsma. Mimetic Spectral Element solution for conservative advection. *Proceedings of ICOSAHOM 2012–2013 (this issue)*, 2012.
12. E. Tonti and Gruppo nazionale per la fisica matematica. *On the formal structure of physical theories*. Istituto de matematica, Politecnico, 1975.
13. A. Yavari. On geometric discretization of elasticity. *Journal of Mathematical Physics*, 49: 1–36, 2008.

A Spectral Method for Optimal Control Problems Governed by the Time Fractional Diffusion Equation with Control Constraints

Xingyang Ye and Chuanju Xu

Abstract In this paper, we study the fractional optimal control problem and its spectral approximation. The problem under investigation consists in finding the optimal solution governed by the time fractional diffusion equation with constraints on the control variable. We construct a suitable weak formulation, study its well-posedness, and design a Galerkin spectral method for its numerical solution. The main contribution of the paper includes: (1) a priori error estimates for the space-time spectral approximation is derived; (2) a projection gradient algorithm is designed to efficiently solve the discrete minimization problem; (3) some numerical experiments are carried out to confirm the efficiency of the proposed method. The obtained numerical results show that the convergence is exponential for smooth exact solutions.

1 Introduction

Let $\Lambda = (-1, 1), I = (0, T), T > 0$. We consider the following linear-quadratic optimal control problem for the control variable q under constraints:

$$\min_q \left\{\frac{1}{2}\int_0^T\int_\Lambda (u(x,t) - \bar{u}(x,t))^2 \mathrm{d}x\mathrm{d}t + \frac{\lambda}{2}\int_0^T\int_\Lambda q^2(x,t)\mathrm{d}x\mathrm{d}t\right\}, \tag{1}$$

X. Ye
School of Science, Jimei University, 361021 Xiamen, China
e-mail: xingyangye@163.com

C. Xu (✉)
School of Mathematical Sciences, Xiamen University, 361005 Xiamen, China
e-mail: cjxu@xmu.edu.cn

M. Azaïez et al. (eds.), *Spectral and High Order Methods for Partial Differential Equations - ICOSAHOM 2012*, Lecture Notes in Computational Science and Engineering 95, DOI 10.1007/978-3-319-01601-6_33,

where λ and $\bar{u}$ are given, u is governed by:

$$\begin{aligned} {}_0\partial_t^\alpha u(x,t) - \partial_x^2 u(x,t) &= f(x,t) + q(x,t), && \forall (x,t) \in \Lambda \times I, \\ u(x,0) &= u_0(x), && \forall x \in \Lambda, \\ u(-1,t) = u(1,t) &= 0, && \forall t \in I, \end{aligned} \tag{2}$$

with ${}_0\partial_t^\alpha$ $(0 < \alpha < 1)$ denoting the left Caputo fractional derivative and q satisfying

$$\int_0^T \int_\Lambda q(x,t)\mathrm{d}x\mathrm{d}t \geq 0. \tag{3}$$

The optimal control problem (1)–(3) has been subject of many research in scientific and engineering computing. Although most research on control problems have been focused on partial differential equations of integer order, we are seeing a growing interest for research on using fractional partial differential equations, which are novel extensions of the traditional models. It has been found that the fractional order model can provide a more realistic description for some kind of complex systems in the fields covering control theory [16], viscoelastic materials [11, 13], anomalous diffusion [3, 5, 10], advection and dispersion of solutes in porous or fractured media [2], and etc. [6, 14, 19].

An approach for the numerical solution of the fractional optimal control problem (FOCP) was first proposed in [1], where the fractional variational principle and the Lagrange multiplier technique were used. Following this idea, Frederico and Torres [8, 9] formulated a Noether-type theorem in the general context and studied fractional conservation laws. In [17], a scheme using eigenfunctions expansion was derived for FOCP in a 2-dimensional distributed system. Also, by means of eigenfunction expansion approach, Özdemir [18] investigated the control problem of a distributed system in cylindrical coordinates.

More recently, Mophou [15] applied the classical control theory to a fractional diffusion equation, involving a Riemann-Liouville fractional time derivative. The existence and uniqueness of the solution were established. Dorville et al. [7] extended the results of [15] to a boundary fractional optimal control with finite observation expressed in terms of the Riemann-Liouville integral of order α. However, none of the above work has studied the error estimates of the approaches.

In this paper we consider the optimal control problem associated to the time fractional diffusion equation (2) with Caputo fractional derivative. Differing from the approach based on the Grünwald-Letnikov or eigenfunctions expansion, we construct a spectral approximation in both space and time directions based on the weak formulation introduced in [12]. We will see that the spectral method shows great advantages over low-order methods in approximating the optimal control problem with control integral constraints. Moreover, as compared to the unconstrained method considered in our previous work [20], the presence of the control constraints here leads to many additional difficulties, one of which is that the constrained problem requires some additional variational inequalities. The purpose of this work is to derive a priori error estimates for the space-time spectral approximation to the underlying problem, and propose an efficient algorithm to solve the discrete control problem.

The outline of the paper is as follows: In the next section we formulate the optimal control problem under consideration and derive the optimality conditions. Section 3 is devoted to the spectral discretization of the optimal problem. In Sect. 4, a priori error estimates for the control, state, and adjoint variables are provided. Finally we carry out, in Sect. 5, some numerical tests to verify the theoretical results.

2 Formulation of the Problem and Optimization

Let c be a generic positive constant. We use the expression $A \lesssim B$ to mean that $A \leq cB$, and use the expression $A \cong B$ to mean that $A \lesssim B \lesssim A$.

Let $\Omega = \Lambda \times I$. For a domain $\mathcal{O}$, which may be Λ, I or Ω, we use $L^2(\mathcal{O})$, $H^s(\mathcal{O})$, and $H_0^s(\mathcal{O})$ to denote the usual Sobolev spaces, equipped with the norms $\|\cdot\|_{0,\mathcal{O}}$ and $\|\cdot\|_{s,\mathcal{O}}$ respectively. For the Sobolev space X with norm $\|\cdot\|_X$, we define the space $H^s(I;X) := \{v| \, \|v(\cdot,t)\|_X \in H^s(I)\}$, endowed with the norm$\|v\|_{H^s(I;X)} := \| \|v(\cdot,t)\|_X \|_{s,I}$. Particularly, when X stands for $H^{\mu}(\Lambda)$ or $H_0^{\mu}(\Lambda)$, the norm of the space $H^s(I;X)$ will be denoted by $\|\cdot\|_{\mu,s,\Omega}$. Hereafter, in cases where no confusion would arise, the domain symbols I, Λ, Ω may be dropped from the notations.

We also introduce the state space $B^s(\Omega) = H^s(I, L^2(\Lambda)) \cap L^2(I, H_0^1(\Lambda))$, $\forall s > 0$, equipped with the norm $\|v\|_{B^s(\Omega)} = (\|v\|^2_{H^s(I,L^2(\Lambda))} + \|v\|^2_{L^2(I,H_0^1(\Lambda))})^{1/2}$.

Now we consider the following weak formulation of the state equation (2): given $q, f \in L^2(\Omega)$, find $u \in B^{\frac{\alpha}{2}}(\Omega)$, such that

$$\mathscr{A}(u,v) = (f+q,v)_\Omega + \left(\frac{u_0(x)t^{-\alpha}}{\Gamma(1-\alpha)}, v\right)_\Omega, \quad \forall v \in B^{\frac{\alpha}{2}}(\Omega), \tag{4}$$

where the bilinear form $\mathscr{A}(\cdot,\cdot)$ is defined by

$$\mathscr{A}(u,v) := \left({}_0^R\partial_t^{\frac{\alpha}{2}} u, \, {}_t^R\partial_T^{\frac{\alpha}{2}} v \right)_\Omega + (\partial_x u, \partial_x v)_\Omega.$$

Here, ${}_0^R\partial_t^{\frac{\alpha}{2}}$ and ${}_t^R\partial_T^{\frac{\alpha}{2}}$ respectively denote the left and right Riemann-Liouville fractional derivative of order $\frac{\alpha}{2}$.

It has been proved [12] that the following continuity and coercivity hold

$$\mathscr{A}(u,v) \lesssim \|u\|_{B^{\frac{\alpha}{2}}(\Omega)} \|v\|_{B^{\frac{\alpha}{2}}(\Omega)}, \quad \mathscr{A}(v,v) \gtrsim \|v\|^2_{B^{\frac{\alpha}{2}}(\Omega)}, \quad \forall u, v \in B^{\frac{\alpha}{2}}(\Omega),$$

and the problem (4) is well-posed.

To formulate the problem we introduce the admissible set K associated to (3) as $K := \{q \in L^2(\Omega) : \int_\Omega q(x,t)\mathrm{d}x\mathrm{d}t \geq 0\}$, and define the cost functional:

$$\mathscr{J}(q,u) := \frac{1}{2}\|u-\bar{u}\|_{0,\Omega}^2 + \frac{\lambda}{2}\|q\|_{0,\Omega}^2, \ (q,u) \in K \times B^{\frac{\alpha}{2}}(\Omega). \tag{5}$$

Then the optimal control problem reads: find $(q^*, u(q^*)) \in K \times B^{\frac{\alpha}{2}}(\Omega)$, such that

$$\mathscr{J}(q^*, u(q^*)) = \min_{(q,u)\in K\times B^{\frac{\alpha}{2}}(\Omega)} \mathscr{J}(q,u) \ \text{ subject to (4).} \tag{6}$$

The well-posedness of the state problem ensures the existence of a control-to-state mapping $q \mapsto u = u(q)$ defined through (4). By means of this mapping we introduce the reduced cost functional $J(q) := \mathscr{J}(q, u(q)),\ q \in L^2(\Omega)$. Then the optimal control problem (6) is equivalent to: find $q^* \in K$, such that

$$J(q^*) = \min_{q\in K} J(q). \tag{7}$$

The first order necessary optimality condition for (7) reads

$$J'(q^*)(\delta q - q^*) \geq 0, \ \ \forall \delta q \in K, \tag{8}$$

where $J'(q^*)(\cdot)$ is the gradient of $J(q)$, defined through the Gâteaux derivative. The convexity of the quadratic functional implies that (8) is also sufficient for optimality.

Lemma 1. *It holds*

$$J'(q)(\delta q) = (\lambda q + z(q), \delta q)_\Omega, \ \ \forall \delta q \in L^2(\Omega), \tag{9}$$

where $z(q) = z$ is the solution of the following adjoint state equation

$$\begin{aligned} {}_t\partial_T^\alpha z(x,t) - \partial_x^2 z(x,t) &= u(x,t) - \bar{u}(x,t), \ \forall (x,t) \in \Omega, \\ z(x,T) &= 0, \qquad\qquad\qquad \forall x \in \Lambda, \\ z(-1,t) = z(1,t) &= 0, \qquad\qquad\qquad \forall t \in I, \end{aligned} \tag{10}$$

with ${}_t\partial_T^\alpha$ being the right Caputo fractional derivative of order α.

Proof. The proof goes along the same lines as Theorem 3.1 in [20]. □

The weak form of (10) reads: find $z \in B^{\frac{\alpha}{2}}(\Omega)$, such that

$$\mathscr{A}(\varphi, z) = (u - \bar{u}, \varphi)_\Omega, \ \ \forall \varphi \in B^{\frac{\alpha}{2}}(\Omega). \tag{11}$$

It can also be proved that (11) admits a unique solution for any given $u \in B^{\frac{\alpha}{2}}(\Omega)$.

In what follows we will need the mapping $q \to u(q) \to z(q)$, where for any given q, $u(q)$ is defined by (4), and once $u(q)$ is known $z(q)$ is defined by (11).

Theorem 1. *Let $(q^*, u(q^*))$ be the solution of the optimal control problem (6) and $z(q^*)$ be the corresponding adjoint state. Then we have*

$$\lambda q^* = \max\{0, \overline{z(q^*)}\} - z(q^*)$$

where $\overline{z(q^*)} = \int_\Omega z(q^*)/\int_\Omega 1$.

Proof. The proof is similar to Theorem 3.1 in [4]. □

3 Space-Time Spectral Discretization

We define the polynomial space $P_M^0(\Lambda) := P_M(\Lambda) \cap H_0^1(\Lambda), S_L := P_M^0(\Lambda) \otimes P_N(I)$, where P_M denotes the space of all polynomials of degree less than or equal to M, L stands for the parameter pair (M, N).

Then we consider the spectral approximation to (4): find $u_L(q) \in S_L$ such that

$$\mathscr{A}(u_L(q), v_L) = (f + q, v_L)_\Omega + \left(\frac{u_0(x)t^{-\alpha}}{\Gamma(1-\alpha)}, v_L\right)_\Omega, \quad \forall v_L \in S_L. \tag{12}$$

The following estimate, derived in [12], will be used in the analysis later on.

Lemma 2. *For any $q \in L^2(\Omega)$, let $u(q)$ be the solution of (4), $u_L(q)$ be the solution of (12). Suppose $u \in H^{\frac{\alpha}{2}}(I; H^\mu(\Lambda)) \cap H^\gamma(I; H_0^1(\Lambda))$, $0 < \alpha < 1, \gamma > 1, \mu \geq 1$, then we have*

$$\begin{aligned}\|u(q) - u_L(q)\|_{B^{\frac{\alpha}{2}}(\Omega)} \lesssim{}& N^{\frac{\alpha}{2}-\gamma}\|u\|_{0,\gamma} + N^{-\gamma}\|u\|_{1,\gamma} + N^{\frac{\alpha}{2}-\gamma}M^{-\mu}\|u\|_{\mu,\gamma} \\ &+ M^{-\mu}\|u\|_{\mu,\frac{\alpha}{2}} + M^{1-\mu}\|u\|_{\mu,0}.\end{aligned} \tag{13}$$

Similar to the continuous case, we introduce the semidiscrete reduced cost functional $J_L : L^2(\Omega) \to \mathbb{R}$:

$$J_L(q) := \mathscr{J}(q, u_L(q)), \ q \in L^2(\Omega), \tag{14}$$

where $u_L(q)$ is given by (12). Then we consider the following auxiliary optimal problem: find $q^* \in K$, such that

$$J_L(q^*) = \min_{q \in K} J_L(q). \tag{15}$$

The solution q^* of above problem fulfills the first order optimality condition

$$J_L'(q^*)(\delta q - q^*) \geq 0, \quad \forall \delta q \in K, \tag{16}$$

where

$$J_L'(q)(\phi) = (\lambda q + z_L(q), \phi)_\Omega, \ \forall q, \phi \in K, \tag{17}$$

with $z_L(q) \in S_L$ being the solution of the semidiscrete adjoint problem:

$$\mathscr{A}(\varphi_L, z_L(q)) = (u_L(q) - \bar{u}, \varphi_L)_\Omega, \quad \forall \varphi_L \in S_L. \tag{18}$$

Now we consider the approximation of the control space to obtain the fully discrete optimal control problem. To this end, we introduce the finite dimensional subspace for the control variable: $K_L = K \cap (P_M(\Lambda) \otimes P_N(I))$. Then the full discrete optimal control problem reads: find $q_L^* \in K_L$, such that

$$J_L(q_L^*) = \min_{q_L \in K_L} J_L(q_L), \tag{19}$$

where $J_L(\cdot)$ is defined in (14). The unique solution of (19), q_L^*, satisfies:

$$J_L'(q_L^*)(\delta q - q_L^*) \geq 0, \ \forall \delta q \in K_L. \tag{20}$$

Remark 1. Although the polynomial degree used to approximate the control variable may be different from those for the discretization of the state variable, we choose to use the same degree pair (M, N) for simplification of the notation.

4 A Priori Error Estimates

In order to carry out an error analysis for the spectral approximation (19), we first recall two results to be used in what follows.

Lemma 3 ([20]). *For all $p, q \in L^2(\Omega)$, we have*

$$J_L'(p)(p-q) - J_L'(q)(p-q) \geq \lambda \|p-q\|_{0,\Omega}^2. \tag{21}$$

Lemma 4 ([20]). *Let $q \in L^2(\Omega)$ be a given control. Suppose $z(q) \in B^{\frac{\alpha}{2}}(\Omega)$ is the continuous adjoint state determined by (11) and $z_L(q)$ is the solution of (18). Then*

$$\|z(q) - z_L(q)\|_{B^{\frac{\alpha}{2}}(\Omega)} \lesssim \|u(q) - u_L(q)\|_{0,\Omega} + \inf_{\forall \varphi_L \in S_L} \|z(q) - \varphi_L\|_{B^{\frac{\alpha}{2}}(\Omega)}. \tag{22}$$

We are now in a position to derive one of the main results of this paper.

Lemma 5. *Let $q^* \in K$ be the solution of the continuous optimization problem (7), $q_L^* \in K_L$ be the solution of its discrete counterpart (19). Suppose $q^* \in L^2(I; H^\mu(\Lambda)) \cap H^\gamma(I; L^2(\Lambda))$, $\gamma > 1, \mu \geq 1$, then it holds*

$$\|q^* - q_L^*\|_{0,\Omega} \leq N^{-\gamma} \|q^*\|_{0,\gamma} + M^{-\mu} \|q^*\|_{\mu,0} + \|z(q^*) - z_L(q^*)\|_{0,\Omega}. \tag{23}$$

Proof. It follows from (21), (8) and (20) that for any $p_L \in K_L$,

$$
\begin{aligned}
&\lambda \|q^* - q_L^*\|_{0,\Omega}^2 \\
&\leq J_L'(q^*)(q^* - q_L^*) - J_L'(q_L^*)(q^* - q_L^*) \\
&= J_L'(q^*)(q^* - q_L^*) - J'(q^*)(q^* - q_L^*) + J'(q^*)(q^* - q_L^*) - J_L'(q_L^*)(q^* - q_L^*) \\
&\leq J_L'(q^*)(q^* - q_L^*) - J'(q^*)(q^* - q_L^*) - J_L'(q_L^*)(q^* - p_L) \\
&= (z_L(q^*) - z(q^*), q^* - q_L^*)_\Omega + (z_L(q_L^*) + \lambda q_L^*, p_L - q^*)_\Omega \\
&\leq c(\delta) \|z_L(q^*) - z(q^*)\|_{0,\Omega}^2 + \delta \|q^* - q_L^*\|_{0,\Omega}^2 + (z_L(q_L^*) + \lambda q_L^*, p_L - q^*)_\Omega,
\end{aligned}
\tag{24}
$$

where δ is an arbitrary small positive number, $c(\delta)$ is a constant dependent on δ. Furthermore, for the last term in the above estimate, we have

$$
\begin{aligned}
&(z_L(q_L^*) + \lambda q_L^*, p_L - q^*)_\Omega \\
&= (z(q^*) + \lambda q^*, p_L - q^*)_\Omega + (\lambda q_L^* - \lambda q^*, p_L - q^*)_\Omega \\
&\quad + (z_L(q_L^*) - z_L(q^*), p_L - q^*)_\Omega + (z_L(q^*) - z(q^*), p_L - q^*)_\Omega \\
&\leq (z(q^*) + \lambda q^*, p_L - q^*)_\Omega + \lambda\delta \|q^* - q_L^*\|_{0,\Omega}^2 + (\lambda + 2)C(\delta) \|p_L - q^*\|_{0,\Omega}^2 \\
&\quad + \delta \|z_L(q_L^*) - z_L(q^*)\|_{0,\Omega}^2 + \delta \|z_L(q^*) - z(q^*)\|_{0,\Omega}^2 .
\end{aligned}
\tag{25}
$$

Notice that $z_L(q_L^*) - z_L(q^*)$ solves

$$
\mathscr{A}(\varphi_L, z_L(q_L^*) - z_L(q^*)) = (u_L(q_L^*) - u_L(q^*), \varphi_L)_\Omega, \quad \forall \varphi_L \in S_L, \tag{26}
$$

and $u_L(q_L^*) - u_L(q^*)$ satisfies

$$
\mathscr{A}(u_L(q_L^*) - u_L(q^*), v_L) = (q_L^* - q^*, v_L)_\Omega, \quad \forall v_L \in S_L. \tag{27}
$$

Thus taking $\varphi_L = z_L(q_L^*) - z_L(q^*)$ in (26) and $v_L = u_L(q_L^*) - u_L(q^*)$ in (27) gives

$$
\|z_L(q_L^*) - z_L(q^*)\|_{B^{\frac{\alpha}{2}}(\Omega)} \leq c_1 \|u_L(q_L^*) - u_L(q^*)\|_{B^{\frac{\alpha}{2}}(\Omega)} \leq c_1 \|q^* - q_L^*\|_{0,\Omega} . \tag{28}
$$

Then plugging (25) and (28) into (24) yields

$$
\begin{aligned}
\lambda \|q^* - q_L^*\|_{0,\Omega}^2 \leq{}& (z(q^*) + \lambda q^*, p_L - q^*)_\Omega + c_2\delta \|q^* - q_L^*\|_{0,\Omega}^2 + (\lambda + 2)c(\delta) \|p_L - q^*\|_{0,\Omega}^2 \\
&+ (\delta + c(\delta)) \|z_L(q^*) - z(q^*)\|_{0,\Omega}^2 ,
\end{aligned}
$$

where $c_2 = 1 + \lambda + c_1$. Now by taking $\delta = \frac{\lambda}{2c_2}$, we obtain, $\forall p_L \in K_L$,

$$\|q^* - q_L^*\|_{0,\Omega}^2 \lesssim (z(q^*) + \lambda q^*, p_L - q^*)_\Omega + \|p_L - q^*\|_{0,\Omega}^2 + \|z_L(q^*) - z(q^*)\|_{0,\Omega}^2. \tag{29}$$

Let Π_N and Π_M be the standard L^2-orthogonal projectors defined in I and Λ, respectively. Then, it holds

$$(q^* - \Pi_N \Pi_M q^*, r_L)_\Omega = 0, \quad \forall r_L \in P_M(\Lambda) \otimes P_N(I),$$

and in particular

$$(q^* - \Pi_N \Pi_M q^*, 1)_\Omega = 0,$$

that is

$$\int_\Omega \Pi_N \Pi_M q^* \mathrm{d}x\mathrm{d}t = \int_\Omega q^* \mathrm{d}x\mathrm{d}t \geq 0.$$

This means $\Pi_N \Pi_M q^* \in K_L$. Thus by taking $p_L = \Pi_N \Pi_M q^*$ in (29), we get

$$\begin{aligned} &\|q^* - q_L^*\|_{0,\Omega}^2 \\ &\lesssim (z(q^*) + \lambda q^*, \Pi_N \Pi_M q^* - q^*)_\Omega + N^{-2\gamma} \|q^*\|_{\gamma,0}^2 + M^{-2\mu} \|q^*\|_{0,\mu}^2 + \|z_L(q^*) - z(q^*)\|_{0,\Omega}^2. \end{aligned} \tag{30}$$

Next, it follows from Theorem 1 that

$$z(q^*) + \lambda q^* = \max\{0, \overline{z(q^*)}\} = const,$$

and hence

$$(z(q^*) + \lambda q^*, \Pi_N \Pi_M q^* - q^*)_\Omega = 0. \tag{31}$$

Finally, (23) results from (30) and (31). □

Using the above Lemmas and following the same lines as the proof of Theorem 4.1 in [20], we obtain the main result concerning the approximation errors.

Theorem 2. *Suppose q^* and q_L^* are respectively the solutions of the continuous optimization problem (7) and its discrete counterpart (19), $u(q^*)$ and $u_L(q_L^*)$ are the state solutions of (4) and (12) associated to q^* and q_L^* respectively, and $z(q^*)$ and $z_L(q_L^*)$ are the associated solutions of (11) and (18) respectively.*

If $q^* \in L^2(I; H^\mu(\Lambda)) \cap H^\gamma(I; L^2(\Lambda))$ *and* $u(q^*), z(q^*) \in H^{\frac{\alpha}{2}}(I; H^\mu(\Lambda)) \cap H^\gamma(I; H_0^1(\Lambda))$, $0 < \alpha < 1, \gamma > 1$ *and* $\mu \geq 1$, *then the following estimate holds:*

$$
\begin{aligned}
&\|q^* - q_L^*\|_{0,\Omega} + \|u(q^*) - u_L(q_L^*)\|_{B^{\frac{\alpha}{2}}(\Omega)} + \|z(q^*) - z_L(q_L^*)\|_{B^{\frac{\alpha}{2}}(\Omega)} \\
&\lesssim N^{-\gamma}\|q^*\|_{0,\gamma} + M^{-\mu}\|q^*\|_{\mu,0} + N^{\frac{\alpha}{2}-\gamma}(\|u(q^*)\|_{0,\gamma} + \|z(q^*)\|_{0,\gamma}) \\
&\quad + N^{-\gamma}(\|u(q^*)\|_{1,\gamma} + \|z(q^*)\|_{1,\gamma}) + N^{\frac{\alpha}{2}-\gamma}M^{-\mu}(\|u(q^*)\|_{\mu,\gamma} + \|z(q^*)\|_{\mu,\gamma}) \\
&\quad + M^{-\mu}(\|z(q^*)\|_{\mu,\frac{\alpha}{2}} + \|u(q^*)\|_{\mu,\frac{\alpha}{2}}) + M^{1-\mu}(\|u(q^*)\|_{\mu,0} + \|z(q^*)\|_{\mu,0}).
\end{aligned}
$$

5 Optimization Algorithm and Numerical Results

We carry out in this section a series of numerical experiments and present some results to validate the obtained error estimates. We first propose below a projection gradient optimization algorithm to solve the optimization problems.

Projection gradient optimization algorithm Choose an initial control $q_L^{(0)}$, and set $k = 0$.

(a) Solve problems

$$
\mathscr{A}\left(u_L(q_L^{(k)}), v_L\right) = \left(f + q_L^{(k)}, v_L\right)_\Omega + \left(\frac{u_0(x)t^{-\alpha}}{\Gamma(1-\alpha)}, v_L\right)_\Omega, \qquad \forall v_L \in S_L, \tag{32}
$$

$$
\mathscr{A}\left(\varphi_L, z_L(q_L^{(k)})\right) = \left(u_L(q_L^{(k)}) - \bar{u}, \varphi_L\right)_\Omega, \qquad \forall \varphi_L \in S_L. \tag{33}
$$

Let $d_L^{(k)} = z_L(q_L^{(k)}) + \lambda q_L^{(k)}$;

(b) Solve problems

$$
\mathscr{A}(\tilde{u}_L^{(k)}, v_L) = (d_L^{(k)}, v_L)_\Omega, \qquad \forall v_L \in S_L, \tag{34}
$$

$$
\mathscr{A}(\varphi_L, \tilde{z}_L^{(k)}) = (\tilde{u}_L^{(k)}, \varphi_L)_\Omega, \qquad \forall \varphi_L \in S_L, \tag{35}
$$

and set $\tilde{d}_L^{(k)} = \tilde{z}_L^{(k)} + \lambda d_L^{(k)}$, $\rho_k = \frac{(d_L^{(k)}, d_L^{(k)})_\Omega}{(\tilde{d}_L^{(k)}, d_L^{(k)})_\Omega}$;

(c) Update: $q_L^{(k+\frac{1}{2})} = q_L^{(k)} - \rho_k d_L^{(k)}$, $q_L^{(k+1)} = -\min\left\{0, \overline{q_L^{(k+\frac{1}{2})}}\right\} + q_L^{(k+\frac{1}{2})}$;

(d) If $\left\|d_L^{(k)}\right\| \leq$tolerance, then take $q_L^* = q_L^{(k+1)}$ and solve problems (12) and (18) to get $u_L(q_L^*)$ and $z_L(q_L^*)$;

Else, set $k = k + 1$, repeat (a)–(d).

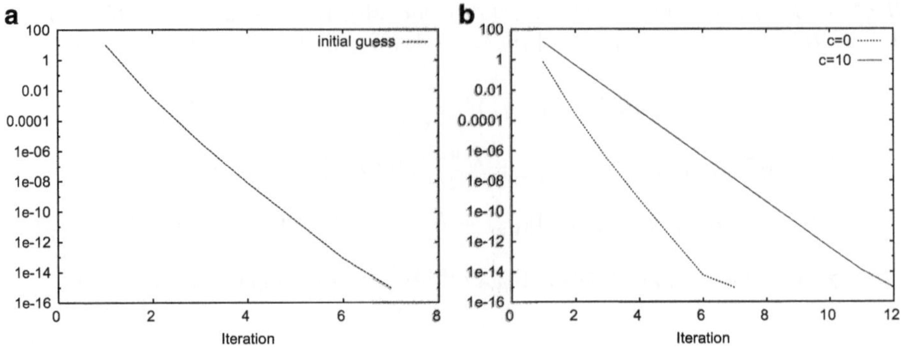

Fig. 1 Convergence history of the gradient of the objective function. (**a**) $q^{(0)} = 15q^*$. (**b**) $q^{(0)} = c$

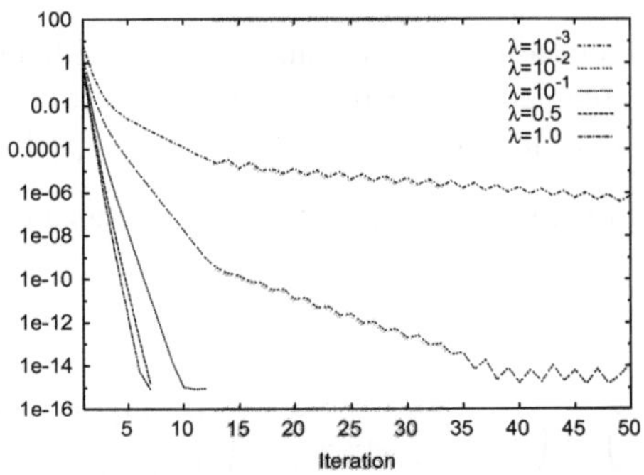

Fig. 2 Impact of λ on the convergence rate of the gradient of the cost functional

Numerical results Let $T = 1$ and consider problem (6) with the exact solutions:

$$u(q^*) = \sin \pi x \cos \pi t,\ z(q^*) = \sin \pi x \sin \pi (1-t),\ \lambda q^* = \max\{0, \overline{z(q^*)}\} - z(q^*).$$

In the first test, we investigate the impact of the initial guess on the convergence of the projection gradient optimization algorithm. We start by considering $q^{(0)} = 15q^*$. In Fig. 1a, we present the convergence history of the gradient of the objective function as a function of the iteration number with $M = 20, N = 20, \alpha = 0.5, \lambda = 1$. We see that the iterative method converges within eight iterations. We then take $q^{(0)}$ to be constant c with $c = 0$ or 10, and repeat the same computation as the previous test. The result is given in Fig. 1b. These results seem to tell that the initial guess has no significant effects on the convergence of the projection gradient iterative algorithm.

We then study the effect of the regularization parameter λ on the convergence rate of the optimization algorithm. In Fig. 2 we plot the convergence history versus the iteration number with $M = N = 18, \alpha = 0.5$, and $q^{(0)} = 0$ for several values of λ ranging from 0 to 1. It is observed that the algorithm has better convergence

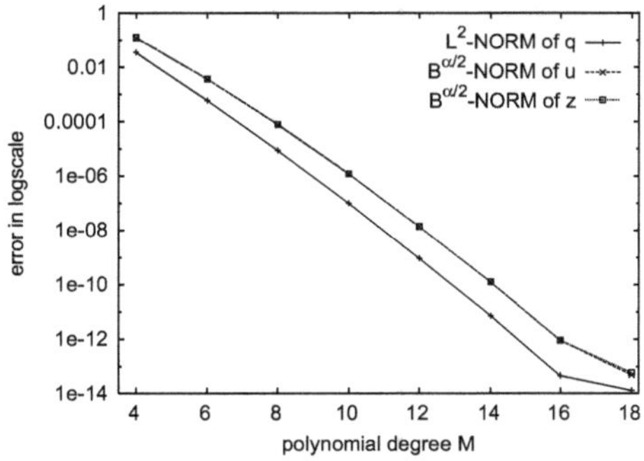

Fig. 3 Errors of q, u and z versus M with $N = 20, \alpha = 0.4$

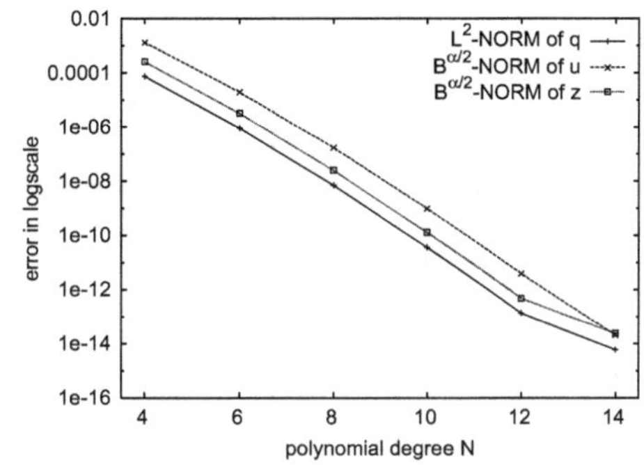

Fig. 4 Errors of q, u and z versus N with $M = 20, \alpha = 0.1$

property for $\lambda = 1$. The convergence slows down as λ decreases. In particular, the algorithm fails to converge with $\lambda = 0$.

In what follows we fix $q^{(0)} = 0$ and $\lambda = 1$ to investigate the error behavior of the numerical solution. In Fig. 3 we plot the errors as functions of the polynomial degrees M with $\alpha = 0.4, N = 20$. As expected, the errors show an exponential decay. The errors versus N with $M = 20$ are shown in Fig. 4. The error curves indicate that the convergence in time is also exponential.

6 Concluding Remarks

We have presented an efficient optimization algorithm for the fractional control problem based on the spectral approximation. A priori error estimates for the numerical solution are derived. Some numerical experiments have been carried out to confirm the theoretical results. However there are many important issues needed to be addressed. For example, we can consider more complicated control problems and constraint sets. Besides, although our analysis and algorithm are designed for the optimization of the distributed control problem, we hope that they are generalizable to a greater variety of situations such as minimization problems associated to boundary conditions, diffusion coefficient, and so on.

Acknowledgements The work of X. Ye was partially supported by the Science Foundation of Jimei University (Grant No. 201303250002), Natural Science Foundation and the Higher College Special Project of Scientific Research of Fujian Province, China (Grant No. 2012J01013 and JK2012025). The work of C. Xu was partially supported by National NSF of China (Grant numbers 11071203 and 91130002).

References

1. O. Agrawal. A general formulation and solution scheme for fractional optimal control problems. *Nonlinear Dynam*, 38(1):323–337, 2004.
2. D.A. Benson, S.W. Wheatcraft, and M.M. Meerschaert. The fractional-order governing equation of lévy motion. *Water Resour. Res.*, 36(6):1413–1423, 2000.
3. J.P. Bouchaud and A. Georges. Anomalous diffusion in disordered media: Statistical mechanisms, models and physical applications. *Phys. Rep.*, 195(4–5):127–293, 1990.
4. Y.P Chen, N.Y Yi, and W.B Liu. A Legendre-Galerkin spectral method for optimal control problems governed by elliptic equations. *SIAM J. Numer. Anal.*, 46(5):2254–2275, 2008.
5. M. Dentz, A. Cortis, H. Scher, and B. Berkowitz. Time behavior of solute transport in heterogeneous media: Transition from anomalous to normal transport. *Adv. Water Resources*, 27(2):155–173, 2004.
6. K. Diethelm. *The Analysis of Fractional Differential Equations*. Springer-Verlag, Berlin, 2010.
7. R. Dorville, G.M. Mophou, and V.S. Valmorin. Optimal control of a nonhomogeneous dirichlet boundary fractional diffusion equation. *Comput. Math. Appl.*, 62(3):1472–1481, 2011.
8. G. Frederico and D. Torres. Fractional conservation laws in optimal control theory. *Nonlinear Dynam*, 53(3):215–222, 2008.
9. G. Frederico and D. Torres. Fractional optimal control in the sense of caputo and the fractional noethers theorem. *Int. Math. Forum*, 3(10):479–493, 2008.
10. I. Goychuk, E. Heinsalu, M. Patriarca, G. Schmid, and P. Hänggi. Current and universal scaling in anomalous transport. *Phys. Rev. E*, 73(2):020101, 2006.
11. R.C. Koeller. Application of fractional calculus to the theory of viscoelasticity. *J. Appl. Mech.*, 51:299–307, 1984.
12. X.J Li and C.J Xu. A space-time spectral method for the time fractional diffusion equation. *SIAM J. Numer. Anal.*, 47(3):2108–2131, 2009.
13. F. Mainardi. Fractional diffusive waves in viscoelastic solids. *Nonlinear Waves in Solids*, 93–97, 1995.
14. K. Miller and B. Ross. *An Introduction to the Fractional Calculus and Fractional Differential Equations*. Wiley, New York, 1993.
15. G.M. Mophou. Optimal control of fractional diffusion equation. *Comput. Math. Appl.*, 61(1):68–78, 2011.
16. A. Oustaloup. *La dérivation non entière: théorie, synthèse et applications*. Hermes, Paris, 1995.
17. N. Ozdemir, O.P. Agrawal, B.B. Iskender, and D. Karadeniz. Fractional optimal control of a 2-dimensional distributed system using eigenfunctions. *Nonlinear Dynam.*, 55(3):251–260, 2009.
18. N. Özdemir, D. Karadeniz, and B.B. İskender. Fractional optimal control problem of a distributed system in cylindrical coordinates. *Phys. Lett. A*, 373(2):221–226, 2009.
19. I. Podlubny. *Fractional differential equations*. Acad. Press, New York, 1999.
20. Xingyang Ye and Chuanju Xu. Spectral optimization methods for the time fractional diffusion inverse problem. *Numerical Mathematics: Theory, Methods and Applications*, to appear.

Two-Phase Flow Solved by High Order Discontinuous Galerkin Method

J.S.B. van Zwieten, D.R. van der Heul, R.H.A. IJzermans, R.A.W.M. Henkes, and C. Vuik

Abstract In this article we present a discretisation of a one-dimensional, hyperbolic model for two-phase pipe flow based on a Discontinuous Galerkin Finite Element Method with a viscous regularisation to suppress the Gibbs phenomenon.

1 Introduction

The goal of this project is to develop an accurate and efficient numerical scheme for solving a one-dimensional model for two-phase flow. Such a model is relevant to various applications in the oil and gas industry, such as the design and analysis of flow lines and wellbores. Due to the length of such pipeline systems (>100 km) and the very large aspect ratio of the domain it is only feasible to use a one-dimensional model.

Several one-dimensional hyperbolic models for multiphase flow are described in literature varying in complexity. Such models are typically derived by averaging of the three-dimensional conservation laws over the cross sectional area of the pipe. A common feature of those models is a nonconservative term, which is responsible for the transfer of momentum between phases. This term is a direct result of the averaging. Although this term seems to be nonphysical, the sum of all momentum equations does satisfy conservation, as expected. Because of this, special care has to be taken in discretising these models. For the present work we select one of these

J.S.B. van Zwieten (✉) · D.R. van der Heul · C. Vuik
Delft Institute of Applied Mathematics, TU Delft, The Netherlands
e-mail: j.s.b.vanzwieten@tudelft.nl; d.r.vanderheul; c.vuik@tudelft.nl

R.H.A. IJzermans · R.A.W.M. Henkes
Pipelines, Flow Assurance and Subsea Systems, P&T, Shell, Amsterdam, The Netherlands
e-mail: rutger.ijzermans@shell.com; ruud.henkes@shell.com

M. Azaïez et al. (eds.), *Spectral and High Order Methods for Partial Differential Equations - ICOSAHOM 2012*, Lecture Notes in Computational Science and Engineering 95, DOI 10.1007/978-3-319-01601-6_34,

models, but the proposed discretisation method is applicable to many of the models without significant adjustments.

Hyperbolic two-phase models are commonly discretised using a Finite Volume Method (FVM). The crucial and most difficult part is the design of a numerical flux, such that the dissipation of the scheme is as little as possible and computationally efficient at the same time. A numerical flux is often tailor made for a certain model and not applicable to variants or extensions of that model.

In this article we propose the Discontinuous Galerkin Finite Element Method (DG-FEM) as an alternative for a FVM. The DG-FEM shares the local conservation property of a FVM but is easily extended to very high order. The amount of dissipation added to the discrete system by the numerical flux reduces with increasing order of the basis functions [3]. For this reason we apply in this article one of the most simple numerical fluxes, Rusanov (Local Lax-Friedrichs) [1], which would be unacceptable for a FVM, together with eight Legendre basis functions per element. The only information needed for Rusanov stabilisation is (an approximation to) the largest eigenvalue of the system, hence the dependence on a particular model is practically eliminated.

Unfortunately, a higher order method introduces another problem. Without additional measures the discrete solutions suffer from the Gibbs phenomenon, which may lead to an unstable method and nonphysical solutions such as negative mass density and can reduce the accuracy in a large area around a discontinuity.

Several techniques for eliminating the Gibbs phenomenon have been proposed. A popular method is to reduce the order of the polynomial basis locally, in the neighbourhood of discontinuities, which will increase the amount of dissipation as the element jumps will in general be larger. The reduction of the order can optionally be combined with mesh refinement in order to prevent a severe loss of accuracy.

In this article we apply, instead, a technique described in [6]. We add an artificial viscous term to the original model and discretise the whole using the local Discontinuous Galerkin method. The amount of viscosity added to the system is based on a shock sensor: in smooth regions no viscosity is added, in the neighbourhood of a discontinuity the amount of viscosity varies with the size of the oscillations. The shock sensor and algorithm for obtaining the amount of viscosity can be tuned with several parameters, which appear to have optimal values depending on the problem under consideration.

We believe that by following this route we will be able to discretise without significant changes a more sophisticated, three-phase model, without the need of computing numerically the eigenvalues and -vectors or constructing a less dissipative numerical flux.

This article is organised as follows. In Sect. 2 we state the model used as basis for our discretisation. In Sect. 3 we repeat the theory on the DLM measure needed to discretise nonconservative hyperbolic systems and discuss the chosen integration path. We state a weak formulation based on the work of [7]. In Sect. 4 we discuss the stabilisation applied to element edges. In Sect. 5 we apply the viscous regularisation by adding a viscous term to the two-phase flow model and rederive a weak formulation for this altered system. We discuss the shock sensor used for our simulations. In Sect. 6 we apply the discretisation to two test problems commonly used in the literature on two-phase models: a Large Relative Velocity shock tube and the water faucet test problem.

2 Model

We use the two-phase flow model as described in [5]. The model originates from the three-dimensional conservation of mass and momentum. Averaging over the pipe cross section for each phase separately yields the following four equations, two for each phase,

$$\frac{\partial}{\partial t}\alpha_k \rho_k + \frac{\partial}{\partial x}\alpha_k \rho_k u_k = 0, \tag{1}$$

$$\frac{\partial}{\partial t}\alpha_k \rho_k u_k + \frac{\partial}{\partial x}\left(\alpha_k \rho_k u_k^2 + \alpha_k\left(p - p_{\text{int}}\right)\right) + \alpha_k \frac{\partial p_{\text{int}}}{\partial x} = \alpha_k \rho_k g. \tag{2}$$

Here, subscript $k \in \{L, G\}$ denotes a phase (L : liquid, G : gas), α_k is the volume fraction of phase k, ρ_k the density, u_k the velocity, p the pressure and g the gravity force. In this model the pressure p is assumed to be constant per cross section, which leads to a conditionally hyperbolic system. The interface pressure correction term p_{int} is added to ensure hyperbolicity,

$$p - p_{\text{int}} = 1.2\frac{\alpha_G \alpha_L \rho_G \rho_L}{\rho_G \alpha_L + \rho_L \alpha_G}\left(u_G - u_L\right)^2. \tag{3}$$

The model is closed by a relation for the volume conservation

$$\alpha_L + \alpha_G = 1 \tag{4}$$

and by defining the equations of state for each phase,

$$\rho_k = \rho_{k0} + \frac{p - p_{k0}}{a_k^2}. \tag{5}$$

We may write the set of differential equations concisely as

$$\frac{\partial q_i}{\partial t}(x,t) + \sum_{j \in \mathscr{I}} f_{ij}(q(x,t))\frac{\partial q_j}{\partial x}(x,t) + s_i(q(x,t)) = 0, \quad \forall i \in \mathscr{I}, \tag{6}$$

where $q := [\alpha_1\rho_1, \alpha_2\rho_2, \alpha_1\rho_1u_1, \alpha_2\rho_2u_2]^T$ is the vector of conserved quantities and $s(q) := [0, 0, \alpha_1\rho_1 g, \alpha_2\rho_2 g]^T$ is the source term and $\mathscr{I} := \{0, 1, 2, 3\}$ is an index set. We define the nonconservative flux f_{ij} as

$$f_{ij} := \frac{\partial F_{\text{c};i}}{\partial q_j} + f_{\text{n};ij} \quad \forall i, j \in \mathscr{I} \tag{7}$$

where $F_c := [\alpha_1\rho_1u_1, \alpha_2\rho_2u_2, \alpha_1\rho_1u_1^2 + \alpha_1(p - p_{int}), \alpha_2\rho_2u_2^2 + \alpha_2(p - p_{int})]^T$ is the conservative part of the flux and

$$f_n = \begin{bmatrix} 0 & 0 & 0 & 0 \\ 0 & 0 & 0 & 0 \\ \alpha_1 \frac{\partial p_{int}}{\partial(\alpha_1\rho_1)} & \alpha_1 \frac{\partial p_{int}}{\partial(\alpha_2\rho_2)} & 0 & 0 \\ \alpha_2 \frac{\partial p_{int}}{\partial(\alpha_1\rho_1)} & \alpha_2 \frac{\partial p_{int}}{\partial(\alpha_2\rho_2)} & 0 & 0 \end{bmatrix} \tag{8}$$

the nonconservative part of the flux.

Note that every quantity can be written in terms of the conserved quantities q together with the closure relations (4) and (5), e.g. the pressure p can be determined by solving a quadratic equation involving $\alpha_1\rho_1 = q_1$ and $\alpha_2\rho_2 = q_2$ originating from (4) and (5).

3 Weak Formulation

We follow the approach of [7] to obtain a weak formulation for the nonconservative hyperbolic PDE. The weak formulation uses a path integral, based on the following definition and theorem, to approximate the nonconservative product in the PDE at points where the weak solution jumps.

Definition 1 (Integration paths, [4]). A Lipschitz continuous path $\phi : [0, 1] \times \mathbb{R}^n \times \mathbb{R}^n \to \mathbb{R}^n$ is called an integration path if it satisfies the following properties:

$$\phi_i(0; q, v) = q_i \quad \text{and} \quad \phi_i(1; q, v) = v_i \quad \forall i \in \mathscr{I}, \forall q, v \in \mathbb{R}^n, \tag{9}$$

$$\phi_i(\tau; q, q) = q_i \quad \forall i \in \mathscr{I}, \forall q \in \mathbb{R}^n, \tag{10}$$

$$\left| \frac{\partial \phi_i}{\partial \tau}(\tau; q, v) \right| \le K\,|q_i - v_i| \quad \forall i \in \mathscr{I}, \forall q, v \in \mathbb{R}^n, \tau \text{ a.e.} \in [0, 1]. \tag{11}$$

Theorem 1 (DLM measure, [4]). *Let $q : (a, b) \to \mathbb{R}^n$ be a function of bounded variation and $f : \mathbb{R}^n \to \mathbb{R}^n$ be a continuous function. Then there exists a unique real-valued bounded Borel measure μ on (a, b) characterised by the two following properties:*

1. If q is continuous on a Borel set $B \subset (a, b)$, then

$$\mu_i(B; f, q) = \int_B \sum_{j \in \mathscr{I}} f_{ij}(q) \frac{\partial q_i}{\partial x}\, d\lambda(x), \quad \forall i \in \mathscr{I}, \tag{12}$$

where λ is the Borel measure.

2. *If q is discontinuous at a point $x \in (a,b)$, then*

$$\mu_i(\{x\}; f, q) = \int_0^1 \sum_{j \in \mathscr{I}} f_{ij}(\phi(\tau; q(x^-), q(x^+))) \frac{\partial \phi_j}{\partial \tau}(\tau; q(x^-), q(x^+))\, d\lambda(\tau), \quad \forall i \in \mathscr{I}, \tag{13}$$

where $q(x^-) := \lim_{y \nearrow x} q(y)$ and $q(x^+) = \lim_{y \searrow x} q(y)$.

We obtain a weak formulation by multiplying the PDE (6) with a test function v and integrating the result using two different measures, a Lebesgue measure for the time derivative and source term and a DLM measure for the terms involving spatial derivatives, yielding

$$\int_\Omega v(x) \left(\frac{\partial q_i}{\partial t}(x,t) + s_i(q(x,t)) \right) d\lambda(x) + \int_\Omega v(x)\, d\mu_i(x; f, q) = 0, \quad \forall i \in \mathscr{I}. \tag{14}$$

The DG-FEM formulation follows from defining a set $\mathscr{E}$ of open, connected, nonoverlapping elements such that $\cup\mathscr{E}$ is dense in Ω, a broken polynomial space Q and assuming $q, v \in Q$. Let $\mathscr{D}$ denote the set of element edges, including any boundaries. Substituting the continuous (12) and discontinuous (13) definitions of the DLM measure yields the semi-discrete DG-FEM formulation

$$\sum_{E \in \mathscr{E}} \int_E v(x) \left(\frac{\partial q_i}{\partial t}(x,t) + s_i(q(x,t)) + \sum_{j \in \mathscr{I}} f_{ij}(q(x,t)) \frac{\partial q_j}{\partial x}(x,t) \right) d\lambda(x) +$$
$$\sum_{x \in \mathscr{D}} v(x) \int_0^1 \sum_{j \in \mathscr{I}} f_{ij}(\phi(\tau; q(x^-,t), q(x^+,t))) \frac{\partial \phi_j}{\partial \tau}(\tau; q(x^-,t), q(x^+,t)))\, d\lambda(\tau) = 0, \quad \forall i \in \mathscr{I}. \tag{15}$$

Since the DLM measure is nonzero on element edges, the trial function v needs to have a (single) value at these points. We define this value by comparing the scheme with conventional DG-FEM schemes for conservative hyperbolic PDE's. Let $[\![q]\!] := x \mapsto q(x^+) - q(x^-)$ denote the jump of q at x and $\{\!\{q\}\!\} := x \mapsto \frac{1}{2}(q(x^+) + q(x^-))$ the average. Assuming the nonconservative part of f to be zero, $f_\mathrm{n} = 0$, the DG-FEM formulation (15) becomes

$$\sum_{E \in \mathscr{E}} \int_E v(x) \left(\frac{\partial q_i}{\partial t}(x,t) + s_i(q(x,t)) + \frac{\partial}{\partial x} F_{\mathrm{c},i}(q(x,t)) \right) d\lambda(x)$$
$$+ \sum_{x \in \mathscr{D}} v(x)\, [\![F_{\mathrm{c},i}(q(\cdot,t))]\!]\,(x) = 0, \quad \forall i \in \mathscr{I}. \tag{16}$$

By asserting $v = \{\{v\}\}$ and applying integration by parts to the flux term in the element integral we obtain an unstable, conservative DG-FEM scheme with central fluxes:

$$\sum_{E\in\mathscr{E}} \int_E v(x) \left(\frac{\partial q_i}{\partial t}(x,t) + s_i(q(x,t)) - \frac{\partial v}{\partial x}(x) F_{c,i}(q(x,t)) \right) d\lambda(x) \\ - \sum_{x\in\mathscr{D}} [\![v]\!](x) \{\!\{F_{c,i}(q(\cdot,t))\}\!\}(x) = 0, \quad \forall i \in \mathscr{I}. \quad (17)$$

4 Stabilisation of Edge Flux

Due to the central approximation of the flux at element edges the semi-discrete ODE (15) is unstable. There are various methods to stabilise the ODE, mostly originating from Finite Volume Methods, varying in the amount of added dissipation. In terms of computational effort the cheapest methods add the largest amount of dissipation, which often severely reduces the accuracy of low-order Finite Volume Methods. A high-order DG-FEM discretisation suffers far less from the dissipation [3]. For this reason we choose one of the cheapest stabilisation methods: Rusanov.

In [1] Rusanov stabilisation is derived for a nonconservative FVM. The equivalent of this stabilisation for a DG-FEM scheme is given by the following term, to be added to the right hand side of the semi-discrete ODE (15):

$$\mathrm{stab}_i := -\frac{1}{2} \sum_{x\in\mathscr{D}} C(q(x^-,t), q(x^+,t)) [\![v]\!](x) [\![q_i(\cdot,t)]\!](x), \quad \forall i \in \mathscr{I}. \quad (18)$$

The function C determines locally, i.e. at each interface, the amount of added viscosity and should be larger than the absolute eigenvalues of the flux function f at the interface.

5 Stabilisation of Element Flux

Strong oscillations may occur in a large area (spreading across multiple elements) around a discontinuity, possibly leading to an unstable situation, for instance when a solution becomes nonphysical. We apply a technique described in [6] to suppress these oscillations.

We add a viscous term to the original PDE (6) as follows:

$$\frac{\partial u_i}{\partial t} + \sum_j f_{ij}(u) \frac{\partial u_j}{\partial x} + s_i(u) = \frac{\partial}{\partial x} \left(\epsilon \frac{\partial u_i}{\partial x} \right), \quad \forall i \in \mathscr{I}, \quad (19)$$

where ϵ is a parameter controlled by a shock sensor. As noted in [3] discretisation of this PDE with the DG-FEM as described above by regarding $\epsilon \frac{\partial u_i}{\partial x}$ as a flux function may lead to an inconsistent scheme. The *local DG* [2] method is introduced to circumvent this. The PDE (19) is written as a (larger) system of first order derivatives by introducing the intermediate variable w_i,

$$\frac{\partial u_i}{\partial t} + \sum_j f_{ij}(u)\frac{\partial u_j}{\partial x} + s_i(u) = \frac{\partial \epsilon w_i}{\partial x}, \quad \forall i \in \mathscr{I}, \tag{20}$$

where w_i is defined as

$$w_i - \frac{\partial u_i}{\partial x} = 0, \quad \forall i \in \mathscr{I}. \tag{21}$$

There are other choices possible for this splitting, e.g. the viscosity could be incorporated in w_i, however, we observed that system (20), (21) produces the most stable and sharp results.

We discretise the viscous system (20), (21) in a similar way as described above. The PDE's are multiplied by a test function $v \in Q$ and integrated over the domain using the DLM measure for all spatial derivatives, including the split viscous terms, and the Lebesgue measure for the remaining terms:

$$\int_\Omega v(x)\left(\frac{\partial q_i}{\partial t}(x,t) + s_i(q(x,t),x,t)\right) d\lambda(x) + \int_\Omega v(x)\, d\mu_i(x; f, q)$$
$$= \int_\Omega v(x)\, d\mu_i(x; \delta, \epsilon w) + \text{stab}_i, \quad \forall i \in \mathscr{I}, \tag{22}$$

$$\int_\Omega v(x) w_i(x,t)\, d\lambda(x) - \int_\Omega v(x)\, d\mu_i(x; \delta, u) = 0, \quad \forall i \in \mathscr{I}. \tag{23}$$

Note that for conservative fluxes, in this case $\frac{\partial \epsilon w_i}{\partial x}$ and $\frac{\partial u_i}{\partial x}$, the DLM measure on discontinuous points is equivalent to a central flux. No additional stabilisation is required since the derivatives combined represent a second order derivative.

The amount of viscosity ϵ is determined by a shock sensor. We use the shock sensor as described in [6] with a minor modification. The shock sensor is applied to the sum of the internal energy for each phase, projected on Q, denoted by z. The highest mode of z at element E is measured against $z - z_{E1}$, the quantity with zero average value:

$$S_E = \frac{\int_E \hat{z}_{EP}^2 \psi_{EP}^2(x)\, d\lambda(x)}{\int_E (z - z_{E1})^2(x)\, d\lambda(x)}. \tag{24}$$

We use the same algorithm for obtaining a viscosity as described in [6] using the above shock sensor S_E.

6 Numerical Results

We apply the proposed discretisation to two test problems commonly used to verify discrete two-phase pipe flow models. We do this mainly to show that the solutions converge to the correct entropy solutions. It is not expected that for the chosen test problems, which develop shocks, the high order accuracy of the discretisation comes to light.

6.1 *Shock Tube*

The *large relative velocity shock tube* test problem is a Riemann problem with the following initial data: the liquid volume fraction jumps from 0.71 to 0.7 and the gas velocity from 65 to 50 ms^{-1}, the pressure is constant at $2.65 \cdot 10^5$ Pa and the liquid velocity is 1 ms^{-1}. There is no gravity force. (See for example [5] and the references therein.) This Riemann problem generates four shocks, two of them moving at relatively high velocities, the other two with low velocities.

Figure 1 shows the solution of the system at $t = 0.1$ s, discretised with the DG-FEM method as described in this paper with 6 and 24 elements, both having 8 basis functions per element. For comparison results obtained using the first order Roe FVM as described in [5] is shown using 192 elements, the same amount of DOFs as for DG with 24 elements.

All three solutions attain the same shock speeds and intermediate levels in the eyeball-norm. Both DG-schemes show no signs of the Gibbs phenomenon on the global scale. Zooming in on the edges does reveal a slight oscillation. The size of the oscillations is related to the amount of viscosity added to the system and can be further reduced at the cost of less sharp shocks.

6.2 *Water Faucet*

The *water faucet test problem* consists of a vertical pipe, 12 m long, filled with a mixture of water (0.8 volume fraction) and air. Water flows initially with 10 ms^{-1} downwards, the air is at rest. At the top of the pipe the conditions are the same as the initial conditions. The bottom the pipe is at a constant pressure of 10^5 Pa. Under influence of gravity (10 ms^{-2}) the liquid will accelerate and the liquid fraction will decrease by conservation of mass. (See for example [5] and the references therein.)

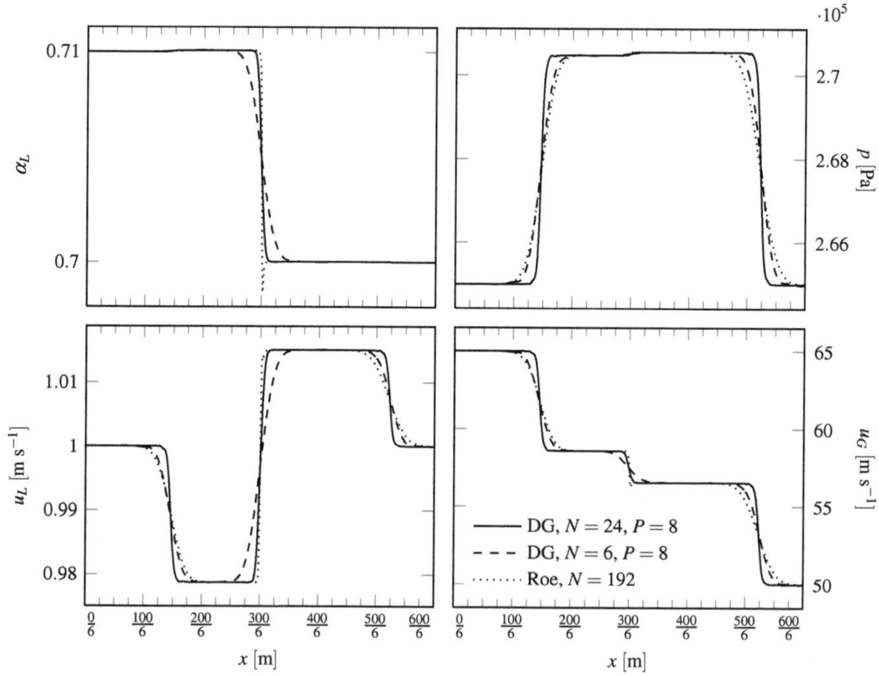

Fig. 1 Results for the shock tube test problem at $t = 0.6$ s with 24 and 6 elements, all using 8 Legendre basis functions. *Top-left*: volume fraction water, *top-right*: pressure, *bottom-left*: velocity water, *bottom-right*: velocity air

The exact solution for this problem with incompressible liquid and gas phase is a single shock moving downwards, leaving a steady, contracted water column behind (see [5] and the references therein). Since the pressure variation is small, this solution is considered to be a very good approximation to the system with compressible phases.

Figure 2 shows the solution of the system at $t = 0.6$ s, discretised with the DG-FEM method as described in this paper with 6 and 24 elements, both having 8 basis functions per element. For comparison results obtained using the first order Roe FVM as described in [5] is shown using 192 elements, the same amount of DOFs as for DG with 24 elements. Also the approximate solution is shown, which is based on the exact solution of the incompressible problem for the volume fraction and liquid velocity and a numerical solution of the Roe scheme with 1,200 DOFs for the pressure and gas velocity.

The results are similar to the shock tube test problem. The Gibbs phenomenon is completely suppressed for both DG solutions, even on a small scale there are no wiggles present. The DG solutions converge to the approximate solution.

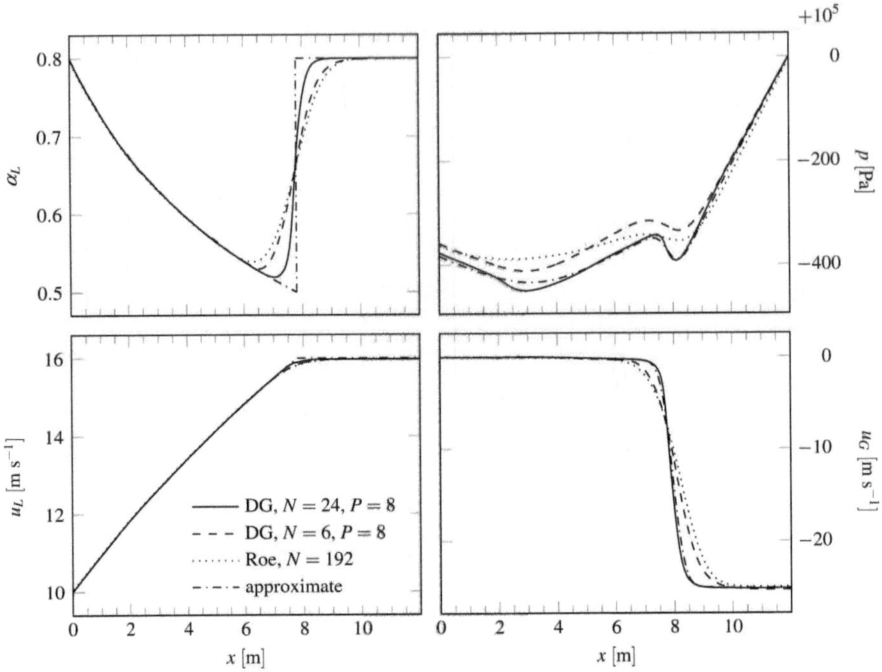

Fig. 2 Results for the water faucet test problem at $t = 0.6$ s with 24 and 6 elements, all using 8 Legendre basis functions. *Top-left*: volume fraction water, *top-right*: pressure, *bottom-left*: velocity water, *bottom-right*: velocity air

7 Conclusions

We have presented a discretisation of a one-dimensional, nonconservative, hyperbolic, two-phase pipe flow model using a high-order Discontinuous Galerkin Finite Element Method which avoids the costly, numerical evaluation of the eigenstructure of the system. The Gibbs phenomenon is suppressed by using a viscous regularisation in the neighbourhood of discontinuities, which are identified by a shock sensor. We have shown that the numerical solutions converge to the correct entropy solutions.

References

1. Castro, M., Pardo, A., Parés, C., Toro, E.: On some fast well-balanced first order solvers for nonconservative systems. Mathematics of Computation **79**(271), 1427–1472 (2010)
2. Cockburn, B., Shu, C.: The local discontinuous galerkin method for time-dependent convection-diffusion systems. SIAM Journal on Numerical Analysis **35**(6), 2440–2463 (1998)

3. Cockburn, B., Shu, C.: Runge-kutta discontinuous galerkin methods for convection-dominated problems. Journal of Scientific Computing **16**(3), 173–261 (2001)
4. Dal Maso, G., LeFloch, P., Murat, F.: Definition and weak stability of nonconservative products. Journal de Mathématiques Pures et Appliquées **74**(6), 483–548 (1995)
5. Evje, S., Flåtten, T.: Hybrid flux-splitting schemes for a common two-fluid model. Journal of Computational Physics **192**(1), 175–210 (2003)
6. Persson, P., Peraire, J.: Sub-cell shock capturing for discontinuous galerkin methods. In: Proceedings of the 44th AIAA Aerospace Sciences Meeting and Exhibit. American Institute of Aeronautics and Astronautics (2006)
7. Rhebergen, S., Bokhove, O., Van der Vegt, J.: Discontinuous galerkin finite element methods for hyperbolic nonconservative partial differential equations. Journal of Computational Physics **227**(3), 1887–1922 (2008)

Editorial Policy

1. Volumes in the following three categories will be published in LNCSE:

i) Research monographs
ii) Tutorials
iii) Conference proceedings

Those considering a book which might be suitable for the series are strongly advised to contact the publisher or the series editors at an early stage.

2. Categories i) and ii). Tutorials are lecture notes typically arising via summer schools or similar events, which are used to teach graduate students. These categories will be emphasized by Lecture Notes in Computational Science and Engineering. **Submissions by interdisciplinary teams of authors are encouraged.** The goal is to report new developments – quickly, informally, and in a way that will make them accessible to non-specialists. In the evaluation of submissions timeliness of the work is an important criterion. Texts should be well-rounded, well-written and reasonably self-contained. In most cases the work will contain results of others as well as those of the author(s). In each case the author(s) should provide sufficient motivation, examples, and applications. In this respect, Ph.D. theses will usually be deemed unsuitable for the Lecture Notes series. Proposals for volumes in these categories should be submitted either to one of the series editors or to Springer-Verlag, Heidelberg, and will be refereed. A provisional judgement on the acceptability of a project can be based on partial information about the work: a detailed outline describing the contents of each chapter, the estimated length, a bibliography, and one or two sample chapters – or a first draft. A final decision whether to accept will rest on an evaluation of the completed work which should include

- at least 100 pages of text;
- a table of contents;
- an informative introduction perhaps with some historical remarks which should be accessible to readers unfamiliar with the topic treated;
- a subject index.

3. Category iii). Conference proceedings will be considered for publication provided that they are both of exceptional interest and devoted to a single topic. One (or more) expert participants will act as the scientific editor(s) of the volume. They select the papers which are suitable for inclusion and have them individually refereed as for a journal. Papers not closely related to the central topic are to be excluded. Organizers should contact the Editor for CSE at Springer at the planning stage, see *Addresses* below.

In exceptional cases some other multi-author-volumes may be considered in this category.

4. Only works in English will be considered. For evaluation purposes, manuscripts may be submitted in print or electronic form, in the latter case, preferably as pdf- or zipped ps-files. Authors are requested to use the LaTeX style files available from Springer at http://www.springer.com/authors/book+authors/helpdesk?SGWID=0-1723113-12-971304-0 (Click on Templates → LaTeX → monographs or contributed books).

For categories ii) and iii) we strongly recommend that all contributions in a volume be written in the same LaTeX version, preferably LaTeX2e. Electronic material can be included if appropriate. Please contact the publisher.

Careful preparation of the manuscripts will help keep production time short besides ensuring satisfactory appearance of the finished book in print and online.

5. The following terms and conditions hold. Categories i), ii) and iii):

Authors receive 50 free copies of their book. No royalty is paid.
Volume editors receive a total of 50 free copies of their volume to be shared with authors, but no royalties.

Authors and volume editors are entitled to a discount of 33.3 % on the price of Springer books purchased for their personal use, if ordering directly from Springer.

6. Springer secures the copyright for each volume.

Addresses:

Timothy J. Barth
NASA Ames Research Center
NAS Division
Moffett Field, CA 94035, USA
barth@nas.nasa.gov

Michael Griebel
Institut für Numerische Simulation
der Universität Bonn
Wegelerstr. 6
53115 Bonn, Germany
griebel@ins.uni-bonn.de

David E. Keyes
Mathematical and Computer Sciences
and Engineering
King Abdullah University of Science
and Technology
P.O. Box 55455
Jeddah 21534, Saudi Arabia
david.keyes@kaust.edu.sa

and

Department of Applied Physics
and Applied Mathematics
Columbia University
500 W. 120 th Street
New York, NY 10027, USA
kd2112@columbia.edu

Risto M. Nieminen
Department of Applied Physics
Aalto University School of Science
and Technology
00076 Aalto, Finland
risto.nieminen@aalto.fi

Dirk Roose
Department of Computer Science
Katholieke Universiteit Leuven
Celestijnenlaan 200A
3001 Leuven-Heverlee, Belgium
dirk.roose@cs.kuleuven.be

Tamar Schlick
Department of Chemistry
and Courant Institute
of Mathematical Sciences
New York University
251 Mercer Street
New York, NY 10012, USA
schlick@nyu.edu

Editor for Computational Science
and Engineering at Springer:
Martin Peters
Springer-Verlag
Mathematics Editorial IV
Tiergartenstrasse 17
69121 Heidelberg, Germany
martin.peters@springer.com

Lecture Notes in Computational Science and Engineering

1. D. Funaro, *Spectral Elements for Transport-Dominated Equations.*
2. H.P. Langtangen, *Computational Partial Differential Equations.* Numerical Methods and Diffpack Programming.
3. W. Hackbusch, G. Wittum (eds.), *Multigrid Methods V.*
4. P. Deuflhard, J. Hermans, B. Leimkuhler, A.E. Mark, S. Reich, R.D. Skeel (eds.), *Computational Molecular Dynamics: Challenges, Methods, Ideas.*
5. D. Kröner, M. Ohlberger, C. Rohde (eds.), *An Introduction to Recent Developments in Theory and Numerics for Conservation Laws.*
6. S. Turek, *Efficient Solvers for Incompressible Flow Problems.* An Algorithmic and Computational Approach.
7. R. von Schwerin, ***M**ulti **B**ody **S**ystem **SIM**ulation.* Numerical Methods, Algorithms, and Software.
8. H.-J. Bungartz, F. Durst, C. Zenger (eds.), *High Performance Scientific and Engineering Computing.*
9. T.J. Barth, H. Deconinck (eds.), *High-Order Methods for Computational Physics.*
10. H.P. Langtangen, A.M. Bruaset, E. Quak (eds.), *Advances in Software Tools for Scientific Computing.*
11. B. Cockburn, G.E. Karniadakis, C.-W. Shu (eds.), *Discontinuous Galerkin Methods.* Theory, Computation and Applications.
12. U. van Rienen, *Numerical Methods in Computational Electrodynamics.* Linear Systems in Practical Applications.
13. B. Engquist, L. Johnsson, M. Hammill, F. Short (eds.), *Simulation and Visualization on the Grid.*
14. E. Dick, K. Riemslagh, J. Vierendeels (eds.), *Multigrid Methods VI.*
15. A. Frommer, T. Lippert, B. Medeke, K. Schilling (eds.), *Numerical Challenges in Lattice Quantum Chromodynamics.*
16. J. Lang, *Adaptive Multilevel Solution of Nonlinear Parabolic PDE Systems.* Theory, Algorithm, and Applications.
17. B.I. Wohlmuth, *Discretization Methods and Iterative Solvers Based on Domain Decomposition.*
18. U. van Rienen, M. Günther, D. Hecht (eds.), *Scientific Computing in Electrical Engineering.*
19. I. Babuška, P.G. Ciarlet, T. Miyoshi (eds.), *Mathematical Modeling and Numerical Simulation in Continuum Mechanics.*
20. T.J. Barth, T. Chan, R. Haimes (eds.), *Multiscale and Multiresolution Methods.* Theory and Applications.
21. M. Breuer, F. Durst, C. Zenger (eds.), *High Performance Scientific and Engineering Computing.*
22. K. Urban, *Wavelets in Numerical Simulation.* Problem Adapted Construction and Applications.

23. L.F. Pavarino, A. Toselli (eds.), *Recent Developments in Domain Decomposition Methods.*

24. T. Schlick, H.H. Gan (eds.), *Computational Methods for Macromolecules: Challenges and Applications.*

25. T.J. Barth, H. Deconinck (eds.), *Error Estimation and Adaptive Discretization Methods in Computational Fluid Dynamics.*

26. M. Griebel, M.A. Schweitzer (eds.), *Meshfree Methods for Partial Differential Equations.*

27. S. Müller, *Adaptive Multiscale Schemes for Conservation Laws.*

28. C. Carstensen, S. Funken, W. Hackbusch, R.H.W. Hoppe, P. Monk (eds.), *Computational Electromagnetics.*

29. M.A. Schweitzer, *A Parallel Multilevel Partition of Unity Method for Elliptic Partial Differential Equations.*

30. T. Biegler, O. Ghattas, M. Heinkenschloss, B. van Bloemen Waanders (eds.), *Large-Scale PDE-Constrained Optimization.*

31. M. Ainsworth, P. Davies, D. Duncan, P. Martin, B. Rynne (eds.), *Topics in Computational Wave Propagation.* Direct and Inverse Problems.

32. H. Emmerich, B. Nestler, M. Schreckenberg (eds.), *Interface and Transport Dynamics.* Computational Modelling.

33. H.P. Langtangen, A. Tveito (eds.), *Advanced Topics in Computational Partial Differential Equations.* Numerical Methods and Diffpack Programming.

34. V. John, *Large Eddy Simulation of Turbulent Incompressible Flows.* Analytical and Numerical Results for a Class of LES Models.

35. E. Bänsch (ed.), *Challenges in Scientific Computing - CISC 2002.*

36. B.N. Khoromskij, G. Wittum, *Numerical Solution of Elliptic Differential Equations by Reduction to the Interface.*

37. A. Iske, *Multiresolution Methods in Scattered Data Modelling.*

38. S.-I. Niculescu, K. Gu (eds.), *Advances in Time-Delay Systems.*

39. S. Attinger, P. Koumoutsakos (eds.), *Multiscale Modelling and Simulation.*

40. R. Kornhuber, R. Hoppe, J. Périaux, O. Pironneau, O. Wildlund, J. Xu (eds.), *Domain Decomposition Methods in Science and Engineering.*

41. T. Plewa, T. Linde, V.G. Weirs (eds.), *Adaptive Mesh Refinement – Theory and Applications.*

42. A. Schmidt, K.G. Siebert, *Design of Adaptive Finite Element Software.* The Finite Element Toolbox ALBERTA.

43. M. Griebel, M.A. Schweitzer (eds.), *Meshfree Methods for Partial Differential Equations II.*

44. B. Engquist, P. Lötstedt, O. Runborg (eds.), *Multiscale Methods in Science and Engineering.*

45. P. Benner, V. Mehrmann, D.C. Sorensen (eds.), *Dimension Reduction of Large-Scale Systems.*

46. D. Kressner, *Numerical Methods for General and Structured Eigenvalue Problems.*

47. A. Boriçi, A. Frommer, B. Joó, A. Kennedy, B. Pendleton (eds.), *QCD and Numerical Analysis III.*

48. F. Graziani (ed.), *Computational Methods in Transport.*

49. B. Leimkuhler, C. Chipot, R. Elber, A. Laaksonen, A. Mark, T. Schlick, C. Schütte, R. Skeel (eds.), *New Algorithms for Macromolecular Simulation.*

50. M. Bücker, G. Corliss, P. Hovland, U. Naumann, B. Norris (eds.), *Automatic Differentiation: Applications, Theory, and Implementations.*

51. A.M. Bruaset, A. Tveito (eds.), *Numerical Solution of Partial Differential Equations on Parallel Computers.*

52. K.H. Hoffmann, A. Meyer (eds.), *Parallel Algorithms and Cluster Computing.*

53. H.-J. Bungartz, M. Schäfer (eds.), *Fluid-Structure Interaction.*

54. J. Behrens, *Adaptive Atmospheric Modeling.*

55. O. Widlund, D. Keyes (eds.), *Domain Decomposition Methods in Science and Engineering XVI.*

56. S. Kassinos, C. Langer, G. Iaccarino, P. Moin (eds.), *Complex Effects in Large Eddy Simulations.*

57. M. Griebel, M.A Schweitzer (eds.), *Meshfree Methods for Partial Differential Equations III.*

58. A.N. Gorban, B. Kégl, D.C. Wunsch, A. Zinovyev (eds.), *Principal Manifolds for Data Visualization and Dimension Reduction.*

59. H. Ammari (ed.), *Modeling and Computations in Electromagnetics: A Volume Dedicated to Jean-Claude Nédélec.*

60. U. Langer, M. Discacciati, D. Keyes, O. Widlund, W. Zulehner (eds.), *Domain Decomposition Methods in Science and Engineering XVII.*

61. T. Mathew, *Domain Decomposition Methods for the Numerical Solution of Partial Differential Equations.*

62. F. Graziani (ed.), *Computational Methods in Transport: Verification and Validation.*

63. M. Bebendorf, *Hierarchical Matrices.* A Means to Efficiently Solve Elliptic Boundary Value Problems.

64. C.H. Bischof, H.M. Bücker, P. Hovland, U. Naumann, J. Utke (eds.), *Advances in Automatic Differentiation.*

65. M. Griebel, M.A. Schweitzer (eds.), *Meshfree Methods for Partial Differential Equations IV.*

66. B. Engquist, P. Lötstedt, O. Runborg (eds.), *Multiscale Modeling and Simulation in Science.*

67. I.H. Tuncer, Ü. Gülcat, D.R. Emerson, K. Matsuno (eds.), *Parallel Computational Fluid Dynamics 2007.*

68. S. Yip, T. Diaz de la Rubia (eds.), *Scientific Modeling and Simulations.*

69. A. Hegarty, N. Kopteva, E. O'Riordan, M. Stynes (eds.), *BAIL* 2008 – *Boundary and Interior Layers.*

70. M. Bercovier, M.J. Gander, R. Kornhuber, O. Widlund (eds.), *Domain Decomposition Methods in Science and Engineering XVIII.*

71. B. Koren, C. Vuik (eds.), *Advanced Computational Methods in Science and Engineering.*

72. M. Peters (ed.), *Computational Fluid Dynamics for Sport Simulation.*

73. H.-J. Bungartz, M. Mehl, M. Schäfer (eds.), *Fluid Structure Interaction II - Modelling, Simulation, Optimization.*

74. D. Tromeur-Dervout, G. Brenner, D.R. Emerson, J. Erhel (eds.), *Parallel Computational Fluid Dynamics 2008.*

75. A.N. Gorban, D. Roose (eds.), *Coping with Complexity: Model Reduction and Data Analysis.*

76. J.S. Hesthaven, E.M. Rønquist (eds.), *Spectral and High Order Methods for Partial Differential Equations.*

77. M. Holtz, *Sparse Grid Quadrature in High Dimensions with Applications in Finance and Insurance.*

78. Y. Huang, R. Kornhuber, O.Widlund, J. Xu (eds.), *Domain Decomposition Methods in Science and Engineering XIX.*

79. M. Griebel, M.A. Schweitzer (eds.), *Meshfree Methods for Partial Differential Equations V.*

80. P.H. Lauritzen, C. Jablonowski, M.A. Taylor, R.D. Nair (eds.), *Numerical Techniques for Global Atmospheric Models.*

81. C. Clavero, J.L. Gracia, F.J. Lisbona (eds.), *BAIL 2010 – Boundary and Interior Layers, Computational and Asymptotic Methods.*

82. B. Engquist, O. Runborg, Y.R. Tsai (eds.), *Numerical Analysis and Multiscale Computations.*

83. I.G. Graham, T.Y. Hou, O. Lakkis, R. Scheichl (eds.), *Numerical Analysis of Multiscale Problems.*

84. A. Logg, K.-A. Mardal, G. Wells (eds.), *Automated Solution of Differential Equations by the Finite Element Method.*

85. J. Blowey, M. Jensen (eds.), *Frontiers in Numerical Analysis - Durham 2010.*

86. O. Kolditz, U.-J. Gorke, H. Shao, W. Wang (eds.), *Thermo-Hydro-Mechanical-Chemical Processes in Fractured Porous Media - Benchmarks and Examples.*

87. S. Forth, P. Hovland, E. Phipps, J. Utke, A. Walther (eds.), *Recent Advances in Algorithmic Differentiation.*

88. J. Garcke, M. Griebel (eds.), *Sparse Grids and Applications.*

89. M. Griebel, M.A. Schweitzer (eds.), *Meshfree Methods for Partial Differential Equations VI.*

90. C. Pechstein, *Finite and Boundary Element Tearing and Interconnecting Solvers for Multiscale Problems.*

91. R. Bank, M. Holst, O. Widlund, J. Xu (eds.), *Domain Decomposition Methods in Science and Engineering XX.*

92. H. Bijl, D. Lucor, S. Mishra, C. Schwab (eds.), *Uncertainty Quantification in Computational Fluid Dynamics.*

93. M. Bader, H.-J. Bungartz, T. Weinzierl (eds.), *Advanced Computing.*

94. M. Ehrhardt, T. Koprucki (eds.), *Advanced Mathematical Models and Numerical Techniques for Multi-Band Effective Mass Approximations.*

95. M. Azaïez, H. El Fekih, J.S. Hesthaven (eds.), *Spectral and High Order Methods for Partial Differential Equations ICOSAHOM 2012.*

For further information on these books please have a look at our mathematics catalogue at the following URL: www.springer.com/series/3527

Monographs in Computational Science and Engineering

1. J. Sundnes, G.T. Lines, X. Cai, B.F. Nielsen, K.-A. Mardal, A. Tveito, *Computing the Electrical Activity in the Heart.*

For further information on this book, please have a look at our mathematics catalogue at the following URL: www.springer.com/series/7417

Texts in Computational Science and Engineering

1. H. P. Langtangen, *Computational Partial Differential Equations.* Numerical Methods and Diffpack Programming. 2nd Edition
2. A. Quarteroni, F. Saleri, P. Gervasio, *Scientific Computing with MATLAB and Octave.* 3rd Edition
3. H. P. Langtangen, *Python Scripting for Computational Science.* 3rd Edition
4. H. Gardner, G. Manduchi, *Design Patterns for e-Science.*
5. M. Griebel, S. Knapek, G. Zumbusch, *Numerical Simulation in Molecular Dynamics.*
6. H. P. Langtangen, *A Primer on Scientific Programming with Python.* 3rd Edition
7. A. Tveito, H. P. Langtangen, B. F. Nielsen, X. Cai, *Elements of Scientific Computing.*
8. B. Gustafsson, *Fundamentals of Scientific Computing.*
9. M. Bader, *Space-Filling Curves.*
10. M. Larson, F. Bengzon, *The Finite Element Method: Theory, Implementation and Applications.*

For further information on these books please have a look at our mathematics catalogue at the following URL: www.springer.com/series/5151

MIX
Papier aus verantwortungsvollen Quellen
Paper from responsible sources
FSC® C105338

If you have any concerns about our products,
you can contact us on
ProductSafety@springernature.com

In case Publisher is established outside the EU,
the EU authorized representative is:
Springer Nature Customer Service Center GmbH
Europaplatz 3, 69115 Heidelberg, Germany

Printed by Libri Plureos GmbH
in Hamburg, Germany